Experimental Design
and Model Choice

Helge Toutenburg

Experimental Design and Model Choice

The Planning and Analysis of Experiments
with Continuous or Categorical Response

With Contributions by Sabina Illi

With 28 Figures

Springer-Verlag Berlin Heidelberg GmbH

Prof. Dr. Dr. Helge Toutenburg
University of Munich
Institute of Statistics
Akademiestr. 1
D-80799 Munich
Germany

Die Deutsche Bibliothek - CIP-Einheitsaufnahme

Toutenburg, Helge:
Experimental design and model choice : the planning and
analysis of experiments with continuous or categorical response
/ Helge Toutenburg. With contributions by Sabina Illi. -

ISBN 978-3-642-52500-1   ISBN 978-3-642-52498-1 (eBook)
DOI 10.1007/978-3-642-52498-1

ISBN 978-3-642-52500-1

© Springer-Verlag Berlin Heidelberg 1995
Originally published by Physica-Verlag Heidelberg in 1995

The use of general descriptive names, registered names, trademarks, etc. in this publication
does not imply, even in the absence of a specific statement, that such names are exempt from
the relevant protective laws and regulations and therefore free for general use.

SPIN 10505620   88/2202 - 5 4 3 2 1 0 - Printed on acid-free paper

Preface

This book is a translation of the up-dated version of my German textbook that was written parallel to my lecture "Design of Experiments" which I held at the University of Munich.

It was often called to my attention by statisticians in the pharmaceutical industry, that there is a need for a summarizing and standardized representation of the design and analysis of experiments, that includes the different aspects of classical theory for continuous response and of modern procedures for categorical and especially correlated response, as well as more complex designs, as for example cross-over and repeated measures.

My staff and graduate students played an essential part in the preparation of the manuscript. They wrote the text in well-tried precision (Andreas Fieger), worked out examples (Andreas Fieger, Christian Kastner), and prepared several sections in this book (Ulrike Feldmeier, Sabina Illi, Elke Ortmann, Andrea Schöpp, Irmgard Strehler, Christian Kastner, Oliver Loch).

I am appreciative of the efforts of those who assisted in the preparation of the English version. In particular, I would like to thank Sabina Illi and Oliver Loch, as well as V.K. Srivastava (University of Lucknow, India) for their careful reading of the English version and invaluable suggestions.

Of course not all aspects could be taken into account. Especially the development in the field of generalized linear models is so dynamic, that it is hard to include all current tendencies. In order to keep up with this development, the book contains more recent methods for the analysis of clusters.

Finally, I would like to thank Dr. Werner A. Müller of the Physica-Verlag for his cooperation and the confidence in this book.

Helge Toutenburg   Munich, March 1995

Contents

1 Introduction

1.1 Data, Variables, and Random Processes

Many processes that occur in nature, engineering sciences and biomedical or pharmaceutical experiments cannot be characterized by theoretical or even mathematical models.

The analysis of such processes, especially the study of cause-effect relationships, may be carried out by drawing inferences from a finite number of samples. One important goal now consists in designing sampling experiments that are productive, cost effective, and provide a sufficient data base in a qualitative sense. Statistical methods of experimental design aim at improving and optimizing the effectiveness and productivity of empirically conducted experiments.

An almost unlimited capacity of hard- and software facilities suggests nowadays almost unlimited quality of information. It is often overlooked, however, that large numbers of data do not necessarily coincide with a large contents of information. Basically, it is desirable to collect data that contains a high level of information, i.e., *information-rich-data*. Statistical methods of experimental design offer a possibility to increase the proportion of such information-rich-data.

As data serves to understand, as well as to control processes, we may formulate several basic ideas of experimental design:

- Selection of appropriate variables

- Determination of the optimal range of input values

- Determination of the optimal process regime under restrictions or marginal conditions specific for the process under study (for example: pressure, temperature, toxicity).

Examples:

(a) Let the response variable Y denote the flexibility of a plastic that is used in dental medicine to prepare a set of dentures. Let the input variable X denote the proportion of silan (%). A suitably designed experiment should

　(i) confirm that the flexibility increases if the proportion of silan increases.

(ii) find out the optimal dose of silan, that leads to an appropriate increase of flexibility (cf. Table 1.1).

PMMA 2.2 Vol% quartz without silan	PMMA 2.2 Vol% quartz with silan
98.47	106.75
106.20	111.75
100.47	96.67
98.72	98.70
91.42	118.61
108.17	111.03
98.36	90.92
92.36	104.62
80.00	94.63
114.43	110.91
104.99	104.62
101.11	108.77
102.94	98.97
103.95	98.78
99.00	102.65
106.05	
$\bar{x} = 100.42$	$\bar{y} = 103.91$
$s_x^2 = 7.9^2$	$s_y^2 = 7.6^2$
$n = 16$	$m = 15$

Table 1.1: Flexibility of PMMA with and without silan

(b) In metallurgy, the effect of two competing methods (oil, A – or salt water, B) to harden a given alloy had to be investigated. Some metallic pieces were hardened by method A and some by method B. In both samples, the average hardness $\bar{x}_A$ and $\bar{x}_B$ was calculated and interpreted as a measure to assess the effect of the respective method (cf. Montgomery, 1976, p.1).

In both examples, the following questions are of interest:

- Are all explaining factors incorporated that affect flexibility or hardness?

- How many workpieces have to be subjected to treatment such that possible differences are statistically significant?

- What is the smallest difference between average treatment effects that can be described as being substantially?

- Which methods of data analysis should be used?

1.2   Basic Principles of Experimental Design

We shall demonstrate the basic principles of experimental design by the following example of dental medicine. Let us assume that a study is to be planned in the framework of a prophylactic program for children in pre-school age. Answers to the following questions are to be expected:

- Are different intensity levels of instruction in dental care for pre-school children different in their effect?

- Are they substantially different from situations in which no instruction is given at all?

Before we try to answer these questions we have to discuss some topics:

(a) Exact definition of *intensity levels of instruction* in medical care.

Level I: instruction by dentists and parents and
            instruction of the kindergarten teacher by dentists.
Level II: as I, but without instruction of parents.
Level III: instruction by dentists only.
Additionally we define
Level IV: no instruction at all (control group).

(b) How can we measure the effect of the instruction?
    As an appropriate parameter, we chose the increase in caries during the period of observation, expressed by the difference in carious teeth.

Obviously, the most simple design is to give instructions to one child whereas another is left without advice. The criterion to quantify the effect is given by the increase of carious teeth developed during a fixed period.

Treatment	Unit	Increase of carious teeth
A (without instruction)	1 child	increase a
B (with instruction)	1 child	increase b

It would be unreasonable to conclude that instruction will definitely reduce the increase of carious teeth if b is smaller then a, as only one child was observed for each treatment. If more children are investigated and the difference of average effects (a – b) still continues to be large, one may decide that instruction definitely leads to improvements.

One important fact has to be mentioned at this stage. If more than one unit per group is observed, there will be some variability in the outcomes of the experiment in spite of homogeneous experimental conditions. This phenomenon is called *sampling error* or *natural variation*. In what follows, we will establish some basic principles to study the sampling error.

Principle 1: The experiment has to be carried out on several units (children) in order to determine the sampling error.

Principle 2: The units have to be assigned *randomly* to treatments.

In our example, every level of instruction must have the same chance to get assigned. These two principles are essential to determine the sampling error correctly. Additionally, the conditions under which the treatments were given should be comparable, if not identical. Also the units should be similar in structure. This means, for example, that children are of almost the same age, or live in the same area, or show a similar sociological environment. An appropriate setup of a correctly designed trial would consist of four groups, each with four children that have similar characteristics. The four levels of instruction are then randomly distributed to the children such that in the end, all levels are present in every group. This is the reasoning behind

Principle 3: To increase the sensitivity of an experiment, one usually stratifies the units into groups with similar (homogeneous) characteristics. These are called blocks. The criterion to stratify is often given by age, sex, risk-exposure, or sociological factors.

Principle 4: The experiment should be balanced. The number of units assigned to a specific treatment should nearly be the same, i.e., every instruction level occurs equally often among the children. The last principle ensures that every treatment is given as often as the others.

1.3   Scaling of Variables

In general, the applicability of statistical methods depends on the scale in which variables have been measured. Some methods, for example, assume that data may take any value within a given interval, whereas others require only an ordinal or ranked scale. The measurement scale is of particular importance as quality and goodness of statistical methods depend to some extent on it.

Nominal scale (qualitative data)

This is the most simple scale. Each data point belongs uniquely to a specific category. These categories are often coded by numbers that have no real numeric meaning.

Examples:

- classification of patients by sex: two categories *male* and *female* are possible

- classification of patients by blood groups

- increase of carious teeth in a given period. Possible categories: 0 (no increase), 1 (1 additional carious tooth), etc.

- profession

- race

- marital status.

This type of data is called nominal data. The following scale contains substantially more information.

Ordinal or ranked scale (quantitative data)

If we intend to characterize objects according to an ordering, for example grades or ratings, we may use an ordinal or ranked scale. Different categories now symbolize different qualities. Note that this does not mean, that differences between numerical values may be interpreted.

Example: The *index of oral-hygiene* (OHI) may take values 0, 1, 2, and 3. OHI is 0 if teeth are entirely free of dental plaque and OHI is 3 if more than two-thirds of teeth are attacked. The following classification serves as an example for an ordered scale.

Group 1	0–1	excellent hygiene
Group 2	2	satisfactory hygiene
Group 3	3	poor hygiene

Further examples of ordinal-scaled data are

- age groups (< 40, < 50, < 60, ≥ 60 years)

- intensity of a medical treatment (low, average, high dose)

- preference rating of an object (low, average, high)

Metric or interval scale

One disadvantage of a ranked scale consists in the fact, that numerical differences in the data are not liable to interpretation. In order to measure differences we shall use a metric or interval scale with a defined origin and equal scaling units (for example temperature). An interval scale with a natural origin is called ratio scale. Length, time, or weight measurements are examples of such ratio scales. It is convenient to consider interval and ratio scale as one scale.

Examples:

- Resistance to pressure of material

- p_H-value in dental plaque

- Time to produce a workpiece

- Rates of return in per cent

- Price of a good in $

Interval data may be represented by an ordinal scale and ordinal data by a nominal scale. In both situations, there is a loss of information. Obviously, there is no way to transfrom data from a lower into a higher scale.

Advanced statistical techniques are available for ordinal and interval data. A survey is given in table 1.2.

	Appropriate measures	Appropriate test procedures	Appropriate measures of correlation
Nominal scale	absolute and relative frequency- mode	χ^2-test	contingency coefficient
Ranked scale	frequencies, mode, ranks median, quantiles rank variance	χ^2-test, nonparametric methods based on ranks	rank correlation coefficient
Interval scale	frequencies, mode, ranks quantiles, median, skewness, $\bar{x}, s, s^2$	χ^2-test, nonparametric methods, parametric methods (for example under normality) χ^2-, t-, F-test, variance and regression analysis	correlation coefficient

Table 1.2: Measurement scales and related statistics

It should be noted that all types of measurement scales may occur simultaneously if more than one variable is observed from a person or an object.

Examples: Typical data at the registration of a hospital:

- Sex (nominal)

- Deformities: congenital/transmitted/received (nominal)

- Age (interval)

- Order of therapeutic steps (ordinal)

- OHI-index (ordinal)

- Time of treatment (interval)

1.4   Measuring and Scaling in Statistical Medicine

We shall discuss briefly some general measurement problems that are typical for medical data. It turns out that interval, or even ordinal data are more difficult to realize than one might have expected. Some variables are *directly measurable*, for example height, weight, age, or blood pressure of a patient, whereas others may be observed only via *proxy* variables. The latter case is called *indirect measurement*. Results for the variable of interest may only be derived from results of a proxy.

Examples:

- Assessing the health of a patient by measuring the effect of a drug.

- Determining the extention of a cardiac infarction by measuring the concentration of transaminase.

An indirect measurement may be regarded as the sum of the actual effect and an additional random effect. To quantify the actual effect may be problematic. Such an indirect measurement leads to a metric scale if

- the indirect observation is metric

- the actual effect is measurable by a metric variable

- there is a unique relation between both measurement scales.

Unfortunately, the latter case arises, rarely in medicine.

Another problem comes up by introducing *derived scales* which are defined as a function of metric scales. Their statistical treatment is rather difficult and much care has to be taken in order to analyse such data.

Example: Heart defects are usually measured by the ratio

$$\frac{\text{strain duration}}{\text{time of expulsion}}.$$

Even if we assume that both variables in numerator and denominator are normally distributed, we should refrain from analysing the ratio by standard statistical methods as the ratio follows a Cauchy distribution, such that neither expectation nor variance exist.

Another important point is the scaling of an interval scale itself. If measurement units are chosen unnecessarily wide, this may lead to identical values (ties) and therefore to a loss of information.

From our opinion, it should be stressed that real interval scales are rarely to justify, especially in bio-medical experiments.

Furthermore, metric data are often derived by transformations such that parametric assumptions, for example normality, have to be checked carefully.

As a conclusion, statistical methods based on rank or nominal data assume new importance in the analysis of bio-medical data.

1.5 Experimental Design in Biotechnology

Data represent a combination of *signals and noise.* A signal may be defined as the effect a variable takes on a process. Noise, or experimental errors, cover the natural variability in the data or variables.

If a biological, clinical or even chemical trial is repeated several times, one cannot expect that results are identical. Response variables always show some variation that has to be analysed by statistical methods.

There are two main sources of uncontrolled variability. These are given by a pure experimental error and a measurement error in which possible interactions are also subsumed. An *experimental error* is the variability of a response variable under exactly the same experimental conditions. *Measurement errors* describe the variability of a response if repeated measurements are taken.

In practice, the experimental error is usually assumed to be much higher than the measurement error. Additionally, it is often impossible to separate both errors, such that noise may be understood as the sum of both errors. As the measurement error is negligible in relation to the experimental error we have

$$\text{Noise} \quad \approx \quad \text{Experimental error.}$$

One task of experimental design is to separate signals from noise under marginal conditions given by restrictions in material, time, or money.

Example: If a response is influenced by two variables A and B, then one tries to quantify the effect of each variable. If the response is measured only at low or high levels of A and B, then there is no way to isolate their effects. If measurements are taken according to the following combinations, then individual effects may be separated:

- Level: A low, B low

- A low, B high

- A high, B low

- A high, B high.

1.6 Relative Importance of Effects – the Pareto-Principle

The analysis of models of the form

$$\text{Response} = f(X_1, \ldots, X_k),$$

where the X_i symbolize exogeneous influence variables, is subject to several requirements:

- Choice of the link function $f(\cdot)$.

- Choice of factors X_i.

- Consideration of interactions and hierarchical structures.

- Estimation of effects and interpretation of results.

Theory of loglinear regression (Agresti, 1990, Fahrmeir and Hamerle, 1984, Toutenburg, 1992) suggests that a special coding of variables as dummies yields estimates of the effects that are independent of measurement units. Ishihawa (1976) has illustrated this principle by a Pareto-chart. Figure 1.1 shows such a chart in which influence variables and interactions are ordered according to their relative importance.

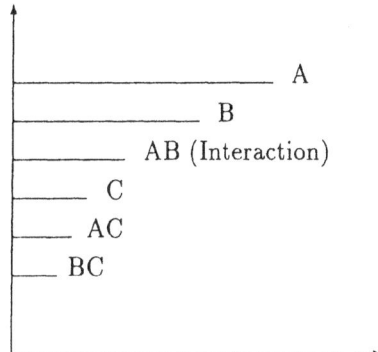

Figure 1.1: Typical Pareto-chart of a model: response = $f(A, B, C)$

1.7 An Alternative Chart in Case of Bivariate Ordinal Cause-Effect Relationships

The results of statistical analyses become undoubtly more apparent if they are accompanied by appropriate graphs and charts. Based on the Pareto-principle, one such chart has been presented in the previous section. It facilitates to find and identify the main effects and interactions. In this section, we will illustrate a method developed by Heumann et al. (1993) if bivariate cause-effect relationships for ordinal data are investigated by loglinear models. Let the response variable Y take two values:

$$Y = \begin{cases} 1 & \text{if response is success} \\ 0 & \text{otherwise.} \end{cases}$$

Let the influence variables A and B have three ordinal factor levels (low, average, high).

The loglinear model is given by

$$\ln(n_{1jk}) = \mu + \lambda_1^{\text{success}} + \lambda_j^A + \lambda_k^B + \lambda_{1j}^{\text{success}/A} + \lambda_{1k}^{\text{success}/B} \qquad (1.1)$$

Data are taken from Table 1.3.

Y		Factor B		
	Factor A	low	average	high
0	low	40	10	20
	average	60	70	30
	high	80	90	70
1	low	20	30	5
	average	60	150	20
	high	100	210	50

Table 1.3: Three-dimensional contingency table

The loglinear model with interactions (1.1)

$$Y \ / \ \text{factor A}, \qquad Y \ / \ \text{factor B}$$

yields the following parameter estimates for the main effects (Table 1.4).

Parameter	Standardized estimate
$Y = 0$	0.257
$Y = 1$	−0.257
Factor A low	−13.982
Factor A average	4.908
Factor A high	14.894
Factor B low	2.069
Factor B average	10.515
Factor B high	−10.057

Table 1.4: Main effects in model (1.1)

The estimated interactions are given in Table 1.5.

The interactions are displayed in Figures 1.2 and 1.3. The effects are shown proportional to the highest effect. Note that a comparison of main effects (shown at the border) and interactions is not possible due to different scaling. Solid circles correspond to a positive, non solid circles to a negative interaction.

Parameter	Standardized estimate
$Y = 0$/Factor A low	3.258
$Y = 0$/Factor A average	-1.963
$Y = 0$/Factor A high	-2.589
$Y = 1$/Factor A low	-3.258
$Y = 1$/Factor A average	1.963
$Y = 1$/Factor A high	2.589
$Y = 0$/Factor B low	1.319
$Y = 0$/Factor B average	-8.258
$Y = 0$/Factor B high	5.432
$Y = 1$/Factor B low	-1.319
$Y = 1$/Factor B average	8.258
$Y = 1$/Factor B high	-5.432

Table 1.5: Estimated interactions

The standardization was calculated according to

$$\text{Area effect}_i = \pi r_i^2 \tag{1.2}$$

with

$$r_i = \sqrt{\frac{\text{Estimation of effect}_i}{\max_i\{\text{Estimation of effect}_i\}}} \cdot r,$$

where r denotes the radius of the maximum effect.

Interpretation: Figure 1.2 shows, that (A low)/failure and (A high)/success are positively correlated, such that a recommendation to control is given by "A high". Analogously, we extract from Figure 1.3 the recommendation "B average".

Note: Interactions are to be assessed only within one figure and not between different figures as standardization is different. A Pareto-chart for the effects of positive response yields the following Figure 1.6, where negative effects are shown as thin- and positive effects are shown as thick lines.

Figures 1.4 and 1.5 show an alternative method to display interactions (graphics produced by MS Excel).

Example 1.1: To illustrate the principle further, we focus our attention on the cause-effect relationship between smoking and tartar. The loglinear model related to the following Table 1.6 is given by

$$\ln(n_{ij}) = \mu + \lambda_i^{\text{Smoking}} + \lambda_j^{\text{Tartar}} + \lambda_{ij}^{\text{Smoking/Tartar}} \tag{1.3}$$

The parameters stand for:

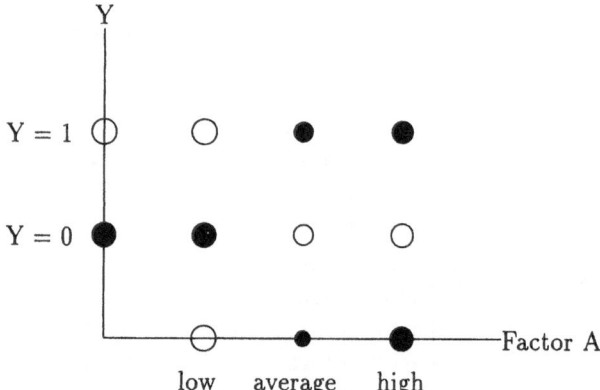

Figure 1.2: Main effects and interactions of factor A

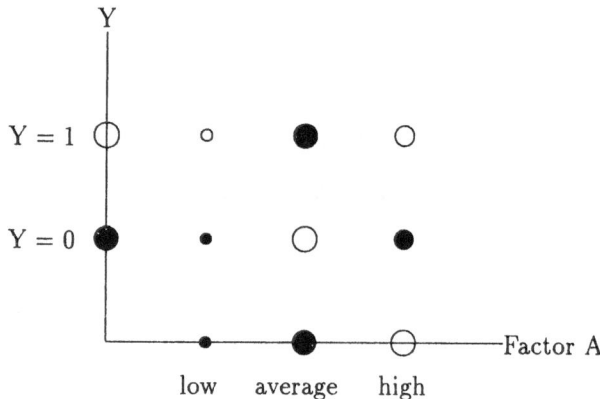

Figure 1.3: Main effects and interactions of factor B

$\lambda_i^{\text{Smoking}}$:   Main effect of the three levels non-smoker, light smoker and heavy smoker

$\lambda_j^{\text{Tartar}}$:   Main effect of the three levels (no/average/high) of tartar

$\lambda_{ij}^{R/Z}$:   Interaction smoking/tartar

Parameter estimates are given in Table 1.7.

Basically, Figure 1.7 shows a diagonal structure of interactions, where positive values are located on the main diagonal. This indicates a positive relationship between tartar and smoking. In the final figure of this section, interactions are displayed as a MS Excel graphic (Figure 1.8).

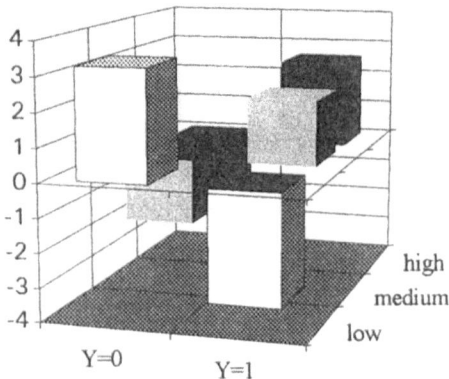

Figure 1.4: Interactions A × Response

		No tartar	Medium tartar	High-level tartar	
	j	1	2	3	$n_{i\cdot}$
i					
Non-smoker	1	284	236	48	568
Smoker, less than 6.5g per day	2	606	983	209	1798
Smoker, more than 6.5g per day	3	1028	1871	425	3324
$n_{\cdot j}$		1918	3090	682	5690

Table 1.6: Contingency table: consumption of tobacco / tartar

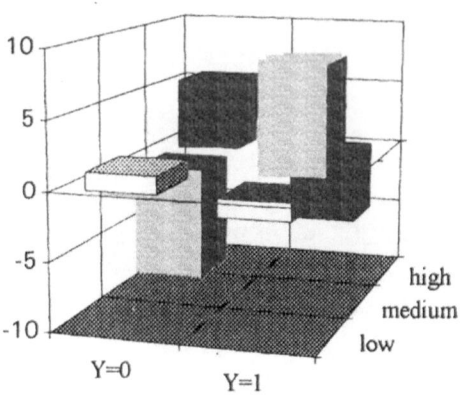

Figure 1.5: Interactions B × Response

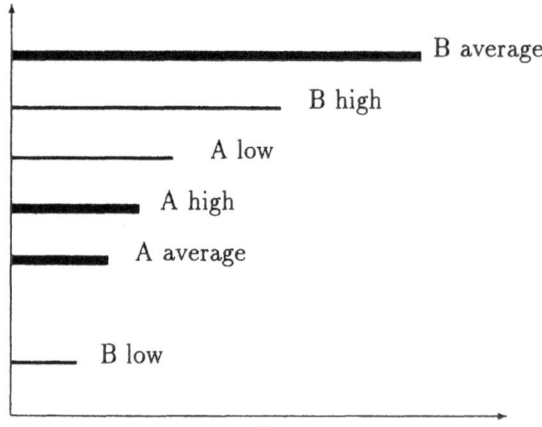

Figure 1.6: Simple Pareto-chart of a loglinear model

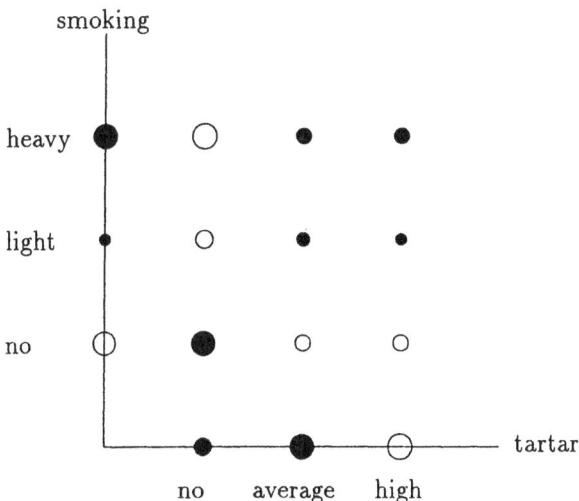

Figure 1.7: Effects in a loglinear model (1.3) displayed proportional to size

Standardized parameter estimates	Effect
−25.93277	smoking(non)
7.10944	smoking(light)
32.69931	smoking(heavy)
11.70939	tartar(no)
23.06797	tartar(average)
−23.72608	tartar(high)
7.29951	smoking(non)/tartar(no)
−3.04948	smoking(non)/tartar(average)
−2.79705	smoking(non)/tartar(high)
−3.51245	smoking(light)/tartar(no)
1.93151	smoking(light)/tartar(average)
1.17280	smoking(light)/tartar(high)
−7.04098	smoking(heavy)/tartar(no)
2.66206	smoking(heavy)/tartar(average)
3.16503	smoking(heavy)/tartar(high)

Table 1.7: Estimations in model (1.3)

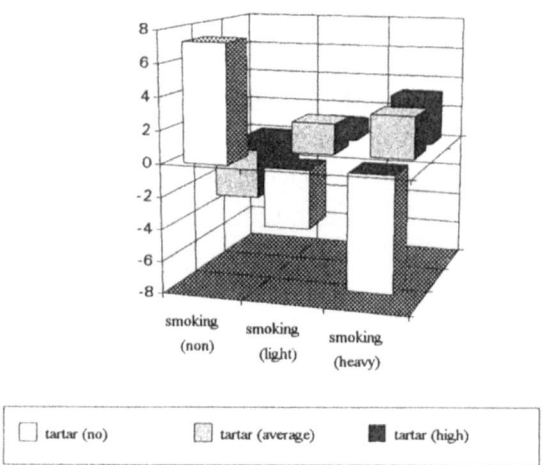

Figure 1.8: Interactions by MS Excel

1.8 A One-Way Factorial Experiment by Example

To illustrate the theory of the preceeding section, we shall consider a typical application of experimental design in agriculture. Let us assume that $n_1 = 10$ and $n_2 = 10$ plants are collected randomly out of n (homogeneous) plants. The first group is subjected to a fertilizer A and the second to a fertilizer B. After the period of vegetation, the weight (response) y of all plants is measured.

Suppose for simplicity that the response variable in the population is distributed according to $Y \sim N(\mu, \sigma^2)$. Then we have for both subpopulations (fertilizers A and B)

$$Y_A \sim N(\mu_A, \sigma^2)$$

and

$$Y_B \sim N(\mu_B, \sigma^2),$$

where the variances are assumed to be equal.

These assumptions are comprised by the following one-way factorial model, where the factor fertilizer is imposed on two levels A and B. For the actual response values we have

$$y_{ij} = \mu_i + \epsilon_{ij} \qquad (i = 1, 2 \quad j = 1, \ldots, n_i) \tag{1.4}$$

with

$$\epsilon_{ij} \sim N(0, \sigma^2)$$

and ϵ_{ij} independent, for all $i \neq j$. The null hypothesis is given by

$$H_0 : \mu_1 = \mu_2 \qquad (\text{i.e., } H_0 \colon \mu_A = \mu_B).$$

The alternative hypothesis is

$$H_1 : \mu_1 \neq \mu_2.$$

The one-way analysis of variance is equivalent to testing the equality of expected values of two samples by the t-test under normality. The test statistic in case of independent samples of size n_1 and n_2 is given by

$$t_{n_1+n_2-2} = \frac{\bar{x} - \bar{y}}{s} \sqrt{\frac{n_1 \cdot n_2}{n_1 + n_2}}, \tag{1.5}$$

where

$$s^2 = \frac{\sum_{i=1}^{n_1}(x_i - \bar{x})^2 + \sum_{j=1}^{n_2}(y_j - \bar{y})^2}{n_1 + n_2 - 2} \tag{1.6}$$

is the pooled estimate of the variance (experimental error). H_0 will be rejected, if

$$|t| > t_{n_1+n_2-2;1-\frac{\alpha}{2}}, \tag{1.7}$$

where $t_{n_1+n_2-2;1-\frac{\alpha}{2}}$ stands for the $(1 - \frac{\alpha}{2})$-quantile of the $t_{n_1+n_2-2}$-distribution. Assume the data from Table 1.8 was observed.

i	Fertilizer A x_i	$(x_i - \bar{x})^2$	Fertilizer B y_i	$(y_i - \bar{y})^2$
1	4	1	5	1
2	3	4	4	4
3	5	0	6	0
4	6	1	7	1
5	7	4	8	4
6	6	1	7	1
7	4	1	5	1
8	7	4	8	4
9	6	1	5	1
10	2	9	5	1
$\sum$	50	26	60	18

Table 1.8: One-way factorial experiment with two independent distributions

We calculate $\bar{x} = 5$, $\bar{y} = 6$ and

$$s^2 = \frac{26 + 18}{10 + 10 - 2} = \frac{44}{18} = 1.56^2,$$

$$t_{18} = \frac{5 - 6}{1.56} \sqrt{\frac{100}{20}} = -1.43,$$

$$t_{18;0.975} = 2.10,$$

such that $H_0 : \mu_A = \mu_B$ cannot be rejected.

The underlying assumption of the above test is that both subpopulations can be characterized by identical distributions which may differ only in location. This assumption should be checked carefully, as (insignificant) differences may have their reason in inhomogeneous populations. This inhomogeneity leads to an increase of the experimental error and makes it difficult to detect different factor effects.

Pairwise comparisons (paired *t*-test)

Another experimental setup that arises frequently in the analysis of biomedical data is given if two factor levels are subjected consecutively to the same object or person. After the first treatment a *wash-out period* is established, in which the response variable is traced back to its original level.

Consider, for example, two alternative pesticides A and B which should reduce lice attack on plants. Each plant is treated initially by method A before the concentration of lice is measured. Then, after some time, each plant is treated by method B and again the concentration is measured. The underlying statistical model is given by

$$y_{ij} = \mu_i + \beta_j + \epsilon_{ij} \qquad \begin{cases} i = 1, 2 \\ j = 1, \ldots, J \end{cases} \tag{1.8}$$

where

y_{ij} : concentration in plant j after treatment i
μ_i : effect of treatment i
β_j : effect of jth replication

and ϵ_{ij} is the experimental error. A comparison of the treatments is possible by inspecting the individual differences

$$d_j = y_{1j} - y_{2j} \qquad j = 1, \ldots, J \tag{1.9}$$

of concentrations on one specific plant. We derive

$$\begin{aligned} \mu_d := \mathrm{E}(d_j) &= \mathrm{E}(y_{1j} - y_{2j}) \\ &= \mu_1 + \beta_j - \mu_2 - \beta_j \\ &= \mu_1 - \mu_2. \end{aligned}$$

Testing $\mathrm{H}_0 : \mu_1 = \mu_2$ is therefore equivalent to testing for significance of $\mathrm{H}_0 : \mu_d = 0$. In this situation the paired t-test for one sample may be applied assuming $d_i \sim N(0, \sigma_d^2)$:

$$t_{n-1} = \frac{\bar{d}}{s_d} \sqrt{n} \tag{1.10}$$

with

$$s_d^2 = \frac{\sum(d_i - \bar{d})^2}{n - 1} \quad .$$

H_0 is rejected if

$$|t_{n-1}| > t_{n-1;1-\frac{\alpha}{2}}.$$

Let us assume that the data shown in Table (1.9) was observed (i.e., the same data as in Table (1.8)): We get

$$\bar{d} = -1,$$

$$s_d^2 = \frac{8}{9} = 0.94^2,$$

$$t_9 = \frac{-1}{0.94} \sqrt{10} = -3.36,$$

$$t_{9;0.975} = 2.26,$$

such that $\mathrm{H}_0 : \mu_1 = \mu_2$ (i.e., $\mu_A = \mu_B$) is rejected, which confirms that method A is superior to B.

j	y_{1j}	y_{2j}	d_j	$(d_j - \bar{d})^2$
1	4	5	-1	0
2	3	4	-1	0
3	5	6	-1	0
4	6	7	-1	0
5	7	8	-1	0
6	6	7	-1	0
7	4	5	-1	0
8	7	8	-1	0
9	6	5	1	4
10	2	5	-3	4
$\sum$			-10	8

Table 1.9: Pairwise experimental design

If we compare the two experimental designs – completely randomized design and randomized block design – a loss in degrees of freedom becomes apparent in the latter design. The respective confidence intervals are given by

$$(\bar{x} - \bar{y}) \pm t_{18;0.975}\, s\sqrt{\frac{n_1 + n_2}{n_1 n_2}},$$

$$-1 \pm 2.10 \cdot 1.56\sqrt{\frac{20}{100}},$$

$$-1 \pm 1.46,$$

$$[-2.46, +0.46],$$

and

$$\bar{d} \pm t_{9;0.975}\frac{s_d}{\sqrt{n}},$$

$$-1 \pm 2.26\frac{0.94}{\sqrt{10}},$$

$$-1 \pm 0.67,$$

$$[-1.67; -0.33].$$

We observe a higher concentration in the second experiment. A comparison of the respective variances $s^2 = 1.56^2$ and $s_d^2 = 0.94^2$ indicates that a reduction of the experimental error to $\frac{0.94}{1.56} \cdot 100 = 60\%$ was achieved by blocking.

Note that these positive effects of blocking depend on the homogeneity of variances within each block. In Chapter 4 we will discuss this topic in detail.

1.9   Exercises and Questions

1.9.1 Describe the basic principles of experimental design.

1.9.2 Why are control groups useful?

1.9.3 To what type of scaling does the following data belong:

- Male/female

- Catholic, Protestant

- Pressure

- Temperature

- Tax category

- Small car, car in the middle range, luxury limousine

- Age

- Lenght of stay of a patient in a clinical trial

- University degrees?

1.9.4 What is the difference between direct and indirect measurements?

1.9.5 What are ties and their consequences in a set of data?

1.9.6 What is a Pareto-chart?

1.9.7 Describe problems occuring in experimental setups with paired observations.

2 Comparison of Two Samples

2.1 Introduction

Problems of comparing two samples arise frequently in medicine, sociology, agriculture, engineering, or marketing. The data may have been generated by observation or may be the outcomes of a controlled experiment. In the latter case, randomization plays a crucial role in gaining information about possible differences in the samples which may be due to a specific factor. Randomization means, for example, that in a controlled clinical trial there is a constant chance for every patient of getting a specific treatment. The idea of a blind, double blind, or even triple blind setup of the experiment is that neither patient, nor clinician, nor statistician know what treatment has been given. This should exclude possible biases in the response variable, that would be induced by such knowledge. It becomes clear that careful planning is indispensible to achieve valid results.

Another problem in the framework of a clinical trial may consist in the fact of a systematic effect of a subgroup of patients, for example males and females. If such a situation is to be expected, one should stratify the sample into homogeneous subgroups. Such a strategy proves to be useful in planned experiments as well as observational studies.

Another experimental setup is given by a *matched-pair design*. Subgroups then contain only one individual and pairs of subgroups are compared w.r.t. different treatments. This procedure requires pairs to be homogeneous w.r.t. all possible factors that may exhibit an influence on the response variable and is thus limited to very special situations.

2.2 Paired *t*-Test and Matched-Pair Design

In order to illustrate the basic reasoning of a matched-pair design, consider an experiment, the structure of which is given by Table 2.1.

We consider the linear model already given in (1.8). Assuming that

$$d_i \overset{i.i.d.}{\sim} N(\mu_d, \sigma_d^2), \tag{2.1}$$

Pair	Treatment 1	Treatment 2	Difference
1	y_{11}	y_{21}	$y_{11} - y_{21} = d_1$
2	y_{12}	y_{22}	$y_{12} - y_{22} = d_2$
$\vdots$	$\vdots$	$\vdots$	$\vdots$
n	y_{1n}	y_{2n}	$y_{1n} - y_{2n} = d_n$
			$\bar{d} = \frac{\sum d_i}{n}$

Table 2.1: Response in a matched-pair design

the best linear unbiased estimator of $\bar{d}$ of μ_d is distributed as

$$\bar{d} \sim N(\mu_d, \frac{\sigma_d^2}{n}) \quad . \tag{2.2}$$

An unbiased estimator of σ_d^2 is given by

$$s_d^2 = \frac{\sum_{i=1}^{n}(d_i - \bar{d})^2}{n-1} \sim \frac{\sigma_d^2}{n-1} \chi_{n-1}^2 \tag{2.3}$$

such that under H_0: $\mu_d = 0$ the ratio

$$t = \frac{\bar{d}}{s_d}\sqrt{n} \tag{2.4}$$

is distributed according to a (central) t-distribution.
A two-sided test for H_0: $\mu_d = 0$ vs. H_1: $\mu_d \neq 0$ rejects H_0, if

$$|t| > t_{n-1;1-\alpha}(\text{two-sided}) = t_{n-1;1-\frac{\alpha}{2}} \quad . \tag{2.5}$$

A one-sided test H_0: $\mu_d = 0$ vs. H_1: $\mu_d > 0$ $(\mu_d < 0)$ rejects H_0 in favour of H_1: $\mu_d > 0$, if

$$t > t_{n-1;1-\alpha} \quad . \tag{2.6}$$

H_0 is rejected in favour of H_1: $\mu_d < 0$, if

$$t < -t_{n-1;1-\alpha} \quad . \tag{2.7}$$

Necessary Sample Size and Power of the Test

We consider a test of H_0 vs. H_1 for a distribution with an unknown parameter θ. Obviously, there are four possible situations, two of which lead to a correct decision. The probability

$$P_\theta(\text{reject } H_0 | \ H_0 \text{ true}) = P_\theta(H_1 | H_0) \leq \alpha \quad \text{for all} \quad \theta \in H_0 \tag{2.8}$$

is called probability of a *type I error*. α is to be fixed before the experiment. Usually, $\alpha = 0.05$ is a reasonable choice.

| | Real situation | |
Decision	H_0 true	H_0 false
H_0 accepted	correct decision	false decision
H_0 rejected	false decision	correct decision

Table 2.2: Test decisions

The probability

$$P_\theta(\text{accept } H_0 \mid H_0 \text{ false}) = P_\theta(H_0|H_1) \geq \beta \quad \text{for all} \quad \theta \in H_1 \qquad (2.9)$$

is called probability of a *type II error*. Obviously, this probability depends on the true value of θ such that the function

$$G(\theta) = P_\theta(\text{reject } H_0) \qquad (2.10)$$

is called *power* of the test. Generally, a test to a given α aims to fix the type II error at a defined level or beyond. Equivalently, we could say that the power should reach or even exceed a given value. Moreover the following rules apply:

(i) the power rises as the sample size n increases, keeping α and the parameters under H_1 fixed,

(ii) the power rises and therefore β decreases as α increases, keeping n and the parameters under H_1 fixed,

(iii) the power rises as the difference δ between the parameters under H_0 and under H_1 increases.

We keep in mind that the power of a test depends on the difference δ, the type I error, the sample size n and on the hypothesis being one- or two-sided. Changing from a one-sided to a two-sided problem reduces the power.

The comparison of means in a matched-pair design yields the following relationship. Consider a one-sided test (H_0: $\mu_d = \mu_0$ vs. H_1: $\mu_d = \mu_0 + \delta$, $\delta > 0$) and a given α. To start with, we assume σ_d^2 to be known. We now try to derive the sample size n that is required to achieve a fixed power of $1 - \beta$ for a given α and known σ_d^2. This means that we have to settle n in a way that H_0: $\mu_d = \mu_0$ with fixed α is accepted with probability β, although the true parameter is $\mu_d = \mu_0 + \delta$. We define

$$u := \frac{\bar{d} - \mu_0}{\sigma_d/\sqrt{n}} \quad .$$

Then under H_1: $\mu_d = \mu_0 + \delta$ we have

$$\tilde{u} = \frac{\bar{d} - (\mu_0 + \delta)}{\sigma_d/\sqrt{n}} \sim N(0,1) \quad . \qquad (2.11)$$

$\tilde{u}$ and u are related as follows

$$u = \tilde{u} + \frac{\delta}{\sigma_d}\sqrt{n} \sim N(\frac{\delta}{\sigma_d}\sqrt{n}, 1) \quad . \tag{2.12}$$

The null hypothesis H_0: $\mu_d = \mu_0$ is accepted erroneously if the test statistic u has a value of $u \leq u_{1-\alpha}$.

The probability for this case should be $\beta = P(H_0|H_1)$. So we get

$$\begin{aligned} \beta &= P(u \leq u_{1-\alpha}) \\ &= P(\tilde{u} \leq u_{1-\alpha} - \frac{\delta}{\sigma_d}\sqrt{n}) \end{aligned}$$

and therefore

$$u_\beta = u_{1-\alpha} - \frac{\delta}{\sigma_d}\sqrt{n} \quad ,$$

which yields

$$n \geq \frac{(u_{1-\alpha} - u_\beta)^2 \sigma_d^2}{\delta^2} \tag{2.13}$$

$$= \frac{(u_{1-\alpha} + u_{1-\beta})^2 \sigma_d^2}{\delta^2} \quad . \tag{2.14}$$

For application in practice, we have to estimate σ_d^2 in (2.13). If we estimate σ_d^2 using the sample variance, we also have to replace $u_{1-\alpha}$ and $u_{1-\beta}$ by $t_{n-1;1-\alpha}$ and $t_{n-1;1-\beta}$, respectively. The value of δ is the difference of expectations of the two parameter ranges, which is either known or estimated using the sample.

2.3 Comparison of Means in Independent Groups

2.3.1 Two Sample t-Test

We already discussed the two sample problem in Section 1.8. Now we consider the two independent samples

$$\begin{aligned} A &: & x_1, \ldots, x_{n_1} &,& x_i &\sim N(\mu_A, \sigma_A^2) \\ B &: & y_1, \ldots, y_{n_2} &,& y_i &\sim N(\mu_B, \sigma_B^2) \quad . \end{aligned}$$

Assuming $\sigma_A^2 = \sigma_B^2 = \sigma^2$, we may apply the linear model (1.4). To compare the two groups A and B we test the hypothesis H_0: $\mu_A = \mu_B$ using the statistic (1.5), i.e., $t_{n_1+n_2-2} = \frac{\bar{x}-\bar{y}}{s}\sqrt{\frac{n_1 n_2}{n_1+n_2}}$. In practical applications, we have to check the assumption that $\sigma_A^2 = \sigma_B^2$.

2.3.2 Testing H_0: $\sigma_A^2 = \sigma_B^2 = \sigma^2$

Under H_0, the two independent sample variances

$$s_x^2 = \frac{1}{n_1 - 1} \sum_{i=1}^{n_1} (x_i - \bar{x})^2$$

and

$$s_y^2 = \frac{1}{n_2 - 1} \sum_{i=1}^{n_2} (y_i - \bar{y})^2$$

follow a χ^2-distribution with $n_1 - 1$ and $n_2 - 1$ degrees of freedom, respectively, and their ratio follows an F-distribution

$$F = \frac{s_x^2}{s_y^2} \sim F_{n_1-1, n_2-1} \quad . \tag{2.15}$$

Decision

Two-sided: $H_0: \sigma_A^2 = \sigma_B^2$ vs. $H_1: \sigma_A^2 \neq \sigma_B^2$
H_0 is rejected if

$$\left. \begin{array}{l} F > F_{n_1-1, n_2-1; 1-\frac{\alpha}{2}} \\ \text{or} \\ F < F_{n_1-1, n_2-1; \alpha/2} \end{array} \right\} \tag{2.16}$$

with

$$F_{n_1-1, n_2-1; \alpha/2} = \frac{1}{F_{n_1-1, n_2-1; 1-\frac{\alpha}{2}}} \quad . \tag{2.17}$$

One-sided: $H_0: \sigma_A^2 = \sigma_B^2$ vs. $H_1: \sigma_A^2 > \sigma_B^2$
If

$$F > F_{n_1-1, n_2-1; 1-\alpha} \tag{2.18}$$

then H_0 is rejected.

Remark: The larger sample variance always acts as the numerator of F in the case of one-sided testing.

Example 2.1: Using the data set of Table 1.8, we want to test $H_0: \sigma_A^2 = \sigma_B^2$. In Table 1.8, we find the values $n_1 = n_2 = 10$, $s_A^2 = \frac{26}{9}$ and $s_B^2 = \frac{18}{9}$. This yields

$$F = \frac{26}{18} = 1.44 < 3.18 = F_{9,9;0.95}$$

so that we cannot reject the null hypothesis $H_0: \sigma_A^2 = \sigma_B^2$ vs. $H_1: \sigma_A^2 > \sigma_B^2$ according to (2.18). Therefore, our analysis in Section 1.8 was correct.

2.3.3 Comparison of Means in the Case of Unequal Variances

If $H_0: \sigma_A^2 = \sigma_B^2$ is not valid, we are up against the so-called Behrens-Fisher-problem, which has no exact solution. For practical use, the following correction of the test statistic according to *Welch* gives sufficiently good results

$$t = \frac{|\bar{x} - \bar{y}|}{\sqrt{\dfrac{s_x^2}{n_1} + \dfrac{s_y^2}{n_2}}} \sim t_v \tag{2.19}$$

with degrees of freedom approximated with

$$v = \frac{\left(\frac{s_x^2}{n_1} + \frac{s_y^2}{n_2}\right)^2}{\frac{(s_x^2/n_1)^2}{n_1+1} + \frac{(s_y^2/n_2)^2}{n_2+1}} - 2 \qquad (2.20)$$

(v is rounded). We have $\min(n_1 - 1, n_2 - 1) < v < n_1 + n_2 - 2$.

Example 2.2: In material-testing, two normal variables A and B were examined. The sample parameters are summarized as follows:

$$\bar{x} = \quad 27.99 \quad , \quad s_x^2 = \quad 5.98^2 \quad , \quad n_1 = \quad 9$$
$$\bar{y} = \quad 1.92 \quad , \quad s_y^2 = \quad 1.07^2 \quad , \quad n_2 = \quad 10$$

The sample variances are not equal:

$$F = \frac{5.98^2}{1.07^2} = 31.23 > 3.23 = F_{8,9;0.95} \quad .$$

Therefore, we have to use *Welch's test* to compare the means.

$$t_v = \frac{|27.99 - 1.92|}{\sqrt{\frac{5.98^2}{9} + \frac{1.07^2}{10}}} = 12.89$$

with $v \approx 9$ degrees of freedom. The critical value of $t_{9;0.975} = 2.26$ is exceeded and we reject H_0: $\mu_A = \mu_B$.

2.3.4   Transformations of Data to Assure Homogeneity of Variances

We know from experience that the two sample t-test is more sensitive to discrepencies in the homogeneity of variances than to deviations from the assumption of normal distribution. The two sample t-test usually reaches the level of significance if the assumption of normal distribution is not fully justified but sample sizes are large enough ($n_1, n_2 > 20$) and the homogeneity of variances is valid. This result is based on the central limit theorem. Analogously, deviations from variance homogeneity can have severe effects on the level of significance.

The following transformations may be used to avoid inhomogeneity of variances

- logarithmic transformation $\ln(x_i)$, $\ln(y_i)$

- logarithmic transformation $\ln(x_i + 1)$, $\ln(y_i + 1)$, esp. if x_i and y_i have zero values or if $0 \leq x_i, y_i \leq 10$ (Woolson, 1987, p. 171).

2.3.5 Necessary Sample Size and Power of the Test

The necessary sample size to achieve a desired power of the two sample t-test is derived like in the paired t-test problem. Let $\delta = \mu_A - \mu_B > 0$ be the one-sided alternative to be tested against H_0: $\mu_A = \mu_B$ with $\sigma_A^2 = \sigma_B^2 = \sigma^2$. Then with $n_2 = a \cdot n_1$ (if $a = 1$ then $n_1 = n_2$), the minimum sample size to preserve a power of $1 - \beta$ (cf. (2.13)) is given by

$$n_1 = \sigma^2 (1 + \frac{1}{a})(u_{1-\alpha} + u_{1-\beta})^2 / \delta^2 \tag{2.21}$$

and

$$n_2 = a \cdot n_1 \quad \text{with } n_1 \text{ from (2.21).}$$

2.3.6 Comparison of Means Without Prior Testing H_0: $\sigma_A^2 = \sigma_B^2$; Cochran-Cox Test for Independent Groups

There are several alternative methods to be used instead of the two sample t-test in case of unequal variances. The test of Cochran and Cox (1957) uses a statistic which approximately follows a t-distribution. The Cochran-Cox test is conservative compared to the usually used t-test. Substantially, this fact is due to the special number of degrees of freedom that has to be used. The degrees of freedom of this test are a weighted average of $n_1 - 1$ and $n_2 - 1$. In the balanced case ($n_1 = n_2 = n$) the Cochran-Cox test has $n - 1$ degrees of freedom compared to $2(n - 1)$ degrees of freedom used in the two sample t-test. The test statistic

$$t_{c-c} = \frac{\bar{x} - \bar{y}}{s_{(\bar{x}-\bar{y})}} \tag{2.22}$$

with

$$s_{(\bar{x}-\bar{y})}^2 = \frac{s_x^2}{n_1} + \frac{s_y^2}{n_2}$$

has critical values at

two-sided:
$$t_{c-c(1-\alpha/2)} = \frac{\frac{s_x^2}{n_1} t_{n_1-1;1-\alpha/2} + \frac{s_y^2}{n_2} t_{n_2-1;1-\alpha/2}}{s_{(\bar{x}-\bar{y})}^2} \tag{2.23}$$

one-sided:
$$t_{c-c(1-\alpha)} = \frac{\frac{s_x^2}{n_1} t_{n_1-1;1-\alpha} + \frac{s_y^2}{n_2} t_{n_2-1;1-\alpha}}{s_{(\bar{x}-\bar{y})}^2} \tag{2.24}$$

The null hypothesis is rejected if $|t_{c-c}| > t_{c-c}(1 - \alpha/2)$ (two-sided), resp., $t_{c-c} > t_{c-c}(1 - \alpha)$ (one-sided, H_1: $\mu_A > \mu_B$).

Example 2.3: (Example 2.2 continued)
We test H_0: $\mu_A = \mu_B$ using the one-sided Cochran-Cox-test. With

$$\begin{aligned} s_{(\bar{x}-\bar{y})}^2 &= \frac{5.98^2}{9} + \frac{1.07^2}{10} \\ &= 3.97 + 0.11 = 4.08 = 2.02^2 \end{aligned}$$

and (one-sided)

$$t_{c-c(1-\alpha)} = \frac{3.97 \cdot 1.86 + 0.11 \cdot 1.83}{4.08}$$
$$= 1.86 \quad,$$

we get $t_{c-c} = \frac{27.99 - 1.92}{2.02} = 12.91 > 1.86$, so that H_0 has to be rejected.

2.4 Wilcoxon's Sign-Rank-Test in the Matched-Pair Design

Wilcoxon's test for differences of pairs is the nonparametric analogon to the paired t-test. This test can be applied to continuous (not necessarily normal distributed) response. The test allows to check whether the differences $y_{1i} - y_{2i}$ of paired observations (y_{1i}, y_{2i}) are symmetrically distributed with median M $=$ 0.

In the two-sided test problem, the hypothesis is given by

$$\left.\begin{array}{l} H_0: \quad M = 0 \quad \text{or, equivalently,} \quad H_0: \quad P(Y_1 < Y_2) = 0.5 \\ \text{versus} \\ H_1: \quad M \neq 0 \end{array}\right\} \qquad (2.25)$$

and in the one-sided test problem

$$H_0: \quad M \leq 0 \quad \text{versus} \quad H_1: \quad M > 0 \quad . \qquad (2.26)$$

Assuming $Y_1 - Y_2$ being distributed symmetrically, the relation $f(-d) = f(d)$ holds for each value of the difference $D = Y_1 - Y_2$, with $f(\cdot)$ denoting the density function of the difference variable. Therefore, we can expect under H_0 that the ranks of absolute differences $|d|$ are equally distributed amongst negative and positive differences. We put the absolute differences in ascending order and note the sign of each difference $d_i = y_{1i} - y_{2i}$. Then, we sum over the ranks of absolute difference with positive sign (or analogously with negative sign) and get the following statistic (cf. Büning and Trenkler, 1978, p. 187)

$$W^+ = \sum_{i=1}^{n} Z_i R(|d_i|) \qquad (2.27)$$

with

$$\begin{array}{rcl} d_i &=& y_{1i} - y_{2i} \quad, \\ R(|d_i|) &:& \text{rank of } |d_i|, \\ Z_i &=& \left\{\begin{array}{lcl} 1 &:& d_i > 0 \\ 0 &:& d_i < 0 \end{array}\right. \end{array} \qquad (2.28)$$

We also could sum over the ranks of negative differences (W^-) and get the relationship $W^+ + W^- = n(n+1)/2$.

Exact Distribution of W^+ under H_0

The term W^+ can also be expressed as

$$W^+ = \sum_{i=1}^{n} i Z_{(i)} \quad \text{with} \quad Z_{(i)} = \begin{cases} 1 & : \quad D_j > 0 \\ 0 & : \quad D_j < 0 \end{cases} \tag{2.29}$$

In this case D_j denotes the difference for which $r(|D_j|) = i$ for given i. Under H_0: $M = 0$ the variable W^+ is symmetrically distributed with center

$$\mathrm{E}(W^+) = \mathrm{E}(\sum_{i=1}^{n} i\, Z_{(i)}) = \frac{n(n+1)}{4}.$$

The sample space may be regarded as a set L of all n-tuples built of 1 or 0. L itself consists of 2^n elements and each of it has probability $1/2^n$ under H_0. Hence, we get

$$P(W^+ = w) = \frac{a(w)}{2^n} \tag{2.30}$$

with $a(w)$: number of possibilities to assign $+$ signs to the numbers from 1 to n in a manner that leads to the sum w.

Example: Let $n = 4$. The exact distribution of W^+ under H_0 can be found in the last column of the table.

w	Tuple of Ranks	$a(w)$	$P(W^+ = w)$
10	(1 2 3 4)	1	1/16
9	(2 3 4)	1	1/16
8	(1 3 4)	1	1/16
7	(1 2 4) , (3 4)	2	2/16
6	(1 2 3) , (2 4)	2	2/16
5	(1 4) , (2 3)	2	2/16
4	(1 3) , (4)	2	2/16
3	(1 2) , (3)	2	2/16
2	(2)	1	1/16
1	(1)	1	1/16
0		1	1/16

$\sum$: $16/16 = 1$

E.g., $P(W^+ \geq 8) = 3/16$.

Testing

Test A: H_0: $M = 0$ is rejected versus H_1: $M \neq 0$, if $W^+ \leq w_{\alpha/2}$ or $W^+ \geq w_{1-\alpha/2}$.
Test B: H_0: $M \leq 0$ is rejected versus H_1: $M > 0$, if $W^+ \geq w_{1-\alpha}$.

The exact critical values can be found in tables (e.g., Table H, p. 373 in Büning and Trenkler, 1978). For large sample sizes $(n > 20)$ we can use the following approximation

$$Z = \frac{W^+ - E(W^+)}{\sqrt{Var(W^+)}} \overset{H_0}{\sim} N(0,1) \quad ,$$

i.e.,

$$Z = \frac{W^+ - \dfrac{n(n+1)}{4}}{\sqrt{\dfrac{n(n+1)(2n+1)}{24}}} . \tag{2.31}$$

For both tests, H_0 is rejected if $|Z| > u_{1-\alpha/2}$ resp. $Z > u_{1-\alpha}$.

Ties

Ties may occur as *zero-differences* $(d_i = y_{1i} - y_{2i} = 0)$ and/or as *compound-differences* $(d_i = d_j$ for $i \neq j)$. Depending on the type of ties, we use one of the following tests

- zero-differences test

- compound-differences test

- zero- plus compound-differences test.

The following methods are comprehensively described in Lienert (1986, p. 327-332).

Zero-Differences Test

(a) Sample reduction method of Wilcoxon and Hemelrijk:
This method is used if the sample size is large enough $(n > 10)$ and the percentage of ties is less than 10% $(t_0/n \leq 1/10,$ with t_0 denoting the number of zero-differences.)
Zero-differences are excluded from the sample and the test is conducted using the remaining $n_0 = n - t_0$ pairs.

(b) Pratt's partial-rank randomization method:
This method is used for small sample sizes with more than 10% of zero-differences.
The zero-differences are included during the association of ranks but are excluded from the test statistic. The exact distribution of W_0^+ under H_0 is calculated for the remaining n_0 signed ranks. The probabilities of rejection are given by

– Test A (two-sided)

$$P_0' = \frac{2A_0' + a_0'}{2n_0}$$

– Test B (one-sided)

$$P_0' = \frac{A_0' + a_0'}{2n_0}$$

Here A_0' denotes the number of orderings which give $W_0^+ > w_0$ and a_0' the number of orderings, which give $W_0^+ = w_0$.

(c) Cureton's asymptotic version of the partial-rank randomization test:
This test is used for large sample sizes and many zero-differences ($t_0/n > 0.1$). The test statistic is given by

$$Z_{W_0} = \frac{W_0^+ - \mathrm{E}\left(W_0^+\right)}{\sqrt{\mathrm{Var}\left(W_0^+\right)}}$$

with

$$\mathrm{E}\left(W_0^+\right) = \frac{n(n+1) - t_0(t_0+1)}{4}$$

$$\mathrm{Var}\left(W_0^+\right) = \frac{n(n+1)(2n+1) - t_0(t_0+1)(2t_0+1)}{24}$$

Under H_0, the statistic Z_{W_0} follows asymptotically the standard normal distribution.

Compound-Differences Test

(a) Shared-ranks randomization method:
In small samples and for any percentage of compound-differences we assign averaged ranks to the compound-differences. The exact distribution and one and two-sided critical values are calculated as shown in 1.b.

(b) Approximated compound-differences test:
If we have a larger sample ($n > 10$) and a small percentage of compound-differences ($t/n \leq 1/5$ with t = number of compound-differences), then we assign averaged ranks to the compounded values. The test statistic is calculated and tested as usual.

(c) Asymptotic sign rank test corrected for ties:
This method is useful for large samples with $t/n > 1/5$.
In equation (2.32) we replace $\mathrm{Var}(W^+)$ by a corrected variance (due to the association of ranks) $\mathrm{Var}(W_{corr.}^+)$, given by

$$\mathrm{Var}\left(W_{corr.}^+\right) = \frac{n(n+1)(2n+1)}{24} - \sum_{j=1}^{r} \frac{t_j^3 - t_j}{48} \quad ,$$

with r denoting the number of groups of ties and t_j the number of ties in the jth group ($1 \le j \le r$). Unbounded observations are regarded as groups of size 1. If there are no ties, then $r = n$ and $t_j = 1$ for all j, e.g., the correction term becomes zero.

Zero- plus Compound-Differences Test

These tests are used if there are both zero- and compound-differences.

(a) Pratt's randomization method:
   For small samples which are cleared up for zeros ($n_0 \le 10$), we proceed like in 1.b. but additionally assign averaged ranks to the compound-differences.

(b) Cureton's approximation method:
   In larger zero-cleared samples the test statistic is calculated analogously to (a). The expectation $\mathrm{E}\,(W_0^+)$ equals that in 1.c. and is given by

$$\mathrm{E}\,(W_0^+) = \frac{n(n+1) - t_0(t_0+1)}{4} \quad .$$

The variance in 1.c. has to be corrected due to ties and is given by

$$\mathrm{Var}_{\,corr.}(W_0^+) = \frac{n(n+1)(2n+1) - t_0(t_0+1)(2t_0+1)}{24} - \sum_{j=1}^{r} \frac{t_j^3 - t_j}{48} \quad .$$

Finally the test statistic is given by

$$Z_{W_0 - corr.} = \frac{W_0^+ - \mathrm{E}\,(W_0^+)}{\sqrt{\mathrm{Var}_{\,corr.}(W_0^+)}} \quad . \tag{2.32}$$

2.5 Rank Test for Homogeneity of Wilcoxon, Mann and Whitney

We consider two independent continuous random variables X and Y with unkown distribution or non-normal distribution. We would like to test whether the samples of the two variables are samples of the same population (homogeneity). The so-called U-test of Wilcoxon, Mann and Whitney is a rank test. It is the non parametric analogon of the t-test and is used if the assumptions for the use of the t-test are not justified or called in question. The relative efficiancy of the U-Test compared to the t-test is about 95% in case of normally distributed variables. The U-test is often used as a quick test or as a control if the test statistic of the t-test gives values close to the critical values.

The hypothesis to be tested is H_0: The probability P to observe a value from the first population X that is greater than any given value of the population Y is equal to 0.5. The two-sided alternative is H_1: $P \ne 0.5$. The one-sided alternative H_1: $P > 0.5$ means that X is *stochastically greater than* Y.

We combine the observations of the samples $(x_1, \ldots, x_m)$ and $(y_1, \ldots, y_n)$ in ascending order of ranks and note for each rank the sample it belongs to. Let R_1 and R_2 denote the sum of ranks of the X-sample and Y-sample, respectively. The test statistic U is the smaller one of the values U_1 and U_2:

$$U_1 = m \cdot n + \frac{m(m+1)}{2} - R_1 \qquad (2.33)$$

$$U_2 = m \cdot n + \frac{n(n+1)}{2} - R_2 \quad , \qquad (2.34)$$

with $U_1 + U_2 = m \cdot n$ (control).

H_0 is rejected if $U \leq U(m, n; \alpha)$ (Table 2.3 contains some values for $\alpha = 0.05$ (one-sided) and $\alpha = 0.10$ (two-sided)).

					n				
m	2	3	4	5	6	7	8	9	10
4	–	0	1						
5	0	1	2	4					
6	0	2	3	5	7				
7	0	2	4	6	8	11			
8	1	3	5	8	10	13	15		
9	1	4	6	9	12	15	18	21	
10	1	4	7	11	14	17	20	24	27

Table 2.3: Critical values of the U-test ($\alpha = 0.05$ one-sided, $\alpha = 0.10$ two-sided)

In case of m and $n \geq 8$, the excellent approximation

$$u = \frac{U - \frac{m \cdot n}{2}}{\sqrt{\frac{m \cdot n(m+n+1)}{12}}} \sim N(0, 1) \qquad (2.35)$$

is used. For $|u| > u_{1-\alpha/2}$ the hypothesis H_0 is rejected (type I error α two-sided and $\alpha/2$ one-sided).

Example 2.4: We test the equality of means of the two series of measurements given in Table 2.4 using the U-test. Let variable X be the flexibility of PMMA with silan and variable Y be the flexibility of PMMA without silan. We put the $(16 + 15)$ values of both series in ascending order, apply ranks and calculate the sums of ranks $R_1 = 231$ and $R_2 = 265$ (Table 2.5).
Then we get

$$U_1 = 16 \cdot 15 + \frac{16(16+1)}{2} - 231 = 145$$

$$U_2 = 16 \cdot 15 + \frac{15(15+1)}{2} - 265 = 95$$

$$U_1 + U_2 = 240 = 16 \cdot 15 \quad .$$

PMMA 2.2 Vol% quartz without silan	PMMA 2.2 Vol% quartz with silan
98.47	106.75
106.20	111.75
100.74	96.67
98.72	98.70
91.42	118.61
108.17	111.03
98.36	90.92
92.36	104.62
80.00	94.63
114.43	110.91
104.99	104.62
101.11	108.77
102.94	98.97
103.95	98.78
99.00	102.65
106.05	
$\bar{x} = 100.42$	$\bar{y} = 103.91$
$s_x^2 = 7.9^2$	$s_y^2 = 7.6^2$
$n = 16$	$m = 15$

Table 2.4: Flexibility of PMMA with and without silan (cf. Toutenburg et al., 1991, p.100)

Since $m = 16$ and $n = 15$ (both sample sizes ≥ 8), we calculate the test statistic according to (2.35) with $U = U_2$ being the smaller of the two values of U:

$$u = \frac{95 - 120}{\sqrt{\frac{240(16+15+1)}{12}}} = -\frac{25}{\sqrt{640}} = -0.99 \quad ,$$

and therefore $|u| = 0.99 < 1.96 = u_{1-0.05/2} = u_{0.975}$.

The null hypothesis is not rejected (type I error 5% and 2.5% using two- and one-sided alternative, respectively). The exact critical value of U is $U(16, 15, 0.05_{\text{two-sided}}) = 70$ (Tables in Sachs, 1974, p. 232), i.e., the decision is the same (H_0 is not rejected).

Correction of the U-Statistic in Case of Equal Ranks

If observations occur more than once in the combined and ordered samples $(x_1, \ldots, x_m)$ and $(y_1, \ldots, y_n)$, we assign an averaged rank to each of them. The

Rank	1	2	3	4	5	6	7	8	9
Observation	80.00	90.92	91.42	92.36	94.63	96.67	98.36	98.47	98.70
Variable	X	Y	X	X	Y	Y	X	X	Y
Sum of ranks X	1		+3	+4			+7	+8	
Sum of ranks Y		2			+5	+6			+9

Rank	10	11	12	13	14	15	16	17
Observation	98.72	98.78	98.97	99.00	100.47	101.11	102.65	102.94
Variable	X	Y	Y	X	X	X	Y	X
Sum of ranks X	+10	+11		+13	+14	+15		+17
Sum of ranks Y			+12				+16	

Rank	18	19	20	21	22	23	24
Observation	103.95	104.62	104.75	104.99	106.05	106.20	106.75
Variable	X	Y	Y	X	X	X	Y
Sum of ranks X	+18			+21	+22	+23	
Sum of ranks Y		+19	+20				+24

Rank	25	26	27	28	29	30	31
Observation	108.17	108.77	110.91	111.03	111.75	114.43	118.61
Variable	X	Y	Y	Y	Y	X	Y
Sum of ranks X	+25					+30	
Sum of ranks Y		+26	+27	+28	+29		+31

Table 2.5: Computing the sums of ranks (Example 2.3, cf. Table 2.4)

corrected U-test is given by (with $m + n = S$)

$$
u = \frac{U - \dfrac{m \cdot n}{2}}{\sqrt{[\dfrac{m \cdot n}{S(S-1)}][\dfrac{S^3 - S}{12} - \sum_{i=1}^{r} \dfrac{t_i^3 - t_i}{12}]}}
\tag{2.36}
$$

The number of groups of equal obervations (ties) is r and t_i denotes the number of equal observations in each group.

Example 2.5: We compare the time that two dentists B and C need to manufacture an inlay (Table 4.1). At first, we combine the two samples in ascending order (Table 2.6).

Observation	19.5	31.5	31.5	33.5	37.0	40.0	43.5	50.5	53.0	54.0
Dentist	C	C	C	B	B	C	B	C	C	B
Rank	1	2.5	2.5	4	5	6	7	8	9	10
Observation	56.0	57.0	59.5	60.0	62.5	62.5	65.5	67.0	75.0	
Dentist	B	B	B	B	C	C	B	B	B	
Rank	11	12	13	14	15.5	15.5	17	18	19	

Table 2.6: Association of ranks (cf. Table 4.1)

We have $r = 2$ groups with equal data

group 1 : twice a value of 31.5; $t_1 = 2$
group 2 : twice a value of 62.5; $t_2 = 2$.

The correction term then is

$$\sum_{i=1}^{2} \frac{t_i^3 - t_i}{12} = \frac{2^3 - 2}{12} + \frac{2^3 - 2}{12} = 1 \quad .$$

The sums of ranks are given by

$$\begin{aligned} R_1 \ (\text{dentist B}) &= 4 + 5 + \cdots + 19 = 130 \\ R_2 \ (\text{dentist C}) &= 1 + 2.5 + \cdots + 15.5 = 60 \quad, \end{aligned}$$

and according to (2.33), we get

$$U_1 = 11 \cdot 8 + \frac{11(11 + 1)}{2} - 130 = 24$$

and according to (2.34)

$$\begin{aligned} U_2 &= 11 \cdot 8 + \frac{8(8 + 1)}{2} - 60 = 64 \quad, \\ U_1 + U_2 &= 88 = 11 \cdot 8 \quad (\text{control}). \end{aligned}$$

With $S = m + n = 11 + 8 = 19$ and with $U = U_1$ the test statistic (2.36) becomes

$$u = \frac{24 - 44}{\sqrt{[\frac{88}{19 \cdot 18}][\frac{19^3 - 19}{12} - 1]}} = -1.65 \quad,$$

and therefore $|u| = 1.65 < 1.96 = u_{1-0.05/2}$.

The null hypothesis H_0: *Both dentists need the same time to make an inlay is not rejected.* Both samples can be regarded as homogeneous and may be combined in a single sample for further evaluation.

We now assume the working time to be normally distributed. Hence, we can apply the the t-test and get:

dentist B : $\bar{x} = 55.27$ $s_x^2 = 12.74^2$ $n_1 = 11$
dentist C : $\bar{y} = 43.88$ $s_y^2 = 15.75^2$ $n_2 = 8$

(see Table 4.1).
The test statistic (2.15) is given by

$$F_{10,7} = \frac{15.75^2}{12.74^2} = 1.53 < 3.15 = F_{10,7;0.95} \quad,$$

the hypothesis of equal variance is not rejected. To test the hypothesis H_0: $\mu_x = \mu_y$ the test statistic (1.5) is used. The pooled sample variance is calculated according to (1.6) and gives $s^2 = (10 \cdot 12.74^2 + 7 \cdot 15.75^2)/17 = 14.06^2$. We now can evaluate the test statistic (1.5) and get:

$$t_{17} = \frac{55.27 - 43.88}{14.06} \sqrt{\frac{11 \cdot 8}{11 + 8}} = 1.74 < 2.11 = t_{17;0.95}(\text{two-sided})$$

As before, the null hypothesis is not rejected.

2.6 Comparison of Two Groups with Categorical Response

In the previous sections, the comparisons in the matched-pair design and in designs with two independent groups were based on the assumption of continuous response. Now, we want to compare two groups with categorical response. The distributions (binomial, multinomial, and Poisson distribution) and the maximum likelihood estimation are discussed in detail in Chapter 10.

To start with, we first focus on binary response, e.g., to recover/not to recover from an illness, success/no success in a game, scoring more/less than a given level.

2.6.1 McNemar's Test and Matched-Pair Design

In case of binary response we use the codings 0 and 1, so that the pairs in a matched design are one of the tuples of response $(0,0)$, $(0,1)$, $(1,0)$ or $(1,1)$. The observations are summarized in a 2×2 table.

		Group 1		
		0	1	Sum
Group 2	0	a	c	$a + c$
	1	b	d	$b + d$
	Sum	$a + b$	$c + d$	$a + b + c + d = n$

The null hypothesis is H_0: $p_1 = p_2$, where p_i is the probability $P(1|\text{group } i)$ $(i = 1, 2)$. The test is based on the relative frequencies $h_1 = (c + d)/n$ and $h_2 = (b + d)/n$ for response 1, which differ in b and c (these are the frequencies for disconcordant results $(0,1)$ and $(1,0)$).

Under H_0, the values of b and c are expected to be equal, or analogously, the expression $b - (b + c)/2$ is expected to be zero. For a given value of $b + c$, the number of disconcordant pairs follows a binomial distribution with the parameter $p = 1/2$ (probability to observe a disconcordant pair $(0,1)$ or $(1,0)$). As a result, we get $\mathrm{E}[(0,1)\text{-response}] = (b + c)/2$ and $\mathrm{Var}[(0,1)\text{-response}] = (b + c) \cdot \frac{1}{2} \cdot \frac{1}{2}$ (analogously, this holds symmetrically for $[(1,0)\text{-response}]$).

The following ratio then has expectation 0 and variance 1:

$$\frac{b - (b + c)/2}{\sqrt{(b + c) \cdot 1/2 \cdot 1/2}} = \frac{b - c}{\sqrt{b + c}} \overset{H_0}{\sim} (0, 1)$$

and follows the standard normal distribution for reasonable large $(b + c)$ due to the central limit theorem. This approximation can be used for $(b + c) \geq 20$. For continuity correction, the absolute value of $|b - c|$ is decreased by 1. Finally, we get the following test statistic:

$$Z = \frac{(b - c) - 1}{\sqrt{b + c}} \quad \text{if } b \geq c \tag{2.37}$$

$$Z = \frac{(b - c) + 1}{\sqrt{b + c}} \quad \text{if } b < c \tag{2.38}$$

Critical values are the quantiles of the cumulated binomial distribution $B(b + c, \frac{1}{2})$ in case of a small sample size. For larger samples (i.e., $b + c \geq 20$), we choose the quantiles of the standard normal distribution. The test statistic of McNemar is the product of the two Z-statistics given above. It is used for a two-sided test problem in case of $b + c \geq 20$ and follows a χ^2-distribution:

$$Z^2 = \frac{(|b - c| - 1)^2}{b + c} \sim \chi_1^2 \quad . \tag{2.39}$$

Example 2.6: A clinical experiment is used to examine two different teeth cleaning techniques in their effect on oral hygiene. The response is coded binary: reduction of tartar yes/no. The patients are stratified into matched-pairs according to sex, actual teeth cleaning technique and age. We assume the following outcome of the trial:

		Group 1		
		0	1	Sum
	0	10	50	60
Group 2	1	70	80	150
	Sum	80	130	210

We test H_0: $p_1 = p_2$ versus H_1: $p_1 \neq p_2$. Since $b + c = 70 + 50 > 20$, we choose the McNemar statistic

$$Z^2 = \frac{(|70 - 50| - 1)^2}{70 + 50} = \frac{19^2}{120} = 3.01 < 3.84 = \chi_{1;0.95}^2$$

and do not reject H_0.

Remark: Modifications of the McNemar test can be constructed similarly to sign tests. Let n be the number of nonzero differences in the response of the pairs and let T_+ and T_- be the number of positive and negative differences, respectively. Then the test statistic, analogously to the Z-statistics (2.37) and (2.38), is given by

$$Z = \frac{(T_+/n - 1/2) \pm n/2}{1/\sqrt{4n}} \quad , \tag{2.40}$$

in which we use $+n/2$ if $T_+/n < 1/2$ and $-n/2$ if $T_+/n \geq 1/2$. The null hypothesis is H_0: $\mu_d = 0$. Depending on the sample size ($n \geq 20$ or $n < 20$) we use the quantiles of the normal or binomial distribution.

2.6.2 Fisher's Exact Test for Two Independent Groups

Regarding two independent groups of size n_1 and n_2 with binary response, we get the following 2×2-table

	Group 1	Group 2	
1	a	c	$a+c$
0	b	d	$b+d$
	n_1	n_2	n

The relative frequencies of response 1 are $\hat{p}_1 = a/n_1$ and $\hat{p}_2 = c/n_2$. The null hypothesis is H_0: $p_1 = p_2 = p$. In this contingency table, we identify the cell with the weakest cell count and calculate the probability for this and all other tables with an even smaller cell count in the weakest cell. In doing so, we have to assure that the marginal sums keep constant.

Assume $(1, 1)$ to be the weakest cell. Under H_0, we have for response 1 in both groups (for given n, n_1, n_2 and p):

$$P((a+c)|n, p) = \binom{n}{a+c} p^{a+c}(1-p)^{n-(a+c)} \quad ,$$

for group 1 and response 1:

$$P(a|(a+b), p) = \binom{a+b}{a} p^a (1-p)^b \quad ,$$

for group 2 and response 1:

$$P(c|(c+d), p) = \binom{c+d}{c} p^c (1-p)^d \quad .$$

Since the two groups are independent, the joint probability is given by

$$P(\text{group 1} = a \quad \wedge \quad \text{group 2} = c) = \binom{a+b}{a} p^a (1-p)^b \binom{c+d}{c} p^c (1-p)^d$$

and the conditional probability of a and c (for given marginal sum $a + c$) is

$$P(a, c|a+c) = \binom{a+b}{a}\binom{c+d}{c} / \binom{n}{a+c}$$

$$= \frac{(a+b)!(c+d)!(a+c)!(b+d)!}{n!} \cdot \frac{1}{a!b!c!d!} \quad .$$

Hence, the probability to observe the given table or a table with an even smaller count in the weakest cell is

$$P = \frac{(a+b)!(c+d)!(a+c)!(b+d)!}{n!} \cdot \sum_i \frac{1}{a_i! b_i! c_i! d_i!} \quad ,$$

with summation over all cases i with $a_i \leq a$. If $P < 0.05$ (one-sided) or $2P < 0.05$ (two-sided) holds, then the hypothesis H_0: $p_1 = p_2$ is rejected.

Example 2.7: We compare two independent groups of subjects receiving either type A or B of an implanted denture and observe whether it is lost during the healing process (eight weeks after implantation). The data are

		A	B	
loss	yes	2	8	10
	no	10	4	14
		12	12	24

The two tables with weaker count in the (yes$|A$) cell are

1	9
11	3

and

0	10
12	2

and therefore we get

$$P = \frac{10!14!12!12!}{24!}\left(\frac{1}{2!8!10!4!} + \frac{1}{1!9!11!3!} + \frac{1}{0!10!12!2!}\right) = 0.018$$

$$\left.\begin{array}{l}\text{one-sided test:} \quad P = 0.018 \\ \text{two-sided test:} \quad 2P = 0.036\end{array}\right\} < 0.05$$

Decision: H_0: $p_1 = p_2$ is rejected in both of the two cases. The risk of loss is significantly higher for B than for A.

Recurrence Relation

Instead of using tables, we also can use the following recurrence relation (cited by Sachs, 1974, p. 280):

$$P_{i+1} = \frac{a_i d_i}{b_{i+1} c_{i+1}} P_i$$

In our example we get

$$P = P_1 + P_2 + P_3$$
$$P_1 = \frac{10!14!12!12!}{24!} \frac{1}{2!8!10!4!}$$
$$= 0.0166$$
$$P_2 = \frac{2 \cdot 4}{11 \cdot 9} P_1 = 0.0013$$
$$P_3 = \frac{1 \cdot 3}{12 \cdot 10} P_2 = 0.0000 \quad ,$$

and therefore $P = 0.0179 \approx 0.0180$.

2.7 Exercises and Questions

2.7.1 What are the differences between the paired t-test and the two sample t-test (degrees of freedom, power)?

2.7.2 Consider two samples with $n_1 = n_2$, $\alpha = 0.05$ and $\beta = 0.05$ in a matched-pair design and in a design of two independent groups. What is the minimum sample size needed to achieve a power of 0.95, assuming $\sigma^2 = 1$ and $\delta^2 = 4$.

2.7.3 Apply Wilcoxon's sign rank test for a matched-pair design to the following table

Student	Before	After
1	17	25
2	18	45
3	25	37
4	12	10
5	19	21
6	34	27
7	29	29

Table: Scorings of students who took a cup of coffee either before or after a lecture.

Does treatment B (coffee before) significantly influence the score?

2.7.4 For a comparison of two independent samples X : leaf-length of strawberries with manuring A and Y : manuring B, the normal distribution is put in question. Test H_0: $\mu_X = \mu_Y$ using the homogeneity test of Wilcoxon, Mann and Whitney.

A	B
37	45
49	51
51	62
62	73
74	87
89	45
44	33
53	
17	

Note that there are ties.

2.7.5 Recode the response in Table 2.4 into binary response with
flexibility < 100 : 0
flexibility ≥ 100 : 1
and apply Fisher's exact test for H_0: $p_1 = p_2$ ($p_i = P(1|\text{group } i)$).

2.7.6 Considering Exercise 2.7.3, we assume that the response has been binary
recoded according to scoring higher/lower than average: 1/0. A sample of
$n = 100$ shows the following outcome:

		Before		
		0	1	
	0	20	25	45
After	1	15	40	55
		35	65	100

Test for H_0: $p_1 = p_2$ using McNemar's test.

3 The Linear Regression Model

3.1 Descriptive Linear Regression

The main focus in this chapter will be the linear regression model and its basic principle of estimation. We introduce the fundamental method of *Least Squares* by looking at the least squares geometry and discuss some of its algebraic properties.

In empirical work, it is quite often appropriate to specify the relation between two sets of data by a simple linear function.

Let Y denote the dependent variable which is related to a set of K independent variables $X_1, \ldots X_K$ by a function f. As both sets comprise T observations on each variable, it is convenient to use the following notation:

$$(\boldsymbol{y}, \boldsymbol{X}) = \begin{pmatrix} y_1 & x_{11} & \cdots & x_{K1} \\ \vdots & \vdots & & \vdots \\ y_T & x_{1T} & \cdots & x_{KT} \end{pmatrix} = (\boldsymbol{y} \, \boldsymbol{x}_{(1)} \ldots \boldsymbol{x}_{(K)}) = \begin{pmatrix} y_1 \, \boldsymbol{x}'_1 \\ \vdots \\ y_T \, \boldsymbol{x}'_T \end{pmatrix} . \quad (3.1)$$

If we choose f to be linear, we intend to obtain both a good overall fit of the model and easy mathematical tractability. This seems to be realistic as almost every specification of f suffers from the exclusion of important variables or the inclusion of unimportant variables. Additionally, even a correct set of variables is often measured with at least some error such that a correct functional relationship between $\boldsymbol{y}$ and $\boldsymbol{X}$ will most unlikely be precise. On the other hand, the linear approach may serve as a suitable approximation to several nonlinear functional relationships.

If we assume Y to be *generated* additively by a linear combination of the independent variables, we may write

$$Y = X_1\beta_1 + \ldots + X_K\beta_K . \quad (3.2)$$

The β's in (3.2) are unknown (scalar-valued) coefficients. The magnitude of the β's indicates their importance in explaining Y.

An obvious goal of empirical regression analysis consists in finding those values for $\beta_1, \ldots, \beta_K$ which minimize the differences

$$e_t := y_t - \boldsymbol{x}_t'\boldsymbol{\beta}, \quad (t = 1, \ldots, T)$$

where $\boldsymbol{\beta}' = (\beta_1, \ldots, \beta_K)$. The e_t's are called *residuals* and play an important role in regression analysis. In general, we cannot expect that $e_t = 0$ will hold for all $t = 1, \ldots, T$. Accordingly, the residuals are incorporated into the linear approach upon setting

$$y_t = \boldsymbol{x}_t'\boldsymbol{\beta} + e_t \quad (t = 1, \ldots, T). \tag{3.3}$$

This may be summarized in matrix notation by

$$\boldsymbol{y} = \boldsymbol{X}\boldsymbol{\beta} + \boldsymbol{e}. \tag{3.4}$$

Obviously, a successful choice for $\boldsymbol{\beta}$ is indicated by small values of e_t, $t = 1, \ldots, T$. Thus, there are quite a few principles conceivable by which the quality of an actual choice for $\boldsymbol{\beta}$ may be evaluated.

Among others, the following measures have been proposed:

$$\sum_{t=1}^{T} |e_t|, \quad \max_t |e_t|, \quad \sum_{t=1}^{T} e_t^2 = \boldsymbol{e}'\boldsymbol{e}. \tag{3.5}$$

Whereas the first two proposals are subject to either some complicated mathematics or poor statistical properties, the last principle has become widely accepted. It provides the basis for the famous method of least squares.

3.2   The Principle of Ordinary Least Squares (OLS)

Let $\boldsymbol{B}$ be the set of all possible vectors $\boldsymbol{\beta}$. If there is no further information, we have $\boldsymbol{B} = \mathcal{R}^K$ (K-dimensional real Euclidean space). The idea is to find a vector $\boldsymbol{b}' = (b_1, \ldots, b_K)$ from $\boldsymbol{B}$ that minimizes the sum of squared residuals

$$S(\boldsymbol{\beta}) = \sum_{t=1}^{T} e_t^2 = \boldsymbol{e}'\boldsymbol{e} = (\boldsymbol{y} - \boldsymbol{X}\boldsymbol{\beta})'(\boldsymbol{y} - \boldsymbol{X}\boldsymbol{\beta}) \tag{3.6}$$

given $\boldsymbol{y}$ and $\boldsymbol{X}$. A minimum will always exist, as $S(\boldsymbol{\beta})$ is a real-valued convex differentiable function. If we rewrite $S(\boldsymbol{\beta})$ as

$$S(\boldsymbol{\beta}) = \boldsymbol{y}'\boldsymbol{y} + \boldsymbol{\beta}'\boldsymbol{X}'\boldsymbol{X}\boldsymbol{\beta} - 2\boldsymbol{\beta}'\boldsymbol{X}'\boldsymbol{y}' \tag{3.7}$$

and differentiate w.r.t. $\boldsymbol{\beta}$ (by help of A 91 - A 95), we obtain

$$\frac{\partial S(\boldsymbol{\beta})}{\partial \boldsymbol{\beta}} = 2\boldsymbol{X}'\boldsymbol{X}\boldsymbol{\beta} - 2\boldsymbol{X}'\boldsymbol{y}, \tag{3.8}$$

$$\frac{\partial^2 S(\boldsymbol{\beta})}{\partial \boldsymbol{\beta}^2} = 2\boldsymbol{X}'\boldsymbol{X} \quad \text{(nonnegativ definite)}. \tag{3.9}$$

Equating the first derivative to zero yields the *normal equations*

$$X'Xb = X'y. \tag{3.10}$$

The solution of (3.10) is now straightforwardly obtainable by considering a system of linear equations

$$Ax = a , \tag{3.11}$$

where A is an $(n \times m)$-matrix and a an $(n \times 1)$-vector. The $(m \times 1)$-vector x solves the equation. Let A^- be a g-inverse of A (cf. A 62). Then we have

Theorem 3.1 *The linear equation* $Ax = a$ *has a solution if and only if*

$$AA^-a = a . \tag{3.12}$$

If (3.12) holds, then all solutions are given by

$$x = A^-a + (I - A^-A)w , \tag{3.13}$$

where w *is an arbitrary* $(m \times 1)$-*vector.*

Remark: $x = A^-a$ (i.e., (3.13) and $w = 0$) is a particular solution of $Ax = a$.

Proof: Let $Ax = a$ have a solution. Then at least one vector x_0 exists, with $Ax_0 = a$. As $AA^-A = A$ for every g-inverse, we obtain

$$a = Ax_0 = AA^-Ax_0 = AA^-(Ax_0) = AA^-a ,$$

which is just (3.12).

Now let (3.12) be true, i.e., $AA^-a = a$. Then A^-a is a solution of (3.11). Assume now that (3.11) is solvable. To prove (3.13), we have to show that

(i) $A^-a + (I - A^-A)w$ is always a solution of (3.11), (w arbitrary), and that

(ii) every solution x of $Ax = a$ may be represented by (3.13).

Part (i) follows by insertion of the general solution, making also use of $A(I - A^-A) = 0$:

$$A[A^-a + (I - A^-A)w] = AA^-a = a .$$

To prove (ii) we choose $w = x_0$, where x_0 is a solution of the linear equation, i.e., $Ax_0 = a$. Then we have

$$\begin{aligned} A^-a + (I - A^-A)x_0 &= A^-a + x_0 - A^-Ax_0 \\ &= A^-a + x_0 - A^-a \\ &= x_0 , \end{aligned}$$

thus concluding the proof.

We apply this result to our problem, i.e. to (3.10) and check the solvability of the linear equation first.

X is a $(T \times K)$-matrix, thus $X'X$ is a symmetric $(K \times K)$-matrix of rank $(X'X) = p \le K$. (3.10) has a solution if and only if (cf. (3.12))

$$(X'X)(X'X)^- X'y = X'y \; . \tag{3.14}$$

Following the definition of a g-inverse

$$(X'X)(X'X)^-(X'X) = (X'X)$$

we have with A 73

$$X'X(X'X)^- X' = X' \; ,$$

such that (3.14) holds. Thus, the normal equation (3.10) always has a solution. The set of all solutions of (3.10) are by (3.13) of the form

$$b = (X'X)^- X'y + (I - (X'X)^- X'X)w \tag{3.15}$$

where w is an arbitrary $(K \times 1)$-vector.
For the choice $w = 0$, we have with

$$b = (X'X)^- X'y \tag{3.16}$$

a particular solution, which is nonunique as the g-inverse $(X'X)^-$ is nonunique.

In analysis of variance (ANOVA) models, one way to estimate or identify uniquely the effect of the variables in X on the response y is to specify certain restrictions on the parameters (reparametrization).

Corollary: *The set of equations*

$$AXB = C \tag{3.17}$$

with A: $m \times n$, B: $p \times q$, C: $m \times q$ and X: $n \times p$ has a solution X if and only if

$$AA^- CB^- B = C \; , \tag{3.18}$$

where A^- and B^- are arbitrary g-inverses of A and B.

If X is of full rank, i.e., $\mathrm{rank}(X) = p = K$, then we have $(X'X)^- = (X'X)^{-1}$ and the normal equations are uniquely solvable by

$$b = (X'X)^{-1} X'y \; . \tag{3.19}$$

If more generally $\mathrm{rank}(X) = p < K$, then the solutions of the normal equations span the same hyperplane as Xb, i.e., for two solutions b and b^* we have

$$Xb = Xb^* \; . \tag{3.20}$$

This result is easy to prove: If b and b^* are solutions to the normal equations we have

$$X'Xb = X'y \quad \text{and} \quad X'Xb^* = X'y \ .$$

Accordingly, we have for the difference of the above equations

$$X'X(b - b^*) = 0 \ ,$$

which entails by A 72

$$X(b - b^*) = 0 \quad \text{or} \quad Xb = Xb^* \ .$$

Moreover, by (3.20), the two sums of squared errors are given by

$$S(b) = (y - Xb)'(y - Xb) = (y - Xb^*)'(y - Xb^*) = S(b^*) \ .$$

Thus the following Theorem has been proven.

Theorem 3.2 *The vector $\beta = b$ minimizes the sum of squared errors if and only if it is a solution of $X'Xb = X'y$. All solutions are located on the hyperplane Xb.*

The solutions b of the normal equations are called *empirical regression coefficients* or empirical least squares estimates of β. $\hat{y} = Xb$ is called the *empirical regression hyperplane*. An important property of the sum of squared errors $S(b)$ is

$$y'y = \hat{y}'\hat{y} + \hat{e}'\hat{e} \ , \tag{3.21}$$

where $\hat{e}$ denotes the residuals $y - Xb$. The sum of squared observations $y'y$ may be decomposed additively into the sum of squared values $\hat{y}'\hat{y}$ explained by regression and the sum of (unexplained) squared residuals $\hat{e}'\hat{e}$.
We derive (3.21) by premultiplication of (3.10) with b'

$$b'X'Xb = b'X'y$$

and

$$\hat{y}'\hat{y} = (Xb)'(Xb) = b'X'Xb = b'X'y \tag{3.22}$$

according to

$$\begin{aligned} S(b) = \hat{e}'\hat{e} = (y - Xb)'(y - Xb) &= y'y - 2b'X'y + b'X'Xb \\ &= y'y - b'X'y = y'y - \hat{y}'\hat{y} \ . \end{aligned} \tag{3.23}$$

Remark: In analysis of variance, $\hat{y}'\hat{y}$ will be decomposed further into orthogonal components which are related to main- and mixed-effects of treatments.

3.3 Geometric Properties of Ordinary Least Squares Estimation (OLS)

For an $(n \times p)$-matrix $\boldsymbol{X}$, we now define the *column space* $\mathcal{R}(\boldsymbol{X})$ and the *null space* $\mathcal{N}(\boldsymbol{X})$. $\mathcal{R}(\boldsymbol{X})$ is the set of all vectors $\boldsymbol{\Theta}$ such that $\boldsymbol{\Theta} = \boldsymbol{X}\boldsymbol{\beta}$ is fulfilled for all vectors $\boldsymbol{\beta}$ from $\mathcal{R}^p$. $\mathcal{R}(\boldsymbol{X}) = \{\boldsymbol{\Theta} : \boldsymbol{\Theta} = \boldsymbol{X}\boldsymbol{b}\}$ and $\mathcal{N}(X) = \{\boldsymbol{\Phi} : \boldsymbol{X}\boldsymbol{\Phi} = \boldsymbol{0}\}$ are vector spaces. The basic relation between the column space and the null space is given by

$$\mathcal{N}(\boldsymbol{X}) = \mathcal{R}(\boldsymbol{X}')^{\perp} \, . \tag{3.24}$$

Once again, we consider the linear model (3.4), i.e.

$$\boldsymbol{y} = \boldsymbol{X}\boldsymbol{\beta} + \boldsymbol{e} \, ,$$

where $\boldsymbol{X}\boldsymbol{\beta} \in \mathcal{R}(\boldsymbol{X}) = \{\boldsymbol{\Theta} : \boldsymbol{\Theta} = \boldsymbol{X}\tilde{\boldsymbol{\beta}}\}$. If we assume that $\mathrm{rank}(\boldsymbol{X}) = p$, then $\mathcal{R}(\boldsymbol{X})$ is of dimension p. Let $\mathcal{R}(\boldsymbol{X})^{\perp}$ denote the orthogonal complement of $\mathcal{R}(\boldsymbol{X})$ and let $\boldsymbol{X}\boldsymbol{b}$ be denoted by $\boldsymbol{\Theta}_0$ where $\boldsymbol{b}$ is the OLS estimation of $\boldsymbol{\beta}$. Then we have

Theorem 3.3 *The Ordinary Least Squares estimation $\boldsymbol{\Theta}_0$ of $\boldsymbol{X}\boldsymbol{b}$ minimizing*

$$S(\boldsymbol{\beta}) = (\boldsymbol{y} - \boldsymbol{X}\boldsymbol{\beta})'(\boldsymbol{y} - \boldsymbol{X}\boldsymbol{\beta}) = (\boldsymbol{y} - \boldsymbol{\Theta})'(\boldsymbol{y} - \boldsymbol{\Theta}) = \tilde{S}(\boldsymbol{\Theta}) \tag{3.25}$$

for $\boldsymbol{\Theta} \in \mathcal{R}(\boldsymbol{X})$, is given by the orthogonal projection of $\boldsymbol{y}$ on the space $\mathcal{R}(\boldsymbol{X})$.

Proof: As $\mathcal{R}(\boldsymbol{X})$ is of dimension p, an orthonormal basis $\boldsymbol{v}_1, \ldots, \boldsymbol{v}_p$ exists. Furthermore, we may represent the $(T \times 1)$-vector $\boldsymbol{y}$ as

$$\boldsymbol{y} = \sum_{i=1}^{p} a_i \boldsymbol{v}_i + \left(\boldsymbol{y} - \sum_{i=1}^{p} a_i \boldsymbol{v}_i\right) = \boldsymbol{c} + \boldsymbol{d} \tag{3.26}$$

where $a_i = \boldsymbol{y}'\boldsymbol{v}_i$.
As

$$\boldsymbol{v}_j' \boldsymbol{d} = \boldsymbol{v}_j' \boldsymbol{y} - \sum_i a_i \boldsymbol{v}_j' \boldsymbol{v}_i = a_j - \sum_i a_i \delta_{ij} = 0 \tag{3.27}$$

(δ_{ij} denotes the KRONECKER-symbol), we have $\boldsymbol{c} \perp \boldsymbol{d}$, i.e., we have $\boldsymbol{c} \in \mathcal{R}(\boldsymbol{X})$ and $\boldsymbol{d} \in \mathcal{R}(\boldsymbol{X})^{\perp}$, such that $\boldsymbol{y}$ has been decomposed in two orthogonal components. This decomposition is unique as can easily be shown.
We have to show now, that $\boldsymbol{c} = \boldsymbol{X}\boldsymbol{b} = \boldsymbol{\Theta}_0$.
It follows from $\boldsymbol{c} - \boldsymbol{\Theta} \in \mathcal{R}(\boldsymbol{X})$, that

$$(\boldsymbol{y} - \boldsymbol{c})'(\boldsymbol{c} - \boldsymbol{\Theta}) = \boldsymbol{d}'(\boldsymbol{c} - \boldsymbol{\Theta}) = 0 \, . \tag{3.28}$$

Considering $\boldsymbol{y} - \boldsymbol{\Theta} = (\boldsymbol{y} - \boldsymbol{c}) + (\boldsymbol{c} - \boldsymbol{\Theta})$, we get

$$\begin{aligned} \tilde{S}(\boldsymbol{\Theta}) = (\boldsymbol{y} - \boldsymbol{\Theta})'(\boldsymbol{y} - \boldsymbol{\Theta}) &= (\boldsymbol{y} - \boldsymbol{c})'(\boldsymbol{y} - \boldsymbol{c}) + (\boldsymbol{c} - \boldsymbol{\Theta})'(\boldsymbol{c} - \boldsymbol{\Theta}) \\ &\quad + 2(\boldsymbol{y} - \boldsymbol{c})'(\boldsymbol{c} - \boldsymbol{\Theta}) \\ &= (\boldsymbol{y} - \boldsymbol{c})'(\boldsymbol{y} - \boldsymbol{c}) + (\boldsymbol{c} - \boldsymbol{\Theta})'(\boldsymbol{c} - \boldsymbol{\Theta}) \, . \end{aligned} \tag{3.29}$$

$\tilde{S}(\Theta)$ reaches its minimum on $\mathcal{R}(X)$ for the choice $\Theta = c$. As $\tilde{S}(\Theta) = S(\beta)$ we find b to be the optimum $c = \Theta_0 = Xb$.
The OLS estimator Xb of $X\beta$ may also be obtained in a more direct way by using idempotent projection matrices.

Theorem 3.4 *Let P be a symmetric and idempotent matrix of rank p, represen-*
ting the orthogonal projection of $\mathcal{R}^T$ on $\mathcal{R}(X)$.
Then $Xb = \Theta_0 = Py$.

Proof: Following Theorem 3.3, we have

$$\Theta_0 = c = \sum_i a_i v_i = \sum_i v_i(y'v_i)$$

$$= \sum_i v_i(v_i'y)$$

$$= (v_1,\ldots,v_p)(v_1,\ldots,v_p)'y \qquad (3.30)$$

$$= BB'y \qquad [B = (v_1,\ldots,v_p)]$$

$$= Py ,$$

where P is obviously symmetric and idempotent.
We have to make use of the following lemma, which will be stated without proof.
Lemma: *A symmetric and idempotent $T \times T$-matrix P of rank $p \leq T$ represents*
the orthogonal projection matrix of $\mathcal{R}^T$ on a p-dimensional vector space $V =$
$\mathcal{R}(P)$.

(i) **Determination of P if rank $(X) = K$:**
The rows of B constitute an orthonormal basis of $\mathcal{R}(X) = \{\Theta : \Theta = X\beta\}$. But $X = BC$ (C a regular matrix), as the columns of X also form a basis of $\mathcal{R}(X)$.
Thus

$$P = BB' = XC^{-1}C'^{-1}X' = X(C'C)^{-1}X'$$

$$= X(C'B'BC)^{-1}X' \qquad [\text{as } B'B = I] \qquad (3.31)$$

$$= X(X'X)^{-1}X' ,$$

and we finally get

$$\Theta_0 = Py = X(X'X)^{-1}X'y = Xb . \qquad (3.32)$$

(ii) **Determination of P if rank $(X) = p < K$:**
The normal equations have a unique solution, if X is of full column rank K. A method to derive unique solutions if rank $(X) = p < K$ is based on imposing additional linear restrictions, which enable the identification of β.

We introduce only the general strategy by using Theorem 3.4; further details will be given in Section 3.5.

Let R be a $((K - p) \times K)$-matrix with rank $(R) = K - p$ and define the matrix $D = \begin{pmatrix} X \\ R \end{pmatrix}$.

Let r be a known $((K - p) \times 1)$-vector. If rank $(D) = K$, then X and R are *complementary matrices*. The matrix R represents $(K - p)$ additional linear restrictions on β (reparametrization), as it will be assumed that

$$R\beta = r \ . \tag{3.33}$$

Minimization of $S(\beta)$ subject to these exact linear restrictions $R\beta = r$ requires the minimization of the function

$$Q(\beta, \lambda) = S(\beta) + 2\lambda'(R\beta - r) \ , \tag{3.34}$$

here λ stands for a $((K - p) \times 1)$-vector of Lagrangian multipliers. The corresponding normal equations are given by (cf. Theorem A 91 - A 95)

$$\left.\begin{array}{rcl} \dfrac{1}{2}\dfrac{\partial Q(\beta, \lambda)}{\partial \beta} & = & X'X\beta - X'y + R'\lambda \ = \ 0, \\[4mm] \dfrac{1}{2}\dfrac{\partial Q(\beta, \lambda)}{\partial \lambda} & = & R\beta - r \ = \ 0. \end{array}\right\} \tag{3.35}$$

If $r = 0$, we can prove the following Theorem (cf. Seber, 1966, p. 16):

Theorem 3.5 *Under the exact linear restrictions $R\beta = r$ with rank $(R) = K - p$ and rank $(D) = K$ we can state:*

(i) *The orthogonal projection matrix of $\mathcal{R}^T$ on $\mathcal{R}(X)$ is of the form*

$$P = X(X'X + R'R)^{-1}X' \ . \tag{3.36}$$

(ii) *The conditional Ordinary Least Squares estimator of β is given by*

$$b(R, r) = (X'X + R'R)^{-1}(X'y + R'r) \ . \tag{3.37}$$

Proof: We start with the proof of part (i).

From the assumptions we conclude, that for every $\Theta \in \mathcal{R}(X)$ a β exists, such that $\Theta = X\beta$ and $R\beta = r$ are valid. β is unique, as rank $(D) = K$. In other words, for every $\Theta \in \mathcal{R}(X)$, the $((T + K - p) \times 1)$-vector is

$$\begin{pmatrix} \Theta \\ R \end{pmatrix} \in \mathcal{R}(D), \quad \text{therefore} \quad \begin{pmatrix} \Theta \\ r \end{pmatrix} = D\beta \quad (\text{and } \beta \text{ is unique}) \ .$$

If we make use of Theorem 3.4, then we get the projection matrix of $\mathcal{R}^{T+K-p}$ on $\mathcal{R}(D)$ as

$$P^* = D(D'D)^{-1}D' . \tag{3.38}$$

As the projection P^* maps every element of $\mathcal{R}(D)$ onto itself, we have for every $\Theta \in \mathcal{R}(X)$

$$\begin{pmatrix} \Theta \\ r \end{pmatrix} = D(D'D)^{-1}D' \begin{pmatrix} \Theta \\ r \end{pmatrix}$$

$$= \begin{pmatrix} X(D'D)^{-1}X' & X(D'D)^{-1}R' \\ R(D'D)^{-1}X' & R(D'D)^{-1}R' \end{pmatrix} \begin{pmatrix} \Theta \\ r \end{pmatrix} , \tag{3.39}$$

i.e.,

$$\Theta = X(D'D)^{-1}X'\Theta + X(D'D)^{-1}R'r , \tag{3.40}$$

$$r = R(D'D)^{-1}X'\Theta + R(D'D)^{-1}R'r . \tag{3.41}$$

Equations (3.40) and (3.41) hold for every $\Theta \in \mathcal{R}(X)$ and for all $r = R\beta \in \mathcal{R}(R)$. If we choose in (3.33) $r = 0$, then (3.40) and (3.41) specialize to

$$\Theta = X(D'D)^{-1}X'\Theta , \tag{3.42}$$

$$0 = R(D'D)^{-1}X'\Theta . \tag{3.43}$$

From (3.43) it follows that

$$\mathcal{R}(X(D'D)^{-1}R') \perp \mathcal{R}(X) \tag{3.44}$$

and as $\mathcal{R}(X(D'D)^{-1}R') = \{\Theta : \Theta = X\tilde{\beta} \text{ with } \tilde{\beta} = (D'D)^{-1}R'\beta\}$ it holds, that

$$\mathcal{R}(X(D'D)^{-1}R') \subset \mathcal{R}(X) , \tag{3.45}$$

such that finally

$$X(D'D)^{-1}R' = 0 \tag{3.46}$$

(see also Tan, 1971).
The matrices $X(D'D)^{-1}X'$ and $R(D'D)^{-1}R'$ are idempotent (symmetry is evident):

$$\begin{aligned} & X(D'D)^{-1}X'X(D'D)^{-1}X' \\ = & X(D'D)^{-1}(X'X + R'R - R'R)(D'D)^{-1}X' \\ = & X(D'D)^{-1}(X'X + R'R)(D'D)^{-1}X' - X(D'D)^{-1}R'R(D'D)^{-1}X' \\ = & X(D'D)^{-1}X' , \end{aligned}$$

as $D'D = X'X + R'R$ and (3.46) are valid.

The idempotency of $R(D'D)^{-1}R'$ can be shown in a similar way. $D'D$ and $(D'D)^{-1}$ are both positive definite (see Theorem A 39 and A 40). $R(D'D)^{-1}R'$ is positive definite (Theorem A 39 (vi)) and thus regular since rank $(R) = K - p$. But there exists only one idempotent and regular matrix, namely the identity matrix (Theroem A 61 (iii)):

$$R(D'D)^{-1}R' = I \, , \tag{3.47}$$

such that (3.41) is equivalent to $r = r$. As $P = X(D'D)^{-1}X'$ is idempotent, it represents the orthogonal projection matrix of $\mathcal{R}^T$ on a vector space $V \subset \mathcal{R}^T$ (see the Lemma following Theorem 3.4).

With (3.42) we have $\mathcal{R}(X) \subset V$. But also the reverse proposition is true (see Theorem A 25 (iv), (v)):

$$V = \mathcal{R}(X(D'D)^{-1}X') \subset \mathcal{R}(X) \, , \tag{3.48}$$

such that $V = \mathcal{R}(X)$, which proves (i).

Proof of (ii): We will solve the normal equations (3.35). With $R\beta = r$ it also holds that $R'R\beta = R'r$. Inserting the latter identity into the first equation of (3.35) yields

$$(X'X + R'R)\beta = X'y + R'r - R'\lambda \, .$$

Multiplication with $(D'D)^{-1}$ from the left yields:

$$\beta = (D'D)^{-1}(X'y + R'r) - (D'D)^{-1}R'\lambda \, .$$

If we use the second equation of (3.35), (3.46), and (3.47), and then multiply with R from the left we get:

$$R\beta = R(D'D)^{-1}(X'y + R'r) - R(D'D)^{-1}R'\lambda = r - \lambda \, , \tag{3.49}$$

from which $\hat{\lambda} = 0$ follows.

The solution of the normal equations is therefore given by

$$\hat{\beta} = b(R, r) = (X'X + R'R)^{-1}(X'y + R'r) \tag{3.50}$$

which proves (ii).

The conditional OLS estimator (in the sense of being restricted by $R\beta = r$) $b(R, r)$ will be most useful in tackling the problem of multicollinearity which is typical for design matrices in ANOVA-models (see Section 3.5).

3.4 Best Linear Unbiased Estimation

In descriptive regression analysis, the regression coefficient β is allowed to vary freely and is then determined by the method of Least Squares in an algebraical way by using projection matrices. The classical linear regression model now interprets the vector β as a fixed but unknown model parameter. Then, estimation is carried out by minimizing an appropriate risk function. The model and its main assumptions are given as follows:

$$\left.\begin{array}{l} \boldsymbol{y} = \boldsymbol{X}\boldsymbol{\beta} + \boldsymbol{\epsilon} \; , \\ \mathrm{E}\,(\boldsymbol{\epsilon}) = \boldsymbol{0} \; , \qquad \mathrm{E}\,(\boldsymbol{\epsilon}\boldsymbol{\epsilon}') = \sigma^2 \boldsymbol{I} \; , \\ \boldsymbol{X} \;\text{nonstochastic}\; , \;\; \mathrm{rank}\,(\boldsymbol{X}) = K \; . \end{array}\right\} \tag{3.51}$$

As $\boldsymbol{X}$ is assumed to be nonstochastic, $\boldsymbol{X}$ and $\boldsymbol{\epsilon}$ are independent, i.e.,

$$\mathrm{E}\,(\boldsymbol{\epsilon}|\boldsymbol{X}) = \mathrm{E}\,(\boldsymbol{\epsilon}) = \boldsymbol{0} \; , \tag{3.52}$$

$$\mathrm{E}\,(\boldsymbol{X}'\boldsymbol{\epsilon}|\boldsymbol{X}) = \boldsymbol{X}'\mathrm{E}\,(\boldsymbol{\epsilon}) = \boldsymbol{0} \; , \tag{3.53}$$

and

$$\mathrm{E}\,(\boldsymbol{\epsilon}\boldsymbol{\epsilon}'|\boldsymbol{X}) = \mathrm{E}\,(\boldsymbol{\epsilon}\boldsymbol{\epsilon}') = \sigma^2 \boldsymbol{I} \; . \tag{3.54}$$

The rank condition on $\boldsymbol{X}$ means that there are no linear relations between the K regressors $X_1, \ldots, X_K$; especially the inverse matrix $(\boldsymbol{X}'\boldsymbol{X})^{-1}$ exists. Using (3.51) and (3.52) we get the conditional expectation

$$\mathrm{E}\,(\boldsymbol{y}|\boldsymbol{X}) = \boldsymbol{X}\boldsymbol{\beta} + \mathrm{E}\,(\boldsymbol{\epsilon}|\boldsymbol{X}) = \boldsymbol{X}\boldsymbol{\beta} \; , \tag{3.55}$$

and by (3.54) the covariance matrix of $\boldsymbol{y}$ is of the form

$$\mathrm{E}\,[(\boldsymbol{y} - \mathrm{E}\,(\boldsymbol{y}))(\boldsymbol{y} - \mathrm{E}\,(\boldsymbol{y}))'|\boldsymbol{X}] = \mathrm{E}\,(\boldsymbol{\epsilon}\boldsymbol{\epsilon}'|\boldsymbol{X}) = \sigma^2 \boldsymbol{I} \; . \tag{3.56}$$

In what follows, all expected values should be understood as conditional on a fixed matrix $\boldsymbol{X}$.

3.4.1 Linear Estimators

The statistician's task is now to estimate the true but unknown vector β of regression parameters in the model (3.51) on the basis of observations $(\boldsymbol{y}, \boldsymbol{X})$ and assumptions already stated. This will be done by choosing a suitable estimator $\hat{\beta}$ which then will be used to calculate the conditional expectation $\mathrm{E}\,(\boldsymbol{y}|\boldsymbol{X}) = \boldsymbol{X}\boldsymbol{\beta}$ and an estimate for the error variance σ^2. It is common to choose an estimator $\hat{\beta}$ that is linear in $\boldsymbol{y}$, i.e.,

$$\underset{K \times T}{\hat{\beta} =} \;\; \underset{}{\boldsymbol{C}} \;\; \boldsymbol{y} + \;\; \underset{K \times 1}{\boldsymbol{d}} \;\; . \tag{3.57}$$

$\boldsymbol{C}$ and $\boldsymbol{d}$ are nonstochastic matrices, which have been determined by minimizing a suitably chosen risk function in an optimal way.

At first, we have to introduce some definitions.

Definition 3.1 $\hat{\beta}$ *is called homogeneous estimator of β, if $d = 0$; otherwise $\hat{\beta}$ is called inhomogeneous.*

In descriptive regression analysis, we measured the goodness of fit of the model by the sum of squared errors $S(\beta)$. Analogously, we define for the random variable $\hat{\beta}$ the quadratic loss function

$$L(\hat{\beta}, \beta, A) = (\hat{\beta} - \beta)'A(\hat{\beta} - \beta), \tag{3.58}$$

where A is a symmetric and at least nonnegative definite $(K \times K)$-matrix.

Remark: We say that $A \geq 0$ (A nonnegative definite) and $A > 0$ (A positive definite) in accordance to Theorems A 36 - A 38.

Obviously, the loss (3.58) depends on the sample. Thus, we have to consider the average or expected loss over all possible samples. The expected loss of an estimator will be called risk.

Definition 3.2 *The quadratic risk of an estimator $\hat{\beta}$ of β is defined as*

$$R(\hat{\beta}, \beta, A) = \mathrm{E}\,(\hat{\beta} - \beta)'A(\hat{\beta} - \beta). \tag{3.59}$$

The next step now consists in finding an estimator $\hat{\beta}$, that minimizes the quadratic risk function over a class of appropriate functions. Therefore, we have to define a criterion to compare estimators:

Definition 3.3 *($R(A)$-superiority)* *An estimator $\hat{\beta}_2$ of β is called $R(A)$-superior or $R(A)$-improvement over another estimator $\hat{\beta}_1$ of β, if*

$$R(\hat{\beta}_1, \beta, A) - R(\hat{\beta}_2, \beta, A) \geq 0. \tag{3.60}$$

3.4.2   Mean-Square-Error

The quadratic risk is related closely to the matrix-valued criterion of Mean-Square-Error (MSE) of an estimator. The MSE is defined as

$$M(\hat{\beta}, \beta) = \mathrm{E}\,(\hat{\beta} - \beta)(\hat{\beta} - \beta)'\,. \tag{3.61}$$

We will denote the covariance matrix of an estimator $\hat{\beta}$ by $\mathrm{V}\,(\hat{\beta})$:

$$\mathrm{V}\,(\hat{\beta}) = \mathrm{E}\,(\hat{\beta} - \mathrm{E}\,(\hat{\beta}))(\hat{\beta} - \mathrm{E}\,(\hat{\beta}))'. \tag{3.62}$$

If $\mathrm{E}\,(\hat{\beta}) = \beta$, then $\hat{\beta}$ will be called unbiased (for β). If $\mathrm{E}\,(\hat{\beta}) \neq \beta$, then $\hat{\beta}$ is called biased. The difference between $\mathrm{E}\,(\hat{\beta})$ and β is called

$$\mathrm{Bias}(\hat{\beta}, \beta) = \mathrm{E}\,(\hat{\beta}) - \beta. \tag{3.63}$$

If $\hat{\beta}$ is unbiased, then obviously $\text{Bias}(\hat{\beta}, \beta) = 0$.

The following decomposition of the Mean-Square-Error often proves to be useful

$$M(\hat{\beta}, \beta) = \text{E}\left[((\hat{\beta} - \text{E}(\hat{\beta})) + (\text{E}(\hat{\beta}) - \beta)][(\hat{\beta} - \text{E}(\hat{\beta})) + (\text{E}(\hat{\beta}) - \beta)]'\right.$$
$$= \text{V}(\hat{\beta}) + (\text{Bias}(\hat{\beta}, \beta))(\text{Bias}(\hat{\beta}, \beta))', \tag{3.64}$$

i.e., the MSE of an estimator is the sum of the covariance matrix and the squared bias.

MSE-Superiority

As the MSE contains all relevant information about the quality of an estimator, comparisons between different estimators may be made by comparing their MSE-matrices.

Definition 3.4 (MSE-I-criterion) *We consider two estimators $\hat{\beta}_1$ and $\hat{\beta}_2$ of β. Then $\hat{\beta}_2$ is called MSE-superior to $\hat{\beta}_1$ (or $\hat{\beta}_2$ is called MSE-improvement to $\hat{\beta}_1$), if the difference of their MSE-matrices is nonnegative definite, i.e., if*

$$\Delta(\hat{\beta}_1, \hat{\beta}_2) = M(\hat{\beta}_1, \beta) - M(\hat{\beta}_2, \beta) \geq 0. \tag{3.65}$$

MSE-superiority is a local property in the sense that it depends on the particular value of β. The quadratic risk function (3.59) is just a scalar-valued version of the MSE:

$$R(\hat{\beta}, \beta, A) = \text{tr}\{AM(\hat{\beta}, \beta)\}. \tag{3.66}$$

One important connection between $R(A)$- and MSE-superiority has been given by Theobald (1974) and Trenkler (1981):

Theorem 3.6 *Consider two estimators $\hat{\beta}_1$ and $\hat{\beta}_2$ of β. The following two statements are equivalent:*

$$\Delta(\hat{\beta}_1, \hat{\beta}_2) \geq 0, \tag{3.67}$$
$$R(\hat{\beta}_1, \beta, A) - R(\hat{\beta}_2, \beta, A) = \text{tr}\{A\Delta(\hat{\beta}_1, \hat{\beta}_2)\} \geq 0 \tag{3.68}$$

for all matrices of the type $A = aa'$.

Proof: Using (3.65) and (3.66) we get

$$R(\hat{\beta}_1, \beta, A) - R(\hat{\beta}_2, \beta, A) = \text{tr}\{A\Delta(\hat{\beta}_1, \hat{\beta}_2)\}. \tag{3.69}$$

Following Theorem A 43, it holds that $\text{tr}\{A\Delta(\hat{\beta}_1, \hat{\beta}_2)\} \geq 0$ for all matrices $A = aa' \geq 0$ if and only if $\Delta(\hat{\beta}_1, \hat{\beta}_2) \geq 0$.

3.4.3 Best Linear Unbiased Estimation

In (3.57), the matrix C and the vector d are unknown and have to be estimated in an optimal way by minimizing the expectation of the sum of squared errors $S(\hat{\beta})$, namely the risk function

$$r(\beta, \hat{\beta}) = \mathrm{E}\,(y - X\hat{\beta})'(y - X\hat{\beta}) \,. \tag{3.70}$$

Direct calculus yields the following result:

$$\begin{aligned} y - X\hat{\beta} &= X\beta + \epsilon - X\hat{\beta} \\ &= \epsilon - X(\hat{\beta} - \beta) \,, \end{aligned} \tag{3.71}$$

such that

$$\begin{aligned} r(\beta, \hat{\beta}) &= \mathrm{tr}\,\{\mathrm{E}\,(\epsilon - X(\hat{\beta} - \beta))(\epsilon - X(\hat{\beta} - \beta))'\} \\ &= \mathrm{tr}\,\{\sigma^2 I_T + X M(\hat{\beta}, \beta) X' - 2 X \mathrm{E}\,[(\hat{\beta} - \beta)\epsilon']\} \\ &= \sigma^2 T + \mathrm{tr}\,\{X' X M(\hat{\beta}, \beta)\} - 2\mathrm{tr}\,\{X \mathrm{E}\,[(\hat{\beta} - \beta)\epsilon']\} \,. \tag{3.72} \end{aligned}$$

Now, we will specify the risk function $r(\hat{\beta}, \beta)$ for linear estimators, considering *unbiased* estimators only.

Unbiasedness of $\hat{\beta}$ requires that $\mathrm{E}\,(\hat{\beta}|\beta) = \beta$ holds independently of the true β in model (3.51). We will see that this imposes some new restrictions on the matrices to be estimated, i.e.,

$$\begin{aligned} \mathrm{E}\,(\hat{\beta}|\beta) &= C\mathrm{E}\,(y) + d \\ &= C X\beta + d = \beta \quad \text{for all} \quad \beta \,. \tag{3.73} \end{aligned}$$

For the choice $\beta = 0$, we immediately have

$$d = 0 \tag{3.74}$$

and the condition equivalent to (3.73) is

$$C X = I \,. \tag{3.75}$$

Inserting this into (3.71) yields

$$\begin{aligned} y - X\hat{\beta} &= X\beta + \epsilon - X C X\beta - X C\epsilon \\ &= \epsilon - X C\epsilon \,, \tag{3.76} \end{aligned}$$

and (cf. (3.72))

$$\begin{aligned} \mathrm{tr}\,\{X \mathrm{E}\,[(\hat{\beta} - \beta)\epsilon']\} &= \mathrm{tr}\,\{X \mathrm{E}\,(C\epsilon\epsilon')\} \\ &= \sigma^2 \mathrm{tr}\,\{X C\} \\ &= \sigma^2 \mathrm{tr}\,\{C X\} = \sigma^2 \mathrm{tr}\,\{I_K\} = \sigma^2 K \,. \tag{3.77} \end{aligned}$$

Thus we can state the following

Theorem 3.7 *For linear unbiased estimators* $\hat{\beta} = Cy$ *with* $CX = I$ *it holds that* $M(\hat{\beta}, \beta) = \mathrm{V}\,(\hat{\beta}) = \sigma^2 CC'$ *and*

$$r(\hat{\beta}, \beta) = \mathrm{tr}\,\{(X'X)\mathrm{V}\,(\hat{\beta})\} + \sigma^2(T - 2K)\,. \tag{3.78}$$

If we consider the risk functions $r(\hat{\beta}, \beta)$ and $R(\hat{\beta}, \beta, X'X)$, then we may state:

Theorem 3.8 *Let* $\hat{\beta}_1$ *and* $\hat{\beta}_2$ *be two linear unbiased estimators. Then*

$$\begin{aligned} r(\hat{\beta}_1, \beta) - r(\hat{\beta}_2, \beta) &= \mathrm{tr}\,\{(X'X)\,\triangle\,(\hat{\beta}_1, \hat{\beta}_2)\} \\ &= R(\hat{\beta}_1, \beta, X'X) - R(\hat{\beta}_2, \beta, X'X)\,, \end{aligned} \tag{3.79}$$

where $\triangle(\hat{\beta}_1, \hat{\beta}_2) = \mathrm{V}\,(\hat{\beta}_1) - \mathrm{V}\,(\hat{\beta}_2)$, *i.e., the difference of the covariance matrices only.*

Using Theorem 3.7 we get with $CX = I$

$$\begin{aligned} r(\hat{\beta}, \beta) &= \sigma^2(T - 2K) + \mathrm{tr}\,\{X'X\mathrm{V}\,(\hat{\beta})\} \\ &= \sigma^2(T - 2K) + \sigma^2\mathrm{tr}\,\{X'XCC'\}\,. \end{aligned}$$

This function has to be minimized w.r.t C under the restriction

$$CX = \begin{pmatrix} c'_1 \\ \vdots \\ c'_K \end{pmatrix} X = \begin{pmatrix} e'_1 \\ \vdots \\ e'_K \end{pmatrix} = I_K\,,$$

i.e.,

$$\min_{C}[\mathrm{tr}\,\{XCC'X'\}|CX - I = 0]\,.$$

This problem may be reformulated in terms of Lagrangian multipliers as ·

$$\min_{C_i, \lambda_i}\,[\mathrm{tr}\,\{XCC'X'\} - 2\sum_{i=1}^{K}\lambda'_i(c'_iX - e'_i)']\,. \tag{3.80}$$

The $(K \times 1)$-vectors λ_i of Lagrangian multipliers may be comprised in the matrix:

$$\Lambda = \begin{pmatrix} \lambda'_1 \\ \vdots \\ \lambda'_K \end{pmatrix}\,. \tag{3.81}$$

Differentiation of (3.80) w.r.t C and Λ yields (Theorems A 91 – A 95) the normal equations

$$X'XC - \Lambda X' = 0\,, \tag{3.82}$$

$$CX - I = 0\,. \tag{3.83}$$

The matrix $X'X$ is regular since rank $(X) = K$. Premultiplication of (3.82) with $(X'X)^{-1}$ leads to

$$C = (X'X)^{-1}\Lambda X' \, ,$$

from which we have (using (3.83))

$$CX = (X'AX)^{-1}\Lambda(X'X) = I_K \, ,$$

namely

$$\hat{\Lambda} = I_K \, .$$

Therefore, the optimum matrix is

$$\hat{C} = (X'X)^{-1}X' \, .$$

The actual linear unbiased estimator is given by

$$\hat{\beta}_{opt} = \hat{C}y = (X'X)^{-1}X'y \, , \tag{3.84}$$

and coincides with the descriptive or empirical OLS estimator b. The estimator b is unbiased since

$$\hat{C}X = (X'X)^{-1}X'X = I_K \, , \tag{3.85}$$

(see (3.75)) and has the $(K \times K)$-covariance matrix

$$\begin{aligned}
V(b) = V_b &= E(b - \beta)(b - \beta)' \\
&= E\{(X'X)^{-1}X'\epsilon\epsilon'X(X'X)^{-1}\} \\
&= \sigma^2(X'X)^{-1} \, . \tag{3.86}
\end{aligned}$$

The main reason for the popularity of OLS b in contrast to other estimators is obvious, as b possesses the minimum variance property among all members of the class of linear unbiased estimators $\tilde{\beta}$. More precisely:

Theorem 3.9 *Let $\tilde{\beta}$ be an arbitrary linear unbiased estimator of β with covariance matrix $V_{\tilde{\beta}}$ and let a be an arbitrary $(K \times 1)$-vector.*
   Then the following two equivalent statements hold:

  a) The difference $V_{\tilde{\beta}} - V_b$ is always nonnegative definite.

  b) The variance of the linear form $a'b$ is always less than or equal to the variance of $a'b$:

$$a'V_b a \leq a'V_{\tilde{\beta}}a \quad or \quad a'(V_{\tilde{\beta}} - V_b)a \geq 0 \, . \tag{3.87}$$

Proof: The equivalence is a direct consequence from the definition of definiteness. We will prove *a)*.

Let $\tilde{\beta} = \tilde{C}y$ be an arbitrary unbiased estimator. Define without loss of generality

$$\tilde{C} = \hat{C} + D = (X'X)^{-1}X' + D .$$

Unbiasedness of $\tilde{\beta}$ requires that (3.75) is fulfilled:

$$\tilde{C}X = \hat{C}X + DX = I .$$

In view of (3.85) it is necessary that

$$DX = 0 .$$

For the covariance matrix of $\tilde{\beta}$ we get

$$\begin{aligned}
V_{\tilde{\beta}} &= \mathrm{E}(\tilde{C}y - \beta)(\tilde{C}y - \beta)' \\
&= \mathrm{E}(\tilde{C}\epsilon)(\epsilon'\tilde{C}') \\
&= \sigma^2[(X'X)^{-1}X' + D][X(X'X)^{-1} + D'] \\
&= \sigma^2[(X'X)^{-1} + DD'] \\
&= V_b + \sigma^2 DD' \geq V_b .
\end{aligned}$$

Corollary: *Let* $V_{\tilde{\beta}} - V_b \geq 0$. *Denote by* $\mathrm{Var}(b_k)$ *and* $\mathrm{Var}(\tilde{\beta}_k)$ *the maindiagonal elements of* V_b *and* $V_{\tilde{\beta}}$. *Then the following inequality holds for the components of the two vectors* $\tilde{\beta}$ *and* b

$$Var(\tilde{\beta}_i) - Var(b_i) \geq 0 \quad (i = 1, \ldots, K) . \tag{3.88}$$

Proof: From $V_{\tilde{\beta}} - V_b \geq 0$ we have $a'(V_{\tilde{\beta}} - V_b)a \geq 0$ for arbitrary vectors a, such that for the vectors $e_i' = (0 \ldots 010 \ldots 0)$ with 1 at the ith position. Let A be an arbitrary symmetric matrix, such that $e_i' A e_i = a_{ii}$. Then, the ith diagonal element of $V_{\tilde{\beta}} - V_b$ is just (3.88).

The minimum property of b is usually expressed by the fundamental GAUSS-MARKOV-Theorem.

Theorem 3.10 (GAUSS-MARKOV-Theorem.) *Consider the classical linear regression model (3.51). The OLS estimator*

$$b_0 = (X'X)^{-1}X'y \tag{3.89}$$

with covariance matrix

$$V_{b_0} = \sigma^2(X'X)^{-1} \tag{3.90}$$

is the best (homogeneous) linear unbiased estimator of β *in the sense of the two properties of Theorem 3.9.* b_0 *will be also denoted as GAUSS-MARKOV-(GM) estimator.*

Estimation of a Linear Function of β

If we are interested in estimating a linear combination of the components of β, for example linear contrasts in ANOVA-models, then we have to consider

$$d = a'\beta , \tag{3.91}$$

where a is a known $(K \times 1)$-vector. For now, it is sufficient to restrict consideration to linear homogeneous estimators $\tilde{d} = c'y$. Then we have

Theorem 3.11 *In the classical linear regression model (3.51)*

$$\hat{d} = a'b_0 , \tag{3.92}$$

with the variance

$$\mathrm{Var}\,(\hat{d}) = \sigma^2 a'(X'X)^{-1}a = a'V_{b_0}a , \tag{3.93}$$

is the best linear unbiased estimator of $d = a'\beta$.

Proof: Let $\tilde{d} = c'y$ be an arbitrary linear unbiased estimator of d, where c is a $(T \times 1)$-vector. Without loss of generality we set

$$c' = a'(X'X)^{-1}X' + \tilde{c}' .$$

The unbiasedness of $\tilde{d}$ requires that

$$c'X = a' ,$$

i.e.,

$$a'(X'X)^{-1}X'X + \tilde{c}'X = a'$$

and therefore

$$\tilde{c}'X = 0 . \tag{3.94}$$

Using (3.94) we get

$$\begin{aligned}
\tilde{d} - d &= a'\beta + a'(X'X)^{-1}X'\epsilon + \tilde{c}'\epsilon - a'\beta \\
&= a'(X'X)^{-1}X'\epsilon + \tilde{c}'\epsilon = c'\epsilon .
\end{aligned}$$

The variance of $\tilde{d}$ is given by

$$\begin{aligned}
\mathrm{Var}\,(\tilde{d}) &= \mathrm{E}\,(\tilde{d} - d)^2 = c'\mathrm{E}\,(\epsilon\epsilon')c = \sigma^2 c'c \\
&= \sigma^2[a'(X'X)^{-1}X' + \tilde{c}'][X(X'X)^{-1}a + \tilde{c}] \\
&= a'V_{b_0}a + \sigma^2\tilde{c}'\tilde{c} .
\end{aligned}$$

As $\tilde{c}'\tilde{c} \geq 0$, the variance of $\tilde{d}$ will be minimized if $\tilde{c} = 0$. The estimator $c'y = a'(X'X)^{-1}X'y = a'b_0$ is therefore the best estimator among all linear unbiased estimators in the sense of a minimum variance.

3.4.4 Estimation of σ^2

The sum of squares $\hat{\epsilon}'\hat{\epsilon}$ of estimated errors $\hat{\epsilon} = y - \hat{y}$ obviously provides a basis appropriate to estimate σ^2.

In detail we get

$$
\begin{aligned}
\hat{\epsilon} &= y - \hat{y} = X\beta + \epsilon - Xb_0 \\
&= \epsilon - X(X'X)^{-1}X'\epsilon \\
&= (I - X(X'X)^{-1}X')\epsilon \\
&= M\epsilon \,.
\end{aligned}
\tag{3.95}
$$

The matrix M is idempotent by Theorem A 61. As a consequence, the sum of squared errors

$$
\hat{\epsilon}'\hat{\epsilon} = \epsilon' M M \epsilon = \epsilon' M \epsilon
$$

has expectation

$$
\begin{aligned}
\mathrm{E}\,(\hat{\epsilon}'\hat{\epsilon}) &= \mathrm{E}\,(\epsilon' M \epsilon) \\
&= \mathrm{E}\,(\mathrm{tr}\,\{\epsilon' M \epsilon\}) \quad [\text{Theorem A 13 (vi)}] \\
&= \mathrm{E}\,(\mathrm{tr}\,\{M \epsilon' \epsilon\}) \\
&= \mathrm{tr}\,\{M\,\mathrm{E}\,(\epsilon \epsilon')\} \\
&= \sigma^2 \mathrm{tr}\,\{M\} \\
&= \sigma^2 \mathrm{tr}\,\{I_T\} - \sigma^2 \mathrm{tr}\,\{X(X'X)^{-1}X'\} \quad [\text{Theorem A 13 (i)}] \\
&= \sigma^2 \mathrm{tr}\,\{I_T\} - \sigma^2 \mathrm{tr}\,\{(X'X)^{-1}X'X\} \\
&= \sigma^2 \mathrm{tr}\,\{I_T\} - \sigma^2 \mathrm{tr}\,\{I_K\} \\
&= \sigma^2(T - K) \,.
\end{aligned}
\tag{3.96}
$$

An unbiased estimator for σ^2 is then given by

$$
s^2 = \hat{\epsilon}'\hat{\epsilon}(T - K)^{-1} = (y - Xb_0)'(y - Xb_0)(T - K)^{-1} \,.
\tag{3.97}
$$

Hence, an unbiased estimator of V_{b_0} is given by

$$
\hat{V}_{b_0} = s^2(X'X)^{-1} \,.
\tag{3.98}
$$

Bivariate Regression $K = 2$

One important special case of the general linear model with K regressors $X_1, \ldots, X_K$ deserves attention. If there is only one true explanatory variable accompanied by a dummy regressor, i.e., a column of one's, then we speak of the simple linear regresion model

$$
y_t = \alpha + \beta x_t + \epsilon_t \quad (t = 1, \ldots, T) \,.
\tag{3.99}
$$

It is often useful to transform the observations (x_t, y_t) in a way that $(\tilde{x}_t, \tilde{y}_t)$ represent deviations of the sample means $(\bar{x}_t, \bar{y}_t)$:

$$
\tilde{y}_t = y_t - \bar{y}, \quad \tilde{x}_t = x_t - \bar{x} \,.
\tag{3.100}
$$

As

$$E\left(\tilde{y}_t | x_1, \ldots, x_T\right) = \alpha + \beta x_t - (\alpha + \beta \bar{x}) = \beta \tilde{x}_t \, ,$$

we are able to obtain an even simpler form of the model (3.99), while the parameter β remains unchanged, i.e.,

$$\tilde{y}_t = \beta \tilde{x}_t + \tilde{\epsilon}_t \quad (t = 1, \ldots, T) \, . \tag{3.101}$$

Assuming that $\bar{\epsilon} = 1/T \sum \epsilon_t = 0$, we have $\tilde{\epsilon}_t = \epsilon_t$ for all t. The OLS estimator of β and the unbiased estimator of σ^2 are obtained by (3.84) and (3.97) as

$$b = \frac{\sum \tilde{x}_t \tilde{y}_t}{\sum \tilde{x}_t^2} \quad \text{with} \quad \text{Var}\,(b) = \frac{\sigma^2}{\sum \tilde{x}_t^2} \, . \tag{3.102}$$

$$s^2 = (T - 2)^{-1} \sum (\tilde{y}_t - \tilde{x}_t b)^2 \, . \tag{3.103}$$

It is easy to see that the OLS estimator for α is given by

$$\hat{\alpha} = \bar{y} - b\bar{x} \, . \tag{3.104}$$

3.5   Multicollinearity

3.5.1   Extreme Multicollinearity and Estimability

A typical problem in practical work is that there is almost always at least some correlation between the exogeneous variables in X. We speak of extreme multicollinearity if two of more columns in X are linearly dependent. As a consequence, we have rank $(X) < K$ such that one basic assumption of model (3.51) is violated. In this case, no unbiased linear estimators for β exist.

We recall that the condition for unbiasedness is equivalent to $d = 0$ and $CX = I$ (cf. (3.75)). If rank $(X) = p < K$, then CX is of rank p at the most, cf. Theorem A 23(iv), whereas the identity matrix I_K is of rank K. Condition (3.75) is thus never fulfilled.

To prove this result in an alternative way, we may use the corollary following Theorem 3.1.

The condition of unbiasedness is a condition on the matrix C, namely

$$CX = I \, .$$

The latter equation is solvable w.r.t C if and only if (3.17) holds, i.e., $X^- X = I_K$. By help of Theorem A 65 (ii), we know that rank $(X^- X) = $ rank (X) and rank $(X) = p < K$. On the other hand, rank $(I_K) = K$. Thus $(X^- X) = I_K$ cannot be valid so that $CX = I$ is not solvable.

The matrix $(X'X)$ is singular, since rank $(X) < K$ and solutions to the normal equation (3.10) are no longer unique.

We say that the parameter vector β is not estimable in the sense that no linear unbiased estimator exists.

Another aspect of extreme multicollinearity becomes apparent when considering, without loss of generality, for $\boldsymbol{x}_1$ a linear combination consisting of all other columns, i.e.,

$$\boldsymbol{x}_1 = \sum_{k=2}^{K} \alpha_k \boldsymbol{x}_k \ .$$

For an arbitrary scalar $\lambda \neq 0$, we can derive the decomposition

$$\begin{aligned}
\boldsymbol{X}\boldsymbol{\beta} &= \sum_{k=1}^{K} \boldsymbol{x}_k \beta_k = (1-\lambda)\beta_1 \boldsymbol{x}_1 + \sum_{k=2}^{K} (\beta_k + \lambda \alpha_k \beta_1)\boldsymbol{x}_k \\
&= \tilde{\beta}_1 \boldsymbol{x}_1 + \sum_{k=2}^{K} \tilde{\beta}_k \boldsymbol{x}_k = \boldsymbol{X}\tilde{\boldsymbol{\beta}}
\end{aligned} \tag{3.105}$$

where $\tilde{\beta}_1 = (1-\lambda)\beta_1$, $\tilde{\beta}_k = (\beta_k + \lambda \alpha_k \beta_1)$ $(k = 2, \ldots, K)$. This means, that the parameter vectors $\boldsymbol{\beta}$ and $\tilde{\boldsymbol{\beta}}$ with $\boldsymbol{\beta} \neq \tilde{\boldsymbol{\beta}}$ yield the same systematical component $\boldsymbol{X}\boldsymbol{\beta} = \boldsymbol{X}\tilde{\boldsymbol{\beta}}$. Now, the observations $\boldsymbol{y}$ do not depend directly, but over $\boldsymbol{X}\boldsymbol{\beta}$ on $\boldsymbol{\beta}$. The information in $\boldsymbol{y}$ therefore does not allow to distinguish between $\boldsymbol{\beta}$ and $\tilde{\boldsymbol{\beta}}$. The regression coefficients are *not identifiable*, the related models are *observational equivalent*.

Example 3.1: We consider the model

$$y_t = \alpha + \beta x_t + \epsilon_t \quad (t = 1, \ldots, T) \ . \tag{3.106}$$

Exact linear dependence between $\boldsymbol{X}_1 \equiv 1$ and $\boldsymbol{X}_2 = \boldsymbol{X}$ means that $x_1 = \ldots = x_t = a$ (a constant), such that $\sum(x_t - \bar{x})^2 = 0$ and b (3.102) cannot be calculated. Let $\begin{pmatrix} \hat{\alpha} \\ \hat{\beta} \end{pmatrix} = \boldsymbol{C}\boldsymbol{y}$ be a linear homogeneous estimator of $(\alpha, \beta)'$. Unbiasedness requires that (3.75) is fulfilled, such that

$$\begin{pmatrix} \sum c_{1t} & a \sum c_{1t} \\ \sum c_{2t} & a \sum c_{2t} \end{pmatrix} = \begin{pmatrix} 1 & 0 \\ 0 & 1 \end{pmatrix} \ . \tag{3.107}$$

There exists no matrix $\boldsymbol{C}$ and no real-valued $a \neq 0$; $(\alpha, \beta)'$ are not estimable. Since $x_t = a$ for all t, we have $y_t = (\alpha + \beta a) + \epsilon_t$, such that α and β are only jointly estimable as $(\alpha \hat{+} \beta a) = \bar{y}$.

3.5.2 Estimation in Case of Extreme Multicollinearity

We are mainly interested in making use of a prior restriction of the form (3.33) with $\boldsymbol{r} = \boldsymbol{0}$, i.e.,

$$\boldsymbol{0} = \boldsymbol{R}\boldsymbol{\beta} \ . \tag{3.108}$$

Parameter values that are observational equivalent are thus excluded.

The identifiability of $\boldsymbol{\beta}$ is guaranteed if $\boldsymbol{R}\boldsymbol{X} = 0$ and the assumptions of Theorem 3.5 are fulfilled. Following Theorem 3.5, the OLS estimator of $\boldsymbol{\beta}$ is of the form

$$b(\boldsymbol{R}, \boldsymbol{0}) = b(\boldsymbol{R}) = (\boldsymbol{X}'\boldsymbol{X} + \boldsymbol{R}'\boldsymbol{R})^{-1} \boldsymbol{X}'\boldsymbol{y} \ , \tag{3.109}$$

if $r = 0$. Summarizing, we may state: in the classical linear restrictive regression model

$$
\left.
\begin{aligned}
& y = X\beta + \epsilon \\[6pt]
& E\,(\epsilon) = 0\ , \quad E\,(\epsilon\epsilon') = \sigma^2 I\ , \\[6pt]
& X \quad \text{nonstochastic,} \quad \text{rank}\,(X) = p < K\ , \\[6pt]
& 0 = R\beta\ , \quad \text{rank}\,(R) = K - p,\ \ \text{rank}\,(D) = K\ ,
\end{aligned}
\right\}
\tag{3.110}
$$

with $D' = (X', R')$, the following fundamental Theorem is valid

Theorem 3.12 *In model (3.110), the conditional OLS estimator*

$$
b(R) = (X'X + R'R)^{-1}X'y = (D'D)^{-1}X'y \tag{3.111}
$$

with covariance matrix

$$
V_{b(R)} = \sigma^2 (D'D)^{-1}X'X(D'D)^{-1} \tag{3.112}
$$

is the best linear unbiased estimator of β.

Definition 3.5 *A linear estimator $\hat{\beta}$ is called conditionally unbiased under*

$$
\underset{K \times K}{A}\ \beta - \underset{K \times 1}{a}\ = 0\ ,
$$

if

$$
E\,(\hat{\beta} - \beta\,|\,A\beta - a = 0) = 0\,. \tag{3.113}
$$

Proof of Theorem 3.12:
a) $b(R)$ is unbiased:
With $R\beta = 0$ we also have $R'R\beta = 0$ (Theorem A 72, A 73), such that

$$
\begin{aligned}
E\,(b(R)) & = (X'X + R'R)^{-1}X'X\beta \\
& = (X'X + R'R)^{-1}(X'X + R'R)\beta = \beta\ .
\end{aligned}
$$

$b(R)$ fulfills the restriction:

$$
Rb(R) = R(X'X + R'R)^{-1}X'y = 0 \quad \text{(compare (3.46))}\,.
$$

b) We immediately get

$$
b(R) - \beta = (D'D)^{-1}X'\epsilon
$$

and therefore

$$
\begin{aligned}
V_{b(R)} &= \mathrm{E}\{(D'D)^{-1}X'\epsilon\epsilon'X(D'D)^{-1}\} \\
&= \sigma^2(D'D)^{-1}X'X(D'D)^{-1}.
\end{aligned}
$$

c) We now have to prove, that $b(R)$ is the best linear conditionally unbiased estimator of β under the restriction $R\beta = 0$, i.e., the best linear unbiased estimator in model (3.110). (A somewhat different way of proof is given by Tan (1971) who deals with multivariate models using generalized inverses.)
Model (3.110) is then of the form

$$
\begin{pmatrix} y \\ 0 \end{pmatrix} \begin{pmatrix} X \\ R \end{pmatrix} \beta + \begin{pmatrix} \epsilon \\ 0 \end{pmatrix}, \tag{3.114}
$$

or in new symbols $(\tilde{T} = T + K - p)$ of the form

$$
\underset{\tilde{T}\times 1}{\tilde{y}} = \underset{\tilde{T}\times K}{D} \; \underset{K\times 1}{\beta} + \underset{\tilde{T}\times 1}{\tilde{\epsilon}}. \tag{3.115}
$$

We have $\mathrm{E}(\tilde{\epsilon}) = 0$, $\mathrm{E}(\epsilon\epsilon') = V = \begin{pmatrix} \sigma^2 I & 0 \\ 0 & 0 \end{pmatrix}$ and rank $(D) = K$, such that the model is singular. The estimator $b(R)$ is still linear in $\tilde{y}$:

$$
\begin{aligned}
b(R) &= (D'D)^{-1}X'y = (D'D)^{-1}(X'y + R'0) \\[2mm]
&= (D'D)^{-1}D'\tilde{y} = C\tilde{y} \quad (C \text{ is a } K \times \tilde{T}\text{-matrix}).
\end{aligned} \tag{3.116}
$$

Since $b(R)$ is conditionally unbiased, we have

$$
CD = I. \tag{3.117}
$$

Let $\tilde{\beta} = C\tilde{y} + d$ be an arbitrary unbiased estimator of β in model (3.114). Without loss of generality we write

$$
\tilde{C} = C + F \quad \text{with} \quad F = (F_1, F_2), \tag{3.118}
$$

where $C = (D'D)^{-1}D'$ is the matrix from (3.116), F_1 is a $(K \times T)$-matrix and F_2 is a $(K \times (K - p))$-matrix. Unbiasedness of $\tilde{\beta}$ in model (3.114) requires that

$$
\mathrm{E}(\tilde{\beta}) = \tilde{C}D\beta + d = \beta \quad \text{for all} \quad \beta,
$$

from which we have $d = 0$ by choosing $\beta = 0$. A necessary condition for unbiasedness is thus given by

$$
\begin{aligned}
\tilde{C}D\beta &= CD\beta + FD\beta \\
&= CD\beta + F_1X\beta + F_2R\beta \\
&= \beta + F_1X\beta = \beta, \qquad [R\beta = 0 \text{ and } (3.117)]
\end{aligned}
$$

and thus
$$F_1 X = 0 \ . \tag{3.119}$$

It follows that
$$
\begin{aligned}
\tilde{\beta} - \beta &= (C + F)D\beta + (C + F)\tilde{\epsilon} - \beta \\
&= (C + F)\tilde{\epsilon} = \tilde{C}\tilde{\epsilon}
\end{aligned}
$$

and we can express the covariance matrix of $\tilde{\beta}$ in the following form:
$$
\begin{aligned}
V_{\tilde{\beta}} = \mathbf{E}\,(\tilde{\beta} - \beta)(\tilde{\beta} - \beta)' &= \tilde{C}V\tilde{C}' \\
&= (C + F)V(C' + F') \\
&= CVC' + FVF' + FVC' + CVF' \ .
\end{aligned}
$$

Furthermore, we have (with $\mathbf{E}\,(\tilde{\epsilon}\tilde{\epsilon}') = V$, compare (3.115))
$$
\begin{aligned}
CVC' &= V_{b(R)} \\
FVF' &= (F_1, F_2)\begin{pmatrix} \sigma^2 I & 0 \\ 0 & 0 \end{pmatrix}\begin{pmatrix} F_1' \\ F_2' \end{pmatrix} = \sigma^2 F_1 F_1'
\end{aligned}
$$

where $\sigma^2 F_1 F_1'$ is nonnegative definite [Theorem A 41 (v)].
For mixed products it holds
$$
\begin{aligned}
FVC' &= (F_1, F_2)\begin{pmatrix} \sigma^2 I & 0 \\ 0 & 0 \end{pmatrix}\begin{pmatrix} X \\ R \end{pmatrix}(D'D)^{-1} \\
&\quad \tag{3.120} \\
&= F_1 X (D'D)^{-1} = 0 \quad \text{[by (3.119)]}
\end{aligned}
$$

Finally we get
$$V_{\tilde{\beta}} - V_{b(R)} = \sigma^2 F_1 F_1' \geq 0 \tag{3.121}$$

and the asserted optimality of $b(R)$ has been proven. Therefore, $b(R)$ is a GM estimator of β in model (3.114).

3.6   Classical Regression Under Normal Errors

All results obtained so far are valid irrespective of the actual distribution of the random disturbances ϵ, provided that $\mathbf{E}\,(\epsilon) = 0$ and $\mathbf{E}\,(\epsilon\epsilon') = \sigma^2 I$. Now, we shall specify the type of the distribution of ϵ by additionally imposing the following condition: The vector ϵ of random disturbances ϵ_t is distributed according to a T-dimensional normal distribution $N(0, \sigma^2 I)$, i.e., $\epsilon \sim N(0, \sigma^2 I)$.

The probability density of ϵ is given by
$$
\begin{aligned}
f(\epsilon; 0, \sigma^2 I) &= \prod_{t=1}^{T}(2\pi\sigma^2)^{1/2}\exp\left(-\frac{1}{2\sigma^2}\epsilon_t^2\right) \\
&= (2\pi\sigma^2)^{-T/2}\exp\left\{-\frac{1}{2\sigma^2}\sum_{t=1}^{T}\epsilon_t^2\right\} , \tag{3.122}
\end{aligned}
$$

such that its components ϵ_t $(t = 1, \ldots, T)$ are independent and identically distributed as $N(0, \sigma^2)$. (3.122) is a special case of the general T-dimensional normal distribution $N(\boldsymbol{\mu}, \boldsymbol{\Sigma})$. Let $\boldsymbol{\xi} \sim N_T(\boldsymbol{\mu}, \boldsymbol{\Sigma})$, i.e., $\mathrm{E}\,(\boldsymbol{\xi}) = \boldsymbol{\mu}$, $\mathrm{E}\,(\boldsymbol{\xi} - \boldsymbol{\mu})(\boldsymbol{\xi} - \boldsymbol{\mu})' = \boldsymbol{\Sigma}$. Then $\boldsymbol{\xi}$ is normally distributed with density (cf. A 81)

$$f(\boldsymbol{\xi}; \boldsymbol{\mu}, \boldsymbol{\Sigma}) = \{(2\pi)^T |\boldsymbol{\Sigma}|\}^{-1/2} \exp\{-\frac{1}{2}(\boldsymbol{\xi} - \boldsymbol{\mu})'\boldsymbol{\Sigma}^{-1}(\boldsymbol{\xi} - \boldsymbol{\mu})\} . \qquad (3.123)$$

The classical linear regression model under normal errors is given by

$$\left.\begin{aligned} &\boldsymbol{y} = \boldsymbol{X}\boldsymbol{\beta} + \boldsymbol{\epsilon}, \\[1mm] &\boldsymbol{\epsilon} \sim N(\boldsymbol{0}, \sigma^2 \boldsymbol{I}), \\[1mm] &\boldsymbol{X} \quad \text{nonstochastic,} \quad \text{rank}\,(\boldsymbol{X}) = K . \end{aligned}\right\} \qquad (3.124)$$

The Maximum-Likelihood (ML) principle

Definition 3.6 *Let $\boldsymbol{\xi} = (\xi_1, \ldots, \xi_n)'$ be a random variable with density function $f(\boldsymbol{\xi}; \boldsymbol{\Theta})$, where the parameter vector $\boldsymbol{\Theta} = (\Theta_1, \ldots, \Theta_m)'$ is a member of the parameter space Ω comprising all values that are a priori admissible.*
The basic idea of the Maximum-Likelihood principle is to interpret the density $f(\boldsymbol{\xi}; \boldsymbol{\Theta})$ for a specific realisation of the sample $\boldsymbol{\xi}_0$ of $\boldsymbol{\xi}$ as a function of $\boldsymbol{\Theta}$:

$$L(\boldsymbol{\Theta}) = L(\Theta_1, \ldots, \Theta_m) = f(\boldsymbol{\xi}_0; \boldsymbol{\Theta}) .$$

$L(\boldsymbol{\Theta})$ will be denoted as the Likelihood function of $\boldsymbol{\xi}_0$.

The ML principle now postulates to choose a value $\hat{\boldsymbol{\Theta}} \in \Omega$ which maximizes the Likelihood function, i.e.,

$$L(\hat{\boldsymbol{\Theta}}) \geq L(\boldsymbol{\Theta}) \quad \text{for all} \quad \boldsymbol{\Theta} \in \Omega .$$

Note that $\hat{\boldsymbol{\Theta}}$ may not be unique. If we consider all possible samples, then $\hat{\boldsymbol{\Theta}}$ is a function of $\boldsymbol{\xi}$ and thus a random variable itself. We will call it Maximum-Likelihood estimator of $\boldsymbol{\Theta}$.

ML estimation in classical normal regression

Following Theorem A 82, we have for $\boldsymbol{y}$ from (3.51)

$$\boldsymbol{y} = \boldsymbol{X}\boldsymbol{\beta} + \boldsymbol{\epsilon} \sim N(\boldsymbol{X}\boldsymbol{\beta}, \sigma^2 \boldsymbol{I}) , \qquad (3.125)$$

such that the Likelihood function of $\boldsymbol{y}$ is given by

$$L(\boldsymbol{\beta}, \sigma^2) = (2\pi\sigma^2)^{-T/2} \exp\left\{-\frac{1}{2\sigma^2}(\boldsymbol{y} - \boldsymbol{X}\boldsymbol{\beta})'(\boldsymbol{y} - \boldsymbol{X}\boldsymbol{\beta})\right\} . \qquad (3.126)$$

The logarithmic transformation is monotonic. Hence, it is appropriate to maximize $\ln L(\boldsymbol{\beta}, \sigma^2)$ instead of $L(\boldsymbol{\beta}, \sigma^2)$, as the maximizing argument remains unchanged:

$$\ln L(\boldsymbol{\beta}, \sigma^2) = -\frac{T}{2}\ln(2\pi\sigma^2) - \frac{1}{2\sigma^2}(\boldsymbol{y} - \boldsymbol{X}\boldsymbol{\beta})'(\boldsymbol{y} - \boldsymbol{X}\boldsymbol{\beta}) \ . \qquad (3.127)$$

If there are no a-priori-restrictions on the parameters, then the parameter space is given by $\boldsymbol{\Omega} = \{\boldsymbol{\beta}; \sigma^2 : \boldsymbol{\beta} \in \mathcal{R}^K; \sigma^2 > 0\}$. We derive the ML estimators of $\boldsymbol{\beta}$ and σ^2 by equating the first derivatives to zero (Theorems A 91 - A 95)

$$(I) \qquad \frac{\partial \ln L}{\partial \boldsymbol{\beta}} = \frac{1}{2\sigma^2}2\boldsymbol{X}'(\boldsymbol{y} - \boldsymbol{X}\boldsymbol{\beta}) = \boldsymbol{0} \ , \qquad (3.128)$$

$$(II) \quad \frac{\partial \ln L}{\partial \sigma^2} = -\frac{T}{2\sigma^2} + \frac{1}{2(\sigma^2)^2}(\boldsymbol{y} - \boldsymbol{X}\boldsymbol{\beta})'(\boldsymbol{y} - \boldsymbol{X}\boldsymbol{\beta}) = 0 \ . \qquad (3.129)$$

The *Likelihood equations* are given by

$$\left. \begin{array}{ll} (I) & \boldsymbol{X}'\boldsymbol{X}\hat{\boldsymbol{\beta}} = \boldsymbol{X}'\boldsymbol{y} \ , \\[2mm] (II) & \hat{\sigma}^2 = \frac{1}{T}(\boldsymbol{y} - \boldsymbol{X}\hat{\boldsymbol{\beta}})'(\boldsymbol{y} - \boldsymbol{X}\hat{\boldsymbol{\beta}}) \end{array} \right\} \qquad (3.130)$$

Equation (I) is identical to the well known normal equation (3.10). Its solution is unique, as rank $(\boldsymbol{X}) = K$, and we get the unique ML estimator

$$\hat{\boldsymbol{\beta}} = \boldsymbol{b} = (\boldsymbol{X}'\boldsymbol{X})^{-1}\boldsymbol{X}'\boldsymbol{y}. \qquad (3.131)$$

If we compare (II) with the unbiased estimator s^2 (3.97) for σ^2, we immediately see that

$$\hat{\sigma}^2 = \frac{T - K}{T}s^2, \qquad (3.132)$$

such that $\hat{\sigma}^2$ is a biased estimator. The asymptotic expectation is given by (cf. A 99 (i))

$$\lim_{T \to \infty} \mathrm{E}\left(\hat{\sigma}^2\right) = \bar{\mathrm{E}}\left(\hat{\sigma}^2\right) = \mathrm{E}\left(s^2\right) = \sigma^2 \ . \qquad (3.133)$$

Thus we can state

Theorem 3.13 *The Maximum-Likelihood estimator and the OLS estimator of $\boldsymbol{\beta}$ are identical in model (3.125) of classical normal regression. The ML estimator $\hat{\sigma}^2$ of σ^2 is asymptotically unbiased.*

Remark: The Cramér-Rao-bound defines a lower bound (in the sense of definiteness of matrices) for the covariance matrix of unbiased estimators. In the model of normal regression, the Cramér-Rao-bound is given by (Amemiya, 1985, p. 19)

$$\mathrm{V}\,(\tilde{\boldsymbol{\beta}}) \geq \sigma^2(\boldsymbol{X}'\boldsymbol{X})^{-1},$$

where $\tilde{\boldsymbol{\beta}}$ is an arbitrary estimator. The covariance matrix of the ML estimator is just identical to this lower bound, such that $\boldsymbol{b}$ is the best unbiased estimator in the linear regression model under normal errors.

3.7 Testing Linear Hypotheses

In this section, testing procedures are being derived in order to test linear hypotheses in the model (3.125) of classical normal regression. The general linear hypothesis

$$H_0 : \boldsymbol{R\beta} = \boldsymbol{r} \, ; \quad \sigma^2 > 0 \quad \text{arbitrary} \tag{3.134}$$

is usually tested against the alternative

$$H_1 : \boldsymbol{R\beta} \neq \boldsymbol{r} \, ; \quad \sigma^2 > 0 \quad \text{arbitrary} \tag{3.135}$$

where the following will be assumed:

$$\left. \begin{array}{l} \boldsymbol{R} \text{ a } (K - s) \times K\text{-matrix,} \\[2ex] \boldsymbol{r} \text{ a } (K - s) \times 1\text{-vector,} \\[2ex] \text{rank} \, (\boldsymbol{R}) = K - s, \\[2ex] s \in \{0, 1, \ldots, K - 1\}, \\[2ex] \boldsymbol{R}, \boldsymbol{r} \text{ nonstochastic and known.} \end{array} \right\} \tag{3.136}$$

The hypothesis H_0 expresses the fact, that the parameter vector $\boldsymbol{\beta}$ obeys $(K - s)$ exact linear restrictions which are independent, as it is required that rank $(\boldsymbol{R}) = K - s$. The general linear hypothesis (3.134) contains two main special cases:

Case 1: $s = 0$

The $K \times K$-matrix $\boldsymbol{R}$ is regular by assumption (3.136) and we may express H_0 and H_1 in the following form:

$$H_0 : \quad \boldsymbol{\beta} = \boldsymbol{R}^{-1}\boldsymbol{r} = \boldsymbol{\beta}^* \, ; \quad \sigma^2 > 0 \quad \text{arbitrary} \tag{3.137}$$
$$H_1 : \quad \boldsymbol{\beta} \neq \boldsymbol{\beta}^* \, ; \quad \sigma^2 > 0 \quad \text{arbitrary} \tag{3.138}$$

Case 2: $s > 0$

We choose an $s \times K$-matrix $\boldsymbol{G}$ complementary to $\boldsymbol{R}$ such that the $K \times K$-Matrix $\begin{pmatrix} \boldsymbol{G} \\ \boldsymbol{R} \end{pmatrix}$ is regular of rank K. Let

$$\boldsymbol{X} \begin{pmatrix} \boldsymbol{G} \\ \boldsymbol{R} \end{pmatrix}^{-1} = \underset{T \times K}{\tilde{\boldsymbol{X}}} = \left(\underset{T \times s}{\tilde{\boldsymbol{X}}_1}, \underset{T \times (K-s)}{\tilde{\boldsymbol{X}}_2} \right) \quad \text{and}$$

$$\underset{s \times 1}{\tilde{\boldsymbol{\beta}}_1} = \boldsymbol{G\beta}, \qquad \underset{(K-s) \times 1}{\tilde{\boldsymbol{\beta}}_2} = \boldsymbol{R\beta} \, .$$

Then we may write:

$$y = X\beta + \epsilon = X \begin{pmatrix} G \\ R \end{pmatrix}^{-1} \begin{pmatrix} G \\ R \end{pmatrix} \beta + \epsilon$$

$$= \tilde{X} \begin{pmatrix} \tilde{\beta}_1 \\ \tilde{\beta}_2 \end{pmatrix} + \epsilon$$

$$= \tilde{X}_1 \tilde{\beta}_1 + \tilde{X}_2 \tilde{\beta}_2 + \epsilon .$$

The latter model obeys all assumptions (3.51). The hypotheses H_0 and H_1 are thus equivalent to

$$H_0 : \tilde{\beta}_2 = r; \ \tilde{\beta}_1 \text{ and } \sigma^2 > 0 \quad \text{arbitrary} , \tag{3.139}$$

$$H_1 : \tilde{\beta}_2 \neq r; \ \tilde{\beta}_1 \text{ and } \sigma^2 > 0 \quad \text{arbitrary} . \tag{3.140}$$

Ω stands for the whole parameter space (either H_0 or H_1 are valid) and $\omega \subset \Omega$ stands for the subspace in which only H_0 is true, i.e.,

$$\Omega = \{\beta; \sigma^2 : \beta \in E^K, \sigma^2 > 0\},$$
$$\omega = \{\beta; \sigma^2 : \beta \in E^K \text{ and } R\beta = r; \sigma^2 > 0\}. \tag{3.141}$$

As a genuine test statistic, we will use the Likelihood ratio

$$\lambda(y) = \frac{\max_\omega L(\Theta)}{\max_\Omega L(\Theta)}, \tag{3.142}$$

which may be derived in terms of model (3.125) in the following way. $L(\Theta)$ attains its maximum at the ML estimator $\hat{\Theta}$. Let $\Theta = (\beta, \sigma^2)$, then it holds that

$$\max_{\beta, \sigma^2} L(\beta, \sigma^2) = L(\hat{\beta}, \hat{\sigma}^2)$$

$$= (2\pi\hat{\sigma}^2)^{-T/2} \exp\left\{-\tfrac{1}{2\hat{\sigma}^2}(y - X\hat{\beta})'(y - X\hat{\beta})\right\} \tag{3.143}$$

$$= (2\pi\hat{\sigma}^2)^{-T/2} \exp\left\{-\tfrac{T}{2}\right\}$$

and therefore

$$\lambda(y) = \left(\frac{\hat{\sigma}_\omega^2}{\hat{\sigma}_\Omega^2}\right)^{-T/2} , \tag{3.144}$$

where $\hat{\sigma}_\omega^2$ and $\hat{\sigma}_\Omega^2$ are ML estimators of σ^2 under H_0 and in Ω. The random variable $\lambda(y)$ can take values between 0 and 1, as is obvious from (3.142). If H_0 is true, the numerator of $\lambda(y)$ should be greater than the denominator, so that

$\lambda(\boldsymbol{y})$ should be close to one in repeated samples. On the other hand, $\lambda(\boldsymbol{y})$ should be close to zero if H_1 is true. Consider the linear transform of $\lambda(\boldsymbol{y})$:

$$F = \{(\lambda(\boldsymbol{y}))^{-2/T} - 1\}(T - K)(K - s)^{-1}$$

$$= \frac{\hat{\sigma}_\omega^2 - \hat{\sigma}_\Omega^2}{\hat{\sigma}_\Omega^2} \cdot \frac{T - K}{K - s}. \tag{3.145}$$

If $\lambda \to 0$ then $F \to \infty$ and if $\lambda \to 1$ we have $F \to 0$, such that "F close to 0" if H_0 seems to be true and "F sufficiently large" if H_1 is supposed to be true.

In what follows, we will determine F and its distribution for the two special cases of the general linear hypothesis.

Distribution of F

Case 1: $s = 0$

The ML estimators under H_0 (3.137) are given by

$$\hat{\boldsymbol{\beta}} = \boldsymbol{\beta}^* \quad \text{and} \quad \hat{\sigma}_\omega^2 = \frac{1}{T}(\boldsymbol{y} - \boldsymbol{X}\boldsymbol{\beta}^*)'(\boldsymbol{y} - \boldsymbol{X}\boldsymbol{\beta}^*). \tag{3.146}$$

The ML estimators over Ω are available from Theorem 3.13:

$$\hat{\boldsymbol{\beta}} = \boldsymbol{b} \quad \text{and} \quad \hat{\sigma}_\Omega^2 = \frac{1}{T}(\boldsymbol{y} - \boldsymbol{X}\boldsymbol{b})'(\boldsymbol{y} - \boldsymbol{X}\boldsymbol{b}). \tag{3.147}$$

Subsequent modifications then yield:

$$\left.\begin{aligned}
\boldsymbol{b} - \boldsymbol{\beta}^* &= (\boldsymbol{X}'\boldsymbol{X})^{-1}\boldsymbol{X}'(\boldsymbol{y} - \boldsymbol{X}\boldsymbol{\beta}^*), \\[1ex]
(\boldsymbol{b} - \boldsymbol{\beta}^*)'\boldsymbol{X}'\boldsymbol{X} &= (\boldsymbol{y} - \boldsymbol{X}\boldsymbol{\beta}^*)'\boldsymbol{X}, \\[1ex]
\boldsymbol{y} - \boldsymbol{X}\boldsymbol{b} &= (\boldsymbol{y} - \boldsymbol{X}\boldsymbol{\beta}^*) - \boldsymbol{X}(\boldsymbol{b} - \boldsymbol{\beta}^*), \\[1ex]
(\boldsymbol{y} - \boldsymbol{X}\boldsymbol{b})'(\boldsymbol{y} - \boldsymbol{X}\boldsymbol{b}) &= (\boldsymbol{y} - \boldsymbol{X}\boldsymbol{\beta}^*)'(\boldsymbol{y} - \boldsymbol{X}\boldsymbol{\beta}^*) \\
&\quad + (\boldsymbol{b} - \boldsymbol{\beta}^*)'\boldsymbol{X}'\boldsymbol{X}(\boldsymbol{b} - \boldsymbol{\beta}^*) \\
&\quad - 2(\boldsymbol{y} - \boldsymbol{X}\boldsymbol{\beta}^*)'\boldsymbol{X}(\boldsymbol{b} - \boldsymbol{\beta}^*) \\
&= (\boldsymbol{y} - \boldsymbol{X}\boldsymbol{\beta}^*)'(\boldsymbol{y} - \boldsymbol{X}\boldsymbol{\beta}^*) \\
&\quad - (\boldsymbol{b} - \boldsymbol{\beta}^*)'\boldsymbol{X}'\boldsymbol{X}(\boldsymbol{b} - \boldsymbol{\beta}^*).
\end{aligned}\right\} \tag{3.148}$$

It follows that

$$T(\hat{\sigma}_\omega^2 - \hat{\sigma}_\Omega^2) = (\boldsymbol{b} - \boldsymbol{\beta}^*)'\boldsymbol{X}'\boldsymbol{X}(\boldsymbol{b} - \boldsymbol{\beta}^*), \tag{3.149}$$

and we now have the test statistic

$$F = \frac{(\boldsymbol{b} - \boldsymbol{\beta}^*)'\boldsymbol{X}'\boldsymbol{X}(\boldsymbol{b} - \boldsymbol{\beta}^*)}{(\boldsymbol{y} - \boldsymbol{X}\boldsymbol{b})'(\boldsymbol{y} - \boldsymbol{X}\boldsymbol{b})} \cdot \frac{T - K}{K}. \tag{3.150}$$

Numerator:
The following statements hold:

$$b - \beta^* = (X'X)^{-1}X'[\epsilon + X(\beta - \beta^*)] \qquad \text{[by (3.148)]},$$

$$\tilde{\epsilon} = \epsilon + X(\beta - \beta^*) \sim N(X(\beta - \beta^*), \sigma^2 I) \quad \text{[Theorem A 82]},$$

$$X(X'X)^{-1}X' \quad \text{idempotent and of rank } K \qquad \text{[Theorem 3.5]},$$

$$(b - \beta^*)'X'X(b - \beta^*) = \tilde{\epsilon}'X(X'X)^{-1}X'\tilde{\epsilon}$$

$$\sim \sigma^2 \chi_K^2(\sigma^{-2}(\beta - \beta^*)'X'X(\beta - \beta^*)) \qquad \text{[Theorem A 84]}$$

$$\text{and} \quad \sim \sigma^2 \chi_K^2 \text{ under } H_0.$$

Denominator:

$$(y - Xb)'(y - Xb) = (T - K)s^2 = \epsilon'M\epsilon \qquad \text{[by (3.97)]},$$

$$M = I - X(X'X)^{-1}X' \quad \text{idempotent of rank} \quad T - K \qquad \text{[A 61 (vi)]},$$

$$\epsilon'M\epsilon \sim \sigma^2 \chi_{T-K}^2 \qquad \text{[Theorem A 87].}$$

$$(3.151)$$

We have

$$MX(X'X)^{-1}X' = 0 \qquad \text{[Theorem A 61 (vi)]}, \qquad (3.152)$$

such that numerator and denominator are independently distributed (Theorem A 89).

Thus (Theorem A 86), the ratio F exhibits the following properties

- F is distributed as $F_{K,T-K}(\sigma^{-2}(\beta - \beta^*)'X'X(\beta - \beta^*))$ under H_1, and

- F is distributed as central $F_{K,T-K}$ under $H_0 : \beta = \beta^*$.

If we denote by $F_{m,n,1-q}$ the $(1-q)$-quantile of $F_{m,n}$ (i.e., $P(F \leq F_{m,n,1-q}) = 1 - q$), then we may derive a uniformly most powerful test, given a fixed level of significance α (cf. Lehmann, 1986, p. 372):

$$\begin{array}{ll} \text{Region of acceptance of } H_0: & 0 \leq F \leq F_{K,T-K,1-\alpha}, \\ \text{critical area of } H_0: & F > F_{K,T-K,1-\alpha}. \end{array} \qquad (3.153)$$

A selection of critical values is provided in Appendix B.

Case 2: $s > 0$

Next we consider a decomposition of the model in order to determine the ML estimators under H_0 (3.139) and compare them with the corresponding ML estimators over Ω. Let

$$\boldsymbol{\beta}' = \left(\underset{1\times s}{\boldsymbol{\beta}'_1}, \quad \underset{1\times(K-s)}{\boldsymbol{\beta}'_2} \right) \tag{3.154}$$

and, respectively

$$\boldsymbol{y} = \boldsymbol{X}\boldsymbol{\beta} + \boldsymbol{\epsilon} = \boldsymbol{X}_1\boldsymbol{\beta}_1 + \boldsymbol{X}_2\boldsymbol{\beta}_2 + \boldsymbol{\epsilon}. \tag{3.155}$$

We set

$$\tilde{\boldsymbol{y}} = \boldsymbol{y} - \boldsymbol{X}_2\boldsymbol{r}. \tag{3.156}$$

Since rank $(\boldsymbol{X}) = K$, we have

$$\underset{T\times s}{\text{rank } (\boldsymbol{X}_1)} = s, \quad \underset{T\times(K-s)}{\text{rank } (\boldsymbol{X}_2)} = K - s, \tag{3.157}$$

such that the inverse matrices $(\boldsymbol{X}'_1\boldsymbol{X}_1)^{-1}$ and $(\boldsymbol{X}'_2\boldsymbol{X}_2)^{-1}$ exist. The ML estimators under H_0 then are given by

$$\hat{\boldsymbol{\beta}}_2 = \boldsymbol{r}, \quad \hat{\boldsymbol{\beta}}_1 = (\boldsymbol{X}'_1\boldsymbol{X}_1)^{-1}\boldsymbol{X}'_1\tilde{\boldsymbol{y}} \tag{3.158}$$

and

$$\hat{\sigma}^2_\omega = \frac{1}{T}(\tilde{\boldsymbol{y}} - \boldsymbol{X}_1\hat{\boldsymbol{\beta}}_1)'(\tilde{\boldsymbol{y}} - \boldsymbol{X}_1\hat{\boldsymbol{\beta}}_1). \tag{3.159}$$

Separation of b

It can easily be seen that

$$\boldsymbol{b} = (\boldsymbol{X}'\boldsymbol{X})^{-1}\boldsymbol{X}'\boldsymbol{y}$$

$$= \left(\begin{array}{cc} \boldsymbol{X}'_1\boldsymbol{X}_1 & \boldsymbol{X}'_1\boldsymbol{X}_2 \\ \boldsymbol{X}'_2\boldsymbol{X}_1 & \boldsymbol{X}'_2\boldsymbol{X}_2 \end{array} \right)^{-1} \left(\begin{array}{c} \boldsymbol{X}'_1\boldsymbol{y} \\ \boldsymbol{X}'_2\boldsymbol{y} \end{array} \right). \tag{3.160}$$

Making use of the formulae for the inverse of a partitioned matrix, yields (Theorem A 19)

$$\left(\begin{array}{cc} (\boldsymbol{X}'_1\boldsymbol{X}_1)^{-1}[\boldsymbol{I} + \boldsymbol{X}'_1\boldsymbol{X}_2\boldsymbol{D}^{-1}\boldsymbol{X}'_2\boldsymbol{X}_1(\boldsymbol{X}'_1\boldsymbol{X}_1)^{-1}] & -(\boldsymbol{X}'_1\boldsymbol{X}_1)^{-1}\boldsymbol{X}'_1\boldsymbol{X}_2\boldsymbol{D}^{-1} \\ -\boldsymbol{D}^{-1}\boldsymbol{X}'_2\boldsymbol{X}_1(\boldsymbol{X}'_1\boldsymbol{X}_1)^{-1} & \boldsymbol{D}^{-1} \end{array} \right), \tag{3.161}$$

where

$$\boldsymbol{D} = \boldsymbol{X}'_2\boldsymbol{M}_1\boldsymbol{X}_2 \tag{3.162}$$

and

$$\boldsymbol{M}_1 = \boldsymbol{I} - \boldsymbol{X}_1(\boldsymbol{X}'_1\boldsymbol{X}_1)^{-1}\boldsymbol{X}'_1 = \boldsymbol{I} - \boldsymbol{P}_{\boldsymbol{X}_1}. \tag{3.163}$$

M_1 is (analogously to M) idempotent and of rank $T - s$, furthermore, we have $M_1 X_1 = 0$. The $(K - s) \times (K - s)$-matrix

$$D = X_2' X_2 - X_2' X_1 (X_1' X_1)^{-1} X_1' X_2 \qquad (3.164)$$

is symmetric and regular, as the normal equations are uniquely solvable. The components b_1 and b_2 of b are then given by

$$b = \begin{pmatrix} b_1 \\ b_2 \end{pmatrix} = \begin{pmatrix} (X_1' X_1)^{-1} X_1' y - (X_1' X_1)^{-1} X_1' X_2 D^{-1} X_2' M_1 y \\ D^{-1} X_2' M_1 y \end{pmatrix}. \qquad (3.165)$$

Various relations immediately become apparent from (3.165):

$$\left. \begin{aligned} b_2 \quad &= D^{-1} X_2' M_1 y, \\ b_1 \quad &= (X_1' X_1)^{-1} X_1' (y - X_2 b_2), \\ b_2 - r \ &= D^{-1} X_2' M_1 (y - X_2 r) \\ &= D^{-1} X_2' M_1 \tilde{y} \\ &= D^{-1} X_2' M_1 (\epsilon + X_2 (\beta_2 - r)), \end{aligned} \right\} \qquad (3.166)$$

$$\left. \begin{aligned} b_1 - \hat{\beta}_1 \ &= (X_1' X_1)^{-1} X_1' (y - X_2 b_2 - \tilde{y}) \\ &= -(X_1' X_1)^{-1} X_1' X_2 (b_2 - r) \\ &= -(X_1' X_1)^{-1} X_1' X_2 D^{-1} X_2' M_1 \tilde{y}. \end{aligned} \right\} \qquad (3.167)$$

Decomposition of $\hat{\sigma}_\Omega^2$

We write (using symbols u and v)

$$\begin{aligned} (y - X b) \ &= \ (y - X_2 r - X_1 \hat{\beta}_1) \ - \ \left(X_1 (b_1 - \hat{\beta}_1) + X_2 (b_2 - r) \right) \\ &= \qquad\qquad u \qquad\qquad - \qquad\qquad v. \end{aligned}$$
$$(3.168)$$

Thus, we may decompose the ML estimator $T \hat{\sigma}_\Omega^2 = (y - X b)'(y - X b)$ as

$$(y - X b)'(y - X b) = u'u + v'v - 2u'v. \qquad (3.169)$$

We have

$$\begin{aligned} u \ &= \ y - X_2 r - X_1 \hat{\beta}_1 = \tilde{y} - X_1 (X_1' X_1)^{-1} X_1' \tilde{y} = M_1 \tilde{y} \quad (3.170) \\ u'u \ &= \ \tilde{y}' M_1 \tilde{y}, \qquad\qquad\qquad\qquad\qquad\qquad\qquad\qquad\qquad (3.171) \\ v \ &= \ X_1 (b_1 - \hat{\beta}_1) + X_2 (b_2 - r) \\ &= \ -X_1 (X_1' X_1)^{-1} X_1' X_2 D^{-1} X_2' M_1 \tilde{y} \qquad \text{[by (3.166)]} \\ &\quad + X_2 D^{-1} X_2' M_1 \tilde{y} \qquad\qquad\qquad\qquad \text{[by (3.167)]} \\ &= \ M_1 X_2 D^{-1} X_2' M_1 \tilde{y} , \qquad\qquad\qquad\qquad\qquad (3.172) \end{aligned}$$

$$\boldsymbol{v'v} = \boldsymbol{\tilde{y}'M_1X_2D^{-1}X_2'M_1\tilde{y}}$$
$$= (\boldsymbol{b_2 - r})'\boldsymbol{D}(\boldsymbol{b_2 - r}) \,, \tag{3.173}$$

$$\boldsymbol{u'v} = \boldsymbol{v'v} \,. \tag{3.174}$$

Summarizing, we may state

$$(\boldsymbol{y - Xb})'(\boldsymbol{y - Xb}) = \boldsymbol{u'u - v'v}$$
$$\tag{3.175}$$
$$= (\boldsymbol{\tilde{y} - X_1\hat{\beta}_1})'(\boldsymbol{\tilde{y} - X_1\hat{\beta}_1}) - (\boldsymbol{b_2 - r})'\boldsymbol{D}(\boldsymbol{b_2 - r})$$

or,

$$T(\hat{\sigma}_\omega^2 - \hat{\sigma}_\Omega^2) = (\boldsymbol{b_2 - r})'\boldsymbol{D}(\boldsymbol{b_2 - r}) \,. \tag{3.176}$$

Hence, for case 2: $s > 0$, we get

$$F = \frac{(\boldsymbol{b_2 - r})'\boldsymbol{D}(\boldsymbol{b_2 - r})}{(\boldsymbol{y - Xb})'(\boldsymbol{y - Xb})} \frac{T - K}{K - s} \,. \tag{3.177}$$

Distribution of $\boldsymbol{F}$

Numerator:
We use the following relations:

$$\boldsymbol{A = M_1X_2D^{-1}X_2'M_1} \quad \text{is idempotent,}$$

$$\text{rank}(\boldsymbol{A}) = \text{tr}(\boldsymbol{A}) = \text{tr}\{(\boldsymbol{M_1X_2D^{-1}})(\boldsymbol{X_2'M_1})\}$$
$$= \text{tr}\{(\boldsymbol{X_2'M_1})(\boldsymbol{M_1X_2D^{-1}})\} \quad \text{[Theorem A 13 (iv)]}$$
$$= \text{tr}(\boldsymbol{I_{K-s}}) = K - s,$$

$$\boldsymbol{b_2 - r = D^{-1}X_2'M_1\tilde{\epsilon}} \quad \text{[by (3.166)],}$$

$$\boldsymbol{\tilde{\epsilon} = \epsilon + X_2(\beta_2 - r)} \sim N(\boldsymbol{X_2(\beta_2 - r)}, \sigma^2\boldsymbol{I}), \quad \text{[Theorem A 82],}$$

$$(\boldsymbol{b_2 - r})'\boldsymbol{D}(\boldsymbol{b_2 - r}) = \boldsymbol{\tilde{\epsilon}'A\tilde{\epsilon}} \sim \sigma^2\chi_{K-s}^2(\sigma^{-2}(\boldsymbol{\beta_2 - r})'\boldsymbol{D}(\boldsymbol{\beta_2 - r})) \tag{3.178}$$

[Theorem A 84] and

$$\sim \sigma^2\chi_{K-s}^2 \quad \text{under } H_0. \tag{3.179}$$

Denominator:

The denominator is equal in both cases, i.e., with $P_X = X(X'X)^{-1}X'$ we have

$$(y - Xb)'(y - Xb) = \epsilon'(I - P_X)\epsilon \quad \sim \quad \sigma^2\chi^2_{T-K}. \tag{3.180}$$

Since

$$(I-P_X)X = (I-P_X)(X_1, X_2) = ((I-P_X)X_1, (I-P_X)X_2) = (0, 0) \tag{3.181}$$

we find

$$(I - P_X)M_1 = (I - P_X) \tag{3.182}$$

and

$$(I - P_X)A = (I - P_X)M_1X_2D^{-1}X_2'M_1 = 0, \tag{3.183}$$

such that numerator and denominator of F (3.177) are independently distributed [Theorem A 89]. Hence [see also Theorem A 86], the test statistic F is distributed under H$_1$ as $F_{K-s,T-K}(\sigma^{-2}(\beta_2 - r)'D(\beta_2 - r))$ and as central $F_{K-s,T-K}$ under H$_0$.

The region of acceptance of H$_0$ at a level of significance α is then given by

$$0 \leq F \leq F_{K-s,T-K,1-\alpha}. \tag{3.184}$$

Accordingly, the critical area of H$_0$ is given by

$$F > F_{K-s,T-K,1-\alpha}. \tag{3.185}$$

3.8   Analysis of Variance and Goodness of Fit

3.8.1   Bivariate Regression

To illustrate the basic ideas, we shall consider the model (3.99) with a dummy variable 1 and a regressor x:

$$y_t = \beta_0 + \beta_1 x_t + e_t \quad (t = 1, \ldots, T). \tag{3.186}$$

Ordinary Least Squares estimators of $\beta = (\beta_0, \beta_1)'$ are given by:

$$b_1 = \frac{\sum(x_t - \bar{x})(y_t - \bar{y})}{\sum(x_t - \bar{x})^2}, \tag{3.187}$$

$$b_0 = \bar{y} - b_1\bar{x} . \tag{3.188}$$

The best predictor of y on the basis of a given x is

$$\hat{y} = b_0 + b_1x , \tag{3.189}$$

Especially, we have for $x = x_t$

$$\hat{y}_t = b_0 + b_1 x_t$$

$$= \bar{y} + b_1(x_t - \bar{x}) \tag{3.190}$$

(cf. (3.187)).

On the basis of the identity

$$y_t - \hat{y}_t = (y_t - \bar{y}) - (\hat{y}_t - \bar{y}) \tag{3.191}$$

we may express the sum of squared residuals (cf.(3.23)) as

$$S(b) = \sum(y_t - \hat{y}_t)^2 = \sum(y_t - \bar{y})^2 + \sum(\hat{y}_t - \bar{y})^2$$

$$-2\sum(y_t - \bar{y})(\hat{y}_t - \bar{y}).$$

Further manipulation yields

$$\begin{aligned}
\sum(y_t - \bar{y})(\hat{y}_t - \bar{y}) &= \sum(y_t - \bar{y})b_1(x_t - \bar{x}) \quad [\text{cf.}(3.190)] \\
&= b_1^2 \sum(x_t - \bar{x})^2 \quad [\text{cf.}(3.187)] \\
&= \sum(\hat{y}_t - \bar{y})^2. \quad [\text{cf.}(3.190)]
\end{aligned}$$

Thus, we have

$$\sum(y_t - \bar{y})^2 = \sum(y_t - \hat{y}_t)^2 + \sum(\hat{y}_t - \bar{y})^2. \tag{3.192}$$

This relation has already been established in (3.21). The left hand side of (3.192) is called **sum of squares about the mean** or **corrected sum of squares of** Y (i.e., SS (corrected)) or SYY.

The first term on the right hand side describes the deviation: "observation – predicted value" , i.e., the residual sum of squares

$$SS \text{ Residual :} \qquad RSS = \sum(y_t - \hat{y}_t)^2 , \tag{3.193}$$

whereas the second term describes the proportion of variability explained by regression

$$SS \text{ Regression :} \qquad SS_{Reg} = \sum(\hat{y}_t - \bar{y})^2 . \tag{3.194}$$

If all observations y_t are located on a straight line, we obviously have $\sum(y_t - \hat{y}_t)^2 = 0$ and thus $SS(\text{corrected}) = SS_{Reg}$.

Accordingly, the goodness of fit of a regression is measured by the ratio

$$R^2 = \frac{SS_{Reg}}{SS \text{ corrected}} . \tag{3.195}$$

We will discuss R^2 in some detail. The degrees of freedom (df) of the sum of squares are

$$\sum_{t=1}^{T}(y_t - \bar{y})^2 \quad : \quad df = T - 1$$

and

$$\sum_{t=1}^{T}(\hat{y}_t - \bar{y})^2 = b_1^2 \sum(x_t - \bar{x})^2 \quad : \quad df = 1$$

as *one* function in y_t – namely b_1 – is sufficient to calculate SS_{Reg}. In view of (3.192), the degree of freedom for the sum of squares $\sum(y_t - \hat{y}_t)^2$ is just the difference of the other two df's, i.e., $df = T - 2$.

All sums of squares are mutually independently distributed as χ_{df}^2, if the errors e_t are normally distributed. This enables us to establish the following analysis of variance table:

Source of variation	SS	df	Mean Square $(= SS/df)$
Regression	SS Regression	1	MS_{Reg}
Residual	RSS	$T - 2$	$s^2 = \frac{RSS}{T-2}$
Total	SS corrected$=SYY$	$T - 1$	

We will use the following abbreviations:

$$SXX = \sum(x_t - \bar{x})^2 \,, \tag{3.196}$$

$$SYY = \sum(y_t - \bar{y})^2 \,, \tag{3.197}$$

$$SXY = \sum(x_t - \bar{x})(y_t - \bar{y}) \,. \tag{3.198}$$

The sample correlation coefficient may then be written as

$$r_{XY} = \frac{SXY}{\sqrt{SXX}\sqrt{SYY}}. \tag{3.199}$$

Moreover, we have (cf. (3.187))

$$b_1 = \frac{SXY}{SXX} = r_{XY}\sqrt{\frac{SYY}{SXX}}. \tag{3.200}$$

The estimator of σ^2 may be expressed, by using (3.193), as:

$$s^2 = \frac{1}{T-2}\sum \hat{e}_t^2 = \frac{1}{T-2}RSS. \tag{3.201}$$

Various alternative formulations for RSS are in use as well:

$$RSS = \sum(y_t - (b_0 + b_1 x_t))^2$$

$$= \sum[(y_t - \bar{y}) - b_1(x_t - \bar{x})]^2$$

$$= SYY + b_1^2 SXX - 2b_1 SXY$$

$$= SYY - b_1^2 SXX \tag{3.202}$$

$$= SYY - \frac{(SXY)^2}{SXX} . \tag{3.203}$$

Further relations immediately become apparent

$$SS \text{ corrected} = SYY \tag{3.204}$$

and

$$SS_{Reg} = SYY - RSS$$

$$= \frac{(SXY)^2}{SXX} = b_1^2 SXX . \tag{3.205}$$

Checking the adequacy of regression analysis

If model (3.186)

$$y_t = \beta_0 + \beta_1 x_t + \epsilon_t$$

is appropriate, the coefficient b_1 should be significantly different from zero. This is equivalent to the fact, that X and Y are significantly correlated.
Formally, we compare the models (cf. Weisberg, 1980, p. 17)

$$
\begin{aligned}
\text{H}_0 \quad &: \quad y_t = \beta_0 + \epsilon_t \\
\text{H}_1 \quad &: \quad y_t = \beta_0 + \beta_1 x_t + \epsilon_t,
\end{aligned}
$$

by testing $\text{H}_0 : \beta_1 = 0$ against $\text{H}_1 : \beta_1 \neq 0$.
We assume normality of the errors $\epsilon \sim N(\mathbf{0}, \sigma^2 \mathbf{I})$. If we recall (3.164), i.e.,

$$\boldsymbol{D} = \boldsymbol{x}'\boldsymbol{x} - \boldsymbol{x}'\mathbf{1}(\mathbf{1}'\mathbf{1})^{-1}\mathbf{1}'\boldsymbol{x}$$

$$= \sum x_t^2 - \frac{(\sum x_t)^2}{T} = \sum(x_t - \bar{x})^2 = SXX , \tag{3.206}$$

then the Likelihood ratio test (3.177) is given by

$$F_{1,T-2} = \frac{b_1^2 SXX}{s^2}$$

$$= \frac{SS_{Reg}}{RSS} \cdot (T - 2)$$

$$= \frac{MS_{Reg}}{s^2}. \tag{3.207}$$

The coefficient of determination

In (3.195) R^2 has been introduced as a measure of goodness of fit. Using (3.205), we get

$$R^2 = \frac{SS_{Reg}}{SYY} = 1 - \frac{RSS}{SYY}. \tag{3.208}$$

The ratio $\frac{SS_{Reg}}{SYY}$ describes the proportion of variability that is covered by regression in relation to the total variability of y. The right-hand side of the equation is 1 minus the proportion of variability that is not covered by regression.

Definition 3.7 R^2 *(3.208) is called coefficient of determination.*

By using (3.199) and (3.205), we get the basic relation between R^2 and the sample correlation coefficient

$$R^2 = r_{XY}^2. \tag{3.209}$$

Confidence intervals for b_0 and b_1

The covariance matrix of OLS is generally of the form $V_b = \sigma^2(X'X)^{-1} = \sigma^2 S^{-1}$. In model (3.186) we get

$$S = \begin{pmatrix} 1'1 & 1'x \\ 1'x & x'x \end{pmatrix} = \begin{pmatrix} T & T\bar{x} \\ T\bar{x} & \sum x_t^2 \end{pmatrix}, \tag{3.210}$$

$$S^{-1} = \frac{1}{SXX} \begin{pmatrix} \frac{1}{T}\sum x_t^2 & -\bar{x} \\ -\bar{x} & 1 \end{pmatrix} \tag{3.211}$$

and therefore

$$\text{Var}(b_1) = \sigma^2 \frac{1}{SXX} \tag{3.212}$$

$$\text{Var}(b_0) = \frac{\sigma^2}{T} \cdot \frac{\sum x_t^2}{SXX} = \frac{\sigma^2}{T} \frac{\sum x_t^2 - T\bar{x}^2 + T\bar{x}^2}{SXX}$$

$$= \sigma^2 \left(\frac{1}{T} + \frac{\bar{x}^2}{SXX} \right). \tag{3.213}$$

The estimated standard deviations are

$$SE(b_1) = s\sqrt{\frac{1}{SXX}} \tag{3.214}$$

and

$$SE(b_0) = s\sqrt{\frac{1}{T} + \frac{\bar{x}^2}{SXX}} \tag{3.215}$$

with s from (3.201).
Under normal errors $\epsilon \sim N(\mathbf{0}, \sigma^2 \mathbf{I})$ in model (3.186), we have

$$b_1 \sim N\left(\beta_1, \sigma^2 \cdot \frac{1}{SXX}\right). \tag{3.216}$$

Thus it holds that

$$\frac{b_1 - \beta_1}{s}\sqrt{SXX} \quad \sim \quad t_{T-2}. \tag{3.217}$$

Analogously, we get

$$b_0 \quad \sim \quad N\left(\beta_0, \sigma^2 \left(\frac{1}{T} + \frac{\bar{x}^2}{SXX}\right)\right), \tag{3.218}$$

$$\frac{b_0 - \beta_0}{s}\sqrt{\frac{1}{T} + \frac{\bar{x}^2}{SXX}} \sim t_{T-2}. \tag{3.219}$$

This enables us to calculate confidence intervals at level $1 - \alpha$

$$b_0 - t_{T-2,1-\alpha/2} \cdot SE(b_0) \le \beta_0 \le b_0 + t_{T-2,1-\alpha/2} \cdot SE(b_0) \tag{3.220}$$

and

$$b_1 - t_{T-2,1-\alpha/2} \cdot SE(b_1) \le \beta_1 \le b_1 + t_{T-2,1-\alpha/2} \cdot SE(b_1). \tag{3.221}$$

These confidence intervals correspond to the region of acceptance of a two-sided test at the same level:

(i) Testing $H_0 : \beta_0 = \beta_0^*$:
The test statistic is

$$t_{T-2} = \frac{b_0 - \beta_0^*}{SE(b_0)}. \tag{3.222}$$

H_0 is not rejected, if

$$|t_{T-2}| \le t_{T-2,1-\alpha/2}$$

or equivalently if (3.220) holds, with $\beta_0 = \beta_0^*$.

(ii) Testing $\mathbf{H}_0 : \beta_1 = \beta_1^*$:

The test statistic is

$$t_{T-2} = \frac{b_1 - \beta_1^*}{SE(b_1)} \tag{3.223}$$

or equivalently

$$t_{T-2}^2 = F_{1,T-2} = \frac{(b_1 - \beta_1^*)^2}{(SE(b_1))^2}. \tag{3.224}$$

This is identical to (3.207), if $\mathrm{H}_0 : \beta_1 = 0$ is being tested. H_0 will not be rejected, if

$$|t_{T-2}| \leq t_{T-2,1-\alpha/2}$$

or equivalently if (3.221) holds, with $\beta_1 = \beta_1^*$.

3.8.2 Multiple Regression

If we consider more than two regressors, still under the assumption of normality of the errors, we find the methods of analysis of variance to be most convenient in distinguishing the two models $y = 1\beta_0 + X\beta_* + \epsilon = \tilde{X}\beta + \epsilon$ and $y = 1\beta_0 + \epsilon$. In the latter model, we have $\hat{\beta}_0 = \bar{y}$ and the related residual sum of squares is

$$\sum(y_t - \hat{y}_t)^2 = \sum(y_t - \bar{y})^2 = SYY. \tag{3.225}$$

In the former model, $\beta = (\beta_0, \beta_*)'$ will again be estimated by $b = (\tilde{X}'\tilde{X})^{-1}\tilde{X}'y$. The two components of the parameter vector β in the full model may be estimated by

$$b = \begin{pmatrix} \hat{\beta}_0 \\ \hat{\beta}_* \end{pmatrix}, \ \hat{\beta}_* = (X'X)^{-1}X'y, \ \hat{\beta}_0 = \bar{y} - \hat{\beta}_*'\bar{x}. \tag{3.226}$$

Thus, we have (cf. Weisberg, 1980, p. 43)

$$\begin{aligned} RSS &= (y - \tilde{X}b)'(y - \tilde{X}b) \\ &= y'y - b'\tilde{X}'\tilde{X}b \\ &= (y - 1\bar{y})'(y - 1\bar{y}) - \hat{\beta}_*'(X'X)\hat{\beta}_* + T\bar{y}^2. \end{aligned} \tag{3.227}$$

The proportion of variability explained by regression is (cf. (3.205))

$$SS_{Reg} = SYY - RSS \tag{3.228}$$

with RSS from (3.227) and SYY from (3.225). The ANOVA-table is of the form

Source of variation	SS	df	MS
Regression on $X_1, \ldots, X_K$	SS_{Reg}	K	SS_{Reg}/K
Residual	RSS	$T - K - 1$	$s^2 = \frac{RSS}{T-K-1}$
Total	SYY	$T - 1$	

As before, the multiple coefficient of determination

$$R^2 = \frac{SS_{Reg}}{SYY} \tag{3.229}$$

is a measure of the proportion of variability explained by regression of y on $X_1, \ldots, X_K$ in relation to the total variability SYY.
The F-test of

$$\mathrm{H}_0 \; : \; \boldsymbol{\beta}_* = 0$$

versus

$$\mathrm{H}_1 \; : \; \boldsymbol{\beta}_* \neq 0$$

(i.e., $\mathrm{H}_0 : \boldsymbol{y} = \mathbf{1}\beta_0 + \boldsymbol{\epsilon}$ versus $\mathrm{H}_1 : \boldsymbol{y} = \mathbf{1}\beta_0 + \boldsymbol{X}\boldsymbol{\beta}_* + \boldsymbol{\epsilon}$) is based on the test statistic

$$F_{K, T-K-1} = \frac{SS_{Reg}/K}{s^2}. \tag{3.230}$$

Often, it is of interest to test for significance of single components of $\boldsymbol{\beta}$. This type of problem arises, for example, in stepwise model selection, if an optimal subset is selected w.r.t. the coefficient of determination.

Criteria for model choice

Draper and Smith (1966) and Weisberg (1980) have established a variety of criteria to find the right model. We will follow the strategy proposed by Weisberg.

(i) Ad-hoc criteria

Denote by $X_1, \ldots, X_K$ all available regressors and let $\{X_{i1}, \ldots, X_{ip}\}$ be a subset of $p \leq K$ regressors. We denote the residual-sum of squares by RSS_K, respective RSS_p. The parameter vectors are

$$\boldsymbol{\beta} \text{ for } X_1, \cdots, X_K,$$
$$\boldsymbol{\beta}_1 \text{ for } X_{i1}, \cdots, X_{ip},$$
$$\text{and}$$
$$\boldsymbol{\beta}_2 \text{ for } (X_1, \cdots, X_K) \backslash (X_{i1}, \cdots, X_{ip}).$$

A choice between both models can be conducted by testing $\mathrm{H}_0 : \boldsymbol{\beta}_2 = 0$. We apply the F-test, since the hypotheses are nested:

$$F_{(K-p), T-K} = \frac{(RSS_p - RSS_K)/(K-p)}{RSS_K/(T-K)}. \tag{3.231}$$

We prefer the full model against the partial model if $\mathrm{H}_0 : \boldsymbol{\beta}_2 = 0$ is rejected, i.e., if $F > F_{1-\alpha}$ (with degrees of freedom $K - p$ and $T - K$).

Model choice based on an adjusted coefficient of determination

The coefficient of determination (see (3.228) and (3.229))

$$R_p^2 = 1 - \frac{RSS_p}{SYY} \tag{3.232}$$

is inappropriate to compare a model with K and one with $p < K$ since R^2 always increases if an additional regressor is incorporated into the model, irrespective of its values. The full model always has the greatest value of R^2.

Theorem 3.14 *Let* $\mathbf{y} = \mathbf{X}_1\boldsymbol{\beta}_1 + \mathbf{X}_2\boldsymbol{\beta}_2 + \boldsymbol{\epsilon} = \mathbf{X}\boldsymbol{\beta} + \boldsymbol{\epsilon}$ *be a full model and* $\mathbf{y} = \mathbf{X}_1\boldsymbol{\beta}_1 + \boldsymbol{\epsilon}$ *a submodel. Then it holds*

$$R_X^2 - R_{X_1}^2 \geq 0. \tag{3.233}$$

Proof: Let

$$R_X^2 - R_{X_1}^2 = \frac{RSS_{X_1} - RSS_X}{SYY} \, ,$$

such that the assertion (3.233) is equivalent to

$$RSS_{X_1} - RSS_X \geq 0 \, .$$

Since

$$
\begin{aligned}
RSS_X &= (\mathbf{y} - \mathbf{X}\mathbf{b})'(\mathbf{y} - \mathbf{X}\mathbf{b}) \\
&= \mathbf{y}'\mathbf{y} + \mathbf{b}'\mathbf{X}'\mathbf{X}\mathbf{b} - 2\mathbf{b}'\mathbf{X}'\mathbf{y} \\
&= \mathbf{y}'\mathbf{y} - \mathbf{b}'\mathbf{X}'\mathbf{y} \tag{3.234}
\end{aligned}
$$

and, analogously,

$$RSS_{X_1} = \mathbf{y}'\mathbf{y} - \hat{\boldsymbol{\beta}}_1'\mathbf{X}_1'\mathbf{y}$$

where

$$\mathbf{b} = (\mathbf{X}'\mathbf{X})^{-1}\mathbf{X}'\mathbf{y}$$

and

$$\hat{\boldsymbol{\beta}}_1 = (\mathbf{X}_1'\mathbf{X}_1)^{-1}\mathbf{X}_1'\mathbf{y}$$

are OLS estimators in the full model and in the submodel, we have

$$RSS_{X_1} - RSS_X = \mathbf{b}'\mathbf{X}'\mathbf{y} - \hat{\boldsymbol{\beta}}_1'\mathbf{X}_1'\mathbf{y} \, . \tag{3.235}$$

Now we have with (3.160) – (3.166)

$$
\begin{aligned}
\mathbf{b}'\mathbf{X}'\mathbf{y} &= (\mathbf{b}_1', \mathbf{b}_2') \begin{pmatrix} \mathbf{X}_1'\mathbf{y} \\ \mathbf{X}_2'\mathbf{y} \end{pmatrix} \\
&= (\mathbf{y}' - \mathbf{b}_2'\mathbf{X}_2')\mathbf{X}_1(\mathbf{X}_1'\mathbf{X}_1)^{-1}\mathbf{X}_1'\mathbf{y} + \mathbf{b}_2'\mathbf{X}_2'\mathbf{y} \\
&= \hat{\boldsymbol{\beta}}_1'\mathbf{X}_1'\mathbf{y} + \mathbf{b}_2'\mathbf{X}_2'\mathbf{M}_1\mathbf{y} \quad (\text{cf.}(3.175)) \, .
\end{aligned}
$$

Thus, (3.235) becomes

$$RSS_{X_1} - RSS_X = b_2' X_2' M_1 y$$
$$= y' M_1 X_2 D^{-1} X_2' M_1 y \geq 0 , \qquad (3.236)$$

such that (3.233) is proven.

On the basis of Theorem 3.14 we define the statistic:

$$\text{F-change} = \frac{(RSS_{X_1} - RSS_X)/(K - p)}{RSS_X/(T - K)} , \qquad (3.237)$$

which is distributed as $F_{K-p,T-K}$ under H_0: "submodel is valid". In model choice procedures, F-change tests for significance of the change of R^2 by adding further $K - p$ variables to the submodel.

In multiple regression, the appropriate adjustment of the ordinary coefficient of determination is provided by the coefficient of determination adjusted by the degrees of freedom of the multiple model :

$$\bar{R}_p^2 = 1 - \left(\frac{T-1}{T-p} \right) (1 - R_p^2). \qquad (3.238)$$

Remark: If there is no constant β_0 present in the model, then the numerator is T instead of $T - 1$, such that $\bar{R}_p^2$ may possibly take negative values. This disadvantage cannot occur when using the ordinary R^2.

If we consider two models, the smaller of which is assumed to be completely included in the bigger one, and we find the relation

$$\bar{R}_{p+q}^2 < \bar{R}_p^2 ,$$

then the smaller model obviously shows a better goodness of fit.

Further criteria are, for example, Mallows' C_p, (cf. Weisberg, 1980, p.188) or criteria based on the residual-Mean-Square-Error $\hat{\sigma}_p^2 = RSS_p/(T - p)$. There are close relations between these measures.

Confidence regions

As in bivariate regression, there are close relations between the region of acceptance of the F-test and the confidence intervals for β in the multiple linear regression model as well.

Confidence ellipsoids for the whole parameter vector β

Considering (3.150) and (3.153), we get for $\beta^* = \beta$ a confidence ellipsoid at level $1 - \alpha$

$$\frac{(b - \beta)' X' X (b - \beta)}{(y - Xb)'(y - Xb)} \cdot \frac{T - K}{K} \leq F_{K,T-K,1-\alpha}. \qquad (3.239)$$

Confidence ellipsoids for subvectors of β

From (3.177) and (3.185), we have that

$$\frac{(b_2 - \beta_2)'D(b_2 - \beta_2)}{(y - Xb)'(y - Xb)} \cdot \frac{T - K}{K - s} \leq F_{K-s,T-K,1-\alpha} \tag{3.240}$$

is a $(1 - \alpha)$-confidence ellipsoid for β_2.

Further results may be found in Judge et al. (1980), Goldberger (1964), Pollock (1979), Weisberg (1980), and Kmenta (1971).

3.9   The General Linear Regression Model

3.9.1   Introduction

In many applications, it cannot be justified that the response values y_t $t = 1,\ldots,T$ are independent. Consider, for example, a time series with autocorrelated errors or processes typically arising in medicine or sociology, when measurements are being repeated several times on a single person or cluster analysis. We will discuss these types of models at a later stage (Chapter 7: Repeated Measurements, Chapter 8: Cross-Over Design, Chapter 9: Analysis of Categorical Data).

The general linear regression model is of the form

$$\left.\begin{array}{c} y = X\beta + \epsilon, \\[2mm] \mathrm{E}\,(\epsilon) = 0, \quad \mathrm{E}\,(\epsilon\epsilon') = \sigma^2 W, \\[2mm] W \text{ positive definite and known}, \\[2mm] X \text{ nonstochastic, rank}\,(X) = K. \end{array}\right\} \tag{3.241}$$

The first problem is now, that in case of an unknown matrix W the number of additional parameters to be estimated may increase by $T(T + 1)/2$ at the most. This problem cannot be solved on the basis of T observations only. Therefore we assume for the present, that W is known. Furthermore, it is useful to impose several restrictions on W in the sense that $\mathrm{tr}\,(W) = T$ or $w_{ii} = 1$, $i = 1,\ldots,T$.

Aitken estimator

In order to facilitate the estimation in a general linear regression model (3.241), we shall transform the model by methods already introduced in Section 2.6. The matrices W and W^{-1} may be decomposed [see also A 31 (iii)] as

$$W = MM \quad \text{and} \quad W^{-1} = NN \tag{3.242}$$

where $M = W^{1/2}$ and $N = W^{-1/2}$ are non-singular. We transform the model (3.241) by premultiplication with N:

$$Ny = NX\beta + N\epsilon \tag{3.243}$$

and set

$$Ny = \tilde{y}, \quad NX = \tilde{X}, \quad N\epsilon = \tilde{\epsilon}. \tag{3.244}$$

Then it holds

$$E(\tilde{\epsilon}) = E(N\epsilon) = 0, \quad E(\tilde{\epsilon}\tilde{\epsilon}') = E(N\epsilon\epsilon'N) = \sigma^2 I, \tag{3.245}$$

such that the transformed model $\tilde{y} = \tilde{X}\beta + \tilde{\epsilon}$ obeys all assumptions of the classical regression model. The OLS estimator of β in this model is of the form

$$\begin{aligned} b &= (\tilde{X}'\tilde{X})^{-1}\tilde{X}'\tilde{y} \\ &= (X'NN'X)^{-1}X'NN'y \\ &= (X'W^{-1}X)^{-1}X'W^{-1}y. \end{aligned} \tag{3.246}$$

$b = (\tilde{X}'\tilde{X})^{-1}\tilde{X}'\tilde{X}$ is, as we know, identical to the Gauss-Markov estimator in the transformed model. The GM property of b remains also valid in model (3.241):

$$\begin{aligned} b &= S^{-1}XX'W^{-1}y \text{ is unbiased}: \\ E(b) &= (X'W^{-1}X)^{-1}X'W^{-1}E(y) \\ &= (X'W^{-1}X)^{-1}X'W^{-1}X\beta = \beta \end{aligned} \tag{3.247}$$

Moreover, b possesses the smallest variance (in the sense of Theorem 3.9): Let $\tilde{\beta} = \tilde{C}y$ be an arbitrary linear unbiased estimator of β. We set

$$\tilde{C} = \hat{C} + D \tag{3.248}$$

with

$$\hat{C} = S^{-1}X'W^{-1}. \tag{3.249}$$

The unbiasedness of $\tilde{\beta}$ leads to the condition $DX = 0$, such that $\hat{C}WD = 0$. Therefore, we get for the covariance matrix

$$\begin{aligned} V_{\tilde{\beta}} &= E(\tilde{C}\epsilon\epsilon'\tilde{C}') \\ &= \sigma^2(\hat{C} + D)W(\hat{C}' + D') \\ &= \sigma^2\hat{C}W\hat{C}' + \sigma^2 DWD' \\ &= V_b + \sigma^2 DWD', \end{aligned} \tag{3.250}$$

such that $V_{\tilde{\beta}} - V_b = \sigma^2 D'WD$ is nonnegative definite (Theorem A 41 (v)). This result is summarized in

Theorem 3.15 (GAUSS-MARKOV-AITKEN-Theorem) *In the general linear regression model, the generalized OLS estimator*

$$b = (X'W^{-1}X)^{-1}X'W^{-1}y \tag{3.251}$$

with covariance matrix

$$V_b = \sigma^2(X'W^{-1}X)^{-1} = \sigma^2 S^{-1} \tag{3.252}$$

is the best linear unbiased estimator of β.

(We denote b also as AITKEN estimator or GLS estimator). Analogously to the classical model, we estimate σ^2 and V_b by

$$s^2 = (y - Xb)'W^{-1}(y - Xb)(T - K)^{-1} \tag{3.253}$$

and

$$\hat{V}_b = s^2 S^{-1} \ . \tag{3.254}$$

Both estimators are unbiased:

$$\mathrm{E}\,(s^2) = \sigma^2 \quad \text{and} \quad \mathrm{E}\,(\hat{V}_b) = \sigma^2 S^{-1} \ . \tag{3.255}$$

Analogously to Theorem 3.11, we may formulate

Theorem 3.16 *In the general linear regression model*

$$\hat{d} = a'b \ , \tag{3.256}$$

with variance

$$\mathrm{Var}\,(\hat{d}) = \sigma^2 a' S^{-1} a = a' V_b a \ , \tag{3.257}$$

is the best linear unbiased estimator of $d = a'\beta$.

3.9.2 Misspecification of the Covariance Matrix

Assuming the general linear regression model (3.241) and W to be true, we want to examine the influence of a misspecification of the covariance matrix on the estimator of β and σ^2, compared to the GLS estimator b (3.251) and s^2 (3.253). Reasons for the misspecification could be

- the use of the classical OLS estimator because the correlation between the errors ϵ_t was not recognized,

- that the correlation is generally described by a matrix $\tilde{W} \neq W$,

- that the matrix W is unknown and is estimated independent of y from a presample through $\hat{W}$.

In any case we get the estimator

$$\hat{\beta} = (X'AX)^{-1}X'Ay \ , \tag{3.258}$$

with $A \neq W^{-1}$ symmetric, nonstochastic, and with $(X'AX)$ regular. Then we have

$$\mathrm{E}\,(\hat{\beta}) = \beta \ , \tag{3.259}$$

where $\hat{\beta}$ (3.258) is unbiased for every misspecified matrix A (if rank $(X'AX) = K$). For the covariance matrix of $\hat{\beta}$ we get

$$V_{\hat{\beta}} = \sigma^2 (X'AX)^{-1} X'AWAX(X'AX)^{-1} \ . \tag{3.260}$$

The loss of efficiency, due to the use of $\hat{\boldsymbol{\beta}}$ instead of the GLS estimator $\boldsymbol{b} = \boldsymbol{S}^{-1}\boldsymbol{X}'\boldsymbol{W}^{-1}\boldsymbol{y}$ becomes

$$
\begin{aligned}
\boldsymbol{V}_{\hat{\beta}} - \boldsymbol{V}_b \;=\;& \sigma^2[(\boldsymbol{X}'\boldsymbol{A}\boldsymbol{X})^{-1}\boldsymbol{X}'\boldsymbol{A} - \boldsymbol{S}^{-1}\boldsymbol{X}'\boldsymbol{W}^{-1}] \\
& \times \boldsymbol{W}[(\boldsymbol{X}'\boldsymbol{A}\boldsymbol{X})^{-1}\boldsymbol{X}'\boldsymbol{A} - \boldsymbol{S}^{-1}\boldsymbol{X}'\boldsymbol{W}^{-1}]' . \quad (3.261)
\end{aligned}
$$

Following Theorem A 41 (iv), this matrix is nonnegative definite. There is no loss in efficiency if

$$
(\boldsymbol{X}'\boldsymbol{A}\boldsymbol{X})^{-1}\boldsymbol{X}'\boldsymbol{A} = \boldsymbol{S}^{-1}\boldsymbol{X}'\boldsymbol{W}^{-1} \quad \text{or} \quad \hat{\boldsymbol{\beta}} = \boldsymbol{b} \qquad (3.262)
$$

is valid.

If the first row of $\boldsymbol{X}$ is $\boldsymbol{1}$ then we set $\boldsymbol{X} = (\boldsymbol{1}, \boldsymbol{x}_2, \ldots, \boldsymbol{x}_K) = (\boldsymbol{1}, \tilde{\boldsymbol{X}})$.

For this case and if $\boldsymbol{A} = \boldsymbol{I}$, implying the use of the classical OLS estimator $\boldsymbol{b}_0 = (\boldsymbol{X}'\boldsymbol{X})^{-1}\boldsymbol{X}'\boldsymbol{y}$, the following Theorem by McElroy (1967) is valid:

Theorem 3.17 *The OLS estimator $\boldsymbol{b}_0 = (\boldsymbol{X}'\boldsymbol{X})^{-1}\boldsymbol{X}'\boldsymbol{y}$ is GM estimator in the generalized linear regression model if and only if $\boldsymbol{X} = (\boldsymbol{1}\tilde{\boldsymbol{X}})$ and*

$$
\boldsymbol{W} = (1-\rho)\boldsymbol{I} + \rho\boldsymbol{1}\boldsymbol{1}' \qquad (3.263)
$$

with $0 \le \rho < 1$ and $\boldsymbol{1}' = (1, 1, \ldots, 1)$.

In other words, we have

$$
(\boldsymbol{X}'\boldsymbol{X})^{-1}\boldsymbol{X}'\boldsymbol{y} = (\boldsymbol{X}'\boldsymbol{W}^{-1}\boldsymbol{X})^{-1}\boldsymbol{X}'\boldsymbol{W}^{-1}\boldsymbol{y} \qquad (3.264)
$$

for all $\boldsymbol{y}$ if and only if the errors ϵ_t have the same variance σ^2 and equal nonnegative covariances $\sigma^2\rho$. A matrix of this form is called compound symmetric.

Moreover, a loss in efficiency occurs if σ^2 is estimated by an estimator $\hat{\sigma}^2$ that is based on $\hat{\boldsymbol{\beta}}$. Here we have

$$
\begin{aligned}
\hat{\epsilon} = \boldsymbol{y} - \boldsymbol{X}\hat{\boldsymbol{\beta}} \;=\;& (\boldsymbol{I} - \boldsymbol{X}(\boldsymbol{X}'\boldsymbol{A}\boldsymbol{X})^{-1}\boldsymbol{X}'\boldsymbol{A})\epsilon \\
(T-K)\hat{\sigma}^2 \;=\;& \hat{\epsilon}'\hat{\epsilon} \\
\;=\;& \operatorname{tr}\left\{(\boldsymbol{I} - \boldsymbol{X}(\boldsymbol{X}'\boldsymbol{A}\boldsymbol{X})^{-1}\boldsymbol{X}'\boldsymbol{A})\epsilon\epsilon'(\boldsymbol{I} - \boldsymbol{A}\boldsymbol{X}(\boldsymbol{X}'\boldsymbol{A}\boldsymbol{X})^{-1}\boldsymbol{X}')\right\} , \\
\operatorname{E}(\hat{\sigma}^2)(T-K) \;=\;& \sigma^2\operatorname{tr}(\boldsymbol{W} - \boldsymbol{X}(\boldsymbol{X}'\boldsymbol{A}\boldsymbol{X})^{-1}\boldsymbol{X}'\boldsymbol{A}) \\
& + \operatorname{tr}\left\{\sigma^2\boldsymbol{X}(\boldsymbol{X}'\boldsymbol{A}\boldsymbol{X})^{-1}\boldsymbol{X}'\boldsymbol{A}(\boldsymbol{I} - 2\boldsymbol{W}) + \boldsymbol{X}\boldsymbol{V}_{\hat{\beta}}\boldsymbol{X}'\right\} . \quad (3.265)
\end{aligned}
$$

If we choose the standardization $\operatorname{tr}(\boldsymbol{W}) = T$, then the first term in (3.265) becomes $(T-K)$ (Theorem A 13). In the case $\hat{\boldsymbol{\beta}} = (\boldsymbol{X}'\boldsymbol{X})^{-1}\boldsymbol{X}'\boldsymbol{y}$ (i.e., $\boldsymbol{A} = \boldsymbol{I}$) we get

$$
\begin{aligned}
\operatorname{E}(\hat{\sigma}^2) \;=\;& \sigma^2 + \frac{\sigma^2}{T-K}\operatorname{tr}\left[\boldsymbol{X}(\boldsymbol{X}'\boldsymbol{X})^{-1}\boldsymbol{X}'(\boldsymbol{I} - \boldsymbol{W})\right] \\
\;=\;& \sigma^2 + \frac{\sigma^2}{T-K}(K - \operatorname{tr}\left[(\boldsymbol{X}'\boldsymbol{X})^{-1}\boldsymbol{X}'\boldsymbol{W}\boldsymbol{X}\right]) . \quad (3.266)
\end{aligned}
$$

The average bias of the estimator $\hat{\sigma}^2$ which is based on OLS is given by the second term in (3.266). It is to be expected, that the bias will tend to be negative, especially in processes with positive correlation. As a consequence, the variance will be underestimated, leading in turn to a better goodness of fit (cf. several examples in Goldberger, 1964, pp. 288 in cases of heteroscedasticity and first order autoregression).

3.10   Exercises and Questions

3.10.1 Define the principle of least squares.

3.10.2 Given the normal equation $X'X\beta = X'y$, what are the conditions for a unique solution?

3.10.3 Assume rank $(X) = p < K$.
What are the linear restrictions to ensure estimability of β?
Give the definition of the restricted least squares estimator.

3.10.4 Define the matrix-valued Mean-Square-Error of a linear estimator and the MSE-I-superiority.

3.10.5 Let $\hat{\beta} = Cy + d$ be a linear estimator. Give the condition of unbiasedness of $\hat{\beta}$.
What is the best linear unbiased estimator?

3.10.6 What is the relation of the covariance matrices of the best linear unbiased estimator $\hat{\beta}$ and any linear estimator $\tilde{\beta}$?

3.10.7 How can you get an unbiased estimate of σ^2?

3.10.8 Characterize weak and extreme multicollinearity in terms of rank $(X'X)$, unbiasedness of the least squares estimator and identifiability.

3.10.9 Assume $\epsilon \sim N(0, \sigma^2 I)$ and give the ML estimators of β and σ^2.

4 Single-Factor Experiments with Fixed and Random Effects

4.1 Models I and II in the Analysis of Variance

The analysis of variance, which was originally developed by R. A. Fisher for field experiments, is one of the most widely used and one of the most general statistical procedures for testing and analysing data. These procedures require a large amount of computation, especially in case of complicated classifications. For this reason, these procedures are available as software.

We distinguish between two fundamental problems.

Model I (with fixed effects) is used for the *multiple comparison of means* of quantitative normally distributed factors that are observed on fixed selected experimental units. We test the null hypothesis H_0: $\mu_1 = \mu_2 = \ldots = \mu_s$ against the general alternative H_1: *at least two means are different*, i.e., we compare s normally distributed populations with respect to their means. The corresponding F-test is a generalization of the t-test, that compares two normal distributions. In general, this comparison is called *comparison of the effects of treatments*. If *specific* treatments are to be compared, then it is wise not to choose them at random, but to assume them as *fixed*.

Example 4.1: Comparison of the average manufacturing time for an inlay by three different *prespecified* dentists (Table 4.1)

Model II (with random effects) is used for the *decomposition* of the *total variability* produced by the effect of several factors. This total variability (variance) is decomposed into components that reflect the effect of each factor and into a component that can not be explained by the factors, i.e., the error variance. The experimental units are chosen at *random*, as opposed to model I. The treatments are then to be regarded as a random sample from an assumed infinite population. Hence, we have no interest in the treatments chosen at random, but only in the respective proportion of the total variability.

Dentist A	Dentist B	Dentist C
55.5	67.0	62.5
40.0	57.0	31.5
38.5	33.5	31.5
31.5	37.0	53.0
45.5	75.0	50.5
70.0	60.0	62.5
78.0	43.5	40.0
80.0	56.0	19.5
74.5	65.5	
57.5	54.0	
72.0	59.5	
70.0		
48.0		
59.0		
$n_1 = 14$	$n_2 = 11$	$n_3 = 8$
$\bar{x}_1 = 58.57$	$\bar{x}_2 = 55.27$	$\bar{x}_3 = 43.88$
$n = n_1 + n_2 + n_3$		

Table 4.1: Manufacturing time (in minutes) for the making of inlays, measured for three dentists (cf. S.Toutenburg, 1977)

Example 4.2: From a total population, the manufacturing times of (e.g., three) dentists *chosen at random* are to be analysed with respect to their proportion of the total variability.

4.2 One-Way Classification for the Multiple Comparison of Means

Assume we have s samples from s normally distributed populations $N(\mu_i, \sigma^2)$. Furthermore, assume the sample sizes to be n_i and the total sample size to be n

$$\sum_{i=1}^{s} n_i = n. \tag{4.1}$$

The variances σ^2 are *unknown, but equal* in all populations.

Definition 4.1 *If all n_i are equal, then the sampling design (experimental design) is called balanced. Otherwise it is called unbalanced.*

The s different levels of a factor A are called *treatments*. Since only one factor is investigated, we call this type of experimental design *one-way classification*.

Examples:

1. Factor A: plastic PMMA
   s levels: s different concentrations of quartz in PMMA
   s effects: flexibility of the different PMMA materials

2. Factor A: fertilization
   s levels: s different fertilizers (or one fertilizer with s different concentrations of phosphate)
   s effects: output per acre

	Single experiments per level of A 1 2 ... n_i			Sum of the obser- vations per sample	Sample mean
1	y_{11} y_{12} ...		y_{1n_1}	$\sum y_{1j} = Y_1.$	$Y_1./n_1 = y_1.$
2	y_{21} y_{22} ...		y_{2n_2}	$\sum y_{2j} = Y_2.$	$Y_2./n_2 = y_2.$
$\vdots$					
s	y_{s1} y_{s2} ...		y_{sn_s}	$\sum y_{sj} = Y_s.$	$Y_s./n_s = y_s.$
	$n = \sum n_i$			$\sum Y_i. = Y..$	$Y../n = y..$

Table 4.2: Sample design (one-way classification)

The observations of the s samples are arranged according to Table 4.2. A period in the subscript indicates that we summed over this subscript. For example, $y_1.$ is the sum of the first row, $y..$ is the total sum. For the observations y_{ij} we assume the following model

$$y_{ij} = \mu + \alpha_i + \epsilon_{ij} \quad (i = 1, \ldots, s; j = 1, \ldots, n_i) \quad , \tag{4.2}$$

in which μ is the overall mean, α_i is the effect of the ith level of factor A, i.e., the deviation (treatment effect) from the overall mean μ caused by the ith level and ϵ_{ij} is a random error (i.e., random deviation from μ and α_i).
μ and α_i are fixed parameters, the ϵ_{ij} are random. The following assumptions have to hold:

- the errors ϵ_{ij} are independent and identically distributed with mean 0 and variance σ^2.

- the errors are normal, i.e., we have $\epsilon_{ij} \sim N(0, \sigma^2)$

- the following constraint holds

$$\sum \alpha_i n_i = 0. \tag{4.3}$$

In experimental designs, it is important to have equal sample sizes n_i in the groups *(balanced case)*, otherwise the analysis of variance is not *robust* against deviations from the assumptions (normal distribution, equal variances).

Remark: Model I (with fixed effects) assumes that the s treatments are given in advance, i.e., they are *fixed* before the experiment. Hence the α_i are non-stochastic factors. If the s treatments were selected by a random mechanism from a set of possible treatments, then the α_i would be stochastic, i.e., random variables with a distribution. For the analysis of linear models with stochastic parameters the methods of linear models have to be modified. For now, we restrict ourselves to the case with fixed effects. Models with random effects are discussed in Section 4.6.

Completely Randomized Experimental Design

The simplest and least restrictive design (CRD : completely randomized design) consists of assigning the s treatments to the n experimental units in the following manner. We choose n_1 experimental units at random and assign them to treatment $i = 1$. After that, n_2 experimental units are selected from the remaining $n - n_1$ units, once again at random, and are assigned to treatment $i = 2$, etc.. The remaining $n - \sum_{i=1}^{s-1} n_i = n_s$ units receive the sth treatment. This experimental design has the following *advantages* (cf. e.g., Petersen, 1985, p.7)

- Flexibility: The number s of treatments and the amounts n_i are not restricted; in particular, unbalanced designs are allowed. However, balanced design should be preferred, since for these designs the power of the tests is the highest.

- Degrees of freedom: The design provides maximum number of degrees of freedom for the error variance.

- Statistical analysis: The employment of standard procedures is possible in the unbalanced case as well (e.g., in case of missing values due to non-response).

A *disadvantage* of this design arises in case of inhomogeneous experimental units: a decrease in precision of the results. Often however, the experimental units can be grouped into homogeneous subgroups (blocking) with a resulting increase in precision.

4.2.1 Representation as Restrictive Model

The linear model (4.2) can be formulated in matrix notation

$$
\begin{pmatrix} y_{11} \\ \vdots \\ y_{1n_1} \\ \vdots \\ y_{s1} \\ \vdots \\ y_{sn_s} \end{pmatrix} = \begin{pmatrix} 1 & 1 & 0 & \cdots & 0 \\ \vdots & \vdots & \vdots & \vdots & \vdots \\ 1 & 1 & 0 & \cdots & 0 \\ \vdots & \vdots & \vdots & \vdots & \vdots \\ 1 & 0 & \cdots & 0 & 1 \\ \vdots & \vdots & \vdots & \vdots & \vdots \\ 1 & 0 & \cdots & 0 & 1 \end{pmatrix} \begin{pmatrix} \mu \\ \alpha_1 \\ \vdots \\ \alpha_s \end{pmatrix} + \begin{pmatrix} \epsilon_{11} \\ \vdots \\ \epsilon_{1n_1} \\ \vdots \\ \epsilon_{s1} \\ \vdots \\ \epsilon_{sn_s} \end{pmatrix}
$$

i.e.,

$$
y = X\beta + \epsilon, \quad \epsilon \sim N(0, \sigma^2 I) \tag{4.4}
$$

with X of type $n \times (s+1)$ and $\text{rank}(X) = s$. Hence, we have exact multicollinearity. $X'X$ is now singular, and a linear restriction $r = R'\beta$ with $\text{rank}(R) = J = 1$ and $\text{rank}\begin{pmatrix} X \\ R' \end{pmatrix} = s+1$ has to be introduced for the estimation of the $(s+1) \times 1$-vector $\beta' = (\mu, \alpha_1, \ldots, \alpha_s)$ (cf. Theorem 3.5). We choose

$$
r = 0, \quad R' = (0, n_1, \ldots, n_s) \tag{4.5}
$$

and hence

$$
\sum \alpha_i n_i = 0 \tag{4.6}
$$

(cf. (4.3)).

Remark: The estimatability of β is ensured according to Theorem 3.5 for *every* restriction $r = R'\beta$ with $\text{rank}(R') = J = 1$ and $\text{rank}\begin{pmatrix} X \\ R' \end{pmatrix} = s+1$. However, the *selected* restriction (4.6) has the advantage of an interpretation justified by the subject matter, that follows the effect coding of a loglinear model. The parameters α_i are then the deviations from the overall mean μ and hence standardized with respect to μ. Thus, the α_i determine the relative (positive or negative) factors, with which the ith treatment leads to deviations from the overall mean, by their magnitude and their sign.

According to (3.37), the conditional OLS estimate of $\beta' = (\mu, \alpha_1, \ldots, \alpha_s)$ is of the following form

$$
b(R', 0) = (X'X + RR')^{-1}X'y. \tag{4.7}
$$

As we can easily check, the matrix $\begin{pmatrix} X \\ R' \end{pmatrix}$ with X from (4.4) and R' from (4.5) is of full column rank $s + 1$.

Case $s = 2$

We demonstrate the computation of the estimate $b(R', 0)$ for $s = 2$. With the notation $1'_{n_i} = (1, \ldots, 1)$ for the $n_i \times 1$-vector of ones, we obtain the following representation:

$$\underset{n,3}{X} = \begin{pmatrix} 1_{n_1} & 1_{n_1} & 0 \\ 1_{n_2} & 0 & 1_{n_2} \end{pmatrix}, \tag{4.8}$$

$$X'X = \begin{pmatrix} 1'_{n_1} & 1'_{n_2} \\ 1'_{n_1} & 0' \\ 0' & 1'_{n_2} \end{pmatrix} \begin{pmatrix} 1_{n_1} & 1_{n_1} & 0 \\ 1_{n_2} & 0 & 1_{n_2} \end{pmatrix}$$

$$= \begin{pmatrix} n_1 + n_2 & n_1 & n_2 \\ n_1 & n_1 & 0 \\ n_2 & 0 & n_2 \end{pmatrix},$$

$$RR' = \begin{pmatrix} 0 \\ n_1 \\ n_2 \end{pmatrix} \begin{pmatrix} 0 & n_1 & n_2 \end{pmatrix} \tag{4.9}$$

$$= \begin{pmatrix} 0 & 0 & 0 \\ 0 & n_1^2 & n_1 n_2 \\ 0 & n_1 n_2 & n_2^2 \end{pmatrix}.$$

With $n = n_1 + n_2$ we have

$$(X'X + RR') = \begin{pmatrix} n & n_1 & n_2 \\ n_1 & n_1 + n_1^2 & n_1 n_2 \\ n_2 & n_1 n_2 & n_2 + n_2^2 \end{pmatrix},$$

$$|X'X + RR'| = n_1 n_2 n^2,$$

$$(X'X + RR')^{-1} = \frac{1}{n_1 n_2 n^2} \begin{pmatrix} n_1 n_2 (1+n) & -n_1 n_2 & -n_1 n_2 \\ -n_1 n_2 & n_2(n(1+n_2)-n_2) & -n_1 n_2 (n-1) \\ -n_1 n_2 & -n_1 n_2 (n-1) & n_1(n(1+n_1)-n_1) \end{pmatrix},$$

$$\tag{4.10}$$

$$X'y = \begin{pmatrix} 1'_{n_1} & 1'_{n_2} \\ 1'_{n_1} & 0' \\ 0' & 1'_{n_2} \end{pmatrix} \begin{pmatrix} y_1 \\ y_2 \end{pmatrix}$$

$$= \begin{pmatrix} Y_{..} \\ Y_{1.} \\ Y_{2.} \end{pmatrix}. \tag{4.11}$$

Here we have

$$y_1 = \begin{pmatrix} y_{11} \\ \vdots \\ y_{1n_1} \end{pmatrix}, \quad y_2 = \begin{pmatrix} y_{21} \\ \vdots \\ y_{2n_2} \end{pmatrix}$$

$$Y_{1.} = \sum_{i=1}^{n_1} y_{1i} , \quad Y_{2.} = \sum_{i=1}^{n_2} y_{2i}$$
$$Y_{..} = Y_{1.} + Y_{2.} .$$

Finally, we receive the conditional OLS estimate (4.7) for the case $s = 2$ according to

$$b\left((0, n_1, n_2), 0\right) = (X'X + RR')^{-1} X'y$$
$$= \begin{pmatrix} \hat{\mu} \\ \hat{\alpha}_1 \\ \hat{\alpha}_2 \end{pmatrix} = \begin{pmatrix} y_{..} \\ y_{1.} - y_{..} \\ y_{2.} - y_{..} \end{pmatrix} . \qquad (4.12)$$

Proof: The multiplication of (4.10) by rows with (4.11) yields

$$\hat{\mu} = \frac{n_1 n_2 (1+n) Y_{..} - n_1 n_2 Y_{1.} - n_1 n_2 Y_{2.}}{n_1 n_2 n^2}$$
$$= \frac{n Y_{..}}{n^2} = \frac{Y_{..}}{n} = y_{..} ,$$
$$\hat{\alpha}_1 = \frac{-n_1 n_2 Y_{..} + n_2 (n(1+n_2) - n_2) Y_{1.} - n_1 n_2 (n-1) Y_{2.}}{n_1 n_2 n^2}$$
$$= -\frac{Y_{..}}{n^2} + \frac{n + n n_2 - n_2}{n_1 n^2} Y_{1.} - \frac{n-1}{n^2} (Y_{..} - Y_{1.})$$
$$= Y_{1.} \left(\frac{n + n n_2 - n_2 + n n_1 - n_1}{n_1 n^2} \right) - Y_{..} \left(\frac{1 - 1 + n}{n^2} \right)$$
$$= \frac{Y_{1.}}{n_1} - \frac{Y_{..}}{n} = y_{1.} - y_{..}$$

and analogously

$$\hat{\alpha}_2 = y_{2.} - y_{..} .$$

4.2.2 Decomposition of the Error Sum of Squares

With $b(R', 0)$ from (4.12) we receive

$$\hat{y} = Xb(R', 0) = \begin{pmatrix} y_{1.} \mathbf{1}_{n_1} \\ y_{2.} \mathbf{1}_{n_2} \end{pmatrix} . \qquad (4.13)$$

The decomposition (3.192), i.e.,

$$\sum (y_t - \bar{y})^2 = \sum (y_t - \hat{y}_t)^2 + \sum (\hat{y}_t - \bar{y})^2$$

is of the following form in the model (4.4) with the new notation

$$\sum_{i=1}^{s} \sum_{j=1}^{n_i} (y_{ij} - y_{..})^2 = \sum_{i=1}^{s} \sum_{j=1}^{n_i} (y_{ij} - y_{i.})^2 + \sum_{i=1}^{s} n_i (y_{i.} - y_{..})^2 \qquad (4.14)$$

or, written according to (3.193) and (3.194)

$$SS_{Corr} = RSS + SS_{Reg} \tag{4.15}$$

or in the notation of the analysis of variance

$$SS_{Total} = SS_{Within} + SS_{Between}. \tag{4.16}$$

The sum of squares

$$SS_{Within} = \sum\sum(y_{ij} - y_{i\cdot})^2$$

measures the variability within each treatment. On the other hand, the sum of squares

$$SS_{Between} = \sum_{i=1}^{s} n_i(y_{i\cdot} - y_{\cdot\cdot})^2$$

measures the differences in variability between the treatments, that is the actual treatment effects.

Testing the Regression

We consider the linear model

$$y_{ij} = \mu + \alpha_i + \epsilon_{ij} \quad \begin{array}{l}(i = 1,\ldots,s \\ j = 1,\ldots,n_i)\end{array} \tag{4.17}$$

with

$$\sum n_i\alpha_i = 0 \quad . \tag{4.18}$$

Testing the hypothesis

$$H_0: \quad \alpha_1 = \cdots = \alpha_s = 0 \tag{4.19}$$

is equivalent to comparing the models

$$H_0: \quad y_{ij} = \mu + \epsilon_{ij} \tag{4.20}$$

and

$$H_1: \quad y_{ij} = \mu + \alpha_i + \epsilon_{ij} \quad \text{with} \quad \sum n_i\alpha_i = 0 \quad , \tag{4.21}$$

i.e., is equivalent to testing

$$H_0: \quad \alpha_1 = \cdots = \alpha_s = 0 \quad (parameter\ space\ \ \omega) \tag{4.22}$$

against

$$H_1: \quad \alpha_i \neq 0 \text{ for at least two } i \quad (parameter\ space\ \ \Omega). \tag{4.23}$$

In case of an assumed normal distribution $\epsilon_{ij} \sim N(0,\sigma^2)$ for all i,j the corresponding LR test statistic (3.145)

$$F = \frac{\hat{\sigma}_\omega^2 - \hat{\sigma}_\Omega^2}{\hat{\sigma}_\Omega^2}\frac{T - K}{K - s}$$

changes to

$$F = \frac{SS_{Total} - SS_{Within}}{SS_{Within}} \frac{n-s}{s-1} \tag{4.24}$$

$$= \frac{SS_{Between}}{SS_{Within}} \frac{n-s}{s-1} \tag{4.25}$$

$$= \frac{MS_{Between}}{MS_{Within}}. \tag{4.26}$$

Remark: The sum of squares

$$SS_{Between} = \sum_{i=1}^{s} n_i(y_{i\cdot} - y_{\cdot\cdot})^2$$

is named according to the factor, e.g., SS_A if factor A represents a treatment in s different levels. Analogously, we also denote

$$SS_{Within} = \sum_{i=1}^{s}\sum_{j=1}^{n_i}(y_{ij} - y_{i\cdot})^2$$

as SS_{Error} (SSE, error sum of squares).

The sums of squares can also be written in detail as follows:

$$SS_{Total} = \sum_i\sum_j(y_{ij} - y_{\cdot\cdot})^2 = \sum_i\sum_j y_{ij}^2 - ny_{\cdot\cdot}^2 , \tag{4.27}$$

$$SS_{Between} = SS_A = \sum_i\sum_j(y_{i\cdot} - y_{\cdot\cdot})^2 = \sum_i n_i y_{i\cdot}^2 - ny_{\cdot\cdot}^2 , \tag{4.28}$$

$$SS_{Error} = \sum_i\sum_j(y_{ij} - y_{i\cdot})^2 = \sum_i\sum_j y_{ij}^2 - \sum_i n_i y_{i\cdot}^2 \tag{4.29}$$

These formulas make the computation a lot easier (i.e., if calculators are used).

Under the assumption of a normal distribution, the sums of squares have a χ^2-distribution with the corresponding degrees of freedom. The ratios SS/df are called MS. As we will show further on,

$$MS_E = \frac{SS_{Error}}{n-s} \tag{4.30}$$

is an unbiased estimate of σ^2. For the test of hypothesis (4.22) the test statistic (4.26) is used, i.e.,

$$F = \frac{MS_A}{MS_E} = \frac{n-s}{s-1}\frac{SS_A}{SS_{Error}}. \tag{4.31}$$

Under H_0 F has an $F_{s-1,n-s}$-distribution. If

$$F > F_{s-1,n-s;1-\alpha} \tag{4.32}$$

Source of variation	SS	Degrees of freedom	MS	Test statistic F
Between the levels of factor A	$SS_A = \sum\limits_{i=1}^{s} n_i y_{i.}^2 - n y_{..}^2$	$df_A = s-1$	$MS_A = \frac{SS_A}{df_A}$	$\frac{MS_A}{MS_E}$
Within the levels of factor A	$SS_{Error} = \sum\limits_{i}\sum\limits_{j} y_{ij}^2 - \sum\limits_{i} n_i y_{i.}^2$	$df_E = n-s$	$MS_E = \frac{SS_E}{df_E}$	
	$SS_{Total} = \sum\limits_{i}\sum\limits_{j} y_{ij}^2 - n y_{..}^2$	$df_T = n-1$		

Table 4.3: Layout for the analysis of variance – one-way classification

then H_0 is rejected. For the realization of the analysis of variance we use Table 4.3.

Remark: For the derivation of the test statistic (4.31) we used the results of Chapter 3 and those of Section 3.7 in particular. Hence, we did not again prove the independence of the χ^2-distributions in the numerator and denominator of F (4.31).

An alternative proof for the stochastic independence of SS_A and SS_{Error} is based on a theorem by Cochran in the form, as can be found in Montgomery (1976, p.37) for example.

Theorem 4.1 (Theorem by Cochran) *Let $z_i \sim N(0,1)$, $i = 1, \ldots, v$ be independent random variables and assume the following disjunct decomposition*

$$\sum_{i=1}^{v} z_i^2 = Q_1 + Q_2 + \cdots + Q_s \tag{4.33}$$

with $s \le v$. Hence, the $Q_1, \ldots, Q_s$ are independent $\chi_{v_1}^2, \ldots, \chi_{v_s}^2$-distributed random variables if, and only if,

$$v = v_1 + \cdots + v_s \tag{4.34}$$

holds.

Employing this theorem yields the following:

$$(i) \qquad SS_{Total} = \sum_{i=1}^{s}\sum_{j=1}^{n_i} (y_{ij} - y_{..})^2 \tag{4.35}$$

has $n = \sum_{i=1}^{s} n_i$ summands, that have to satisfy one linear restriction ($\sum\sum y_{ij} = ny..$). Hence, SS_{Total} has $n - 1$ degrees of freedom.

$$(ii) \qquad SS_{Within} = SS_{Error} = \sum\sum(y_{ij} - y_{i.})^2 \qquad (4.36)$$

has s linear restrictions $\sum_{j=1}^{n_i} y_{ij} = n_i y_{i.}$ $(i = 1, \ldots, s)$ in case of n summands. Hence, SS_{Within} has $n - s$ degrees of freedom.

$$(iii) \qquad SS_{Between} = SS_A = \sum_{i=1}^{s} n_i(y_{i.} - y..)^2 \qquad (4.37)$$

has s summands, that have to satisfy one linear restriction ($\sum_{i=1}^{s} n_i y_{i.} = ny..$), and thus $SS_{Between}$ has $s - 1$ degrees of freedom. Hence, for the decomposition (4.33) according to

$$SS_{Total} = SS_{Error} + SS_A$$

we have the decomposition (4.34) of the degrees of freedom, i.e.,

$$n - 1 = (n - s) + (s - 1) \quad ,$$

such that according to Theorem 4.1, SS_{Error} and SS_A have independent χ^2-distributions, i.e., their ratio F (4.31) has an F-distribution.

4.2.3 Estimation of σ^2 by MS_{Error}

In (3.97) we derived the statistic

$$s^2 = \frac{1}{T - K}(y - Xb_0)'(y - Xb_0)$$

as an unbiased estimate for σ^2 in the linear model. In our special case of model (4.4) and using

$$\hat{y} = Xb_0 = \begin{pmatrix} y_1.\mathbf{1}_{n_1} \\ y_2.\mathbf{1}_{n_2} \\ \vdots \\ y_s.\mathbf{1}_{n_s} \end{pmatrix} \qquad (4.38)$$

according to (4.13) for $s > 2$, we receive (equating $K = s$, $T = n$):

$$s^2 = \frac{1}{n - s}((\mathbf{y}_1 - y_1.\mathbf{1}_{n_1})', \ldots, (\mathbf{y}_s - y_s.\mathbf{1}_{n_s})') \begin{pmatrix} \mathbf{y}_1 - y_1.\mathbf{1}_{n_1} \\ \vdots \\ \mathbf{y}_s - y_s.\mathbf{1}_{n_s} \end{pmatrix}$$

$$= \frac{1}{n - s}\sum_{i=1}^{s}\sum_{j=1}^{n_i}(y_{ij} - y_{i.})^2 \qquad (4.39)$$

$$= MS_{Error}. \qquad (4.40)$$

Model (4.2) yields

$$y_{i\cdot} = \mu + \alpha_i + \epsilon_{i\cdot} \quad , \qquad \epsilon_{i\cdot} \sim N\left(0, \frac{\sigma^2}{n_i}\right) \tag{4.41}$$

and hence, in analogy to (3.96)

$$\begin{aligned}
\mathrm{E}\left(MS_{Error}\right) &= \frac{1}{n-s}\mathrm{E}\left[\sum\sum(y_{ij} - y_{i\cdot})^2\right] \\
&= \frac{1}{n-s}\mathrm{E}\left[\sum\sum(\epsilon_{ij}^2 + \epsilon_{i\cdot}^2 - 2\epsilon_{ij}\epsilon_{i\cdot})\right] \\
&= \frac{1}{n-s}\sum\sum(\sigma^2 + \frac{\sigma^2}{n_i} - 2\frac{\sigma^2}{n_i}) \\
&= \sigma^2. \tag{4.42}
\end{aligned}$$

Furthermore, it follows from (4.41) with (4.6) that

$$\begin{aligned}
y_{\cdot\cdot} &= \mu + \frac{1}{n}\sum_{i=1}^{s} n_i\alpha_i + \epsilon_{\cdot\cdot} \\
&= \mu + \epsilon_{\cdot\cdot} \quad , \qquad \epsilon_{\cdot\cdot} \sim N\left(0, \frac{\sigma^2}{n}\right) \tag{4.43}
\end{aligned}$$

$$\begin{aligned}
\mathrm{E}\left(\epsilon_{i\cdot}\epsilon_{\cdot\cdot}\right) &= \frac{1}{n_i n}\mathrm{E}\left[\sum_{j=1}^{n_i}\epsilon_{ij}\sum_{i=1}^{s}\sum_{j=1}^{n_i}\epsilon_{ij}\right] \\
&= \frac{\sigma^2}{n} \quad . \tag{4.44}
\end{aligned}$$

Hence

$$y_{i\cdot} - y_{\cdot\cdot} = \alpha_i + \epsilon_{i\cdot} - \epsilon_{\cdot\cdot}, \tag{4.45}$$

$$\mathrm{E}\left(y_{i\cdot} - y_{\cdot\cdot}\right)^2 = \alpha_i^2 + \frac{\sigma^2}{n_i} - \frac{\sigma^2}{n} \tag{4.46}$$

holds, and thus

$$\begin{aligned}
\mathrm{E}\left(MS_A\right) &= \frac{1}{s-1}\sum\sum\mathrm{E}\left(y_{i\cdot} - y_{\cdot\cdot}\right)^2 \\
&= \sigma^2 + \frac{\sum n_i\alpha_i^2}{s-1} \quad . \tag{4.47}
\end{aligned}$$

Hence, under $H_0 : \alpha_1 = \cdots = \alpha_s = 0$ MS_A is an unbiased estimate for σ^2 as well. Thus, if H_0 does not hold, the test statistic F (4.31) has an expectation larger than one.

Example 4.3: The measured manufacturing times for the making of inlays (Table 4.1) represent one-way classified data material. Here, factor A represents the effect of a dentist on the manufacturing times, it has $s = 3$ levels (dentist A, B, C).

We may assume that the assumptions for a normal distribution hold, if we replace the manufacturing times in Table 4.1 by their natural logarithm (the reason for this transformation is that time values usually have a skewed distribution).

j\i	1	2	3	4	5	6	7	8	9	10
(A) 1	4.02	3.69	3.65	3.45	3.82	4.25	4.36	4.38	4.31	4.05
(B) 2	4.20	4.04	3.51	3.61	4.32	4.09	3.77	4.03	4.18	3.99
(C) 3	4.14	3.45	3.45	3.97	3.92	4.14	3.69	2.97		

j\i	11	12	13	14	$Y_i.$		$y_i.$	
(A) 1	4.28	4.25	3.87	4.08	56.46	$= Y_1.$	4.03	$= y_1.$
(B) 2	4.09				43.83	$= Y_2.$	3.98	$= y_2.$
(C) 3					29.73	$= Y_3.$	3.72	$= y_3.$
		n=33			130.02	$= Y..$	3.94	$= y..$

Table 4.4: Logarithms of the manufacturing times from Table 4.1

	SS		df	MS		F	
SS_A	$=$	512.82 - 512.28	2	MS_A	$= 0.27$	F	$= 2.70$
	$=$	0.54					
SS_{Error}	$=$	515.76 - 512.82	30	MS_E	$= 0.10$		
	$=$	2.94					
SS_{Total}	$=$	515.76 - 512.28	32				
	$=$	3.48					

Table 4.5: Analysis of variance table for Example 4.1

The arrangement in Table 4.4 of the measured values is done according to Table 4.1, the analysis is done in Table 4.5. The analysis yields the test statistic $F = 2.70 < 3.32 = F_{2,30;0.95}$ (Table B6). Hence, the null hypothesis *The mean manufacturing times per inlay are equal for all three dentists* is *not* rejected.

Once again we want to point out the difference between model I and II: the above result indicates that the three selected dentists do not differ with respect to their average manufacturing times per inlay. If however, we want to test the effect that the factor *dentist* has on the manufacturing time, then the manufacturing times would have to be measured in a sample of s dentists *selected at random*, and the proportion of the variability due to dentists compared to the total variation would have to be tested. Hence, not the comparison of means is the point of interest, but the decomposition of the total variation into components (model II).

Remark: The above analysis was done on a PC with maximum precision. If calculators are used, and in case of two-digital precision, deviations in the $SS's$ arise, but not in the test decision.

4.3   Comparison of Single Means

4.3.1   Linear Contrasts

The multiple comparison of means, i.e., the test of H_0 (4.22) against H_1 (4.23), has two possible outcomes — acceptance of H_0 (no treatment effect) and rejection of H_0 (treatment effect). In case of the first decision the analysis is finished, although a second run for the proof of an effect with a larger sample size could be done after appropriate power calculations.

If however H_1: $\alpha_i \neq 0$ for at least one i (or equivalently $\mu_i = \mu + \alpha_i \neq \mu + \alpha_j = \mu_j$ for at least one pair $(i, j), i \neq j$) is accepted, i.e., an overall treatment effect is proven, then the main interest lies in finding those populations that caused this overall effect. Hence, in this situation comparisons of pairs or of linear combinations are appropriate, that is we test for example

$$H_0 : \quad \mu_1 = \mu_2$$

against

$$H_1 : \quad \mu_1 \neq \mu_2$$

with the two-sample t-test by comparing $y_1.$ and $y_2.$ according to (1.5). Another possible hypothesis would be, for example $\mu_1 + \mu_2 = \mu_3 + \mu_4$.

These hypotheses stand for one linear constraint $r = \boldsymbol{R}'\boldsymbol{\beta}$ each, with rank$(\boldsymbol{R}') = 1$. In the analysis of variance, a linear combination of means (in the population or in the sample) is called a linear contrast, as long the following assumption holds.

Definition 4.2 *A linear combination*

$$\sum_{i=1}^{a} c_i y_i. = \boldsymbol{c}'\boldsymbol{y}.$$

of means is called linear contrast if

$$\boldsymbol{c}'\boldsymbol{c} \neq 0 \quad and \quad \sum_{i=1}^{a} c_i = 0 \tag{4.48}$$

holds.

Suppose we want to compare s populations with respect to their means, i.e., if we assume

$$y_{ij} \sim N(\mu_i, \sigma^2) \quad i = 1, \ldots, s; j = 1, \ldots, n_i \tag{4.49}$$

with y_{ij} and $y_{i'j}$ independent for $i \neq i'$, then

$$y_{i\cdot} \sim N\left(\mu_i, \frac{\sigma^2}{n_i}\right). \tag{4.50}$$

Denote by

$$\boldsymbol{\mu} = (\mu_1, \ldots, \mu_s)' \tag{4.51}$$

the vector of the s expectations. Then every linear contrast in the expectations can be written as

$$\boldsymbol{c}'\boldsymbol{\mu} \quad \text{with} \quad \sum c_i = 0 \quad \text{and} \quad \boldsymbol{c}'\boldsymbol{c} \neq 0. \tag{4.52}$$

The vector $\boldsymbol{\mu}$ is not to be mistaken for the overall mean μ from (4.4). Hence, the test statistic for testing H_0: $\boldsymbol{c}'\boldsymbol{\mu} = 0$ has the typical form

$$\frac{(\boldsymbol{c}'\boldsymbol{y}.)^2}{\text{Var}(\boldsymbol{c}'\boldsymbol{y}.)} \tag{4.53}$$

with the vector

$$\boldsymbol{y}' = (y_{1\cdot}, \ldots, y_{s\cdot}) \tag{4.54}$$

of the sample means. Thus, because of the independence of the s populations, we have (cf. (4.4))

$$\boldsymbol{c}'\boldsymbol{y}. \sim N\left(\boldsymbol{c}'\boldsymbol{\mu}, \sigma^2 \sum \frac{c_i^2}{n_i}\right) \tag{4.55}$$

and hence under H_0

$$\frac{(\boldsymbol{c}'\boldsymbol{y}.)^2}{\sigma^2 \sum \frac{c_i^2}{n_i}} \sim \chi_1^2. \tag{4.56}$$

As always, MS_{Error} (4.40) is an unbiased estimate of the variance σ^2, hence the test statistic is of the following form

$$t_{n-s}^2 = F_{1,n-s} = \frac{(\boldsymbol{c}'\boldsymbol{y}.)^2}{MS_{Error} \sum \frac{c_i^2}{n_i}} \tag{4.57}$$

if the χ^2-distributions of the numerator and denominator are independent.

Proof of the *F*-distribution of $F_{1,n-s}$ from (4.57)

(i) Denominator:

First, we derive a representation of MS_{Error} as a quadratic form in the total error vector ϵ (cf.(4.4)).

With (4.2) and (4.41) we have

$$y_{ij} - y_{i.} = \epsilon_{ij} - \epsilon_{i.} \quad , \quad (\text{all } i, j)$$

$$\epsilon_i - 1_{n_i}\epsilon_{i.} = \epsilon_i - \frac{1}{n_i}1_{n_i}1'_{n_i}\epsilon_i$$

$$= (I_{n_i} - \frac{1}{n_i}1_{n_i}1'_{n_i})\epsilon_i$$

$$= Q_i\epsilon_i \quad , \tag{4.58}$$

$$\begin{pmatrix} \epsilon_1 \\ \vdots \\ \epsilon_s \end{pmatrix} - \begin{pmatrix} 1_{n_1}\epsilon_{1.} \\ \vdots \\ 1_{n_s}\epsilon_{s.} \end{pmatrix} = \begin{pmatrix} Q_1 & & 0 \\ & \ddots & \\ 0 & & Q_s \end{pmatrix}\epsilon$$

$$= \text{diag}(Q_1, \ldots, Q_s)\epsilon$$

$$= Q\epsilon \quad . \tag{4.59}$$

The matrices $Q_i = I_{n_i} - \frac{1}{n_i}1_{n_i}1'_{n_i}$ are symmetric:

$$Q_i = Q'_i \quad ,$$

hence we have

$$Q = Q' \quad .$$

Furthermore, Q_i is idempotent:

$$Q_i^2 = I_{n_i} + \frac{1}{n_i^2}1_{n_i}1'_{n_i}1_{n_i}1'_{n_i} - \frac{2}{n_i}1_{n_i}1'_{n_i}$$

$$- Q_i$$

with $\text{rank}(Q_i) = \text{tr}(Q_i) = n_i - 1$. Hence, Q is idempotent as well, with $\text{rank}(Q) = \sum \text{rank}(Q_i) = n - s$.

This yields the following representation

$$MS_{Error} = \frac{1}{n-s}\epsilon'Q\epsilon \quad . \tag{4.60}$$

(ii) Numerator:

We have

$$y_. = \begin{pmatrix} y_{1.} \\ \vdots \\ y_{s.} \end{pmatrix} = \begin{pmatrix} \mu + \alpha_1 + \epsilon_{1.} \\ \vdots \\ \mu + \alpha_s + \epsilon_{s.} \end{pmatrix} \quad . \tag{4.61}$$

Under

$$H_0 : \quad c'\mu = c' \begin{pmatrix} \mu + \alpha_1 \\ \vdots \\ \mu + \alpha_s \end{pmatrix} = 0 \tag{4.62}$$

we have

$$c'y_. = c' \begin{pmatrix} \epsilon_{1.} \\ \vdots \\ \epsilon_{s.} \end{pmatrix} = c'\epsilon. \tag{4.63}$$

with

$$
\begin{aligned}
\epsilon_. &= \begin{pmatrix} \frac{1}{n_1}\mathbf{1}'_{n_1} & & \mathbf{0}' \\ & \ddots & \\ \mathbf{0}' & & \frac{1}{n_s}\mathbf{1}'_{n_s} \end{pmatrix} \epsilon \\
&= \operatorname{diag}(D'_1, \ldots, D'_s)\epsilon \\
&= D'\epsilon \quad . \tag{4.64}
\end{aligned}
$$

Hence, the numerator of F (4.57) can also be presented as a quadratic form in ϵ according to

$$\frac{(c'y_.)^2}{\sum \frac{c_i^2}{n_i}} = \frac{1}{\sum \frac{c_i^2}{n_i}} \epsilon' Dcc'D'\epsilon \quad . \tag{4.65}$$

The matrix of this quadratic form is symmetric and idempotent:

$$\left(\frac{1}{\sum \frac{c_i^2}{n_i}} Dcc'D' \right)^2 = \frac{1}{\sum \frac{c_i^2}{n_i}} Dcc'D' \tag{4.66}$$

We check this for $s = 2$. We have

$$
\begin{aligned}
Dcc'D' &= \begin{pmatrix} \frac{1}{n_1}\mathbf{1}_{n_1} & \mathbf{0} \\ \mathbf{0} & \frac{1}{n_2}\mathbf{1}_{n_2} \end{pmatrix} \begin{pmatrix} c_1 \\ c_2 \end{pmatrix} (c_1\ c_2) \begin{pmatrix} \frac{1}{n_1}\mathbf{1}'_{n_1} & \mathbf{0}' \\ \mathbf{0}' & \frac{1}{n_2}\mathbf{1}'_{n_2} \end{pmatrix} \\
&= \begin{pmatrix} \frac{c_1^2}{n_1^2}\mathbf{1}_{n_1}\mathbf{1}'_{n_1} & \frac{c_1 c_2}{n_1 n_2}\mathbf{1}_{n_1}\mathbf{1}'_{n_2} \\ \frac{c_1 c_2}{n_1 n_2}\mathbf{1}_{n_2}\mathbf{1}'_{n_1} & \frac{c_2^2}{n_2^2}\mathbf{1}_{n_2}\mathbf{1}'_{n_2} \end{pmatrix}
\end{aligned}
$$

and hence

$$(Dcc'D')^2 = \left(\frac{c_1^2}{n_1} + \frac{c_2^2}{n_2} \right)(Dcc'D') \quad .$$

From this the idempotence follows (cf.(4.66)). Furthermore, we have (cf. A 61 (ii))

$$\operatorname{rank}\left(\frac{Dcc'D'}{\sum \frac{c_i^2}{n_i}} \right) = \operatorname{tr}\left(\frac{Dcc'D'}{\sum \frac{c_i^2}{n_i}} \right) = 1 \quad ,$$

since $\operatorname{tr}(\mathbf{1}_{n_i}\mathbf{1}'_{n_i}) = n_i$.

(iii) Independence of numerator and denominator

The numerator and denominator of F from (4.57) are quadratic forms in ϵ with idempotent matrices, hence they have a χ_1^2-distribution, or χ_{n-s}^2-distribution respectively. According to Theorem A 88, their ratio has an $F_{1,n-s}$-distribution if

$$\frac{1}{\sum \frac{c_i^2}{n_i}} QDcc'D' = 0. \tag{4.67}$$

As can easily be seen, we have

$$QD = \begin{pmatrix} Q_1 D_1 & & 0 \\ & \ddots & \\ 0 & & Q_s D_s \end{pmatrix} \tag{4.68}$$

and

$$\begin{aligned} Q_i D_i &= (I_{n_i} - \frac{1}{n_i} 1_{n_i} 1'_{n_i}) \frac{1}{n_i} 1_{n_i} \\ &= \frac{1}{n_i} 1_{n_i} - \frac{1}{n_i} 1_{n_i} = 0 \quad . \end{aligned}$$

Hence

$$QD = 0$$

and (4.67) holds.

Since under H_0: $c'\mu = 0$, a linear contrast is invariant to a multiplication with a constant $a \neq 0$:

$$ac'\mu = 0, \quad a\sum c_i = 0, \tag{4.69}$$

it is advisable to eliminate the ambiguity by the standardization

$$c'c = 1. \tag{4.70}$$

Definition 4.3 *A linear contrast $c'\mu$ is normed if $c'c = 1$*

Definition 4.4 *Two linear contrasts $c_1'\mu$ and $c_2'\mu$ are orthogonal if*

$$c_1'c_2 = 0. \tag{4.71}$$

Analogously, a system $(c_1'\mu, \ldots, c_v'\mu)$ of orthogonal contrasts is called an *orthonormal system* if

$$c_i'c_j = \delta_{ij} \tag{4.72}$$

$(i, j = 1, \ldots, v)$ holds, where δ_{ij} is the Kronecker symbol.

The orthogonal contrasts are an essential aid in reducing the number of possible pairwise comparisons to the maximum number of independent hypotheses, and hence in ensuring the testability.

Example 4.4: Assume we have $s = 3$ samples (3 levels of factor A) and let the design be balanced ($n_i = r$). The overall null hypothesis

$$H_0: \quad \mu_1 = \mu_2 = \mu_3 \quad (\text{i.e., } H_0: \alpha_i = 0 \text{ for } i = 1, 2, 3) \tag{4.73}$$

can be written, for example, as

$$H_0: \quad \mu_1 = \mu_2 \quad \text{and} \quad \mu_2 = \mu_3 \quad, \tag{4.74}$$

or with linear contrasts as

$$H_0: \quad \begin{pmatrix} c_1' \\ c_2' \end{pmatrix} \boldsymbol{\mu} = \begin{pmatrix} 0 \\ 0 \end{pmatrix} \tag{4.75}$$

with

$$\begin{aligned}
\boldsymbol{\mu}' &= (\mu_1, \mu_2, \mu_3) \quad \text{and} \\
c_1' &= (1, -1, 0) \quad, \tag{4.76} \\
c_2' &= (0, 1, -1) \quad. \tag{4.77}
\end{aligned}$$

We have $c_1' c_2 = -1$, hence $c_1' \boldsymbol{\mu}$ and $c_2' \boldsymbol{\mu}$ are not orthogonal and the quadratic forms $(c_1' y.)^2$ and $(c_2' y.)^2$ are not stochastically independent. If however, we choose

$$c_1' = (1, -1, 0), \quad c_1' c_1 = 2 \tag{4.78}$$

as before, and

$$c_2' = (1, 1, -2), \quad c_2' c_2 = 6 \quad, \tag{4.79}$$

then $c_1' c_2 = 0$. $c_1' \boldsymbol{\mu} = 0$ means $\mu_1 = \mu_2$ and $c_2' \boldsymbol{\mu} = 0$ means $\frac{\mu_1 + \mu_2}{2} = \mu_3$, so that both contrasts represent $H_0: \mu_1 = \mu_2 = \mu_3$ simultaneously. The test statistic for H_0 (4.75) is then of the form

$$F_{2, n-2} = \left(\frac{r(c_1' y.)^2}{c_1' c_1} + \frac{r(c_2' y.)^2}{c_2' c_2} \right) / MS_{Error}. \tag{4.80}$$

With the contrasts (4.78) and (4.79), we thus have for the hypothesis H_0 (4.73)

$$F_{2, n-2} = \left(\frac{r(y_1. - y_2.)^2}{2} + \frac{r(y_1. + y_2. - 2y_3.)^2}{6} \right) / MS_{Error} \quad. \tag{4.81}$$

4.3.2 Contrasts of the Total Response Values in the Balanced Case

We want to derive an interesting decomposition of the sum of squares SS_A. We assume

- s levels of factor A (treatments)

- $n_i = r$ repetitions per treatment (balanced design)

- $n = rs$ the total number of response values

- $Y_{i\cdot} = \sum_{j=1}^{r} y_{ij}$ the total response of treatment i

- $Y' = (Y_{1\cdot}, \ldots, Y_{s\cdot})$ the vector of the total response values

-
$$SS_A = \frac{1}{r}\sum_{i=1}^{s} Y_{i\cdot}^2 - \frac{1}{rs}\left(\sum_{i=1}^{s} Y_{i\cdot}\right)^2 \tag{4.82}$$

(cf. (4.28) for the balanced case).

Under these assumptions the following rules apply (cf. e.g., Petersen, 1985, p.92)

(i) Let $c_1' Y_\cdot$ be a linear contrast of the total response values. Then

$$S_1^2 = \frac{\left(\sum_{i=1}^{s} c_{1i} Y_{i\cdot}\right)^2}{\left(r \sum c_{1i}^2\right)} = \frac{(c_1' Y_\cdot)^2}{(r c_1' c_1)} \tag{4.83}$$

is a component of SS_A with one degree of freedom. Hence, with

$$c_{1i} Y_{i\cdot} \sim N(0, r\sigma^2 c_{1i}^2),$$
$$c_1' Y_\cdot \sim N(0, r\sigma^2 \sum c_{1i}^2)$$
$$= N(0, r\sigma^2 c_1' c_1)$$

we have

$$\frac{(c_1' Y_\cdot)^2}{r c_1' c_1} = S_1^2 \sim \sigma^2 \chi_1^2 \quad . \tag{4.84}$$

(ii) If $c_2' Y_\cdot$ and $c_1' Y_\cdot$ are orthogonal contrasts, then

$$S_2^2 = \frac{(c_2' Y_\cdot)^2}{(r c_2' c_2)} \tag{4.85}$$

is a component of $SS_A - S_1^2$.

(iii) If $c_1' Y_\cdot, \ldots, c_{s-1}' Y_\cdot$ is a complete system of orthogonal contrasts, then

$$S_1^2 + \ldots + S_{s-1}^2 - SS_A \tag{4.86}$$

holds.

We now have a decomposition of SS_A into $s - 1$ independent sums of squares. In case of a normal distribution, these components have independent χ^2-distributions. This decomposition corresponds to the decomposition of the G^2-statistic in $I \times 2$-contingency tables into $I - 1$ independent, χ^2-distributed G^2-statistics for the analysis of subeffects. In case of a significant overall treatment effect the main subeffects that contributed to the significance can thus be discovered. The significance of the subeffects, i.e., H_0: $c_i' Y = 0$ against H_1: $c_i' Y \neq 0$, is tested with

$$t_{n-s}^2 = F_{1,n-s} = F_{1,s(r-1)} = \frac{S_i^2}{MS_{Error}}. \tag{4.87}$$

i	1	2	3	4	5	6	$Y_{i\cdot}$	$y_{i\cdot}$	s_i
			Repetitions j						
1	4.5	5.0	3.5	3.7	4.8	4.0	25.5	4.25	0.6091
2	3.8	4.0	3.9	4.2	3.6	4.4	23.9	3.98	0.2858
3	3.5	4.5	3.2	2.1	3.5	4.0	20.8	3.47	0.8116
4	3.0	2.8	2.2	3.4	4.0	3.9	19.3	3.22	0.6882
							$Y_{\cdot\cdot} = 89.5$	$y_{\cdot\cdot} = 3.73$	

Table 4.6: Flexibility in dependency of four levels of factor A (additives)

```
   Source        D.F.   Sum of    Mean       F       F
                       Squares  Squares   Ratio    Prob.
Between Groups      3   4.0046   1.3349   3.3687   0.0389
Within Groups      20   7.9250   0.3962
        Total      23  11.9296
```

Table 4.7: Analysis of variance table for Table 4.6 in SPSS format

Variance of Linear Contrasts

If the s samples are independent, then the variance of a linear contrast is computed as follows

(i) **Contrast of the means**
Let $c'y_\cdot = c_1 y_{1\cdot} + \ldots + c_s y_{s\cdot}$, then

$$\mathrm{Var}(c'y_\cdot) = \left(\frac{c_1^2}{n_1} + \ldots + \frac{c_s^2}{n_s} \right) \sigma^2 \tag{4.88}$$

holds in general. In the balanced case ($n_i = r$, $i = 1, \ldots, s$) this expression simplifies to

$$\mathrm{Var}(c'y_\cdot) = \frac{c'c}{r} \sigma^2 \quad . \tag{4.89}$$

(ii) **Contrast of the totals**
Let $c'Y_\cdot = c_1 Y_{1\cdot} + \ldots + c_s Y_{s\cdot}$, then

$$\mathrm{Var}(c'Y_\cdot) = (n_1 c_1^2 + \ldots + n_s c_s^2)\sigma^2 \tag{4.90}$$

holds in general and in the balanced design

$$\mathrm{Var}(c'Y_\cdot) = rc'c\sigma^2 \quad . \tag{4.91}$$

The variance σ^2 of the population is estimated by $\mathrm{MS}_{Error} = s^2$, hence

$$\widehat{\mathrm{Var}}(c'y_\cdot) = s^2 \sum \frac{c_i^2}{n_i} \tag{4.92}$$

and

$$\widehat{\text{Var}}(c'Y.) = s^2 \sum n_i c_i^2 \tag{4.93}$$

are unbiased estimates of $\text{Var}(c'y.)$ and $\text{Var}(c'Y.)$.

Example 4.5: Consider the following balanced experimental design with $r = 6$ repetitions:

Factor A   level 1 :   control group (neither A_1 nor A_2)
            level 2 :   additive A_1
            level 3 :   additive A_2
            level 4 :   additives A_1 and A_2 (combination)

Suppose, response Y is the flexibility of a plastic material, and that we are interested in the most favourable mixture in the sense of a reduction of the flexibility. The data is shown in Table 4.6.
The analysis of variance with SPSS is done via the following commands

```
GET FILE = 'flex.sav' .
EXECUTE .
ANOVA Response BY Treatment (1,4).
/STATISTICS ALL .
```

We receive the analysis of variance table (Table 4.7) according to the layout of Table 4.3 in the SPSS format. The F-test rejects the hypothesis H_0: $\mu_1 = \mu_2 = \mu_3$ with the statistic $F_{3,20} = 3.3687$ (p-value 0.0389). Hence, we can now compare pairs or combinations of treatments. For $s = 4$ levels, systems exist with $s - 1 = 3$ orthogonal contrasts. We consider the following two systems (Tables 4.8 and 4.9).
In both systems the sums of squares S^2 of the contrasts add up to SS_A (SS Between Groups in Table 4.7) according to (4.86). With $MS_{Error} = 0.3962$, the test statistics (4.87) are

Table 4.8	Table 4.9
2.02	1.01
2.61	9.10 *
5.48 *	0.00

The 95%-quantile of the $F_{1,23}$-distribution is 4.15, so that

- the employment of at least one additive compared to the control group is significant (i.e., reduces the flexibility significantly)

- the employment of A_2 (alone or in combination with A_1) reduces the flexibility significantly.

Contrast	Treatment Response $Y_i.$	1 25.5	2 23.9	3 20.8	4 19.3	$c'Y.$	S^2
A_1 against A_2			0	+1	−1	0	3.1	0.8008
A_1 or A_2 against $(A_1$ and $A_2)$			0	−1	−1	2	−6.1	1.0336
A_1 or A_2 or $(A_1$ and $A_2)$ against Control group			−3	+1	+1	+1	−12.5	2.1702
							$\Sigma = 4.0046$

Table 4.8: Orthogonal contrasts and test statistics S^2

Contrast	Treatment Response $Y_i.$	1 25.5	2 23.9	3 20.8	4 19.3	$c'Y.$	S^2
A_1		−1	+1	−1	+1	−3.1	0.4004
A_2		−1	−1	+1	+1	−9.3	3.6038
$A_1 \times A_2$		+1	−1	−1	+1	0.1	0.0004
						$\Sigma = 4.0046$

Table 4.9: Orthogonal contrasts and test statistics S^2

The orthogonal contrasts of the response sums $Y_i.$ make a decomposition of the variability SS_A possible, i.e., of the treatment effect, and hence enable the determination of significant subeffects. With F from (4.57), the orthogonal contrast of means, on the other hand, yield a test statistic for testing differences of treatments according to the linear function of the means given by the contrast.

We demonstrate this with the same systems of orthogonal contrasts as in Tables 4.8 and 4.9. The results are shown in Tables 4.10 and 4.11. We have for example (Table 4.11, first row)

$$
\begin{aligned}
c'y. &= (y_2. + y_4.) - (y_1. + y_3.) \\
&= 3.98 + 3.22 - (4.25 + 3.47) = -0.52 \quad,
\end{aligned}
$$

$$
\begin{aligned}
\widehat{\mathrm{Var}}(c'y.) &= \frac{c'c}{r}s^2 \\
&= \frac{4}{6} \cdot 0.3962 = 0.2641 \\
&= 0.5140^2
\end{aligned}
$$

with $s^2 = MS_{Error} = 0.3962$ from Table 4.7. The test statistic from (4.57) for

$$
H_0 \quad : \quad c'\mu = (\mu_2 + \mu_4) - (\mu_1 + \mu_3) = 0 \quad,
$$

i.e., for $H_0 : (\alpha_2 + \alpha_4) = (\alpha_1 + \alpha_3)$ is now

$$t_{24-4} = t_{20} = \frac{-0.520}{0.514} = -1.002 \quad .$$

The critical value is (Table B5)

$$t_{20;0.95,\text{one-sided}} = -1.73$$

and

$$t_{20;0.95,\text{two-sided}} = \pm 2.09 \quad ,$$

so that H_0 is not rejected. We can see from the Tables 4.10 and 4.11 that the following contrasts are significant:

$$\frac{\mu_2 + \mu_3 + \mu_4}{3} - \mu_1 < 0$$

(the control group has a higher flexibility than the mean of the three treatments),

$$\mu_3 + \mu_4 - (\mu_1 + \mu_2) < 0$$

(A_2 plus (A_1 and A_2) have a lower mean flexibility than control group plus A_1). Commands and output in SPSS: The contrasts from Table 4.11 are called with the command

```
/contrast = -1 1 -1 1
/contrast = -1 -1 1 1
/contrast = 1 -1 -1 1
```

which is inserted into the SPSS procedure.

Treatment	1	2	3	4	$c'y.$	$\text{Var}(c'y.)$	t_{20}
Mean $y_i.$	4.25	3.98	3.47	3.22			
Contrast							
A_1 against A_2	0	+1	−1	0	0.52	0.363^2	1.42
A_1 or A_2 against (A_1 and A_2)	0	−1	−1	2	−1.02	0.629^2	−1.61
A_1 or A_2 or (A_1 and A_2) against control group	−3	+1	+1	+1	−2.08	0.890^2	−2.33 ∗

Table 4.10: Orthogonal contrasts of the means

The obvious question, whether A_2 should be employed alone or in combination with A_1, could be tested with the two-sample t-test according to (1.5). We

Treatment	1	2	3	4	$c'y.$	$\text{Var}(c'y.)$	t_{20}	
Mean $y_i.$	4.25	3.98	3.47	3.22				
Contrast								
A_1	-1	$+1$	-1	$+1$	-0.52	0.514^2	-1.002	
A_2	-1	-1	$+1$	$+1$	-1.54	0.514^2	-2.996	*
$A_1 \times A_2$	$+1$	-1	-1	$+1$	0.02	0.514^2	0.039	

Table 4.11: Orthogonal contrasts of the means

compute with $s_{A_2} = 0.8116$, $s_{A_1 \text{ and } A_2} = 0.6882$ (Table 4.6) the pooled variance (1.6)

$$s^2 = \frac{5(0.8116^2 + 0.6882^2)}{6 + 6 - 2} = 0.7524^2$$

and

$$t_{10} = \frac{\frac{20.8}{6} - \frac{19.3}{6}}{0.7524} \sqrt{\frac{6 \cdot 6}{6 + 6}} = 0.5755 \quad ,$$

so that H_0: $\mu_{A_2} = \mu_{(A_1 \text{ and } A_2)}$ is not rejected ($t_{10,0.95,\text{one-sided}} = 1.81$). Hence, the two treatments A_2 and $(A_1 \text{ and } A_2)$ show no significant difference.

In the next section however, we will integrate this problem of pairwise comparisons in case of s treatments into the multiple test problem. As we will see, this shows that an adjustment of the degrees of freedom, or of the applied quantile respectively, has to be made.

4.4 Multiple Comparisons

4.4.1 Introduction

With the linear and especially with the orthogonal contrasts we have the possibility to test selected linear combinations for significance and thus structure the treatments. Starting point is a rejection of the overall equality $\mu_1 = \ldots = \mu_s$ of the means of the response.

A number of statistical procedures exist for the comparison of single means or of groups of means. These procedures have the following different objectives:

- Comparison of all possible pairs of means (for s levels of A we have $s(s-1)/2$ different pairs)

- Comparison of all $s - 1$ means with a control group selected in advance

- Comparison of all pairs of treatments that were selected in advance

- Comparison of any linear combinations of the means.

These procedures differ, next to their aims, especially with respect to the way in which they control for the type I error. In one case, the error is controlled on a *per comparison basis*, in the other case the error is controlled simultaneously for all comparisons.

A multiple test procedure that conducts every pairwise comparison at a significance level α, i.e., that works per comparison basis, is possible if the group comparisons are already planned at the beginning of the experiment. It is mainly based on the t statistic. If we want to ensure the significance level α simultaneously for all group comparisons of interest, the appropriate multiple test procedure is one that controls the error rate *per experiment basis*.

The decision for one of the two procedures is to be made ahead of the experiment.

4.4.2   Experimentwise Comparisons

The most popular multiple procedures that control the error simultaneously are those of Dunnet (1955) for the comparison of $s-1$ groups with a control group, of Tukey (1953) for all $s(s-1)/2 = \binom{s}{2}$ pairwise comparisons, and of Scheffé (1953) for any linear combinations. The procedures of Tukey and Scheffé should be applied in the explorative phase of an experiment, in order to avoid comparisons that are suggested by the data. A main condition for all multiple procedures is the rejection of H_0: $\mu_1 = \cdots = \mu_s$.

Hint: A detailed representation and rating of the multiple test procedures can be found in Miller (1981) and in Gather and Pigeot (1990).

Procedure by Scheffé

Let $c'\mu$ be any linear contrast of μ and $c'y.$, with $\sum_{i=1}^{a} c_i = 0$ and $y' = (y_1., \ldots, y_s.)$ the corresponding contrast of the vector of means. We then have for all c

$$P(c'y. - \sqrt{S_{1-\alpha}} \leq c'\mu \leq c'y. + \sqrt{S_{1-\alpha}}) = 1 - \alpha \qquad (4.94)$$

with (cf. (4.88))

$$S_{1-\alpha} = MS_{Error}(s-1)(\frac{c_1^2}{n_1} + \cdots + \frac{c_s^2}{n_s})F_{s-1,n-s;1-\alpha} \qquad (4.95)$$

The null hypothesis H_0: $c'\mu = 0$ is rejected if zero is not within the confidence interval. The multiple level is α.

Procedure by Dunnett

Let group $i = 1$ be selected as the control group that is to be compared with the treatments (groups) $i = 2, \ldots, s$. The $(1 - \alpha) \cdot 100\%$-confidence intervals for the $s - 1$ pairwise comparisons "control – treatment" are of the form

$$(y_1. - y_i.) \pm C_{1-\alpha}(s - 1, n - s)s_{\bar{d}_i} \qquad (4.96)$$

with

$$s_{\bar{d}_i} = \sqrt{MS_{Error}(\frac{1}{n_1} + \frac{1}{n_i})} \quad . \tag{4.97}$$

The quantiles $C_{1-\alpha}(s-1, n-s)$ are are given in special tables (one- and two-sided, cf. Woolson, 1987, Tables 13a and 13b, p. 502–503, or Dunnett (1955, 1964)). We show an excerpt for $C_{0.95}(s-1, n-s)$.

	\multicolumn{5}{c}{$s - 1$}				
$n - s$	1	2	3	4	5
5	2.57	3.03	3.39	3.66	3.88
10	2.23	2.57	2.81	2.97	3.11
15	2.13	2.44	2.64	2.79	2.90
20	2.09	2.38	2.57	2.70	2.81

Table 4.12: $C_{0.95}(s-1, n-s)$-quantiles (two-sided)

	\multicolumn{5}{c}{$s - 1$}				
$n - s$	1	2	3	4	5
5	2.02	2.44	2.68	2.85	2.98
10	1.81	2.15	2.34	2.47	2.56
15	1.75	2.07	2.24	2.36	2.44
20	1.72	2.03	2.19	2.30	2.39

Table 4.13: $\tilde{C}_{0.95}(s-1, n-s)$-quantiles (one-sided)

The hypothesis H_0: $\mu_1 = \mu_i$ ($i = 2, \dots, s$) is rejected

- two-sided in favour of H_1: $\mu_1 \neq \mu_i$, if

$$|y_{1\cdot} - y_{i\cdot}| > C_{1-\alpha}(s-1, n-s) \cdot s_{\bar{d}_i} \tag{4.98}$$

- one-sided in favour of H_1: $\mu_1 > \mu_i$, if

$$y_{1\cdot} - y_{i\cdot} > \tilde{C}_{1-\alpha}(s-1, n-s) \cdot s_{\bar{d}_i} \tag{4.99}$$

- one-sided in favour of H_1: $\mu_1 < \mu_i$, if

$$y_{1\cdot} - y_{i\cdot} < -\tilde{C}_{1-\alpha}(s-1, n-s) \cdot s_{\bar{d}_i} \tag{4.100}$$

holds. For all $s - 1$ comparisons the multiple level α is ensured.

Procedure by Tukey

In case of experiments in the explorative phase it is often not possible to fix the set of planned comparisons in advance. Hence, all $s(s-1)/2$ possible pairwise comparisons are done. The two-sided test procedure by Tukey assumes the balanced case $n_i = r$ and controls for the error experimentwise, i.e., for all $s(s-1)/2$ comparisons the multiple level α holds. We compute the confidence intervals

$$(y_{i\cdot} - y_{j\cdot}) \pm T_\alpha \qquad (i > j) \tag{4.101}$$

with

$$T_\alpha = Q_\alpha(s, n-s)\, s_{\bar{d}} \quad , \tag{4.102}$$

$$s_{\bar{d}} = \sqrt{MS_{Error}/r} \quad . \tag{4.103}$$

The quantiles $Q_{1-\alpha}(s, n-s)$ are so-called studentized rank-values, that are given in special tables (cf. e.g., Woolson, 1987, Table 14, pp. 504-505). The set of null hypotheses $H_0(i,j)$: $\mu_i = \mu_j$ $(i > j)$ is rejected in favour of H_1: H_0 incorrect (i.e., $\mu_i \neq \mu_j$ for at least one pair $i > j$), if

$$|y_{i\cdot} - y_{j\cdot}| > T_\alpha \tag{4.104}$$

holds. For all pairs (i,j), $i > j$ with $|y_{i\cdot} - y_{j\cdot}| > T_\alpha$ we have a statistically significant treatment difference.

Bonferroni Method

Suppose, we want to conduct $k \leq s$ comparisons with a multiple level of α at the most. In this situation the Bonferroni method can be applied. This method splits up the risk α into equal parts α/k for the k comparisons. Basis is Bonferroni's inequality.

Let $H_1, \ldots, H_k$ be the confidence intervals for the k comparisons. Denote by $P(H_i)$ the probability that H_i is true (i.e., H_i covers the respective parameter of the ith comparison). Then $P(H_1 \cap \cdots \cap H_k)$ is the probability that all k confidence intervals cover the respective parameters. According to Bonferroni's inequality we have

$$P(H_1 \cap \cdots \cap H_k) \geq 1 - \sum_{i=1}^{k} P(\bar{H}_i) \tag{4.105}$$

where $\bar{H}_i$ is the complementary event to H_i. If $P(\bar{H}_i) = \frac{\alpha}{k}$ is chosen, then the following holds for the simultaneous probability

$$P(H_1 \cap \cdots \cap H_k) \geq 1 - \alpha \tag{4.106}$$

Assume for example, $k \leq s$ contrasts $c_i'\mu$ are to be tested simultaneously. The confidence intervals for $c_i'\mu$ according to the Bonferroni method are then of the following form

$$c_i'y_\cdot \pm t_{n-s; 1-\alpha/2k}\sqrt{MS_{Error}}\sqrt{\frac{c_1^2}{n_1} + \cdots + \frac{c_s^2}{n_s}} \quad . \tag{4.107}$$

The test runs analogously to the procedure by Scheffé, i.e., if (4.107) does not contain the zero, then H_0 is rejected and the respective comparison is significant.

4.4.3 Select Pairwise Comparisons

The "Least significant difference" (LSD)

Suppose we want to compare the means of two selected treatments, i.e., suppose we want to test H_0: $\mu_1 = \mu_2$ against H_1: $\mu_1 \neq \mu_2$. The appropriate test statistic is

$$t_{df} = \frac{y_1. - y_2.}{\sqrt{\widehat{\mathrm{Var}}(y_1. - y_2.)}} \quad , \tag{4.108}$$

where df is the number of degrees of freedom. For $|t| > t_{df;1-\alpha/2}$ we reject H_0, where $t_{df;1-\alpha/2}$ is the two-sided quantile at the α probability level. If H_0 is rejected, then μ_1 is significantly different from μ_2 at the α level. $|t| > t_{df;1-\alpha/2}$ is equivalent with

$$t_{df;1-\alpha/2}\sqrt{\widehat{\mathrm{Var}}(y_1. - y_2.)} < |y_1. - y_2.| \quad . \tag{4.109}$$

Hence, every sample with a difference $|y_1. - y_2.|$ that exceeds $t_{df;1-\alpha/2}\sqrt{\mathrm{Var}(y_1. - y_2.)}$, indicates a significant difference between μ_1 and μ_2. According to (4.109), the left side would be the smallest difference of $y_1.$ and $y_2.$ for which significance would be declared. Thus, we define (df is the number of degrees of freedom of s^2, the pooled variance of the two samples)

$$\begin{aligned} LSD &= t_{df;1-\alpha/2}\sqrt{\widehat{\mathrm{Var}}(y_1. - y_2.)} \\ &= t_{df;1-\alpha/2}\sqrt{s^2(\frac{1}{n_1} + \frac{1}{n_2})} \quad . \end{aligned} \tag{4.110}$$

In the balanced case $(n_1 = n_2 = r)$ we receive

$$LSD = t_{df;1-\alpha/2}\sqrt{\frac{2s^2}{r}} \quad . \tag{4.111}$$

Using the LSD is controversial, especially if it is used for comparisons suggested by the data (largest/smallest sample mean) or if all pairwise comparisons are done without correction of the test level. If the LSD is used for all pairwise comparisons (i.e., for $s(s-1)/2$ comparisons in case of s treatments), then these tests are not independent. Procedures based on the LSD, that ensure the test level due to correctures of the quantiles, exist (HSD, Duncan test). $FPLSD$ and SNK on the other hand, only ensure the global level.

Fisher's Protected LSD (FPLSD)

This procedure starts out with the analysis of variance and tests the global hypothesis H_0 : $\mu_1 = \cdots = \mu_s$ with the statistic $F = \frac{MS_A}{MS_{Error}}$ from (4.31). If F is not significant the procedure stops. If $F > F_{s-1,n-s;1-\alpha}$, i.e., differences of the means are significant, then all pairs of means $y_{i\cdot}$ and $y_{j\cdot}$ $(i \neq j)$ are tested for differences with

$$FPLSD = t_{n-s;1-\alpha/2}\sqrt{MS_{Error}(\frac{1}{n_i} + \frac{1}{n_j})} \quad . \tag{4.112}$$

For $|y_{i\cdot} - y_{j\cdot}| > FPLSD$ we have a significant difference of means. Note that in (4.112) σ^2 is estimated by MS_{Error}. Hence, t now has $n - s$ degrees of freedom (instead of $n_1 + n_2 - 2$ degrees of freedom as in the two-sample case).

Tukey's Honestly Significant Difference (HSD)

This procedure uses the studentized rank-values $Q_{\alpha,(s,n-s)}$ (cf. (4.102)) instead of the t-quantiles and replaces the standard error of the mean by the standard error of the difference (pooled sample). We compute

$$HSD = Q_{\alpha,(s,n-s)}\sqrt{MS_{Error}/r} \quad . \tag{4.113}$$

All differences of pairs $|y_{i\cdot}-y_{j\cdot}|$ $(i < j)$ are compared with HSD. For $|y_{i\cdot}-y_{j\cdot}| > HSD$ we have a significant difference between μ_i and μ_j.

Student-Newman-Keuls Test (SNK)

The SNK is a test in which the difference needed for significance varies with the degree of separation. Suppose we want to compare k means. The sample means are sorted in descending order:

$$y_{(1)\cdot}, \ldots, y_{(k)\cdot} \quad ,$$

where $y_{(i)\cdot}$ is the mean with the ith rank (i.e., $y_{(1)\cdot}$ is the largest, $y_{(k)\cdot}$ the smallest mean). We compute the SNK differences

$$SNK_i = Q_{\alpha,(i,df)}\sqrt{MS_{Error}/r} \qquad (i = 2, \ldots, k) \tag{4.114}$$

with $Q_{\alpha,(i,df)}$ for df degrees of freedom of SS_{Error} and (in succession) $i = 2, 3, \quad , k$ means.

If $|y_{(1)\cdot} - y_{(k)\cdot}| < SNK_k$ then none of the differences of means are significant and the procedure stops.

If $|y_{(1)\cdot} - y_{(k)\cdot}| > SNK_k$ then this (largest) difference is significant. We proceed by testing whether

$$|y_{(2)\cdot} - y_{(k)\cdot}| > SNK_{k-1} \tag{4.115}$$

and

$$|y_{(1)\cdot} - y_{(k-1)\cdot}| > SNK_{k-1} \tag{4.116}$$

holds. If both conditions hold then those differences of the rank-ordered means are tested, where the ranks differ by $k - 3$. This procedure is continued up to the comparison of rank-neighboured means.

Duncan Test

Duncan (1975) modified the procedure FPLSD by computing alternative quantiles. The least significant difference is Bayes-adjusted and reads as follows

$$BLSD = t_B\sqrt{2MS_{Error}/r} \quad . \tag{4.117}$$

The values t_B are given in special tables (Waller and Duncan, 1972) and are printed out in the SPSS procedure.

Hint: A number of multiple test procedures exist that work with other rank-values. These are implemented in the standard software.

Example 4.6: (Continuation of Example 4.5)
Table 4.6 yields

Treatment	1	2	3	4
Rank	1	2	3	4
Mean	4.25	3.98	3.47	3.22

We had $s = 4$, $r = 6$ and $n = 4 \cdot 6 = 24$ as well as $MS_{Error} = 0.3962$ for $n - s = 20$ degrees of freedom (Table 4.7). The hypothesis $H_0 : \mu_1 = \cdots = \mu_4$ was rejected.

Experimentwise Procedures

Procedure by Scheffé The critical value (4.95) of the confidence interval (4.94) for any contrast $c'\mu$ is, with $F_{3,20;0.95} = 3.10$

$$
\begin{aligned}
S_{1-\alpha} &= 0.3962 \cdot 3 \cdot 3.10 \cdot \frac{c'c}{6} \\
&= 0.61 \cdot c'c \quad .
\end{aligned}
$$

We test the complete system of orthogonal contrasts of the means from Table 4.11 and receive

	$c'y$.	$c'c$	$\sqrt{S_{1-\alpha}}$	$c'y. \pm \sqrt{S_{1-\alpha}}$
A_1	-0.52	4	1.57	$[-2.09 , 1.05]$
A_2	-1.54	4	1.57	$[-3.11 , 0.03]$
$A_1 \times A_2$	0.02	4	1.57	$[-1.55 , 1.59]$

The zero lies in all three intervals, hence $H_0: c'\mu = 0$ is never rejected.

Procedure by Dunnett In Example 4.3 level 1 was designed as control group. We conduct the multiple comparison (according to Dunnett) of the control group with the groups 2, 3, 4. The critical limits (4.96) are $(n_i = n_j = 6)$ (cf. Tables 4.12 and 4.13) two-sided:

$$C_{1-\alpha}(3, 20)\sqrt{0.3962 \cdot \frac{2}{6}} = 2.57 \cdot 0.3634 = 0.9340$$

and one-sided:

$$\tilde{C}_{1-\alpha}(3, 20) \cdot 0.3634 = 2.19 \cdot 0.3634 = 0.7958 \quad .$$

For the one-sided tests we receive

$$
\begin{aligned}
y_{1.} - y_{2.} &= 0.27 \\
y_{1.} - y_{3.} &= 0.78 \\
y_{1.} - y_{4.} &= 1.03 * \quad ,
\end{aligned}
$$

and hence, a significant difference between the control group and group 4.

Procedure by Tukey Here all $4 \cdot 3/2 = 6$ possible comparisons are conducted. With $Q_{0.05}(4, 20) = 3.95$ and $s_{\bar{d}} = \sqrt{MS_{Error}/r} = \sqrt{0.3962/6} = 0.2570$ the critical value (cf. (4.102)) is $T_{0.05} = 3.95 \cdot 0.2570 = 1.02$.

(i,j)	$|y_{i.} - y_{j.}|$
(1,2)	0.27
(1,3)	0.78
(1,4)	1.03 *
(2,3)	0.51
(2,4)	0.76
(3,4)	0.25

Again, the difference between treatments 1 and 4 is significant.

Bonferroni Method We conduct the $k = 3$ comparisons from Table 4.10 according to the Bonferroni method. The critical limit from (4.107) for the chosen contrast $c'\mu$ is

$$
t_{20;1-0.05/2\cdot3} \cdot \sqrt{0.3962} \cdot \sqrt{\frac{c'c}{6}} = 2.95 \cdot \frac{0.6294}{2.4495} \cdot \sqrt{c'c}
$$

$$
= 0.7580 \cdot \sqrt{c'c}
$$

Contrast	$c'y$	$c'c$	$0.7580 \cdot \sqrt{c'c}$	Interval (4.107)
1/2	0.52	2	1.0720	$[-0.5520, 1.5920]$
1 or 2/4	−1.02	6	1.8567	$[-2.8767, 0.8367]$
1/2 or 3 or 4	−2.08	12	2.6258	$[-4.7058, 0.6058]$

In the multiple comparison according to Bonferroni no contrast is statistically significant.

Selected Pairwise Comparisons

SNK Test   The studentized ranges $Q_{0.05,(i,df)}$ for $df = 20$ degrees of freedom
are

i	2	3	4
$Q_{0.05,(i,20)}$	2.95	3.57	3.95
SNK_i	0.76	0.92	1.02

This yields the following comparisons

$$|y_{(1)\cdot} - y_{(4)\cdot}| = |4.25 - 3.22|$$
$$= 1.03 > SNK_4 = 1.02 \quad .$$

Hence the largest difference is significant. Thus, we can proceed with the procedure:

$$|y_{(1)\cdot} - y_{(3)\cdot}| = |4.25 - 3.47|$$
$$= 0.78 < SNK_3 = 0.92 \quad ,$$
$$|y_{(2)\cdot} - y_{(4)\cdot}| = |3.98 - 3.22|$$
$$= 0.76 < SNK_3 = 0.92 \quad .$$

Here, the SNK test stops. Hence, the only significant difference is that between
treatment 1 (control group) and treatment 4 (A_1 and A_2). Hence, the treatments
(1,2,3), or (2,3,4) respectively, may be regarded as homogeneous.

SNK in SPSS

The procedure is started with /Ranges = snk

Note: SPSS computes the SNK statistic according to

$$SNK = \sqrt{\frac{MS_{Error}}{2}} \, Q_{\alpha,(i,df)} \sqrt{\frac{1}{n_i} + \frac{1}{n_j}} \quad ; \tag{4.118}$$

for $n_i = n_j = r$ this yields the expression (4.114).
The SPSS printout is of the following form:

```
Multiple Range Test
Student-Newman-Keuls Procedure
Ranges for the .050 level
      2.95   3.57   3.95
The ranges above are table ranges.

The value actually compared with
Mean(J)-Mean(I) is
      .4451 * Range * Sqrt(1/N(I) + 1/N(J))
```

(∗) Denotes pairs of groups significantly
     different at the .050 level

```
              G G G G
              r r r r
              p p p p
              4 3 2 1
Mean    Group
3.22    Grp 4
3.47    Grp 3
3.98    Grp 2
4.25    Grp 1 *
```

Homogeneous Subsets

```
Subset 1
Group  Grp 4  Grp 3  Grp 2
Mean    3.22   3.47   3.98
```

```
Subset 2
Group  Grp 3  Grp 2  Grp 1
Mean    3.47   3.98   4.25
```

Tukey's HSD Test   We compute the HSD (4.113) according to

$$
\begin{aligned}
HSD &= Q_{\alpha,(4,20)}\sqrt{MS_{Error}/6} \\
    &= 3.95 \cdot 0.2569 = 1.01 \quad .
\end{aligned}
$$

The differences of pairs $y_{i\cdot} - y_{j\cdot}$ $(i < j)$ are

$$
\begin{aligned}
y_{1\cdot} - y_{2\cdot} &= 4.25 - 3.98 = 0.27 \\
y_{1\cdot} - y_{3\cdot} &= 0.78 \\
y_{1\cdot} - y_{4\cdot} &= 1.03\; * \\
y_{2\cdot} - y_{3\cdot} &= 0.51 \\
y_{2\cdot} - y_{4\cdot} &= 0.76 \\
y_{3\cdot} - y_{4\cdot} &= 0.25 \quad,
\end{aligned}
$$

hence only $|y_{1\cdot} - y_{4\cdot}| > HSD$ holds.
SPSS call and printout:

```
/Ranges = tukey
```

```
Tukey-HSD Procedure
Ranges for the .050 level
      3.95  3.95  3.95

                  G G G G
                  r r r r
                  p p p p
                  4 3 2 1
Mean    Group
3.22    Grp 4
3.47    Grp 3
3.98    Grp 2
4.25    Grp 1 *
```

Fisher's Protected LSD

The FPLSD (4.112) at the 5% level is

$$t_{20;0.975}\sqrt{0.3962 \cdot \frac{2}{6}} = 2.09 \cdot 0.3634 = 0.76 \quad .$$

With the differences of means calculated above, we receive

```
                  G G G G
                  r r r r
                  p p p p
                  4 3 2 1
Mean    Group
3.22    Grp 4
3.47    Grp 3
3.98    Grp 2 *
4.25    Grp 1 * *
```

The means μ_1 and μ_4 and μ_1 and μ_3 as well as the means μ_2 and μ_4 are significantly different according to this test.

4.5   Regression Analysis of Variance

For the description of the dependence of a variable Y on another (fixed) variable X by a regression model of the form

$$Y = \alpha + \beta X + \epsilon$$

we need pairs of observations (x_i, y_i), $i = 1, \ldots, n$, i.e., for every x-value a y-value is observed.

Consider the following experimental design. For *every* x-value *several* observations of Y are realized:

$$x_i, y_{i1}, \ldots, y_{in_i} .$$

This corresponds to the idea, that a population of y-values belongs to a fixed x-value. The question of interest is whether a dependence exists between the y-samples, represented by their means $y_{i\cdot}$, and the factor X. First, we test whether the populations Y_i have equal means (analysis of variance - multiple comparison of means).
If this hypothesis is rejected, i.e., if we have reason for assuming a functional relationship $Y = f(X)$, then the most simple model, a linear function, is fitted to the means $y_{i\cdot}$:

$$y_{i\cdot} = \alpha + \beta x_i + \epsilon_i \quad (i = 1, \ldots, s) \quad . \tag{4.119}$$

The estimates of α and β are determined, under consideration of the sample sizes n_i, according to the method of weighted least squares, i.e.,

$$\sum_{i=1}^{s} n_i(y_{i\cdot} - \alpha - \beta x_i)^2 \tag{4.120}$$

is minimized with respect to α and β. Let $n = \sum n_i$ be the sum of all observations. The *weighted least squares estimates* are then of the following form

$$\hat{\beta} = \frac{\sum n_i x_i y_{i\cdot} - \frac{1}{n} \sum n_i x_i \sum n_i y_{i\cdot}}{\sum n_i x_i^2 - \frac{1}{n} \left[\sum n_i x_i \right]^2} \tag{4.121}$$

$$\hat{\alpha} = y_{\cdot\cdot} - b\bar{x} \quad , \tag{4.122}$$

where $y_{i\cdot} = \frac{1}{n_i} \sum_j y_{ij}$ is the ith sample mean and $y_{\cdot\cdot} = \frac{1}{n} \sum_i \sum_j y_{ij}$ is the overall mean of all y-values. We receive the estimated means according to

$$\hat{y}_{i\cdot} = \hat{\alpha} + \hat{\beta} x_i \quad . \tag{4.123}$$

We partition the sum of squares SS_A as follows:

$$SS_A = \sum_{i=1}^{s} n_i(y_{i\cdot} - y_{\cdot\cdot})^2 \tag{4.124}$$

$$= \sum_{i=1}^{s} n_i(\hat{y}_{i\cdot} - y_{\cdot\cdot})^2 + \sum_{i=1}^{s} n_i(y_{i\cdot} - \hat{y}_{i\cdot})^2$$

$$= SS_{Model} + SS_{Deviation} \quad .$$

For the degrees of freedom we have

$$df_A = df_M + df_{Deviation} \quad , \tag{4.125}$$

i.e.,

$$(s - 1) = 1 + s - 2 \quad . \tag{4.126}$$

If not only $K = 2$ parameters are to be estimated, but K parameters in general, then

$$df_A = s - 1 \quad , \; df_M = K - 1 \quad , \; df_{Deviation} = s - K \quad . \tag{4.127}$$

The complete *table of the regression analysis of variance* is shown in Table 4.14. As test value for the fit of the model we compute

Source of variation	SS	df	$MS = SS/df$	Test value
Model	SS_M	$K - 1$	MS_M	
				$F = \frac{MS_{Model}}{MS_{Dev}}$
Model deviation	SS_{Dev}	$s - K$	MS_{Dev}	
Between the y-groups	SS_A	$s - 1$	MS_A	$F = \frac{MS_A}{MS_{Error}}$
Within the y-groups	SS_{Error}	$n - s$	MS_{Error}	
Total	SS_{Total}	$n - 1$		

Table 4.14: Table of the regression analysis of variance

$$F = \frac{MS_{Model}}{MS_{Deviation}} . \tag{4.128}$$

If $F > F_{s-1,n-s;1-\alpha}$ the fit of the model is significant at the α level.

Example 4.7: In a study the rate of abrasion of silanized plastic material PMMA was determined, for various levels of the proportion of quartz (Table 4.15).

The null hypothesis H_0 : *All means are equal (i.e., the proportion of quartz has no effect on the rate of abrasion)* is rejected, since the analysis of variance yields the test value

$$F = \frac{MS_A}{MS_{Error}} = 55.80 > 2.74 = F_{3,33;0.95} \tag{4.129}$$

(Table 4.16).

Hence, we fit a linear regression (4.123) to the means y_i. of the $s = 4$ samples. The parameters are computed according to (4.121) and (4.122):

$$\hat{y}_{i\cdot} = 0.0923 - 0.0020 \; x_i \quad (i = 1, \dots, 4) .$$

x [in Vol% quartz]			
$x_1 = 2.2$	$x_2 = 4.5$	$x_3 = 9.3$	$x_4 = 25.6$
0.1420	0.0964	0.0471	0.0451
0.1113	0.0680	0.0585	0.0311
0.1092	0.0964	0.0544	0.0458
0.1298	0.0764	0.0444	0.0534
0.0962	0.0749	0.0575	0.0488
0.0917	0.0813	0.0406	0.0508
0.0800	0.0813	0.0522	0.0440
0.0996	0.0813	0.0525	0.0549
0.1123		0.0570	0.0539
		0.0559	0.0526
$y_{1.} = 0.1080$	$y_{2.} = 0.0820$	$y_{3.} = 0.0520$	$y_{4.} = 0.0480$
$n_1 = 9$	$n_2 = 8$	$n_3 = 10$	$n_4 = 10$
$y_{..} = 0.0710$		$n = 37$	
$\hat{y}_{1.} = 0.0878$	$\hat{y}_{2.} = 0.0831$	$\hat{y}_{3.} = 0.0733$	$\hat{y}_{4.} = 0.0400$

Table 4.15: Data of the rate of abrasion

SS	df	MS	Test value
$SS_M = 0.01340$	1	$MS_M = 0.01340$	$F = 3.02$
$SS_{Dev.} = 0.00886$	2	$MS_{Dev.} = 0.00443$	
$SS_A = 0.02226$	3	$MS_A = 0.00742$	$F = 55.80$
$SS_E = 0.00440$	33	$MS_E = 0.00013$	
$SS_T = 0.02667$	36		

Table 4.16: Table of the regression analysis of variance of the rate of abrasion

These estimated values are shown in Table 4.15. We can now calculate the
partition (4.124) of SS_A (Table 4.16), the test value is

$$F = \frac{MS_{Model}}{MS_{Dev.}} = 3.02 < 18.51 = F_{1,2;0.95} \quad .$$

Hence, the null hypothesis H_0: $\beta = 0$ can not be rejected.

4.6   One-Factorial Models with Random Effects

So far, we discussed models with fixed effects in this chapter. In the introduction
however, we already referred to the difference to models with random effects.

Models with fixed effects for the analysis of treatment effects are the standard
in designed experiments. Models with random effects however, occur in sample
surveys where the grouping categories are random effects.

Examples: Quality control

(i) Fixed effects: The dayly production of five particular machines from an assembly line.

(ii) Random effects: The dayly production of five machines chosen at random, that represent the machines as a class.

The model with random effects is of the same structure as the model (4.2) with fixed effects:

$$y_{ij} = \mu + \alpha_i + \epsilon_{ij} \quad . \tag{4.130}$$

$$(i = 1, \ldots, s; \ j = 1, \ldots, n_i)$$

The meaning of the parameter α_i however has now changed. The α_i are now the random effects of the ith treatment (ith machine). Hence, the α_i are the random variables whose distributions we have to specify. We assume

$$\mathrm{E}(\alpha_i) = 0 \ , \ \mathrm{Var}(\alpha_i) = \sigma_\alpha^2 \tag{4.131}$$

and

$$\mathrm{E}(\epsilon_{ij}\alpha_i) = 0 \ , \ \mathrm{E}(\alpha_i\alpha_j) = 0 \ (i \neq j) \quad . \tag{4.132}$$

Then

$$y_{ij} \sim (\mu, \sigma_\alpha^2 + \sigma^2) \tag{4.133}$$

holds.

In the model with fixed effects the treatment effect A was represented by the parameter estimates $\hat{\alpha}_i$, or $\hat{\mu}_i = \hat{\mu} + \hat{\alpha}_i$, respectively. In the model with random effects a treatment effect can be expressed by the so-called variance components. The variance σ_α^2 is estimated as a component of the entire variance. The absolute or relative size of this component then makes conclusions about the treatment effect possible.

The estimation of the variances σ_α^2 and σ^2 requires no assumptions about the distribution. For the test procedure and the computation of confidence intervals however, we assume the normal distribution, i.e.,

$$\epsilon_{ij} \sim N(0, \sigma^2) \, , \epsilon_{ij} \ \text{independent}$$
$$\alpha_i \sim N(0, \sigma_\alpha^2) \, , \alpha_i \ \text{independent}$$

and hence

$$y_{ij} \sim N(\mu, \sigma_\alpha^2 + \sigma^2) \, . \tag{4.134}$$

Unlike the model with fixed effects, the response values y_{ij} of a level i of the treatment (i.e., of the ith sample) are no longer uncorrelated:

$$\begin{aligned} \mathrm{E}(y_{ij} - \mu)(y_{ij'} - \mu) &= \mathrm{E}(\alpha_i + \epsilon_{ij})(\alpha_i + \epsilon_{ij'}) \\ &= \mathrm{E}(\alpha_i^2) = \sigma_\alpha^2 \quad . \end{aligned} \tag{4.135}$$

On the other hand, the response values of different samples are still uncorrelated $(i \neq i'$, for any $j, j')$:

$$E(y_{ij} - \mu)(y_{i'j'} - \mu) = E(\alpha_i \alpha_{i'}) + E(\epsilon_{ij}\epsilon_{i'j'}) + E(\alpha_i \epsilon_{i'j'}) + E(\alpha_{i'}\epsilon_{ij}) = 0 \quad . \quad (4.136)$$

In case of a normal distribution, uncorrelated can be replaced by independent.

Test of the null hypothesis $H_0 : \sigma_\alpha^2 = 0$ against $H_1 : \sigma_\alpha^2 > 0$
The hypothesis H_0: "no treatment effect" for the two models is:

– fixed effects $H_0 : \alpha_i = 0 \quad \forall i$
– random effects $H_0 : \sigma_\alpha^2 = 0$.

With the results of Section 4.2.3, which we can partly adopt, we have for the model with random effects:

$$E(MS_{Error}) = \sigma^2 ,$$

i.e., $MS_{Error} = \hat{\sigma}^2$ is an unbiased estimate of σ^2. We compute $E(MS_A)$ as follows:

$$
\begin{aligned}
SS_A &= \sum_{i=1}^{s}\sum_{j=1}^{n_i}(y_{i.} - y_{..})^2 , \\
y_{i.} &= \mu + \alpha_i + \epsilon_{i.} , \\
y_{..} &= \mu + \alpha_. + \epsilon_{..} , \\
\alpha_. &= \sum n_i \alpha_i / n , \\
(y_{i.} - y_{..}) &= (\alpha_i - \alpha_.) + (\epsilon_{i.} - \epsilon_{..}) .
\end{aligned}
$$

With (4.131) and (4.132) we have

$$
\begin{aligned}
E(y_{i.} - y_{..})^2 &= E(\alpha_i - \alpha_.)^2 + E(\epsilon_{i.} - \epsilon_{..})^2 , & (4.137) \\
E(\alpha_i - \alpha_.)^2 &= E(\alpha_i^2) + E(\alpha_.^2) - 2E(\alpha_i \alpha_.) \\
&= \sigma_\alpha^2 \left[1 + \frac{\sum n_i^2}{n^2} - 2\frac{n_i}{n} \right] , & (4.138) \\
E(\epsilon_{i.}^2 - \epsilon_{..})^2 &= E(\epsilon_{i.}^2) + E(\epsilon_{..}^2) - 2E(\epsilon_{i.}\epsilon_{..}) \\
&= \frac{\sigma^2}{n_i} + \frac{\sigma^2}{n} - 2\frac{\sigma^2}{n} \\
&= \sigma^2 \left(\frac{1}{n_i} - \frac{1}{n} \right) . & (4.139)
\end{aligned}
$$

Hence

$$
\begin{aligned}
\sum_{j=1}^{n_i} E(y_{i.} - y_{..})^2 &= n_i E(y_{i.} - y_{..})^2 \\
&= \sigma_\alpha^2 \left[n_i + \frac{n_i \sum n_i^2}{n \quad n} - 2\frac{n_i^2}{n} \right] + \sigma^2 (1 - \frac{n_i}{n})
\end{aligned}
$$

and

$$\sum_{i=1}^{s} n_i E(y_{i\cdot} - y_{\cdot\cdot})^2 = \sigma_\alpha^2 \left[n - \frac{\sum n_i^2}{n} \right] + \sigma^2 (s-1) .$$

We receive

(i) in the unbalanced case

$$E(MS_A) = \frac{1}{s-1} E(SS_A) = \sigma^2 + k\sigma_\alpha^2 \qquad (4.140)$$

with

$$k = \frac{1}{s-1}(n - \frac{1}{n}\sum n_i^2) \qquad (4.141)$$

(ii) in the balanced case ($n_i = r$ for all i, $n = r \cdot s$)

$$k = \frac{1}{s-1}(r \cdot s - \frac{1}{r \cdot s} s \cdot r^2) = r \quad , \qquad (4.142)$$

$$E(MS_A) = \sigma^2 + r\sigma_\alpha^2 \quad . \qquad (4.143)$$

This yields the unbiased estimate $\hat{\sigma}_\alpha^2$ of σ_α^2

(i) in the unbalanced case

$$\hat{\sigma}_\alpha^2 = \frac{MS_A - MS_{Error}}{k} \quad , \qquad (4.144)$$

(ii) in the balanced case

$$\hat{\sigma}_\alpha^2 = \frac{MS_A - MS_{Error}}{r} \quad . \qquad (4.145)$$

In case of an assumed normal distribution we have

$$MS_{Error} \sim \sigma^2 \chi_{n-s}^2$$

and

$$MS_A \sim (\sigma^2 + k\sigma_\alpha^2)\chi_{s-1}^2 \quad .$$

The two distributions are independent, hence the ratio

$$\frac{MS_A}{MS_{Error}} \cdot \frac{\sigma^2}{\sigma^2 + k\sigma_\alpha^2}$$

has a central F-distribution under the assumption of equal variances, i.e., under $H_0 : \sigma_\alpha^2 = 0$. Under $H_0 : \sigma_\alpha^2 = 0$ we thus have

$$\frac{MS_A}{MS_{Error}} \sim F_{s-1,n-s} \quad . \qquad (4.146)$$

| Source | SS | df | E(MS) Effects | |
			Fixed	Random
Treatment	SS_A	$s-1$	$\sigma^2 + \frac{\sum n_i \alpha_i^2}{s-1}$	$\sigma^2 + k\sigma_\alpha^2$
Error	SS_{Error}	$n-s$	σ^2	σ^2

Table 4.17: Expectations of MS_A and MS_{Error}

Hence, $H_0 : \sigma_\alpha^2 = 0$ is tested with the same test statistic as $H_0 : \alpha_i = 0$ (all i) in the model with fixed effects. The table of the analysis of variance remains unchanged.

Example 4.8: (Continuation of Example 4.5)
We now regard the design from Table 4.6 as a model with random effects. The null hypothesis $H_0 : \sigma_\alpha^2 = 0$ is tested with the statistic from (4.146). Table 4.7 yields

$$F_{3,20} = \frac{1.3349}{0.3962} = 3.3687 \quad \text{(p-value : 0.0389)} \quad,$$

hence $H_0 : \sigma_\alpha^2 = 0$ is rejected. The estimated components of variance are

$$\hat{\sigma}^2 = MS_{Error} = 0.3962$$

and (cf. (4.145))

$$\hat{\sigma}_\alpha^2 = \frac{1.3349 - 0.3962}{6} = 0.1564 \quad .$$

4.7 Rank Analysis of Variance in the Completely Randomized Design

4.7.1 Kruskal-Wallis Test

The previous models were designed for the case that the response values follow a normal distribution. We now consider the situation that the response is either continuous but not normal or that we have a categorical response. For this data situation, which is often found in practice, we want to conduct the one-factorial comparison of groups. We first discuss the completely randomized design.

The response values are y_{ij} with the two subscripts $i = 1, \ldots, s$ (groups) and $j = 1, \ldots, n_i$ (subscript within the ith group). The data is collected according to the completely randomized design: n_1 units are chosen at random from $n = \sum n_i$ units and are assigned to the treatment (group) 1, etc.. The data structure is shown in Table 4.18.
To begin with, we choose the following linear additive model

$$y_{ij} = \mu_i + \epsilon_{ij} \tag{4.147}$$

Group

	1	2	$\cdots$	s
	y_{11}	y_{21}	$\cdots$	y_{s1}
	$\vdots$	$\vdots$		$\vdots$
	y_{1n_1}	y_{2n_2}	$\cdots$	y_{sn_s}

Table 4.18: Data matrix in the completely randomized design

and assume that

$$\epsilon_{ij} \sim F(0, \sigma^2) \tag{4.148}$$

holds (where F is any continuous distribution). Additionally, we assume that the observations are independent within and between the groups.

The major statistical task is the comparison of the group means μ_i according to

$$H_0: \mu_1 = \cdots = \mu_s \quad \text{against} \quad H_1: \mu_i \neq \mu_j \quad (\text{at least one pair } i, j, \ i \neq j).$$

The tests are based on the comparison of the rank sums of the groups, in analogy to the Wilcoxon test in the two-sample case. The ranking procedure assigns the rank 1 to the smallest value of all s groups, ..., the rank $n = \sum n_i$ to the largest value of all s groups. These ranks R_{ij} replace the original values y_{ij} of the response table 4.18 according to Table 4.19.

Group

	1	2	$\cdots$	s	
	R_{11}	R_{21}		R_{s1}	
	$\vdots$	$\vdots$		$\vdots$	
	R_{1n_1}	R_{2n_2}		R_{sn_s}	
$\sum$	$R_{1\cdot}$	$R_{2\cdot}$	$\cdots$	$R_{s\cdot}$	$R_{\cdot\cdot}$
Mean	$r_{1\cdot}$	$r_{2\cdot}$	$\cdots$	$r_{s\cdot}$	$r_{\cdot\cdot}$

Table 4.19: Rank values for Table 4.18

The rank sums and rank means are

$$R_{i\cdot} = \sum_{j=1}^{n_i} R_{ij}, \quad R_{\cdot\cdot} = \sum_{i=1}^{s} R_{i\cdot} = \frac{n(n+1)}{2}$$

$$r_{i\cdot} = \frac{R_{i\cdot}}{n_i}, \quad r_{\cdot\cdot} = \frac{R_{\cdot\cdot}}{n} = \frac{n+1}{2} .$$

Under the null hypothesis all $n!/n_1! \cdots n_s!$ possible arrangements of the ranks have equal possibility. Hence, for each of these arrangements we can compute a measure for the difference between the groups. One possible measure for the group difference is based on the comparison of the rank means $r_{i\cdot}$.

In analogy to the error sum of squares $SS_A = \sum_{i=1}^{s} n_i(y_{i\cdot} - y_{\cdot\cdot})^2$ (cf.(4.28)) Kruskal and Wallis constructed the following test statistic (Kruskal and Wallis, 1952):

$$H = \frac{12}{n(n+1)} \sum_{i=1}^{s} n_i(r_{i\cdot} - r_{\cdot\cdot})^2$$

$$= \frac{12}{n(n+1)} \sum_{i=1}^{s} \frac{R_{i\cdot}^2}{n_i} - 3(n+1) \quad . \tag{4.149}$$

The test statistic H is a measure for the variance of the sample rank means. For the case of $n_i \leq 5$, tables exist for the exact critical values (cf. e.g., Hollander and Wolfe, 1973, p.294). For $n_i > 5$ $(i = 1, \ldots, s)$, H is approximatively χ_{s-1}^2-distributed.

Correction in Case of Ties

If equal response values y_{ij} arise and mean ranks are assigned, then the following corrected test statistic is used:

$$H_{Corr} = H \left(1 - \frac{\sum_{k=1}^{r}(t_k^3 - t_k)}{n^3 - n}\right)^{-1} \quad . \tag{4.150}$$

Here r is the number of groups with equal ranks and t_k is the number of equal response values within a group. If $H > \chi_{s-1;1-\alpha}^2$, the hypothesis $_0$: $\mu_1 = \cdots = \mu_s$ is rejected in favour of H_1. If H_{Corr} has to be used, the corrected value does not have to be calculated in case of significance of H, due to $H_{Corr} > H$.

Example 4.9: We now compare the manufacturing times from Table 4.1 according to the Kruskal-Wallis test. (Hint: In Example 4.1 the analysis of variance was done with the logarithms of the response values, since a normal distribution of the original values was doubtful. The null hypothesis was not rejected, cf. Table 4.5). The test statistic based on Table 4.20 is

$$H = \frac{12}{33 \cdot 34}[\frac{275.5^2}{14} + \frac{196.0^2}{11} + \frac{89.5^2}{8}] - 3 \cdot 34$$

$$= 4.04 < 5.99 = \chi_{2;0.95}^2 \quad .$$

Since H is not significant we have to compute H_{Corr}. Table 4.20 yields

$r = 4$, $t_1 = 3$ (3 ranks of 3)
         $t_2 = 2$ (2 ranks of 8.5)
         $t_3 = 2$ (2 ranks of 23.5)
         $t_4 = 2$ (2 ranks of 27.5)

Correction term: $1 - \frac{3 \cdot (2^3 - 2) + (3^3 - 3)}{33^3 - 33} = 1 - \frac{42}{35904} = 0.9988$,

$H_{Corr} = 4.045$.

The decision is: the null hypothesis H_0: $\mu_1 = \mu_2 = \mu_3$ is not rejected, the effect "dentist" can not be proven.

Dentist A		Dentist B		Dentist C	
Manufacturing time	Rank	Manufacturing time	Rank	Manufacturing time	Rank
31.5	3	33.5	5	19.5	1
38.5	7	37.0	6	31.5	3
40.0	8.5	43.5	10	31.5	3
45.5	11	54.0	15	40	8.5
48.0	12	56.0	17	50.5	13
55.5	16	57.0	18	53.0	14
57.5	19	59.5	21	62.5	23.5
59.0	20	60.0	22	62.5	23.5
70.0	27.5	65.5	25		
70.0	27.5	67.0	26		
72.0	29	75.0	31		
74.5	30				
78.0	32				
80.0	33				
$n_1 = 14$		$n_2 = 11$		$n_3 = 8$	
$R_{1.} = 275.5$		$R_{2.} = 196.0$		$R_{3.} = 89.5$	
$r_{1.} = 19.68$		$r_{2.} = 17.82$		$r_{3.} = 11.19$	

Table 4.20: Computation of the ranks and rank sums for Table 4.1

4.7.2   Multiple Comparisons

In analogy to the argumentation in Section 4.4, we want to discuss the procedure in case of a rejection of the null hypothesis H_0: $\mu_1 = \cdots = \mu_s$ for ranked data.

Planned Single Comparisons

If we plan a comparison of two particular groups before the data is collected, then the Wilcoxon rank sum test is the appropriate test procedure (cf. Section 2.5). The type I error however, only holds for this particular comparison.

Comparison of All Pairwise Differences

The procedure for comparing all $s(s - 1)/2$ possible pairs (i, j) of differences with $i > j$ dates back to Dunn (1964). It is based on the Bonferroni method and assumes large sample sizes. The following statistics are computed from the differences $r_{i.} - r_{j.}$ of the rank means $(i \neq j, i > j)$

$$z_{ij} = \frac{r_{i.} - r_{j.}}{\sqrt{\frac{n(n+1)}{12}(1/n_i + 1/n_j)}} \ . \tag{4.151}$$

Let $u_{1-\alpha/s(s-1)}$ be the $[1 - \alpha/s(s - 1)]$-quantile of the N(0,1)-distribution. The multiple testing rule that ensures the α-level overall for all $s(s - 1)$ pairwise comparisons is:

$$H_0\colon \mu_i = \mu_j \quad \text{for all } (i,j)\ i > j \tag{4.152}$$

is rejected in favour of

$$H_1\colon \mu_i \neq \mu_j \quad \text{for at least one pair } (i,j),$$

if

$$|z_{ij}| > z_{1-\alpha/s(s-1)} \quad \text{for at least one pair } (i,j), i > j \quad . \tag{4.153}$$

Example 4.10: Table 4.6 shows the response values of the four treatments (i.e., control group, A_1, A_2, $A_1 \cup A_2$) in the balanced randomized design. The analysis of variance, under the assumption of a normal distribution, rejected the null hypothesis $H_0\colon \mu_1 = \cdots = \mu_4$. In the following, we conduct the analysis based on ranked data, i.e., we no longer assume a normal distribution. From Table 4.6 we compute the rank table 4.21 and receive the Kruskal-Wallis statistic

$$\begin{aligned}
H &= \frac{12}{24 \cdot 25 \cdot 6} \sum R_{i\cdot}^2 - 3 \cdot 25 \\
&= \frac{1}{300} \sum (104^2 + 91^2 + 60^2 + 45^2) - 75 \\
&= 7.41 \quad .
\end{aligned}$$

H_0 is not rejected on the 5%-level, due to $7.41 < 7.81 = \chi^2_{3;0.95}$. Hence, the non-parametric analysis stops.

For the demonstration of non-parametric multiple comparisons we now change to the 10%-level. This yields $H = 7.41 > 6.25 = \chi^2_{3;0.90}$. Since H already is significant, H_{Corr} does not have to be calculated. Hence, $H_0\colon \mu_1 = \cdots = \mu_4$ can be rejected on the 10 %-level.

We can now conduct the multiple comparisons of the pairwise differences. The denominator of the test statistic z_{ij} (4.151) is $\sqrt{\frac{24 \cdot 25}{12} \frac{2}{6}} = \sqrt{\frac{50}{3}} = 4.08$.

Comparison	$r_{i\cdot} - r_{j\cdot}$	z_{ij}	
1/2	2.16	0.53	
1/3	7.33	1.80	
1/4	10.83	2.65	*
2/3	5.17	1.27	
2/4	8.67	2.13	
3/4	3.50	0.86	

For $\alpha = 0.10$ we receive $\alpha/s(s-1) = 0.10/12 = 0.0083$, $1 - \alpha/s(s-1) = 0.9916$, $u_{0.9916} = 2.39$. Hence, the comparison 1/4 is significant.

Comparison Control Group – All Other Treatments

If one treatment out of the s treatments is chosen as control group and compared to the other $s - 1$ treatments, then the test procedure is the same, but with the $[u_{1-\alpha/2(s-1)}]$-quantile.

Example 4.11: (Continuation of Example 4.10)
The control group is treatment 1 (no additives). The comparison with the treatments 2 (A_1), 3 (A_2) and 4 ($A_1 \cup A_2$) is done with the test statistics z_{12}, z_{13}, z_{14}. Here we have to use the $u_{1-\alpha/2(s-1)}$-quantile. We receive $1 - 0.10/6 = 0.9833$, $u_{1-0.10/6} = 2.126 \Rightarrow$ the comparisons 1/4 and 2/4 are significant.

Control group		A_1		A_2		$A_1 \cup A_2$	
Value	Rank	Value	Rank	Value	Rank	Value	Rank
4.5	21.5	3.8	12	3.5	8	3.0	4
5.0	24	4.0	16.5	4.5	21.5	2.8	3
3.5	8	3.9	13.5	3.2	5	2.2	2
3.7	11	4.2	19	2.1	1	3.4	6
4.8	23	3.6	10	3.5	8	4.0	16.5
4.0	16.5	4.4	20	4.0	16.5	3.9	13.5
$R_1. = 104$		$R_2. = 91$		$R_3. = 60$		$R_4. = 45$	
$r_1. = 17.33$		$r_2. = 15.17$		$r_3. = 10.00$		$r_4. = 7.50$	

Table 4.21: Rank table for Table 4.6

4.8   Exercises and Questions

4.8.1 Formulate the one-factorial design with $s = 2$ fixed effects for the balanced case as a linear model in the usual coding and in effect coding.

4.8.2 What does the table of the analysis of variance look like in a two-factorial design with fixed effects?

4.8.3 What meaning does the Theorem of Cochran have? Which effects can be tested with it?

4.8.4 In a field experiment three fertilizers are to be tested. The table of the analysis of variance is

	df	MS	F
$SS_A = 50$			
$SS_{Error} =$			
$SS_{Total} = 350$	32		

Name the hypothesis to be tested and the test decision.

4.8.5 Let $c'y.$ be a linear contrast of the means $y_{1.}, \ldots, y_{s.}$. Complete the following:

$$c'y. \sim N(\ ,\ ).$$

The test statistic for testing H_0: $c'\mu = 0$ is

$$? \sim \chi^2_{df}, \quad df = ?$$

4.8.6 How many independent linear contrasts exist for s means? What is a complete system of linear contrasts? Is this system unique?

4.8.7 Let $c'_1 Y., \ldots, c'_{s-1} Y.$ be a complete system of linear contrasts of the total response values $Y. = (Y_{1.}, \ldots, Y_{s.})'$.

Assume that each contrast has the distribution

$$c'_i Y. \sim N(\ ,\ ).$$

Then

$$\frac{(c'_i Y)^2}{?} \sim ?$$

and, if the contrasts are ..., then

$$SS_A =$$

holds.

4.8.8 Let A_1 be a control group and assume that A_2 and A_3 are two medicaments. Name the contrasts for the comparison of
A_1 against A_2 or A_3
A_2 against A_1
A_3 against A_1 ?

4.8.9 Describe the main concern of multiple comparisons and the two methods of comparison.

4.8.10 Assign the experimentwise designed multiple comparisons correctly into the following matrix:

	Scheffé	Dunnett	Tukey	Bonferroni
(i)				
(ii)				
(iii)				
(iv)				

(i) $k \leq s$ comparisons planned in advance
(ii) set of any linear contrasts
(iii) $(s - 1)$ comparisons with a control group
(iv) all $s(s - 1)/2$ comparisons of means

4.8.11 In case of the two-sample t-test (balanced) the critical value is $t_{n-1;1-\alpha}$. In case of the Bonferroni procedure with 3 comparisons the critical value for each single comparison is t ; .

4.8.12 Name the assumptions in the model $y_{ij} = \mu + \alpha_i + \epsilon_{ij}$ with mixed effects. We have $y_{ij} \sim N(\, , \,)$.

Formulate the hypothesis H_0: no treatment effect!

4.8.13 Conduct the rank analysis of variance according to Kruskal-Wallis for the following table (Hint: completely randomized design)

Student A		Student B		Student C	
Points	Rank	Points	Rank	Points	Rank
32		34		38	
39		37		40	
45		42		43	
47		54		48	
53		60		52	
59		75		61	
71				80	
85				95	

5 More Restrictive Designs

5.1 Randomized Block Design

In statistical practice, the experimental units are often not completely homogeneous. Usually, a grouping according to a stratification factor can be observed (clinical population: stratified according to patient's age, degree of disease, etc.). If we have such prior information then a gain in efficiency compared to the completely randomized experiment is possible by grouping into blocks. The experimental units are grouped together in homogeneous groups (blocks) and the treatments are assigned to the experimental units within each block by random. Hence the block effect (differences between the blocks) can now be separated from the experimental error. This leads to a higher precision. The strategy of building blocks should yield a variability within each block that is as small as possible and a variability between blocks that is as high as possible.

The most widely used block design is the randomized block design (RBD). Here s treatments with r repetitions each (i.e., balanced) are assigned to a total of $n = r \cdot s$ experimental units. First, the experimental units are divided into r blocks with s units each in such a way, that the units within each block are as homogeneous as possible. The s treatments are then assigned to the s units at random, so that each treatment occurs only once per block.

Example 5.1: We want to test $s = 3$ treatments A, B, C with $r = 4$ repetitions each in the randomized block design with respect to their effect. Assume the blocking factor to be ordinal scaled (e.g., $r = 4$ levels of intensity of a disease or $r = 4$ age groups).

The block design of the $n = r \cdot s = 12$ experimental units is then of the structure displayed in Table 5.1. The assignment of the $s = 3$ treatments per

Block

I	II	III	IV		I	II	III	IV
1	1	1	1	$\longrightarrow$	A	B	C	B
2	2	2	2	Randomization	B	A	A	C
3	3	3	3		C	C	B	A

Table 5.1: Randomized assignment of treatments per block

block to the 3 units of the $r = 4$ blocks can be done via random numbers. Ranks 1,2, or 3 are assigned to these random numbers and the assignment to the treatments is then done according to a previously specified coding (rank 1: treatment A, rank 2: B, rank 3: C).

Example: Block II in Table 5.1:

Unit	Random number	Rank	Treatment
1	182	2	B
2	037	1	A
3	217	3	C

The structure of the data is shown in Table 5.2, with

Sums
$Y_{i.} = \sum_j y_{ij}$
$Y_{.j} = \sum_i y_{ij}$
$Y_{..} = \sum_i Y_{i.} = \sum_j Y_{.j}$

Means
$y_{i.} = Y_{i.}/s$ Block i
$y_{.j} = Y_{.j}/r$ Treatment j
$y_{..} = Y_{..}/rs$ Total

			Treatment j			
Block i	1	2	$\cdots$	s	Sum	Mean
1	y_{11}	y_{12}	$\cdots$	y_{1s}	$Y_{1.}$	$y_{1.}$
2	y_{21}	y_{22}	$\cdots$	y_{2s}	$Y_{2.}$	$y_{2.}$
$\vdots$	$\vdots$	$\vdots$		$\vdots$	$\vdots$	$\vdots$
r	y_{r1}	y_{r2}	$\cdots$	y_{rs}	$Y_{r.}$	$y_{r.}$
Sum	$Y_{.1}$	$Y_{.2}$	$\cdots$	$Y_{.s}$	$Y_{..}$	
Mean	$y_{.1}$	$y_{.2}$	$\cdots$	$y_{.s}$	$y_{..}$	

Table 5.2: Data table for the randomized block design

Source	SS	df	MS	F
Block	SS_{Block}	$r-1$	MS_{Block}	F_{Block}
Treatment	SS_{Treat}	$s-1$	MS_{Treat}	F_{Treat}
Error	SS_{Error}	$(r-1)(s-1)$	MS_{Error}	
Total	SS_{Total}	$sr-1$		

Table 5.3: Analysis of variance table for the randomized block design

The linear model for the randomized block design (without interaction) is

$$y_{ij} = \mu + \beta_i + \tau_j + \epsilon_{ij} \tag{5.1}$$

with

y_{ij} :   response of the jth treatment in the ith block
μ :   average response of all experimental units (overall mean)
β_i :   additive effect of the ith block
τ_j :   additive effect of the jth treatment
ϵ_{ij} :   random error of the experimental unit, that receives the
    jth treatment in the ith block.

The following assumptions are made:

(i) The blocks are used for error control, hence the β_i are random effects with

$$\beta_i \sim N(0, \sigma_\beta^2) \quad . \tag{5.2}$$

(ii) Assume the treatments to be fixed factors. The τ_j are then fixed effects that represent the deviation from the overall mean μ. Hence the following constraint holds

$$\sum_{j=1}^{s} \tau_j = 0 \quad . \tag{5.3}$$

Remark: If however the treatment effects are to be regarded as random effects, then we assume

$$\tau_j \sim N(0, \sigma_\tau^2) \tag{5.4}$$

and

$$E(\beta_i \tau_j) = 0 \quad \text{(for all } i, j) \tag{5.5}$$

instead of (5.3).

(iii) The ϵ_{ij} are the random errors. Assume

$$\epsilon_{ij} \stackrel{i.i.d.}{\sim} N(0, \sigma^2) \tag{5.6}$$

and

$$E(\epsilon_{ij} \beta_i) = 0 \tag{5.7}$$

as well as

$$E(\epsilon_{ij} \tau_j) = 0 \quad . \tag{5.8}$$

Then

$$\mu_i = \mu + \beta_i \quad \text{is the mean of the } i\text{th block}$$

and

$$\mu_j = \mu + \tau_j \quad \text{is the mean of the } j\text{th treatment.}$$

Decomposition of the Error Sum of Squares

Using the identity

$$y_{ij} - y_{..} = (y_{ij} - y_{i.} - y_{.j} + y_{..}) + (y_{i.} - y_{..}) + (y_{.j} - y_{..}) \quad, \tag{5.9}$$

it can be shown that the following decomposition holds

$$\sum_i \sum_j (y_{ij} - y_{..})^2 \;=\; \sum_i \sum_j (y_{ij} - y_{i.} - y_{.j} + y_{..})^2$$

$$+ \sum_{i=1}^r s(y_{i.} - y_{..})^2$$

$$+ \sum_{j=1}^s r(y_{.j} - y_{..})^2 \quad. \tag{5.10}$$

If the correction term is computed by

$$C = Y_{..}^2/rs \quad, \tag{5.11}$$

then the above sums of squares can be expressed as

$$SS_{Total} \;=\; \sum_i \sum_j (y_{ij} - y_{..})^2 = \sum_i \sum_j y_{ij}^2 - C \tag{5.12}$$

$$SS_{Block} \;=\; s\sum_i (y_{i.} - y_{..})^2 = \frac{1}{s}\sum_i Y_{i.}^2 - C \tag{5.13}$$

$$SS_{Treat} \;=\; r\sum_j (y_{.j} - y_{..})^2 = \frac{1}{r}\sum_j Y_{.j}^2 - C \tag{5.14}$$

$$SS_{Error} \;=\; SS_{Total} - SS_{Block} - SS_{Treat} \quad. \tag{5.15}$$

The F ratios (cf. Table 5.3) are

$$F_{Block} \;=\; \frac{SS_{Block}}{SS_{Error}} \cdot \frac{(r-1)(s-1)}{(r-1)}$$

$$=\; \frac{MS_{Block}}{MS_{Error}} \tag{5.16}$$

and

$$F_{Treat} \;=\; \frac{SS_{Treat}}{SS_{Error}} \cdot \frac{(s-1)(r-1)}{(s-1)}$$

$$=\; \frac{MS_{Treat}}{MS_{Error}} \quad. \tag{5.17}$$

The significance of the treatment effect, i.e., $H_0 : \tau_j = 0 \ (j = 1, \ldots, s)$ for fixed effects and $H_0 : \sigma_\tau^2 = 0$ for random effects, is tested with F_{Treat}.

Testing for Block Effects

Consider the completely randomized design of the model (4.2) for the balanced case ($n_i = r$ for all i) and exchange the rows and columns (i.e., the meaning of i and j) in Table 4.2. If we additionally assume $\alpha_i = \tau_j$, then the following model corresponds with the completely randomized design

$$y_{ij} = \mu + \tau_j + \epsilon_{ij} \tag{5.18}$$

with the constraint $\sum \tau_j = 0$. The subscript $i = 1, \ldots, r$ represents the repetitions of the jth treatment ($j = 1, \ldots, s$). Hence the completely randomized design (5.18) is a nested submodel of the randomized block design (5.1). Testing for significance of the block effect is therefore equivalent to model choice between the complete model (here (5.1)) and a submodel restricted by constraints (H_0 : $\beta_i = 0$).

The appropriate test statistic for this problem was already derived in Section 3.8.2 with F_{Change} (cf. (3.237)). F_{Change} is of the following form:

$$\frac{\text{Error variance (small model)} - \text{Error variance (large model)}}{\text{Error variance (large model)}} \ . \tag{5.19}$$

Applied to our problem we receive for the "large" model (5.1) according to (5.15)

$$SS_{Error(\text{large})} = SS_{Total} - SS_{Block} - SS_{Treat} \ . \tag{5.20}$$

In the "small" model (5.18) we have

$$SS_{Error(\text{small})} = SS_{Total} - SS_{Treat} \ , \tag{5.21}$$

hence F_{Change} is now

$$\frac{SS_{Block}/(r-1)}{SS_{Error(\text{large})}/(r-1)(s-1)} = F_{Block} \ . \tag{5.22}$$

This statistic tests the significance of the transition from the smaller model (completely randomized design) to the larger model (randomized block design) and hence the significance of the block effects.

Estimates and Variances

The unbiased estimate of the jth treatment mean $\mu_j = \mu + \tau_j$ is given by

$$\hat{\mu}_j = \frac{Y_{\cdot j}}{r} = y_{\cdot j} \ . \tag{5.23}$$

The variance of this estimate is

$$\text{Var}(y_{\cdot j}) = \frac{1}{r^2} r \text{Var}(y_{ij}) = \frac{\sigma^2}{r} \quad \text{(for all } j\text{)}. \tag{5.24}$$

The unbiased estimate of the standard deviation of the estimates $y_{\cdot j}$ is then

$$s_{y_{\cdot j}} = \sqrt{MS_{Error}/r} \quad (j = 1, \ldots, s) \quad . \tag{5.25}$$

Hence, the $(1 - \alpha)$-confidence intervals of the jth treatment means are given by

$$y_{\cdot j} \pm t_{(s-1)(r-1),1-\alpha/2}\sqrt{MS_{Error}/r}. \tag{5.26}$$

For the simple comparison of two treatment means we receive an unbiased estimate of their difference by

$$y_{\cdot j_1} - y_{\cdot j_2}$$

with the standard deviation

$$s_{(y_{\cdot j_1} - y_{\cdot j_2})} = \sqrt{2MS_{Error}/r} \quad . \tag{5.27}$$

Hence the $(1 - \alpha)$-confidence intervals for the differences of means are of the form

$$(y_{\cdot j_1} - y_{\cdot j_2}) \pm t_{(s-1)(r-1),1-\alpha/2}\sqrt{2MS_{Error}/r} \quad . \tag{5.28}$$

Hint: Note the admissibility of simple comparisons.

Example 5.2: A physician wants to test the effect of three blood pressure lowering drugs (drug A, drug B, combination A and B) and of a placebo as control group. The 12 patients are assigned into three groups according to their weight. The "difference of the diastolic blood pressure from taking the drug at 6 o'clock am until 6 o'clock pm" is the measured response. The assignment to the treatments is done at random in each block.

The following table shows the measured values from which the table of variance is calculated.

Block	Placebo 1	A 2	B 3	A and B 4	$\sum$	$y_{i\cdot}$
1	5	7	4	12	28	7
2	7	8	6	15	36	9
3	9	9	8	18	44	11
$\sum$	21	24	18	45	108	
$y_{\cdot j}$	7	8	6	15	9	

Table 5.4: Blood pressure differences

We now receive

$$C = Y_{\cdot\cdot}^2/rs = 108^2/12 = 972$$
$$SS_{Total} = 5^2 + \cdots + 18^2 - C$$

$$
\begin{aligned}
&= 1158 - 972 = 186 \\
SS_{Block} &= \frac{1}{4}(28^2 + 36^2 + 44^2) - C \\
&= 1004 - 972 = 32 \\
SS_{Treat} &= \frac{1}{3}(21^2 + 24^2 + 18^2 + 45^2) - C \\
&= 1122 - 972 = 150 \\
SS_{Error} &= 186 - 32 - 150 = 4
\end{aligned}
$$

	SS	df	MS	F
Block	32	2	16	24.00
Treat	150	3	50	75.00
Error	4	6	0.67	
Total	186	11		

The testing of $H_0 : \tau_j = 0 \ (j = 1, \ldots, 4)$ (no treatment effect) with $F_{Treat} = F_{3,6} = 75.00$ leads to a rejection of H_0 $(F_{3,6;0.95} = 4.76)$, hence the treatment effect is significant. The test of the block effect yields significance with $F_{Block} = F_{2,6} = 24.00$ $(F_{2,6;0.95} = 5.14)$, hence the randomized block design is significant compared to the completely randomized design.

Consider the analysis of variance table in the completely randomized design with the same response values as in Table 5.4:

	SS	df	MS	F
Treat	150	3	50	11.11
Error	36	8	4.5	
Total	186	11		

Due to

$$
F = 11.11 > F_{3,8;0.95} = 4.07
$$

the treatment effect here is significant as well.

Treatment means				Standard error
1	2	3	4	$\sqrt{MS_{Error}/r}$
7	8	6	15	$\sqrt{0.67/3} = 0.47$
Confidence intervals				
7 ± 1.15	8 ± 1.15	6 ± 1.15	15 ± 1.15	

(Hint: $t_{6,0.975} = 2.45$, $2.45\sqrt{MS_{Error}/r} = 1.15$)
Confidence intervals for differences of means
(Hint: $t_{6,0.975} \sqrt{2MS_{Error}/r} = 1.63$)

<u>Treatments</u>

1/2	:	-1	$\pm$	1.63	$\Longrightarrow$	$[-2.63 , 0.63]$	
1/3	:	1	$\pm$	1.63	$\Longrightarrow$	$[-0.63 , 2.63]$	
1/4	:	-8	$\pm$	1.63	$\Longrightarrow$	$[-9.63 , -6.37]$	*
2/3	:	2	$\pm$	1.63	$\Longrightarrow$	$[0.37 , 3.63]$	*
2/4	:	7	$\pm$	1.63	$\Longrightarrow$	$[5.37 , 8.63]$	*.
3/4	:	9	$\pm$	1.63	$\Longrightarrow$	$[7.37 , 10.63]$	*

In the simple comparison of means the treatments 1 and 4, 2 and 3, 2 and 4 as well as 3 and 4 differ significantly. Which correct test result would we have received according to Scheffé (Question 5.4.3)?

Example 5.3: $n = 16$ students are tested for $s = 4$ training methods. The students are divided into $r = 4$ blocks according to their previous level of performance and the training methods are then assigned at random within each block. The response is measured as the level of performance on a scale of 1 to 100 points. The results are shown in Table 5.5.

Again, we calculate the sums of squares and test for treatment effect and block effect.

Block	\multicolumn{4}{c}{Training method}	$\sum$	Means
	1	2	3	4		
1	41	53	54	42	190	47.5
2	47	62	58	41	208	52.0
3	55	71	66	58	250	62.5
4	59	78	72	61	270	67.5
$\sum$	202	264	250	202	918	
Means	50.5	66.0	62.5	50.5	57.375	

Table 5.5: Points

$$C = \frac{(918)^2}{16} = 52670.25$$

$$SS_{Total} = 41^2 + \cdots + 61^2 - \frac{(918)^2}{16} = 54524.00 - 52670.25$$
$$= 1853.75$$

$$SS_{Block} = \frac{190^2 + \cdots + 270^2}{4} - \frac{(918)^2}{16} = 53691.00 - 52670.25$$
$$= 1020.75$$

$$SS_{Treat} = \frac{202^2 + \cdots + 202^2}{4} - \frac{(918)^2}{16} = 53451.00 - 52670.25$$
$$= 780.75$$

$$SS_{Error} = 1853.75 - 1020.75 - 780.75$$

$$= 52.25$$

	SS	df	MS	F	
Block	1020.75	3	340.25	58.61	*
Treat	780.75	3	260.25	44.83	*
Error	52.25	9	5.81		
Total	1853.75	15			

Both effects are significant:

$$F_{Treat} = F_{3,9} = 44.83 > 3.86 = F_{3,9;0.95},$$
$$F_{Block} = F_{3,9} = 58.61 > 3.86 = F_{3,9;0.95}.$$

5.2 Latin Squares

In the randomized block design we divided the experimental units into homogeneous blocks according to a blocking factor and hence eliminated the differences among the blocks from the experimental error, i.e., increased the part of the variability explained by a model.

We now consider the case that the experimental units can be grouped with respect to two factors, as in a contingency table. Hence two block effects can be removed from the experimental error. This design is called a Latin square.

If s treatments are to be compared, s^2 experimental units are required. These units are first classified into s blocks with s units each, based on one of the factors (row classification). The units are then classified into s groups with s units each, based on the other factor (column classification). The s treatments are then assigned to the units in such a way that each treatment occurs once, and only once, in each row and column.

Table 5.6 shows a Latin square for the $s = 4$ treatments A, B, C, D, which were assigned to the $n = 16$ experimental units by permutation.

A	B	C	D
B	C	D	A
C	D	A	B
D	A	B	C

Table 5.6: Latin square for $s = 4$ treatments

This arrangement can be varied by randomization, e.g., by first defining the order of the rows by random numbers. We replace the lexicographical order A, B, C, D of the treatments by the numerical order 1, 2, 3, 4.

Row	Random number	Rank
1	131	2
2	079	1
3	284	3
4	521	4

This yields the following row randomization

$$
\begin{array}{cccc}
B & C & D & A \\
A & B & C & D \\
C & D & A & B \\
D & A & B & C
\end{array}
$$

Assume the randomization by columns leads to

Column	Random number	Rank
1	003	1
2	762	4
3	319	3
4	199	2

The final arrangement of the treatments would then be:

$$
\begin{array}{cccc}
B & A & D & C \\
A & D & C & B \\
C & B & A & D \\
D & C & B & A
\end{array}
$$

If a time trend is present, then the Latin square can be applied to separate these effects.

Figure 5.1: Latin square for the elimination of a time trend

5.2.1 Analysis of Variance

The linear model of the Latin square (without interaction) is of the following form:

$$y_{ij(k)} = \mu + \rho_i + \gamma_j + \tau_{(k)} + \epsilon_{ij} \quad . \tag{5.29}$$

$$(i, j, k = 1, \ldots, s)$$

Here $y_{ij(k)}$ is the response of the experimental unit in the ith row and the jth column, subjected to the kth treatment. The parameters are

μ average response (overall mean)

ρ_i ith row effect

γ_j jth column effect

$\tau_{(k)}$ kth treatment effect

ϵ_{ij} experimental error.

We make the following assumptions:

$$\epsilon_{ij} \sim N(0, \sigma^2) \quad , \tag{5.30}$$

$$\rho_i \sim N(0, \sigma_\rho^2) \quad , \tag{5.31}$$

$$\gamma_j \sim N(0, \sigma_\gamma^2) \quad . \tag{5.32}$$

Additionally, we assume all random variables to be mutually independent. For the treatment effects we assume

$$\text{(i) fixed:} \quad \sum_{k=1}^{s} \tau_{(k)} = 0, \tag{5.33}$$

or

$$\text{(ii) random:} \quad \tau_{(k)} \sim N(0, \sigma_\tau^2) \quad , \tag{5.34}$$

respectively. The treatments are distributed over all s^2 experimental units according to the randomization, such that each unit, or rather its response, has to have the subscript (k) in order to identify the treatment. From the data table of the Latin square we obtain the marginal sums

$Y_{i\cdot} = \sum_{j=1}^{s} y_{ij}$ sum of the ith row

$Y_{\cdot j} = \sum_{i=1}^{s} y_{ij}$ sum of the jth column

$Y_{\cdot\cdot} = \sum_i Y_{i\cdot} = \sum_j Y_{\cdot j}$ total response.

For the treatments we calculate

T_k :   sum of the response values of the kth treatment

$m_k = T_k/s$:   average response of the kth treatment

	Treatment				
	1	2	$\cdots$	s	
Sum	T_1	T_2	$\cdots$	T_s	$\sum_{k=1}^{s} T_k = Y_{\cdot\cdot}$
Mean	m_1	m_2	$\cdots$	m_s	$Y_{\cdot\cdot}/s^2 = y_{\cdot\cdot}$

Table 5.7: Sums and means of the treatments

The decomposition of the error sum of squares is as follows: Assume the correction term defined according to

$$C = Y_{\cdot\cdot}^2/s^2 \quad . \tag{5.35}$$

Source	SS	df	MS	F
Rows	SS_{Row}	$s-1$	MS_{Row}	F_{Row}
Columns	SS_{Column}	$s-1$	MS_{Column}	F_{Column}
Treatment	SS_{Treat}	$s-1$	MS_{Treat}	F_{Treat}
Error	SS_{Error}	$(s-1)(s-2)$	MS_{Error}	
Total	SS_{Total}	s^2-1		

Table 5.8: Analysis of variance table for the Latin square

Then we have

$$SS_{Total} = \sum_i \sum_j y_{ij}^2 - C \tag{5.36}$$

$$SS_{Row} = \frac{1}{s} \sum_i Y_{i.}^2 - C \tag{5.37}$$

$$SS_{Column} = \frac{1}{s} \sum_j Y_{.j}^2 - C \tag{5.38}$$

$$SS_{Treat} = \frac{1}{s} \sum_k T_k^2 - C \tag{5.39}$$

$$SS_{Error} = SS_{Total} - SS_{Row} - SS_{Column} - SS_{Treat} \tag{5.40}$$

The MS-values are obtained by dividing the SS-values by their degrees of freedom. The F ratios are MS/MS_{Error} (cf. Table 5.8). The expectations of the MS are shown in Table 5.9

Source	MS	$E(MS)$
Rows	MS_{Row}	$\sigma^2 + s\sigma_\rho^2$
Columns	MS_{Column}	$\sigma^2 + s\sigma_\gamma^2$
Treatment	MS_{Treat}	$\sigma^2 + \frac{s}{s-1}\sum_k \tau_{(k)}^2$
Error	MS_{Error}	σ^2

Table 5.9: $E(MS)$

The null hypothesis H_0: "no treatment effect", i.e., $H_0 : \tau_1 = \cdots = \tau_s = 0$ against $H_1 : \tau_i \neq 0$ for at least one i is tested with

$$F_{Treat} = \frac{MS_{Treat}}{MS_{Error}}. \tag{5.41}$$

Due to the design of the Latin square, the s treatments are repeated s-times each. Hence, treatment effects can be tested for. On the other hand, we cannot always speak of a repetition of rows and columns in the sense of blocks. Hence, F_{Row} and F_{Column} can only serve as indicators for additional effects which yield a reduction of MS_{Error} and thus an increase in precision. Row and column effects would be statistically detectable if repetitions were realized for each cell.

Point and Confidence Estimates of the Treatment Effects

The OLS-estimate of the kth treatment mean $\mu_k = \mu + \tau_{(k)}$ is

$$m_k = T_k/s \qquad (5.42)$$

with the variance

$$\text{Var}(m_k) = \sigma^2/s \qquad (5.43)$$

and the estimated variance

$$\widehat{\text{Var}}(m_k) = MS_{Error}/s \quad . \qquad (5.44)$$

Hence the confidence interval is of the following form

$$m_k \pm t_{(s-1)(s-2);1-\alpha/2} \sqrt{MS_{Error}/s} \quad . \qquad (5.45)$$

In case of a simple comparison of two treatments the difference is estimated by the confidence interval

$$(m_{k_1} - m_{k_2}) \pm t_{(s-1)(s-2);1-\alpha/2} \sqrt{2MS_{Error}/s} \quad . \qquad (5.46)$$

Example 5.4: The effect of $s = 4$ sleeping pills is tested on $s^2 = 16$ persons, who are stratified according to the design of the Latin square, based on the ordinally classified factors body weight and blood pressure. The response to be measured is the prolongation of sleep (in minutes) compared to an average value (without sleeping pills).

Weight

$\longrightarrow$

A 43	B 57	C 61	D 74
B 59	C 63	D 75	A 46
C 65	D 79	A 48	B 64
D 83	A 55	B 67	C 72

Blood pressure $\downarrow$

Table 5.10: Latin square (prolongation of sleep)

Weight

Blood pressure	1	2	3	4	$Y_{i.}$
1	43	57	61	74	235
2	59	63	75	46	243
3	65	79	48	64	256
4	83	55	67	72	277
$Y_{.j}$	250	254	251	256	1011

Medicament	A	B	C	D	Total
Total(T_k)	192	247	261	311	1011
Mean	48.00	61.75	65.25	77.75	63.19

We calculate the sums of squares

$$C = 1011^2/16 = 63882.56$$
$$SS_{Total} = 65939 - C = 2056.44$$
$$SS_{Row} = \frac{1}{4} \cdot 256539 - C = 252.19$$
$$SS_{Column} = \frac{1}{4} \cdot 255553 - C = 5.69$$
$$SS_{Treat} = \frac{1}{4} \cdot 262715 - C = 1796.19$$
$$SS_{Error} = 2056.44 - (252.19 + 5.69 + 1796.19)$$
$$= 2056.44 - 2054.07$$
$$= 2.37$$

Source	SS	df	MS	F	
Rows	252.19	3	84.06	210.15	*
Columns	5.69	3	1.90	4.75	
Treatment	1796.19	3	598.73	1496.83	*
Error	2.37	6	0.40		
Total	2056.44	15			

The critical value is $F_{3,6;0.95} = 4.76$. Hence the row effect (stratification according to blood pressure groups) is significant, the column effect (weight) however, is not significant. The treatment effect is significant as well. The final conclusion should be, that in further clinical tests of the four different sleeping pills the experiment should be conducted according to the randomized block design with the blocking factor "blood pressure groups".

The simple and the multiple tests require SS_{Error} from the model with the main effect treatment

Source	SS	df	MS	F
Treatment	1796.19	3	598.73	27.60 *
Error	260.25	12	21.69	
Total	2056.44	15		

For the simple mean comparisons we obtain $(t_{6;0.975} \cdot \sqrt{2MS_{Error}/4} = 8.07)$

Medicament	Difference	Confidence interval
2/1	13.75	[5.68 , 21.82]
3/1	17.25	[9.18 , 25.32]
4/1	29.75	[21.68 , 37.82]
3/2	3.50	[−4.57 , 11.57]
4/2	16.00	[7.93 , 24.07]
4/3	12.50	[4.43 , 20.57]

Result: In case of the simple test all pairwise mean comparisons except for 3/2 are significant. These tests however are not independent. Hence, we conduct the multiple tests.

Multiple Tests

The multiple test statistics (cf.(4.112)—(4.114)) with the degrees of freedom of the Latin square are

$$FPLSD = t_{s(s-1);1-\alpha/2}\sqrt{2MS_{Error}/s} \ , \tag{5.47}$$

$$HSD = Q_{\alpha,(s,s(s-1))}\sqrt{MS_{Error}/s} \ , \tag{5.48}$$

$$SNK_i = Q_{\alpha,(i,(s-1)(s-2))}\sqrt{MS_{Error}/s} \ . \tag{5.49}$$

Results of the Multiple Tests

Fisher's test:

$$\begin{aligned} FPLSD &= t_{12,0.975}\sqrt{2MS_{Error}/4} \\ &= 2.18\sqrt{21.69/2} \\ &= 7.18 \end{aligned}$$

Hence, the means are different except for μ_2 and μ_3.
HSD-test:
We have $Q_{0.05,(4,12)} = 4.20$, hence

$$HSD = 4.20\sqrt{21.69/4} = 9.78 \ .$$

All the means except 2/3 differ significantly.
SNK-test:
The means ordered according to their size are

$$48.00(A), \ 61.75(B), \ 65.25(C), \ 77.75(D).$$

The studentized rank-values and the SNK_i-values calculated from them are

i	2	3	4
$Q_{0.05,(i,6)}$	3.46	4.34	4.90
SNK_i	8.06	10.11	11.41

For the largest difference (D minus A) we have

$$77.75 - 48 = 29.75 > 11.41 \quad,$$

for the next differences (D minus B) and (C minus A) we receive

$$77.75 - 61.75 \;=\; 16.00 > 10.11 \quad,$$
$$65.25 - 48.00 \;=\; 17.25 > 10.11 \quad,$$

and finally we have

$$
\begin{array}{lll}
(D \text{ minus } C): & 77.75 - 65.25 = \quad 12.50 & > 8.06 \quad, \\
(C \text{ minus } B): & \qquad\qquad\qquad 3.50 & < 8.06 \quad, \\
(B \text{ minus } A): & \qquad\qquad\qquad 13.75 & > 8.06 \quad.
\end{array}
$$

Hence all means except for 2/3 differ significantly.

5.3 Rank Variance Analysis in the Randomized Block Design

5.3.1 Friedman Test

In the randomized block design, the individuals are grouped into blocks and are assigned one of the s treatments, randomized within each block. The essential demand is that each treatment occurs once, and only once, within each block. The layout of the response values is shown in Table 5.2. Once again, we assume the linear additive model (5.1). Furthermore, we assume

$$\epsilon_{ij} \stackrel{i.i.d.}{\sim} F(0, \sigma^2) \tag{5.50}$$

where F is any continuous distribution and does not have to be equal to the normal distribution. The randomization leads to independence of the ϵ_{ij}. Hence, the actual assumption in (5.50) refers to homogeneity of variance.

The hypothesis of interest is H_0: no treatment effect, i.e., we test

$$H_0: \tau_1 = \cdots = \tau_s$$

against

$$H_1: \tau_i \neq \tau_j \quad \text{for at least one } (i,j), \ i \neq j \quad.$$

The test procedure is based on the rank assignment (ranks 1 to s) for the response values, which is to be done separately for each block. Under the null hypothesis each of the $s!$ possible orders per block have the same probability. Analogously, the $(s!)^r$ possible orders of the intra-block ranks have equal possibilities.

If we take the sums of ranks per treatment $j = 1, \ldots, s$ over the r blocks, then they should be almost equal if H_0 holds. The test statistic by Friedman (1937) for testing H_0 compares these rank sums.

| | Treatment | | |
Block	1	$\cdots$	s
1	R_{11}	$\cdots$	R_{s1}
$\vdots$	$\vdots$		$\vdots$
r	R_{1r}	$\cdots$	R_{sr}
Sum	$R_{1.}$	$\cdots$	$R_{s.}$
Mean	$r_{1.}$	$\cdots$	$r_{s.}$

Table 5.11: Rank sums and rank means in the randomized block design

The test statistic by Friedman is

$$Q = \frac{12r}{s(s+1)} \sum_{j=1}^{s}(r_{j.} - r_{..})^2 \tag{5.51}$$

$$= \frac{12}{rs(s+1)} \sum_{j=1}^{s} R_{j.}^2 - 3r(s+1) \quad . \tag{5.52}$$

Here we have

$$R_{j.} = \sum_{i=1}^{r} R_{ji} \quad \text{rank sum of the } j\text{th treatment}$$

$$r_{j.} = R_{j.}/r \quad \text{rank mean of the } j\text{th treatment}$$

$$r_{..} = (s+1)/2 \quad .$$

If H_0 holds, then the differences $r_{i.} - r_{..}$ are almost equal and Q is sufficiently small. If however H_0 does not hold, then Q becomes large.

The test statistic Q is approximately (for r sufficiently large) χ_{s-1}^2-distributed. Hence, H_0: $\tau_1 = \cdots = \tau_s$ is rejected for

$$Q > \chi_{s-1;1-\alpha}^2 \quad .$$

For small values of r ($r < 15$), this approximation is insufficient. In this case exact quantiles are used (cf. Tables in Hollander and Wolfe, 1973, Michaelis, 1971 and Sachs, 1974, p.424). If ties are present, then the correction term

$$C_{corr} = 1 - \sum_{i=1}^{r} \sum_{k=1}^{s_i}(t_{ik}^3 - t_{ik})/rs(s^2 - 1) \tag{5.53}$$

is calculated. Here t_{i1} is the size of the first group of equally large response values, t_{i2} is the size of the second group of equally large response values, etc. in the ith block.

The corrected Friedman statistic is

$$Q_{corr} = \frac{Q}{C_{corr}}. \tag{5.54}$$

The Friedman test is a test of homogeneity. It tests whether the treatment samples could possibly come from the same population.

Example 5.5: (Continuation of Example 5.2) We conduct the comparison of the $s = 4$ treatments, that are arranged in $r = 3$ blocks, according to Table 5.4 with the Friedman test. From Table 5.4 we calculate the table of ranks 5.12

	Placebo	A	B	A and B
Block	1	2	3	4
1	2	3	1	4
2	2	3	1	4
3	2.5	2.5	1	4
Sum	6.5	8.5	3	12
$r_j.$	2.17	2.83	1	4

Table 5.12: Rank table for Table 5.4

The test statistic Q is

$$Q = \frac{12}{3 \cdot 4 \cdot 5}(6.5^2 + 8.5^2 + 3^2 + 12^2) - 3 \cdot 3 \cdot 5$$
$$= \frac{267.5}{5} - 45 = 8.5 \quad .$$

Since we have ties in the third block, we compute

$$C_{corr} = 1 - (2^3 - 2)/3 \cdot 4 \cdot (4^2 - 1)$$
$$= 1 - \frac{1}{30} = 0.97$$

and

$$Q_{corr} = \frac{Q}{C_{corr}} = 8.76 \quad .$$

The exact test yields the 95%-quantile as 7.4. Hence H_0: "homogeneity of the 4 treatments" is rejected.

5.3.2 Multiple Comparisons

We assume that the null hypothesis H_0: $\tau_1 = \cdots = \tau_s$ is rejected by the Friedman test. Analogously to Section 4.7.2, we distinguish between the planned single comparisons, all pairwise comparisons and the comparison of a control group with all other treatments.

Planned Single Comparisons

If the comparison of two selected treatments is planned before the data collection then the Wilcoxon test (cf. Chapter 2) is applied.

Comparison of all Pairwise Differences According to Friedman

The comparison of all $s(s-1)/2$ possible pairs is based on a modification of the Friedman test (cf. Woolson, 1987, p.387).

For each combination (j_1, j_2), $j_1 > j_2$ of treatments we compute the test statistic

$$Z_{j_1,j_2} = \frac{|r_{j_1\cdot} - r_{j_2\cdot}|}{\sqrt{s(s+1)/12r}} \tag{5.55}$$

for testing H_0: $\tau_{j_1} = \tau_{j_2}$ against H_1: $\tau_{j_1} \neq \tau_{j_2}$. All null hypotheses with $Z_{j_1,j_2} > QP_{1-\alpha}(r)$ are rejected and the multiple level is α. Tables for the critical values $QP_{1-\alpha}(r)$ exist (cf. e.g., Woolson, 1987, Table 15, p. 506, Hollander and Wolfe, 1973). For $\alpha = 0.05$ some selected values are

r	2	3	4	5	6	7	8	9	10
$QP_{0.95}(r)$	2.77	3.31	3.63	3.86	4.03	4.17	4.29	4.39	4.47

Example 5.6: (Continuation of Example 5.5) For the differences of the rank means we obtain from Table 5.12 the following Table ($\sqrt{4(4+1)/12 \cdot 3} = \sqrt{20/36} = 0.745$):

Comparison	$\lvert r_{j_1\cdot} - r_{j_2\cdot}\rvert$	Test statistic
1/2	$\lvert 2.17 - 2.83\rvert = 0.66$	0.86
1/3	$\lvert 2.17 - 1.0\rvert = 1.17$	1.57
1/4	$\lvert 2.17 - 4.0\rvert = 1.83$	2.46
2/3	$\lvert 2.83 - 1.0\rvert = 1.83$	2.46
2/4	$\lvert 2.83 - 4.0\rvert = 1.17$	1.57
3/4	$\lvert 1.0 - 4.0\rvert = 3.00$	4.03 *

Result: The treatment B and the combination (A and B) show differences in effect.

Comparison Control Group vs. All Other Treatments

Let $j = 1$ be the subscript of the control group. The test statistic for the multiple comparison of treatment 1 with the $(s-1)$ other treatments is then

$$Z_{1j} = \frac{|r_{1\cdot} - r_{j\cdot}|}{\sqrt{s(s+1)/6r}} \qquad j = 2,\dots,s \quad . \tag{5.56}$$

The two-stage quantiles $QC_{1-\alpha}(s-1)$ are given in special tables (Woolson, 1987, p.507, Hollander and Wolfe, 1973). For $Z_{1j} > QC_{1-\alpha}(s-1)$ the corresponding null hypothesis H_0: "homogeneity of the treatments 1 and j" is rejected. The multiple level α is ensured. In the following table we give a few selected critical values $QC_{0.95}(s-1)$:

$s-1$	1	2	3	4	5
$QC_{0.95}(s-1)$	1.96	2.21	2.35	2.44	2.51

Example 5.7: (Continuation of Example 5.5) The above table of the $|r_{j1.} - r_{j2.}|$ yields the following results for the comparison 'placebo against A, B, and combination':

$$
\left.
\begin{array}{rl}
1/2 \quad : \quad Z_{12} = \dfrac{0.66}{\sqrt{4.5/6.3}} = 0.63 \\[2mm]
1/3 \quad : \quad Z_{13} = \dfrac{1.17}{\sqrt{20/18}} = 1.11 \\[2mm]
1/4 \quad : \quad Z_{14} = \dfrac{1.83}{\sqrt{20/18}} = 1.74
\end{array}
\right\} < 2.35 \quad .
$$

Hence no comparison is significant.

5.4 Excercises and Questions

5.4.1 Describe the strategy of building blocks (homogeneity/heterogeneity). Does the experimental error diminish or increase in case of blocking?

5.4.2 How can it be shown that the completely randomized design is a submodel of the randomized block design?
How can the block effect be tested?
Name the correct F test for treatment effect in the following table:

	SS		MS	F
Block	20	3		
Treatment	60	3		
Error	10	9		
Total	90	15		

5.4.3 Conduct a comparison of means according to Scheffé and Bonferroni for the Example 5.2 (Table 5.4). Compare the results with those from Example 5.2 for the simple comparisons.

5.4.4 A Latin square is to test the effect of $s = 3$ eating habits of decathletes, who are classified according to the ordinally classified factors sprinting speed and strength. Test for block effects and for treatment effect (measured in points).

Speed

$\longrightarrow$

Strength $\downarrow$

A		B		C	
	40		50		80
C		A		B	
	50		45		65
B		C		A	
	70		70		60

Points above an average value

5.4.5 In Example 5.4.4, conduct the experimentwise multiple tests.

5.4.6 Conduct the Friedman test for Table 5.5. Define training method 1 as the control group and conduct the multiple comparison with the three other training methods.

6 Multifactor Experiments

6.1 Elementary Definitions and Principles

In practice, for most designed experiments it can be assumed that the response Y is not only dependent on a single variable but on a whole group of prognostic factors. If these variables are continuous, their influence on the response is taken into account by so-called factor levels. These are ranges (e.g., low, medium, high) that classify the continuous variables as ordinal variables. In Sections 1.7 and 1.8, we have already cited examples for designed experiments where the dependence of a response on two factors was to be examined.

Designs of experiments that analyse the response for all possible combinations of two or more factors are called *factorial experiments* or *cross-classification*. Suppose that we have s factors $A_1, \ldots, A_s$ with $r_1, \ldots, r_s$ factor levels. The complete factorial design then requires $r = \Pi r_i$ observations for one trial. This shows that it is important to restrict the number of factors as well as the number of their levels.

For factorial experiments, two elementary models are distinguished — models with and without interaction. Assume the situation of two factors A and B with two factor levels each, i.e., A_1, A_2 and B_1, B_2.

The change in response produced by a change in the level of a factor is called the main effect of this factor. Considering Table 6.1, the main effect of factor A can be interpreted as the difference between the average response of the two factor levels A_1 and A_2:

$$\lambda_A = \frac{60}{2} - \frac{40}{2} = 10 \quad .$$

Similarly, the main effect of B is

$$\lambda_B = \frac{70}{2} - \frac{30}{2} = 20 \quad .$$

The effects of A at the two levels of B are

$$\text{for } B_1: \quad 20 - 10 = 10, \quad \text{for } B_2: \quad 40 - 30 = 10 \, ,$$

and hence identical for both levels of B. For the effect of B we have

$$\text{for } A_1: \quad 30 - 10 = 20, \quad \text{for } A_2: \quad 40 - 20 = 20 \, ,$$

Factor B

	B_1	B_2	Σ
A_1	10	30	40
A_2	20	40	60
Σ	30	70	100

Factor A

Table 6.1: Two-factorial experiment without interaction

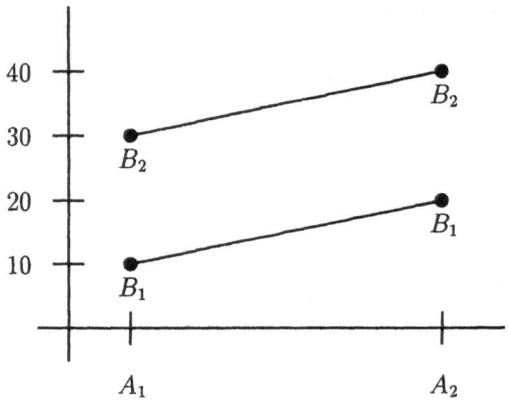

Figure 6.1: Two-factorial experiment without interaction

so that no effect dependent on A can be seen. The response lines are parallel. The analysis of Table 6.2 however leads to the following effects:

$$\text{main effect } \lambda_A = \frac{80-40}{2} = 20 \quad,$$

$$\text{main effect } \lambda_B = \frac{90-30}{2} = 30 \quad,$$

effects of A

$$\text{for } B_1: \quad 20-10 = 10, \quad \text{for } B_2: \quad 60-30 = 30 \quad,$$

effects of B

$$\text{for } A_1: \quad 30-10 = 20, \quad \text{for } A_2: \quad 60-20 = 40 \,.$$

Here the effects depend on the levels of the other factor, the interaction effect amounts to 20. The response lines are no longer parallel (Figure 6.2).

Remark: The term factorial experiment describes the completely crossed combination of the factors (treatments) and not the design of experiment. Factorial

Factor B

	B_1	B_2	Σ
A_1	10	30	40
A_2	20	60	80
Σ	30	90	120

Factor A

Table 6.2: Two-factorial experiment with interaction

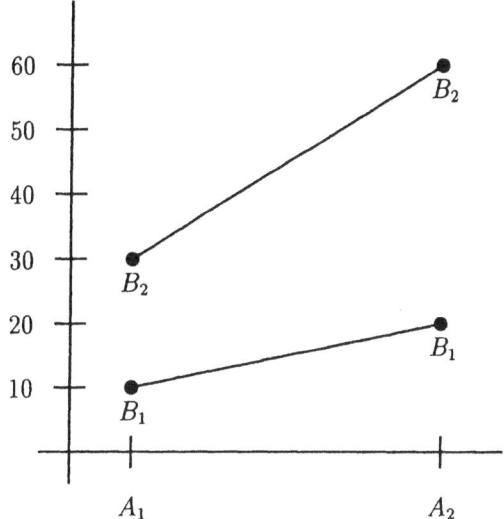

Figure 6.2: Two-factorial experiment with interaction

experiments may be realized as completely randomized designs of experiments, as latin squares etc..

The factorial experiment should be used

- in pilot-studies that analyse the statistical relevance of possible covariables

- for the determination of bivariate interaction

- for the determination of possible rank orders of the factors related to their influence on the response.

Compared to experiments with a single factor, the factorial experiment has the advantage that main effects may be estimated with the same precision, but with a smaller sample size.

Assume that we want to estimate the main effects A and B as in the above examples. The following one factor experiment with two repetitions would be appropriate (cf. Montgomery, 1976, p.124):

$A_1B_1^{(1)}$	$A_1B_2^{(1)}$
$A_2B_1^{(1)}$	

$A_1B_1^{(2)}$	$A_1B_2^{(2)}$
$A_2B_1^{(2)}$	

$$n = 3 + 3 = 6 \text{ observations}$$

estimation of λ_A : $\dfrac{1}{2}\left[(A_2B_1^{(1)} - A_1B_1^{(1)}) + (A_2B_1^{(2)} - A_1B_1^{(2)})\right]$,

estimation of λ_B : $\dfrac{1}{2}\left[(A_1B_1^{(1)} - A_1B_2^{(1)}) + (A_1B_1^{(2)} - A_1B_2^{(2)})\right]$.

Estimation of the effects with same precision is achieved by the factorial experiment

A_1B_1	A_1B_2
A_2B_1	A_2B_2

with only $n = 4$ observations according to

$$\lambda_A = \frac{1}{2}\left[(A_2B_1 - A_1B_1) + (A_2B_2 - A_1B_2)\right]$$

and

$$\lambda_B = \frac{1}{2}\left[(A_1B_2 - A_1B_1) + (A_2B_2 - A_1B_1)\right]$$.

Additionally, the factorial experiment reveals existing interaction and hence leads to an adequate model.

If a present interaction is neglected or not revealed, a serious misinterpretation of the main effects may be the consequence. In principle, if significant interaction is present, then the main effects are of secondary importance since the effect of one factor on the response can no longer be segregated from the other factor.

6.2 Two-Factor Experiments with Interaction (Models with Fixed Effects)

Suppose that there are a levels of factor A and b levels of factor B. For each combination (i,j), r replicates are realized and the design is a completely randomized design. Hence the number of observations equals $N = rab$. The response is described by the linear model

$$y_{ijk} = \mu + \alpha_i + \beta_j + (\alpha\beta)_{ij} + \epsilon_{ijk} ,$$
$$(i = 1,\ldots,a; \; j = 1,\ldots,b; \; k = 1,\ldots,r) . \tag{6.1}$$

Here we have

y_{ijk} : response to the ith level of A and the jth level of B in the kth replicate
μ : overall mean
α_i : effect of the ith level of factor A
β_j : effect of the jth level of factor B
$(\alpha\beta)_{ij}$: effect of the interaction of the combination (i, j)
ϵ_{ijk} : random error.

The following assumption is made for the random vector $\boldsymbol{\epsilon}' = (\epsilon_{111}, \ldots, \epsilon_{abr})$

$$\boldsymbol{\epsilon} \sim N(\mathbf{0}, \sigma^2 \boldsymbol{I}) \quad . \tag{6.2}$$

For the fixed effects, we have the following constraints:

$$\sum_{i=1}^{a} \alpha_i = 0 \quad , \tag{6.3}$$

$$\sum_{j=1}^{b} \beta_j = 0 \quad , \tag{6.4}$$

$$\sum_{i=1}^{a} (\alpha\beta)_{ij} = \sum_{j=1}^{b} (\alpha\beta)_{ij} = 0 \quad . \tag{6.5}$$

Remark: If the randomized block design is chosen as the design of experiment, the model (6.1) additionally contains the (additive) block effects ρ_k as random effects with $\rho_k \sim N(0, \sigma_\rho^2)$.

	B					
A	1	2	$\cdots$	b	$\sum$	Means
1	$Y_{11\cdot}$	$Y_{12\cdot}$	$\cdots$	$Y_{1b\cdot}$	$Y_{1\cdot\cdot}$	$y_{1\cdot\cdot}$
2	$Y_{21\cdot}$	$Y_{22\cdot}$	$\cdots$	$Y_{2b\cdot}$	$Y_{2\cdot\cdot}$	$y_{2\cdot\cdot}$
$\vdots$	$\vdots$	$\vdots$		$\vdots$	$\vdots$	$\vdots$
a	$Y_{a1\cdot}$	$Y_{a2\cdot}$	$\cdots$	$Y_{ab\cdot}$	$Y_{a\cdot\cdot}$	$y_{a\cdot\cdot}$
$\sum$	$Y_{\cdot 1\cdot}$	$Y_{\cdot 2\cdot}$	$\cdots$	$Y_{\cdot b\cdot}$	$Y_{\cdots}$	$y_{\cdots}$
Means	$y_{\cdot 1\cdot}$	$y_{\cdot 2\cdot}$	$\cdots$	$y_{\cdot b\cdot}$		

Table 6.3: Table of the total response values in the $A \times B$-design

Ordinary Least Squares Estimation of the Parameters

The score function (3.6) in model (6.1) is as follows

$$S(\boldsymbol{\theta}) = \sum_i \sum_j \sum_k (y_{ijk} - \mu - \alpha_i - \beta_j - (\alpha\beta)_{ij})^2 \tag{6.6}$$

under the constraints (6.3) - (6.5).

Source	SS	df	MS	F
Factor A	SS_A	$a-1$	MS_A	F_A
Factor B	SS_B	$b-1$	MS_B	F_B
Interaction				
$A \times B$	$SS_{A \times B}$	$(a-1)(b-1)$	$MS_{A \times B}$	$F_{A \times B}$
Error	SS_{Error}	$N - ab$	MS_{Error}	
		$= ab(r-1)$		
Total	SS_{Total}	$N-1$		

Table 6.4: Analysis of variance table in the $A \times B$-design with interaction

Here

$$\theta' = (\mu, \alpha_1, \ldots, \alpha_a, \beta_1, \ldots, \beta_b, (\alpha\beta)_{11}, \ldots, (\alpha\beta)_{ab}) \qquad (6.7)$$

is the vector of the unknown parameters. The normal equations, taking the restrictions (6.3) – (6.5) into consideration, can easily be derived:

$$-\frac{1}{2}\frac{\partial S(\theta)}{\partial \mu} = \sum\sum\sum(y_{ijk} - \mu - \alpha_i - \beta_j - (\alpha\beta)_{ij})$$

$$= Y_{...} - N\mu = 0 \qquad (6.8)$$

$$-\frac{1}{2}\frac{\partial S(\theta)}{\partial \alpha_i} = Y_{i..} - br\alpha_i - br\mu = 0 \quad (i \text{ fixed}) \qquad (6.9)$$

$$-\frac{1}{2}\frac{\partial S(\theta)}{\partial \beta_j} = Y_{.j.} - ar\beta_j - ar\mu = 0 \quad (j \text{ fixed}) \qquad (6.10)$$

$$-\frac{1}{2}\frac{\partial S(\theta)}{\partial (\alpha\beta)_{ij}} = Y_{ij.} - r\mu - r\alpha_i - r\beta_j - (\alpha\beta)_{ij} = 0 \quad (i,j \text{ fixed}) \quad . \quad (6.11)$$

We now obtain the OLS estimates under the constraints (6.3) – (6.5), that is the conditional OLS estimates

$$\hat{\mu} = Y_{...}/N = y_{...} \qquad (6.12)$$

$$\hat{\alpha}_i = \frac{Y_i}{br} - \hat{\mu} = y_{i..} - y_{...} \qquad (6.13)$$

$$\hat{\beta}_j = \frac{Y_{.j.}}{ar} - \hat{\mu} = y_{.j.} - y_{...} \qquad (6.14)$$

$$(\hat{\alpha\beta})_{ij} = \frac{Y_{ij.}}{r} - \hat{\mu} - \hat{\alpha}_i - \hat{\beta}_j = y_{ij.} - y_{i..} - y_{.j.} + y_{...} \qquad (6.15)$$

The correction term is defined as

$$C = Y_{...}^2/N \qquad (6.16)$$

with $N = a\,b\,r$. The sums of squares can now be expressed as follows

$$SS_{Total} = \sum\sum\sum(y_{ijk} - y_{...})^2$$

$$= \sum\sum\sum y_{ijk}^2 - C \tag{6.17}$$

$$SS_A = \frac{1}{br}\sum_i Y_{i..}^2 - C \tag{6.18}$$

$$SS_B = \frac{1}{ar}\sum_j Y_{.j.}^2 - C \tag{6.19}$$

$$SS_{A\times B} = \frac{1}{r}\sum_i\sum_j Y_{ij.}^2 - \frac{1}{br}\sum_i Y_{i..}^2 - \frac{1}{ar}\sum_j Y_{.j.}^2 + C$$

$$= \left[\frac{1}{r}\sum_i\sum_j Y_{ij.}^2 - C\right] - SS_A - SS_B \tag{6.20}$$

$$SS_{Error} = SS_{Total} - SS_A - SS_B - SS_{A\times B}$$

$$= SS_{Total} - \left[\frac{1}{r}\sum_i\sum_j Y_{ij.}^2 - C\right] \quad . \tag{6.21}$$

Remark: The sum of squares between the $a\cdot b$ sums of response $Y_{ij.}$ is also called $SS_{Subtotal}$, i.e.,

$$SS_{Subtotal} = \frac{1}{r}\sum_i\sum_j Y_{ij.}^2 - C \quad . \tag{6.22}$$

Hint: In order to ensure that the interaction effect is detectable (and hence $(\alpha\beta)_{ij}$ can be estimated), at least $r = 2$ replicates have to be realized for each combination (i, j). Otherwise, the interaction effect is included in the error and cannot be seperated.

Test Procedure

The model (6.1) with interaction is called a *saturated model*. The model without interaction is

$$y_{ijk} = \mu + \alpha_i + \beta_j + \epsilon_{ijk} \tag{6.23}$$

and is called the *independence model*.

First, the hypothesis $H_0 : (\alpha\beta)_{ij} = 0$ (for all (i, j)) against $H_1 : (\alpha\beta)_{ij} \neq 0$ (for at least one pair (i, j)) is tested. This corresponds to the model choice *submodel (6.23) compared to the complete model (6.1)* according to our likelihood-ratio test strategy in Chapter 3. The interpretation of inferences obtained from the factorial experiment depends on the result of this test.

H_0 is rejected if

$$F_{A\times B} = \frac{MS_{A\times B}}{MS_{Error}} > F_{(a-1)(b-1),ab(r-1);1-\alpha} \tag{6.24}$$

The interaction effects are significant in case of a rejection of H_0. The main effects are of no importance, no matter whether they are significant or not.

If however H_0 is rejected, then the test results for H_0 : $\alpha_i = 0$ against H_1 : $\alpha_i \neq 0$ (for at least one i) with $F_A = \frac{MS_A}{MS_{Error}}$ and for H_0 : $\beta_j = 0$ against H_1 : $\beta_j \neq 0$ (for at least one j) with $F_B = \frac{MS_B}{MS_{Error}}$ are of importance for the interpretation in model (6.23). If only one factor effect is significant (e.g., A), then the model is reduced further to a balanced one-factor model with a factor levels and br replicates each:

$$y_{ijk} = \mu + \alpha_i + \epsilon_{ijk} \quad . \tag{6.25}$$

Example 6.1: The influence of two factors A (fertilizer) and B (irrigation) on the yield of a type of grain is to be analysed in a pilot study. The factors A and B are applied in two levels (low, high) and $r = 2$ replicates each. Hence, we have $a = b = r = 2$ and $N = abr = 8$. The experimental units (plants) are assigned to the treatments at random. From Tables 6.5 and 6.6, we calculate:

$$
\begin{aligned}
C &= 77.6^2/8 = 752.72 \\
SS_{Total} &= 866.92 - C = 114.20 \\
SS_A &= \frac{1}{4}(39.6^2 + 38.0^2) - C \\
&= 753.04 - 752.72 = 0.32 \\
SS_B &= \frac{1}{4}(26.4^2 + 51.2^2) - C \\
&= 892.60 - 752.72 = 76.88 \\
SS_{Subtotal} &= \frac{1}{2}(17.8^2 + 21.8^2 + 8.6^2 + 29.4^2) - C \\
&= 865.20 - 752.72 = 112.48 \\
SS_{A \times B} &= SS_{Subtotal} - SS_A - SS_B = 35.28 \\
SS_{Error} &= 114.20 - 35.28 - 0.32 - 76.88 \\
&= 1.72
\end{aligned}
$$

B

		1		2	
A	1	8.6	9.2	10.4	11.4
	2	4.7	3.9	14.1	15.3

Table 6.5: Response values

(Hint: $F_{1,4;0.95} = 7.71$)
Result: The test for interaction leads to a rejection of H_0 : *no interaction* with $F_{1,4} = 82.05$. However, a reduction to a an experiment with a single factor is not possible, in spite of the non-significant main effect A.

$$B$$

		1	2	$\sum$
A	1	17.8	21.8	39.6
	2	8.6	29.4	38.0
	$\sum$	26.4	51.2	77.6

Table 6.6: Total response

Source	SS	df	MS	F	
A	0.32	1	0.32	0.74	
B	76.88	1	76.88	178.79	*
$A \times B$	35.28	1	35.28	82.05	*
$Error$	1.72	4	0.43		
$Total$	114.20	7			

Table 6.7: Analysis of variance table for Example 6.1

6.3 Two-Factor Experiments in Effect Coding

In the above section, we have derived the parameter estimates of the components of θ (6.7) by minimizing the error sum of squares under the linear restrictions $\sum_i \alpha_i = 0$, $\sum_j \beta_j = 0$ and $\sum_i (\alpha\beta)_{ij} = \sum_j (\alpha\beta)_{ij} = 0$. This corresponds to the conditional OLS estimate $b(R)$ from (3.111).

We now want to achieve a reduction in the number of parameters. This is done by an alternative parametrization that includes the restrictions already in the model. The result is a set of parameters that corresponds to a design matrix of full column rank. The parameter estimation is now achieved by the OLS estimate b_0. For this purpose we use the so-called *effect coding* of categories. The

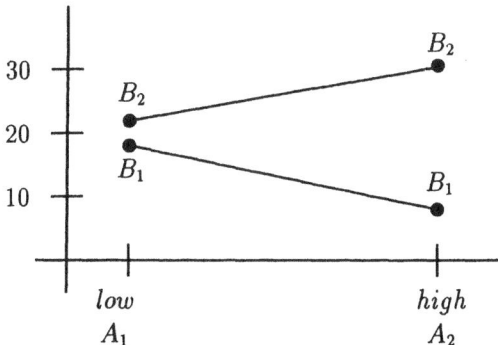

Figure 6.3: Interaction in Example 6.1

effect coding for a factor A at $a=3$ categories (levels) is as follows

$$x_i^A = \begin{cases} 1 & \text{for category } i \quad (i = 1, \ldots, a - 1) \\ -1 & \text{for category } a \\ 0 & \text{else} \end{cases}$$

so that

$$\alpha_a = -\sum_{i=1}^{a-1} \alpha_i \quad , \tag{6.26}$$

or, expressed differently

$$\sum_{i=1}^{a} \alpha_i = 0 \quad . \tag{6.27}$$

Example: Assume factor A has $a = 3$ levels A_1: low, A_2: medium, A_3: high. The original link of design and parameters is as follows

$$\begin{matrix} \text{low:} \\ \text{medium:} \\ \text{high:} \end{matrix} \begin{pmatrix} 1 & 0 & 0 \\ 0 & 1 & 0 \\ 0 & 0 & 1 \end{pmatrix} \begin{pmatrix} \alpha_1 \\ \alpha_2 \\ \alpha_3 \end{pmatrix} \quad \text{and } \alpha_1 + \alpha_2 + \alpha_3 = 0.$$

If effect coding is applied, we obtain

$$\begin{matrix} \text{low:} \\ \text{medium:} \\ \text{high:} \end{matrix} \begin{pmatrix} 1 & 0 \\ 0 & 1 \\ -1 & -1 \end{pmatrix} \begin{pmatrix} \alpha_1 \\ \alpha_2 \end{pmatrix} \quad .$$

Case $a = b = 2$

In case of a linear model with two two-level prognostic factors A and B, we have for fixed k $(k = 1, \ldots, r)$ the following parametrization (cf. Toutenburg, 1992, p.255):

$$\begin{pmatrix} y_{11k} \\ y_{12k} \\ y_{21k} \\ y_{22k} \end{pmatrix} = \begin{pmatrix} 1 & 1 & 1 & 1 \\ 1 & 1 & -1 & -1 \\ 1 & -1 & 1 & -1 \\ 1 & -1 & -1 & 1 \end{pmatrix} \begin{pmatrix} \mu \\ \alpha_1 \\ \beta_1 \\ (\alpha\beta)_{11} \end{pmatrix} + \begin{pmatrix} \epsilon_{11k} \\ \epsilon_{12k} \\ \epsilon_{21k} \\ \epsilon_{22k} \end{pmatrix} \quad . \tag{6.28}$$

Here we get the constraints immediately:

$$\begin{aligned} \alpha_1 + \alpha_2 &= 0 &\Rightarrow& \quad \alpha_2 = -\alpha_1 \\ \beta_1 + \beta_2 &= 0 &\Rightarrow& \quad \beta_2 = -\beta_1 \\ (\alpha\beta)_{11} + (\alpha\beta)_{12} &= 0 &\Rightarrow& \quad (\alpha\beta)_{12} = -(\alpha\beta)_{11} \\ (\alpha\beta)_{11} + (\alpha\beta)_{21} &= 0 &\Rightarrow& \quad (\alpha\beta)_{21} = -(\alpha\beta)_{11} \\ (\alpha\beta)_{21} + (\alpha\beta)_{22} &= 0 &\Rightarrow& \quad (\alpha\beta)_{22} = -(\alpha\beta)_{21} = (\alpha\beta)_{11} \quad . \end{aligned}$$

Of the originally 9 parameters, only 4 remain in the model. The others are calculated from these equations. The following notation is used:

$$\underset{r,4}{\boldsymbol{X}_{11}} = (\mathbf{1}_r \quad \mathbf{1}_r \quad \mathbf{1}_r \quad \mathbf{1}_r)$$

$$\underset{r,4}{\boldsymbol{X}_{12}} = (\mathbf{1}_r \quad \mathbf{1}_r \quad -\mathbf{1}_r \quad -\mathbf{1}_r)$$

$$\underset{r,4}{\boldsymbol{X}_{21}} = (\mathbf{1}_r \quad -\mathbf{1}_r \quad \mathbf{1}_r \quad -\mathbf{1}_r)$$

$$\underset{r,4}{\boldsymbol{X}_{22}} = (\mathbf{1}_r \quad -\mathbf{1}_r \quad -\mathbf{1}_r \quad \mathbf{1}_r)$$

$$\underset{4,4r}{\boldsymbol{X}'} = (\boldsymbol{X}'_{11} \quad \boldsymbol{X}'_{12} \quad \boldsymbol{X}'_{21} \quad \boldsymbol{X}'_{22})$$

$$\boldsymbol{\theta}'_0 = (\mu, \ \alpha_1, \ \beta_1, \ (\alpha\beta)_{11})$$

$$\boldsymbol{y}_{ij} = \begin{pmatrix} y_{ij1} \\ \vdots \\ y_{ijr} \end{pmatrix}, \quad \epsilon_{ij} = \begin{pmatrix} \epsilon_{ij1} \\ \vdots \\ \epsilon_{ijr} \end{pmatrix}$$

$$\boldsymbol{y} = \begin{pmatrix} \boldsymbol{y}_{11} \\ \boldsymbol{y}_{12} \\ \boldsymbol{y}_{21} \\ \boldsymbol{y}_{22} \end{pmatrix}, \quad \epsilon = \begin{pmatrix} \epsilon_{11} \\ \epsilon_{12} \\ \epsilon_{21} \\ \epsilon_{22} \end{pmatrix} .$$

In case of $a = b = 2$ and r replicates, and considering the restrictions (6.3), (6.4), (6.5), the two-factorial model (6.1) can alternatively be expressed in effect coding:

$$y = X\boldsymbol{\theta}_0 + \epsilon \quad . \tag{6.29}$$

The OLS estimate of $\boldsymbol{\theta}_0$ is

$$\hat{\boldsymbol{\theta}}_0 = (\boldsymbol{X}'\boldsymbol{X})^{-1}\boldsymbol{X}'\boldsymbol{y} \quad .$$

We now calculate $\hat{\boldsymbol{\theta}}_0$:

$$\underset{4,4}{\boldsymbol{X}'\boldsymbol{X}} = \boldsymbol{X}'_{11}\boldsymbol{X}_{11} + \boldsymbol{X}'_{12}\boldsymbol{X}_{12} + \boldsymbol{X}'_{21}\boldsymbol{X}_{21} + \boldsymbol{X}'_{22}\boldsymbol{X}_{22}$$

$$= 4r\boldsymbol{I}_4 \quad ,$$

$$\boldsymbol{X}'\boldsymbol{y} = \begin{pmatrix} Y_{...} \\ Y_{1..} - Y_{2..} \\ Y_{.1.} - Y_{.2.} \\ (Y_{11.} + Y_{22.}) - (Y_{12.} + Y_{21.}) \end{pmatrix}$$

$$= \begin{pmatrix} Y_{...} \\ 2Y_{1..} - Y_{...} \\ 2Y_{.1.} - Y_{...} \\ (Y_{11.} + Y_{22.}) - (Y_{12.} + Y_{21.}) \end{pmatrix} . \tag{6.30}$$

With $(X'X)^{-1} = \frac{1}{4r}I$, the OLS estimate $\hat{\theta}_0 = (X'X)^{-1}X'y$ can be written in detail as (cf. (6.12) – (6.15))

$$\begin{pmatrix} \hat{\mu} \\ \hat{\alpha}_1 \\ \hat{\beta}_1 \\ \hat{(\alpha\beta)}_{11} \end{pmatrix} = \begin{pmatrix} y_{...} \\ y_{1..} - y_{...} \\ y_{.1.} - y_{...} \\ y_{11.} - y_{1..} - y_{.1.} + y_{...} \end{pmatrix} . \tag{6.31}$$

The first three relations in (6.31) can easily be detected. The transition from the fourth row in (6.30) to the fourth row in (6.31) however has to be proven in detail.

With $a = b = 2$, we have

$$y_{11.} - y_{1..} - y_{.1.} + y_{...} =$$
$$= \frac{Y_{11.}}{r} - \left[\frac{Y_{11.}}{br} + \frac{Y_{12.}}{br}\right] - \left[\frac{Y_{11.}}{ar} + \frac{Y_{21.}}{ar}\right] + \frac{Y_{11.} + Y_{12.} + Y_{21.} + Y_{22.}}{abr}$$
$$= \frac{Y_{11.}}{r}\left(1 - \frac{1}{b} - \frac{1}{a} + \frac{1}{ab}\right) - \frac{Y_{12.}}{br}\left(1 - \frac{1}{a}\right) - \frac{Y_{21.}}{ar}\left(1 - \frac{1}{b}\right) + \frac{Y_{22.}}{abr}$$
$$= \frac{Y_{11.}}{r}\left(\frac{ab - a - b + 1}{ab}\right) + \frac{Y_{22.}}{abr} - \frac{Y_{12.}}{abr}(a - 1) - \frac{Y_{21.}}{abr}(b - 1)$$
$$= \frac{1}{4r}[(Y_{11.} + Y_{22.}) - (Y_{12.} + Y_{21.})] \quad .$$

Remark: Here we wish to point out an important characteristic of the effect coding. First, we write the matrix X in a different form:

$$X = \begin{pmatrix} X_{11} \\ X_{12} \\ X_{21} \\ X_{22} \end{pmatrix} = \begin{pmatrix} 1_r & 1_r & 1_r & 1_r \\ 1_r & 1_r & -1_r & -1_r \\ 1_r & -1_r & 1_r & -1_r \\ 1_r & -1_r & -1_r & 1_r \end{pmatrix}$$

$$= \begin{pmatrix} \underset{4r,1}{\boldsymbol{x}_\mu} & \underset{4r,1}{\boldsymbol{x}_{\alpha_1}} & \underset{4r,1}{\boldsymbol{x}_{\beta_1}} & \underset{4r,1}{\boldsymbol{x}_{(\alpha\beta)_{11}}} \end{pmatrix}$$

so that

$$\boldsymbol{x}'_\mu \boldsymbol{x}_\mu = \boldsymbol{x}'_{\alpha_1}\boldsymbol{x}_{\alpha_1} = \boldsymbol{x}'_{\beta_1}\boldsymbol{x}_{\beta_1} = \boldsymbol{x}'_{(\alpha\beta)_{11}}\boldsymbol{x}_{(\alpha\beta)_{11}} = 4r,$$
$$\boldsymbol{x}'_\mu \boldsymbol{x}_{\alpha_1} = \boldsymbol{x}'_\mu \boldsymbol{x}_{\beta_1} = \boldsymbol{x}'_\mu \boldsymbol{x}_{(\alpha\beta)_{11}} = 0,$$
$$\boldsymbol{x}'_{\alpha_1}\boldsymbol{x}_{\beta_1} = \boldsymbol{x}'_{\alpha_1}\boldsymbol{x}_{(\alpha\beta)_{11}} = 0,$$
$$\boldsymbol{x}'_{\beta_1}\boldsymbol{x}_{(\alpha\beta)_{11}} = 0 \quad .$$

Hence, as we mentioned before, the following holds

$$
X'X = \begin{pmatrix} x'_\mu \\ x'_{\alpha_1} \\ x'_{\beta_1} \\ x_{(\alpha\beta)_{11}} \end{pmatrix} \begin{pmatrix} x_\mu & x_{\alpha_1} & x_{\beta_1} & x_{(\alpha\beta)_{11}} \end{pmatrix} = 4r I_4 \quad .
$$

The vectors that belong to different effect groups $(\mu, \alpha, \beta, (\alpha\beta))$ are orthogonal. This property remains true in general for effect coding.

General Case: $a > 2, b > 2$

In the general case of a two-factorial model with interaction with
factor A : a levels,
factor B : b levels,
the parameter vector (after taking the constraints into account, that is, in effect coding) is as follows

$$
\theta'_0 = (\mu, \alpha_1, \ldots, \alpha_{a-1}, \beta_1, \ldots, \beta_{b-1}, (\alpha\beta)_{1,1}, \ldots, (\alpha\beta)_{a-1,b-1}) \tag{6.32}
$$

and the design matrix is

$$
X = \begin{pmatrix} x_\mu & X_\alpha & X_\beta & X_{(\alpha\beta)} \end{pmatrix} \quad . \tag{6.33}
$$

Here the column vectors of a submatrix are orthogonal to the column vectors of every other submatrix, e.g.,

$$
X'_\alpha X_\beta = 0 \quad .
$$

The matrix $X'X$ is now block-diagonal

$$
X'X = \mathrm{diag}\left(x'_\mu x_\mu,\ X'_\alpha X_\alpha,\ X'_\beta X_\beta,\ X'_{(\alpha\beta)} X_{(\alpha\beta)} \right)
$$

so that

$$
(X'X)^{-1} = \mathrm{diag}\left((x'_\mu x_\mu)^{-1},\ (X'_\alpha X_\alpha)^{-1},\ (X'_\beta X_\beta)^{-1},\ (X'_{(\alpha\beta)} X_{(\alpha\beta)})^{-1} \right) \tag{6.34}
$$

and the OLS estimate $\hat\theta_0$ can be written as

$$
\hat\theta_0 = \begin{pmatrix} \hat\mu \\ \hat\alpha \\ \hat\beta \\ \widehat{(\alpha\beta)} \end{pmatrix} = \begin{pmatrix} (x'_\mu x_\mu)^{-1} x'_\mu y \\ (X'_\alpha X_\alpha)^{-1} X'_\alpha y \\ (X'_\beta X_\beta)^{-1} X'_\beta y \\ (X'_{(\alpha\beta)} X_{(\alpha\beta)})^{-1} X'_{(\alpha\beta)} y \end{pmatrix} \quad . \tag{6.35}
$$

For the covariance matrix of $\hat\theta$, we get a block-diagonal structure as well

$$
V(\hat\theta) = \sigma^2 \begin{pmatrix} (x'_\mu x_\mu)^{-1} & 0 & 0 & 0 \\ 0 & (X'_\alpha X_\alpha)^{-1} & 0 & 0 \\ 0 & 0 & (X'_\beta X_\beta)^{-1} & 0 \\ 0 & 0 & 0 & (X'_{(\alpha\beta)} X_{(\alpha\beta)})^{-1} \end{pmatrix} . \tag{6.36}
$$

This shows that the estimation vectors $\hat{\mu}, \hat{\alpha}, \hat{\beta}, (\widehat{\alpha\beta})$ are uncorrelated and independent in case of normal errors. From this it follows that the estimates $\hat{\mu}, \hat{\alpha}$ and $\hat{\beta}$ in model (6.1) with interaction and the estimates in the independence model (6.23) are identical. Hence, the estimates for one parameter group — e.g., the main effects of factor B — are always the same, no matter whether the other parameters are contained in the model or not.

In case of rejection of H_0: $(\alpha\beta)_{ij} = 0$, σ^2 is estimated by

$$MS_{Error} = \frac{SS_{Error}}{N - ab} = \frac{1}{N - ab}(SS_{Total} - SS_A - SS_B - SS_{A\times B})$$

(cf. Table 6.4 and (6.21)). If H_0 is not rejected, then the independence model (6.23) holds and we have

$$SS_{Error} = SS_{Total} - SS_A - SS_B$$

for $N - 1 - (a - 1) - (b - 1) = N - a - b + 1$ degrees of freedom.

The model (6.1) with interaction corresponds to the parameter space Ω, according to our notation in Chapter 3. The independence model is the submodel of the parameter space $\omega \subset \Omega$. With (3.176) we have

$$\hat{\sigma}_\omega - \hat{\sigma}_\Omega^2 \geq 0 \quad . \tag{6.37}$$

Applied to our problem, we find

$$\hat{\sigma}_\Omega^2 = \frac{SS_{Total} - SS_A - SS_B - SS_{A\times B}}{N - ab} \tag{6.38}$$

and

$$\hat{\sigma}_\omega^2 = \frac{SS_{Total} - SS_A - SS_B}{N - ab + (a - 1)(b - 1)} \quad . \tag{6.39}$$

Interpretation: In the independence model σ^2 is estimated by (6.39). Hence, the confidence intervals of the parameter estimates $\hat{\mu}, \hat{\alpha}, \hat{\beta}$ are larger when compared with those obtained from the model with interaction. On the other hand, the parameter estimates themselves (which correspond to the center points of the confidence intervals) stay unchanged. Thus, the precision of the estimates $\hat{\mu}, \hat{\alpha}, \hat{\beta}$ decreases. Simultaneously the test statistics change so that in case of a rejection of the saturated model (6.1), tests of significance for μ, α, β based on the analysis of variance table for the independence model are to be carried out.

Case $a = 2, b = 3$

Considering the constraints (6.3) – (6.5), the model in effect coding is as follows

$$\begin{pmatrix} y_{11} \\ y_{12} \\ y_{13} \\ y_{21} \\ y_{22} \\ y_{23} \end{pmatrix} = \begin{pmatrix} 1_r & 1_r & 1_r & 0 & 1_r & 0 \\ 1_r & 1_r & 0 & 1_r & 0 & 1_r \\ 1_r & 1_r & -1_r & -1_r & -1_r & -1_r \\ 1_r & -1_r & 1_r & 0 & -1_r & 0 \\ 1_r & -1_r & 0 & 1_r & 0 & -1_r \\ 1_r & -1_r & -1_r & -1_r & 1_r & 1_r \end{pmatrix} \begin{pmatrix} \mu \\ \alpha_1 \\ \beta_1 \\ \beta_2 \\ (\alpha\beta)_{11} \\ (\alpha\beta)_{12} \end{pmatrix} + \begin{pmatrix} \epsilon_{11} \\ \epsilon_{12} \\ \epsilon_{13} \\ \epsilon_{21} \\ \epsilon_{22} \\ \epsilon_{23} \end{pmatrix} \tag{6.40}$$

Here we once again find the constraints immediately:

$$
\begin{aligned}
\alpha_1 + \alpha_2 &= 0 &\implies& \quad \alpha_2 = -\alpha_1 \\
\beta_1 + \beta_2 + \beta_3 &= 0 &\implies& \quad \beta_3 = -\beta_1 - \beta_2 \\
(\alpha\beta)_{11} + (\alpha\beta)_{21} &= 0 &\implies& \quad (\alpha\beta)_{21} = -(\alpha\beta)_{11} \\
(\alpha\beta)_{12} + (\alpha\beta)_{22} &= 0 &\implies& \quad (\alpha\beta)_{22} = -(\alpha\beta)_{12} \\
(\alpha\beta)_{13} + (\alpha\beta)_{23} &= 0 &\implies& \quad (\alpha\beta)_{23} = -(\alpha\beta)_{13} \\
(\alpha\beta)_{11} + (\alpha\beta)_{12} + (\alpha\beta)_{13} &= 0 &\implies& \quad (\alpha\beta)_{13} = -(\alpha\beta)_{11} - (\alpha\beta)_{12} \\
(\alpha\beta)_{21} + (\alpha\beta)_{22} + (\alpha\beta)_{23} &= 0 &\implies& \quad (\alpha\beta)_{23} = -(\alpha\beta)_{21} - (\alpha\beta)_{22} \\
& & & \qquad\quad = (\alpha\beta)_{11} + (\alpha\beta)_{12}
\end{aligned}
$$

so that of the originally 12 parameters only 6 remain in the model:

$$
\boldsymbol{\theta}_0' = (\mu, \alpha_1, \beta_1, \beta_2, (\alpha\beta)_{11}, (\alpha\beta)_{12}) \quad . \tag{6.41}
$$

We now take advantage of the orthogonality of the submatrices and apply (6.35) for the determination of the OLS estimates. We thus have

$$
\begin{aligned}
\hat{\mu} = (\boldsymbol{x}_\mu' \boldsymbol{x}_\mu)^{-1} \boldsymbol{x}_\mu' \boldsymbol{y} &= \frac{1}{6r} Y_{...} = y_{...} \\
\hat{\alpha}_1 = (\boldsymbol{x}_\alpha' \boldsymbol{x}_\alpha)^{-1} \boldsymbol{x}_\alpha' \boldsymbol{y} &= \frac{1}{6r}(Y_{1..} - Y_{2..}) \\
&= \frac{1}{6r}(2Y_{1..} - Y_{...}) \\
&= y_{1..} - y_{...} \quad , \\
\begin{pmatrix} \hat{\beta}_1 \\ \hat{\beta}_2 \end{pmatrix} &= \left(\boldsymbol{X}_\beta' \boldsymbol{X}_\beta \right)^{-1} \boldsymbol{X}_\beta' \boldsymbol{y} \\
&= \begin{pmatrix} 4r & 2r \\ 2r & 4r \end{pmatrix}^{-1} \begin{pmatrix} Y_{11.} - Y_{13.} + Y_{21.} - Y_{23.} \\ Y_{12.} - Y_{13.} + Y_{22.} - Y_{23.} \end{pmatrix} \\
&= \frac{1}{6r} \begin{pmatrix} 2 & -1 \\ -1 & 2 \end{pmatrix} \begin{pmatrix} Y_{.1.} - Y_{.3.} \\ Y_{.2.} - Y_{.3.} \end{pmatrix} \\
&= \frac{1}{6r} \begin{pmatrix} 2Y_{.1.} - Y_{.2.} - Y_{.3.} \\ 2Y_{.2.} - Y_{.1.} - Y_{.3.} \end{pmatrix} \\
&= \begin{pmatrix} y_{.1.} - y_{...} \\ y_{.2.} - y_{...} \end{pmatrix} \quad ,
\end{aligned}
$$

since for instance

$$
\begin{aligned}
\frac{1}{6r}(2Y_{.1.} - Y_{.2.} - Y_{.3.}) &= \frac{3Y_{.1.} - Y_{...}}{6r} \\
&= y_{.1.} - y_{...} \quad ,
\end{aligned}
$$

$$
\begin{pmatrix} \widehat{(\alpha\beta)}_{11} \\ \widehat{(\alpha\beta)}_{12} \end{pmatrix} = \frac{1}{6r} \begin{pmatrix} 2 & -1 \\ -1 & 2 \end{pmatrix} \begin{pmatrix} Y_{11\cdot} - Y_{13\cdot} - Y_{21\cdot} + Y_{23\cdot} \\ Y_{12\cdot} - Y_{13\cdot} - Y_{22\cdot} + Y_{23\cdot} \end{pmatrix}
$$

$$
= \frac{1}{6r} \begin{pmatrix} 2Y_{11\cdot} - Y_{13\cdot} - 2Y_{21\cdot} + Y_{23\cdot} - Y_{12\cdot} + Y_{22\cdot} \\ -Y_{11\cdot} - Y_{13\cdot} + Y_{21\cdot} + Y_{23\cdot} + 2Y_{12\cdot} - 2Y_{22\cdot} \end{pmatrix}
$$

$$
= \begin{pmatrix} y_{11\cdot} - y_{1\cdot\cdot} - y_{\cdot1\cdot} + y_{\cdots} \\ y_{12\cdot} - y_{1\cdot\cdot} - y_{\cdot2\cdot} + y_{\cdots} \end{pmatrix} .
$$

Example 6.2: A designed experiment is to analyse the effect of different concentrations of phosphate in a combination fertilizer (factor B) on the yield of two types of beans (factor A). A factorial experiment with two factors and fixed effects is chosen:

factor A : A_1 : type of beans I
             A_2 : type of beans II
factor B : B_1 : no phosphate
             B_2 : 10 % per unit
             B_3 : 30 % per unit

Hence, in case of the two-factor approach we have the 6 treatments $A_1 B_1$, $A_1 B_2$, $A_1 B_3$, $A_2 B_1$, $A_2 B_2$, $A_2 B_3$. In order to be able to estimate the error variance, the treatments have to be repeated. Here we choose the completely randomized design of experiment with 4 replicates each. The response values are summarized in Table 6.8.

	B_1	B_2	B_3	Sum
A_1	15	18	22	
	17	19	29	
	14	20	31	
	16	21	35	
Sum	62	78	117	257
A_2	13	17	18	
	9	19	22	
	8	18	24	
	12	18	23	
Sum	42	72	87	201
Sum	104	150	204	458

Table 6.8: Response in the $A \times B$ – design (Example 6.2)

We calculate the sums of squares ($a = 2$, $b = 3$, $r = 4$, $N = 3 \cdot 3 \cdot 4 = 24$):

$$
\begin{aligned}
C &= Y_{\cdots}^2/N = 458^2/24 = 8740.17 \\
SS_{Total} &= (15^2 + 17^2 + \cdots + 23^2) - C
\end{aligned}
$$

$$= 9672 - C = 931.83$$
$$SS_A = \frac{1}{3 \cdot 4}(257^2 + 201^2) - C$$
$$= 8870.83 - C = 130.66$$
$$SS_B = \frac{1}{2 \cdot 4}(104^2 + 150^2 + 204^2) - C$$
$$= 9366.50 - C = 626.33$$
$$SS_{Subtotal} = \frac{1}{4}(62^2 + 78^2 + \cdots + 87^2) - C$$
$$= 9533.50 - C = 793.33$$
$$SS_{A \times B} = SS_{Subtotal} - SS_A - SS_B$$
$$= 36.34$$
$$SS_{Error} = SS_{Total} - SS_{Subtotal} = 138.50$$

	SS	df	MS	F
Factor A	130.66	1	130.66	16.99 *
Factor B	626.33	2	313.17	40.72 *
A × B	36.34	2	18.17	2.36
Error	138.50	18	7.69	
Total	931.83	23		

Table 6.9: Analysis of variance table for Table 6.8

The test strategy starts by testing H$_0$: *no interaction*. The test statistic is

$$F_{A \times B} = F_{2,18} = \frac{18.17}{7.69} = 2.36 \quad .$$

The critical value is

$$F_{2,18;0.95} = 3.55 \quad .$$

Hence, the interaction is not significant at the 5%-level.

	SS	df	MS	F
Factor A	130.66	1	130.66	14.95 *
Factor B	626.33	2	313.17	35.83 *
Error	174.84	20	8.74	
Total	931.83	23		

Table 6.10: Analysis of variance table for Table 6.8 after omitting the interaction (independence model)

The test for significance of the main effects and the interaction effect in Table 6.9 is based on the model (6.1) with interaction. The test statistics for H$_0$: $\alpha_i = 0$,

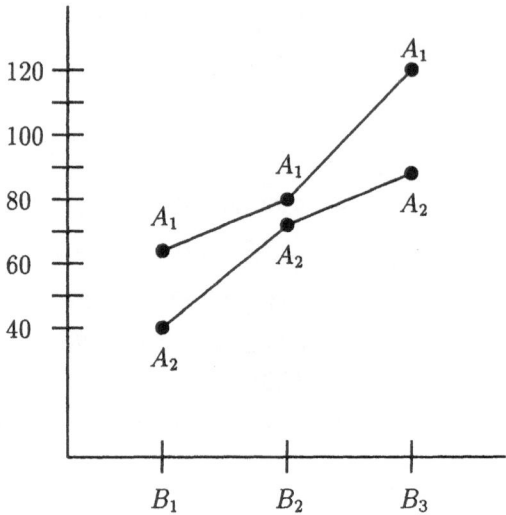

Figure 6.4: Interaction type × fertilization (not significant)

H_0: $\beta_i = 0$ and H_0: $(\alpha\beta)_{ij} = 0$ are independent. We did not reject H_0: $(\alpha\beta)_{ij} = 0$ (cf. Figure 6.4). This leads us back to the independence model (6.23) and we test the significance of the main effects according to Table 6.10. Here both effects are significant as well.

6.4 Two-Factorial Experiment with Block Effects

We now realize the factorial design with the factors A (at a levels) and B (at b levels) as a randomized block design with ab observations for each block (Table 6.11). The appropriate linear model with interaction is then of the following form

$$y_{ijk} = \mu + \alpha_i + \beta_j + \rho_k + (\alpha\beta)_{ij} + \epsilon_{ijk} \qquad (6.42)$$
$$(i = 1, \ldots, a, \; j = 1, \ldots, b, \; k = 1, \ldots, r)$$

Here ρ_k $(k = 1, \ldots, r)$ is the ith block effect and the constraints $\sum_{k=1}^{r} \rho_k = 0$ for fixed effects hold. The other parameters are the same as in model (6.1). In case of random block effects we assume $\rho' = (\rho_1, \ldots, \rho_r) \sim N(0, \sigma_\rho^2 I)$ and $E(\epsilon\rho') = 0$. Let

$$Y_{ij\cdot} = \sum_{k=1}^{r} y_{ijk} \qquad (6.43)$$

be the total response of the factor combination over all r blocks. The error sum of squares SS_{Total} (6.17), SS_A (6.18), SS_B (6.19), $SS_{A \times B}$ (6.20) remain unchanged. For the additional block effect, we calculate

$$SS_{Block} = \frac{1}{ab} \sum_{k=1}^{r} Y_{\cdot\cdot k}^2 - C \quad . \qquad (6.44)$$

The sum of squares SS_{Error} is now

$$SS_{Error} = SS_{Total} - SS_A - SS_B - SS_{A \times B} - SS_{Block} \quad . \qquad (6.45)$$

		Factor B			
Factor A	1	2	$\cdots$	b	Sum
1	$Y_{11\cdot}$	$Y_{12\cdot}$	$\cdots$	$Y_{1b\cdot}$	$Y_{1\cdot\cdot}$
2	$Y_{21\cdot}$	$Y_{22\cdot}$	$\cdots$	$Y_{2b\cdot}$	$Y_{2\cdot\cdot}$
$\vdots$	$\vdots$	$\vdots$		$\vdots$	$\vdots$
a	$Y_{a1\cdot}$	$Y_{a2\cdot}$	$\cdots$	$Y_{ab\cdot}$	$Y_{a\cdot\cdot}$
Sum	$Y_{\cdot1\cdot}$	$Y_{\cdot2\cdot}$	$\cdots$	$Y_{\cdot b\cdot}$	$Y_{\cdots}$

Table 6.11: Two-factorial randomized block design

The analysis of variance is shown in Table 6.12.

The interpretation of the model with block effects is done in the same manner as for the model without block effects. In case of at least one significant interaction, it is not possible to interpret the main effects — including the block effect — seperately.

If H_0: $(\alpha\beta)_{ij} = 0$ is not rejected, then an independence model with the three main effects (A, B, and block) holds, if these effects are significant. Compared to model (6.23), the parameter estimates $\hat{\alpha}$ and $\hat{\beta}$ are more precise, due to the reduction of the variance achieved by the block effect.

Example 6.3: The experiment in Example 6.2 is now designed as a randomized block design with $r = 4$ blocks. The response values are shown in Table 6.13 and the total response is given in the Tables 6.14 and 6.15.
We calculate (with $C = 8740.17$)

$$
\begin{aligned}
SS_{Block} &= \frac{1}{2 \cdot 3}(103^2 + 115^2 + 115^2 + 125^2) - C \\
&= 8780.67 - C = 40.50
\end{aligned}
$$

Source	SS	df	MS	F
A	SS_A	$a - 1$	MS_A	F_A
B	SS_B	$b - 1$	MS_B	F_B
$A \times B$	$SS_{A \times B}$	$(a-1)(b-1)$	$MS_{A \times B}$	$F_{A \times B}$
Block	SS_{Block}	$r - 1$	MS_{Block}	F_{Block}
Error	SS_{Error}	$(r-1)(ab-1)$	MS_{Error}	
Total	SS_{Total}	$rab - 1$		

Table 6.12: Analysis of variance table in the $A \times B$-design (6.42) with interaction and block effects

and

$$SS_{Error} = 98.00 \quad .$$

The analysis of variance table (Table 6.16) shows that with $F_{2,15;0.95} = 3.68$ the interaction effect is once again not significant. In the reduced model

$$y_{ijk} = \mu + \alpha_i + \beta_j + \rho_k + \epsilon_{ijk} \tag{6.46}$$

we test the main effects (Table 6.17).

I	II	III	IV
A_2B_2	A_1B_1	A_1B_3	A_2B_1
17	17	31	12
A_1B_3	A_2B_3	A_2B_1	A_1B_2
22	22	8	21
A_1B_1	A_1B_2	A_1B_2	A_2B_3
15	19	20	23
A_2B_1	A_2B_2	A_2B_2	A_1B_3
13	19	18	35
A_1B_2	A_2B_1	A_1B_1	A_2B_2
18	9	14	18
A_2B_3	A_1B_3	A_2B_3	A_1B_1
18	29	24	16

Table 6.13: Randomized block design and response in the 2×3 factor experiment

Block	I	II	III	IV	Sum
Response total	103	115	115	125	458

Table 6.14: Total response $Y_{..k}$ per block

	B_1	B_2	B_3	
A_1	62	78	117	257
A_2	42	72	87	201
	104	150	204	458

Table 6.15: Total response $Y_{ij.}$ for each factor combination (Example 6.3)

Because of $F_{3,17;0.95} = 3.20$, the block effect is not significant. Hence we return to the model (6.23) with the two main effects A and B which are significant according to Table 6.10.

Source	SS	df	MS	F	
Factor A	130.66	1	130.66	20.01	*
Factor B	626.33	2	313.17	47.96	*
$A \times B$	36.34	2	18.17	2.78	
Block	40.50	3	13.50	2.07	
Error	98.00	15	6.53		
Total	931.83	23			

Table 6.16: Analysis of variance table in model (6.42)

Source	SS	df	MS	F	
Factor A	130.66	1	130.66	16.54	*
Factor B	626.33	2	313.17	39.64	*
Block	40.50	3	13.50	1.71	
Error	134.34	17	7.90		
Total	931.83	23			

Table 6.17: Analysis of variance table in model (6.46)

6.5 Two-Factorial Model with Fixed Effects – Confidence Intervals and Elementary Tests

In a two-factorial experiment with fixed effects there are three different types of means: for A-levels, B-levels and $A \times B$-levels. In case of a non-random block effect, the fourth type of means is that of the blocks. In the following, we assume fixed block effects.

(i) Factor A

The means of the A-levels are

$$y_{i\cdot\cdot} = \frac{1}{br} \sum_{j=1}^{b} \sum_{k=1}^{r} y_{ijk} \sim N(\mu + \alpha_i, \frac{\sigma^2}{br}) \quad . \tag{6.47}$$

The variance σ^2 is estimated by $s^2 = MS_{Error}$ with df degrees of freedom. Here MS_{Error} is computed from the model which holds after testing for interaction and block effects.

The confidence intervals for $\mu + \alpha_i$ are now of the following form ($t_{df,1-\alpha/2}$: two-sided quantile)

$$y_{i\cdot\cdot} \pm t_{df,1-\alpha/2} \sqrt{\frac{s^2}{br}} \quad . \tag{6.48}$$

The standard error of the difference between two A-levels is $\sqrt{2s^2/br}$, so that the test statistic for H_0: $\alpha_{i_1} = \alpha_{i_2}$ is of the following form

$$t_{df} = \frac{y_{i_1\cdot\cdot} - y_{i_2\cdot\cdot}}{\sqrt{2s^2/br}} \quad . \tag{6.49}$$

(ii) Factor B

Similarly, we have

$$y_{\cdot j\cdot} = \frac{1}{ar}\sum_{i=1}^{a}\sum_{k=1}^{r} y_{ijk} \sim N(\mu + \beta_j, \frac{\sigma^2}{ar}) \quad . \tag{6.50}$$

The $(1-\alpha)$ confidence interval for $\mu + \beta_j$ is

$$y_{\cdot j\cdot} \pm t_{df,1-\alpha/2}\sqrt{\frac{s^2}{ar}} \tag{6.51}$$

and the test statistic for the comparison of means $(H_0\!:\!\beta_{j_1} = \beta_{j_2})$ is

$$t_{df} = \frac{y_{\cdot j_1\cdot} - y_{\cdot j_2\cdot}}{\sqrt{2s^2/ar}} \quad . \tag{6.52}$$

(iii) Factor $A \times B$

Here we have

$$y_{ij\cdot} = \frac{1}{r}\sum_{k=1}^{r} y_{ijk} \sim N(\mu + \alpha_i + \beta_j + (\alpha\beta)_{ij}, \frac{\sigma^2}{r}) \quad . \tag{6.53}$$

The $(1-\alpha)$ confidence interval for $\mu + \alpha_i + \beta_j + (\alpha\beta)_{ij}$ is

$$y_{ij\cdot} \pm t_{df,1-\alpha/2}\sqrt{s^2/r} \tag{6.54}$$

and the test statistic for the comparison of two $A \times B$-effects is

$$t_{df} = \frac{y_{i_1 j_1\cdot} - y_{i_2 j_2\cdot}}{\sqrt{2s^2/r}} \quad . \tag{6.55}$$

The significance of single effects is tested by

  (i) H_0: $\mu + \alpha_i = \mu_0$

$$t_{df} = \frac{y_{i\cdot\cdot} - \mu_0}{\sqrt{s^2/br}} \tag{6.56}$$

  (ii) H_0: $\mu + \beta_j = \mu_0$

$$t_{df} = \frac{y_{\cdot j\cdot} - \mu_0}{\sqrt{s^2/ar}} \tag{6.57}$$

  (iii) H_0: $\mu + \alpha_i + \beta_j + (\alpha\beta)_{ij} = \mu_0$

$$t_{df} = \frac{y_{ij\cdot} - \mu_0}{\sqrt{s^2/r}} \tag{6.58}$$

Here the statements in Section 4.4 about elementary and multiple tests hold.

Example 6.4: (Examples 6.2 and 6.3 continued) The test procedure leads to non-significant interaction and block effects. Hence, the independence model holds. From the appropriate analysis of variance table 6.10 we take

$$s^2 = 8.74 \quad \text{for} \quad df = 20.$$

From Table 6.8 we obtain the means of the two levels A_1 and A_2 and of the three levels B_1, B_2, and B_3:

$$A_1 : \quad y_{1..} \;=\; \frac{257}{3 \cdot 4} = 21.42$$

$$A_2 : \quad y_{2..} \;=\; \frac{201}{3 \cdot 4} = 16.75$$

$$B_1 : \quad y_{.1.} \;=\; \frac{104}{2 \cdot 4} = 13.00$$

$$B_2 : \quad y_{.2.} \;=\; \frac{150}{2 \cdot 4} = 18.75$$

$$B_3 : \quad y_{.3.} \;=\; \frac{204}{2 \cdot 4} = 25.50 \quad .$$

(i) A-levels confidence intervals

$$A_1 : \quad 21.42 \pm t_{20;0.975}\sqrt{8.74/3 \cdot 4} \;=\; 21.42 \pm 2.09 \cdot 0.85$$
$$=\; 21.42 \pm 1.78$$
$$\Rightarrow [19.64; 23.20]$$

$$A_2 : \quad 16.75 \pm 1.78$$
$$\Rightarrow [14.97; 18.53]$$

Test for $H_0: \alpha_1 = \alpha_2$ against $H_1: \alpha_1 > \alpha_2$

$$t_{20} \;=\; \frac{21.42 - 16.75}{\sqrt{2 \cdot 8.74/3 \cdot 4}} = \frac{4.67}{1.21} = 3.86$$
$$>\; 1.73 = t_{20;0.95} \text{ (one-sided)}$$

$\Rightarrow H_0$ is rejected.

(ii) B-levels

Confidence intervals

With $t_{20;0.975}\sqrt{8.74/2 \cdot 4} = 2.09 \cdot 1.05 = 2.19$, we obtain

$$B_1 : \quad 13.00 \pm 2.19 \quad \Rightarrow \quad [10.81; 15.19]$$
$$B_2 : \quad 18.75 \pm 2.19 \quad \Rightarrow \quad [16.56; 20.94]$$
$$B_3 : \quad 25.50 \pm 2.19 \quad \Rightarrow \quad [23.31; 27.69]$$

The pairwise comparisons of means reject the hypothesis of identity.

6.6 Two-Factorial Model with Random or Mixed Effects

The first part of Chapter 6 has assumed the effects of A and B to be fixed. This means that the factor levels of A and B are specified before the experiment and hence, the conclusions of the analysis of variance are only valid for these factor levels. Alternative designs allow the factors A and B to act randomly (model with random effects) or keep one factor fixed and choose the other factor by random (model with mixed effects).

6.6.1 Model with Random Effects

We assume that the levels of both factors A and B are chosen at random from populations A and B. The inferences will then be valid about all levels in the (two-dimensional) population. The response values in the model with random effects (or components of variance model) are

$$y_{ijk} = \mu + \alpha_i + \beta_j + (\alpha\beta)_{ij} + \epsilon_{ijk}$$
$$(i = 1, \ldots, a, \ j = 1, \ldots, b, \ k = 1, \ldots, r), \tag{6.59}$$

where α_i, β_j $(\alpha\beta)_{ij}$ are random variables independent of each other and of ϵ_{ijk}. We assume

$$\left.\begin{array}{l} \boldsymbol{\alpha} = (\alpha_1, \ldots, \alpha_a)' \sim N(\mathbf{0}, \sigma_\alpha^2 \, \boldsymbol{I}) \\ \boldsymbol{\beta} = (\beta_1, \ldots, \beta_b)' \sim N(\mathbf{0}, \sigma_\beta^2 \, \boldsymbol{I}) \\ (\boldsymbol{\alpha\beta}) = ((\alpha\beta)_{11}, \ldots, (\alpha\beta)_{ab})' \sim N(\mathbf{0}, \sigma_{\alpha\beta}^2 \, \boldsymbol{I}) \\ \boldsymbol{\epsilon} = (\epsilon_1, \ldots, \epsilon_{abr})' \sim N(\mathbf{0}, \sigma^2 \, \boldsymbol{I}) \end{array}\right\} . \tag{6.60}$$

In matrix notation, the covariance structure is as follows

$$\mathrm{E}\left(\begin{array}{c} \boldsymbol{\alpha} \\ \boldsymbol{\beta} \\ (\boldsymbol{\alpha\beta}) \\ \boldsymbol{\epsilon} \end{array}\right)(\boldsymbol{\alpha}, \boldsymbol{\beta}, (\boldsymbol{\alpha\beta}), \boldsymbol{\epsilon})' = \left(\begin{array}{cccc} \sigma_\alpha^2 \boldsymbol{I} & \mathbf{0} & \mathbf{0} & \mathbf{0} \\ \mathbf{0} & \sigma_\beta^2 \boldsymbol{I} & \mathbf{0} & \mathbf{0} \\ \mathbf{0} & \mathbf{0} & \sigma_{\alpha\beta}^2 \boldsymbol{I} & \mathbf{0} \\ \mathbf{0} & \mathbf{0} & \mathbf{0} & \sigma^2 \boldsymbol{I} \end{array}\right) .$$

Hence the variance of the response values is

$$\mathrm{Var}(y_{ijk}) = \sigma_\alpha^2 + \sigma_\beta^2 + \sigma_{\alpha\beta}^2 + \sigma^2 \ . \tag{6.61}$$

σ_α^2, σ_β^2, $\sigma_{\alpha\beta}^2$, σ^2 are called *variance components*. The hypotheses that we are interested in testing are H_0: $\sigma_\alpha^2 = 0$, H_0: $\sigma_\beta^2 = 0$ and H_0: $\sigma_{\alpha\beta}^2 = 0$.

The formulas for the decomposition of the variance SS_{Total} into SS_A, SS_B, $SS_{A\times B}$, and SS_{Error} and for the calculation of the variance remain unchanged, that is, all sums of squares are calculated as in the fixed effects case. However, to form the test statistics we must examine the expectation of the appropriate mean squares. We have

$$\begin{aligned} SS_A \ &= \ \frac{1}{br} \sum_{i=1}^{a} (Y_{i\cdot\cdot} - Y_{\cdots})^2 \\ &= \ \sum_{i=1}^{a} \sum_{j=1}^{b} \sum_{k=1}^{r} (y_{i\cdot\cdot} - y_{\cdots})^2 \ . \end{aligned} \tag{6.62}$$

With $\alpha. = \frac{1}{a}\sum_{i=1}^{a}\alpha_i$, $\beta. = \frac{1}{b}\sum_{j=1}^{b}\beta_j$, $(\alpha\beta)_{i.} = \frac{1}{b}\sum_{j=1}^{b}(\alpha\beta)_{ij}$, and $(\alpha\beta).. = \frac{1}{ab}\sum\sum(\alpha\beta)_{ij}$, we compute from model (6.59)

$$y_{i..} = \mu + \alpha_i + \beta. + (\alpha\beta)_{i.} + \epsilon_{i..}$$
$$y_{...} = \mu + \alpha. + \beta. + (\alpha\beta).. + \epsilon_{...}$$

so that

$$y_{i..} - y_{...} = (\alpha_i - \alpha.) + [(\alpha\beta)_{i.} - (\alpha\beta)..] + (\epsilon_{i..} - \epsilon_{...}) \quad . \tag{6.63}$$

Because of the mutual independence of the random effects and of the error, we have

$$E(y_{i..} - y_{...})^2 = E(\alpha_i - \alpha.)^2 + E[(\alpha\beta)_{i.} - (\alpha\beta)..]^2 + E(\epsilon_{i..} - \epsilon_{...})^2 \quad . \tag{6.64}$$

For the three components, we observe that

$$
\begin{aligned}
E(\alpha_i - \alpha.)^2 &= E(\alpha_i^2) + E(\alpha_.^2) - 2E(\alpha_i\alpha.) \\
&= \sigma_\alpha^2[1 + \frac{1}{a} - \frac{2}{a}] \\
&= \sigma_\alpha^2[1 - \frac{1}{a}] = \sigma_\alpha^2(\frac{a-1}{a}) \tag{6.65} \\
E[(\alpha\beta)_{i.} - (\alpha\beta)..]^2 &= E[(\alpha\beta)_{i.}^2] + E[(\alpha\beta)_.^2] - 2E[(\alpha\beta)_{i.}(\alpha\beta)..] \\
&= \sigma_{\alpha\beta}^2[\frac{1}{b} + \frac{1}{ab} - \frac{2}{ab}] \\
&= \sigma_{\alpha\beta}^2\left(\frac{a-1}{ab}\right) \tag{6.66} \\
E(\epsilon_{i..} - \epsilon_{...})^2 &= E(\epsilon_{i..}^2) + E(\epsilon_{...}^2) - 2E(\epsilon_{i..}\epsilon_{...}) \\
&= \sigma^2[\frac{1}{br} + \frac{1}{abr} - \frac{2}{abr}] \\
&= \sigma^2\left(\frac{a-1}{abr}\right) \quad , \tag{6.67}
\end{aligned}
$$

whence we find (cf. (6.62) and (6.64))

$$
\begin{aligned}
E(MS_A) &= \frac{1}{a-1}E(SS_A) \\
&= \sigma^2 + r\sigma_{\alpha\beta}^2 + br\sigma_\alpha^2 \quad . \tag{6.68}
\end{aligned}
$$

Similarly, we find

$$E(MS_B) = \sigma^2 + r\sigma_{\alpha\beta}^2 + ar\sigma_\beta^2 \tag{6.69}$$
$$E(MS_{A\times B}) = \sigma^2 + r\sigma_{\alpha\beta}^2 \tag{6.70}$$
$$E(MS_{Error}) = \sigma^2 \quad . \tag{6.71}$$

Estimation of the Variance Components

The estimates $\hat{\sigma}^2$, $\hat{\sigma}_\alpha^2$, $\hat{\sigma}_\beta^2$ and $\hat{\sigma}_{\alpha\beta}^2$ of the variance components σ^2, σ_α^2, σ_β^2 and $\sigma_{\alpha\beta}^2$ are computed from the equating system (6.68) – (6.71) in its sample version, that is, from the system

$$\left.\begin{array}{rlll}
MS_A &= br\hat{\sigma}_\alpha^2 & + r\hat{\sigma}_{\alpha\beta}^2 & + \hat{\sigma}^2 \\
MS_B &= ar\hat{\sigma}_\beta^2 & + r\hat{\sigma}_{\alpha\beta}^2 & + \hat{\sigma}^2 \\
MS_{A\times B} &= & r\hat{\sigma}_{\alpha\beta}^2 & + \hat{\sigma}^2 \\
MS_{Error} &= & & \hat{\sigma}^2
\end{array}\right\} \tag{6.72}$$

i.e.,

$$\begin{pmatrix} MS_A \\ MS_B \\ MS_{A\times B} \\ MS_{Error} \end{pmatrix} = \begin{pmatrix} br & 0 & r & 1 \\ 0 & ar & r & 1 \\ 0 & 0 & r & 1 \\ 0 & 0 & 0 & 1 \end{pmatrix} \begin{pmatrix} \hat{\sigma}_\alpha^2 \\ \hat{\sigma}_\beta^2 \\ \hat{\sigma}_{\alpha\beta}^2 \\ \hat{\sigma}^2 \end{pmatrix} \quad .$$

The coefficient matrix of this linear inhomogeneous system is of triangular shape with its determinant as

$$abr^3 \neq 0 \quad .$$

This yields the unique solution

$$\hat{\sigma}^2 = MS_{Error} \tag{6.73}$$

$$\hat{\sigma}_{\alpha\beta}^2 = \frac{1}{r}(MS_{A\times B} - MS_{Error}) \tag{6.74}$$

$$\hat{\sigma}_\beta^2 = \frac{1}{ar}(MS_B - MS_{A\times B}) \tag{6.75}$$

$$\hat{\sigma}_\alpha^2 = \frac{1}{br}(MS_A - MS_{A\times B}) \quad . \tag{6.76}$$

Testing of Hypotheses about the Variance Components

(i) H_0: $\sigma_{\alpha\beta}^2 = 0$

From the system (6.68) – (6.71) of the expectations of the MS's it can be seen that for H_0: $\sigma_{\alpha\beta}^2 = 0$ (no interaction) we have $E(MS_{A\times B}) = \sigma^2$. Hence the test statistic is of the form

$$F_{A\times B} = \frac{MS_{A\times B}}{MS_{Error}} \quad . \tag{6.77}$$

If H_0: $\sigma_{\alpha\beta}^2 = 0$ does not hold (that is, H_0 is rejected in favor of H_1: $\sigma_{\alpha\beta}^2 \neq 0$), we have $E(MS_{A\times B}) > E(MS_{Error})$. Hence H_0 is rejected if

$$F_{A\times B} > F_{(a-1)(b-1),ab(r-1);1-\alpha} \tag{6.78}$$

holds.

(ii) H_0: $\sigma_\alpha^2 = 0$

The comparison of $E(MS_A)$ (6.68) and $E(MS_{A\times B})$ (6.70) shows that both expectations are identical under H_0: $\sigma_\alpha^2 = 0$ but $E(MS_A) > E(MS_{A\times B})$ holds in case of H_1: $\sigma_\alpha^2 \neq 0$. The test statistic is then

$$F_A = \frac{MS_A}{MS_{A\times B}} \tag{6.79}$$

and H_0 is rejected if

$$F_A > F_{a-1,(a-1)(b-1);1-\alpha} \tag{6.80}$$

holds.

(iii) H_0: $\sigma_\beta^2 = 0$

Similarly, the test statistic for H_0: $\sigma_\beta^2 = 0$ against H_1: $\sigma_\beta^2 \neq 0$ is

$$F_B = \frac{MS_B}{MS_{A\times B}} \quad, \tag{6.81}$$

and H_0 is rejected if

$$F_B > F_{b-1,(a-1)(b-1);1-\alpha} \tag{6.82}$$

holds.

Source	SS	df	MS	F
Factor A	SS_A	$df_A = a - 1$	$MS_A = \frac{SS_A}{df_A}$	$F_A = \frac{MS_A}{MS_{A\times B}}$
Factor B	SS_B	$df_B = b - 1$	$MS_B = \frac{SS_B}{df_B}$	$F_B = \frac{MS_B}{MS_{A\times B}}$
Interaction $A \times B$	$SS_{A\times B}$	$df_{A\times B} = (a-1)(b-1)$	$MS_{A\times B} = \frac{SS_{A\times B}}{df_{A\times B}}$	$F_{A\times B} = \frac{MS_{A\times B}}{MS_{Error}}$
Error	SS_{Error}	$df_{Error} = ab(r-1)$	$MS_{Error} = \frac{SS_{Error}}{df_{Error}}$	
Total	SS_{Total}	$df_{Total} = abr - 1$		

Table 6.18: Analysis of variance table (two-factorial with interaction and random effects)

	SS	df	MS	F
A	130.66	1	130.66	$F_A = \frac{130.66}{18.17} = 7.19$
B	626.33	2	313.17	$F_B = \frac{313.17}{18.17} = 17.24$
$A \times B$	36.34	2	18.17	$F_{A\times B} = \frac{18.17}{7.69} = 2.36$
Error	138.50	18	7.69	
Total	931.83	23		

Table 6.19: Analysis of variance table for Table 6.8 in case of random effects

Remark: In the random effects model the test statistics F_A and F_B are formed with $MS_{A\times B}$ in the denominator. In the model with fixed effects, we have MS_{Error} in the denominator.

Example 6.5: We now consider the experiment in Example 6.2 as a two-factorial experiment with random effects. For this, we assume that the two types of beans (factor A) are chosen at random from a population, instead of being fixed effects. Similarly, we assume that the three phosphate fertilizers are chosen at random from a population. We assume the same response values as in Table 6.8 and adopt the first three columns from Table 6.9 for our analysis (Table 6.19). The estimated variance components are

$$
\hat{\sigma}^2 = 7.69
$$
$$
\hat{\sigma}^2_{\alpha\beta} = \frac{1}{4}(18.17 - 7.69) = 2.62
$$
$$
\hat{\sigma}^2_{\beta} = \frac{1}{2 \cdot 4}(313.17 - 18.17) = 36.88
$$
$$
\hat{\sigma}^2_{\alpha} = \frac{1}{3 \cdot 4}(130.66 - 18.17) = 9.37 \quad .
$$

The three variance components $\sigma^2_{\alpha\beta}$, σ^2_{α}, and σ^2_{β} are not significant at the 5%-level (critical values: $F_{1,2;0.95} = 18.51$, $F_{2,2;0.95} = 19.00$, $F_{2,18;0.95} = 3.55$).

Owing to the non-significance of $\sigma^2_{\alpha\beta}$, we return to the independence model. The analysis of variance table of this model is identical with Table 6.10 so that the two variance components σ^2_{α} and σ^2_{β} are significant.

6.6.2   Mixed Model

We now consider the situation where one factor (e.g., A) is fixed and the other factor (B) is random. The appropriate linear model in the *standard version by Scheffé* (1956, 1959) is

$$
y_{ijk} = \mu + \alpha_i + \beta_j + (\alpha\beta)_{ij} + \epsilon_{ijk}
$$
$$
(i = 1, \ldots, a,\ j = 1, \ldots, b,\ k = 1, \ldots, r)
$$
(6.83)

with the following assumptions

$$
\alpha_i\ :\ \text{fixed effect},\quad \sum_{i=1}^{a} \alpha_i = 0 \tag{6.84}
$$

$$
\beta_j\ :\ \text{random effect},\quad \beta_j \stackrel{i.i.d.}{\sim} N(0, \sigma^2_\beta) \tag{6.85}
$$

$$
(\alpha\beta)_{ij}\ :\ \text{random effect},\quad (\alpha\beta)_{ij} \stackrel{i.i.d.}{\sim} N(0, \frac{a-1}{a}\sigma^2_{\alpha\beta}) \tag{6.86}
$$

$$
\sum_{i=1}^{a} (\alpha\beta)_{ij} = (\alpha\beta)_{.j} = 0 \quad (j = 1, \ldots, b) \quad . \tag{6.87}
$$

We assume that the random variable groups β_j, $(\alpha\beta)_{ij}$, and ϵ_{ijk} are mutually independent, that is, we have $\mathrm{E}(\beta_j(\alpha\beta)_{ij}) = 0$ etc.. As in the above models, we have $\mathrm{E}(\epsilon) = \sigma^2 \mathbf{I}$.

The last assumption (6.87) means that the interaction effects between two different A-levels are correlated. For all $j = 1, \ldots, b$, we have

$$\text{Cov}[(\alpha\beta)_{i_1j}, (\alpha\beta)_{i_2j}] = -\frac{1}{a}\sigma^2_{\alpha\beta} \quad (i_1 \neq i_2) \quad , \tag{6.88}$$

but

$$\text{Cov}[(\alpha\beta)_{i_1j_1}, (\alpha\beta)_{i_2j_2}] = 0 \quad (j_1 \neq j_2, \text{ any } i_1, i_2) \quad . \tag{6.89}$$

For $a = 3$, we provide a short outline of the proof. Using (6.87), we obtain

$$
\begin{aligned}
\text{Cov}[(\alpha\beta)_{1j}, (\alpha\beta)_{2j}] &= \text{Cov}[(\alpha\beta)_{1j}, [-(\alpha\beta)_{1j} - (\alpha\beta)_{3j}]] \\
&= -\text{Var}(\alpha\beta)_{1j} - \text{Cov}[(\alpha\beta)_{1j}, (\alpha\beta)_{3j}] \quad ,
\end{aligned}
$$

whence

$$\text{Cov}[(\alpha\beta)_{1j}, (\alpha\beta)_{2j}] + \text{Cov}[(\alpha\beta)_{1j}, (\alpha\beta)_{3j}] = -\text{Var}(\alpha\beta)_{1j} = -\frac{3-1}{3}\sigma^2_{\alpha\beta} \quad .$$

Since $\text{Cov}[(\alpha\beta)_{i_1j}, (\alpha\beta)_{i_2j}]$ is identical for all pairs, (6.88) holds. If $a = b = 2$ and $r = 1$, then the model (6.83) with all assumptions has a 4-dimensional normal distribution

$$
\begin{pmatrix} y_{11} \\ y_{21} \\ y_{12} \\ y_{22} \end{pmatrix} \sim N \left(\begin{pmatrix} \mu + \alpha_1 \\ \mu + \alpha_2 \\ \mu + \alpha_1 \\ \mu + \alpha_2 \end{pmatrix}, \begin{pmatrix} \tilde{\sigma}^2 & \sigma^2_* & 0 & 0 \\ \sigma^2_* & \tilde{\sigma}^2 & 0 & 0 \\ 0 & 0 & \tilde{\sigma}^2 & \sigma^2_* \\ 0 & 0 & \sigma^2_* & \tilde{\sigma}^2 \end{pmatrix} \right) \tag{6.90}
$$

with

$$
\begin{aligned}
\text{Var}(y_{ij}) &= \tilde{\sigma}^2 = \sigma^2_\beta + \sigma^2_{\alpha\beta}\frac{a-1}{a} + \sigma^2 \tag{6.91} \\
&= (\sigma^2_{\alpha\beta} + \sigma^2) + \sigma^2_* \quad ,
\end{aligned}
$$

using the identity $\sigma^2_* = \sigma^2_\beta - \frac{1}{a}\sigma^2_{\alpha\beta}$. The covariance matrix (6.90) can now be written as

$$\boldsymbol{\Sigma} = ((\sigma^2_{\alpha\beta} + \sigma^2)\mathbf{I}_2 + \sigma^2_* \boldsymbol{J}_2) \otimes \boldsymbol{I}$$

where $\otimes$ is the Kronecker-product (cf. A 100). However, the first matrix has a compound symmetrical structure (3.263) so that the parameter estimates of the fixed effects are computed according to the OLS method (cf. Theorem 3.17):

$$
\begin{aligned}
r = 1: \quad &\hat{\mu} = y_{..} \quad \text{and} \quad \hat{\alpha}_i = y_{i.} - y_{..} \\
r > 1: \quad &\hat{\mu} = y_{...} \quad \text{and} \quad \hat{\alpha}_i = y_{i..} - y_{...} \quad .
\end{aligned}
$$

Expectations of the MS's

The specification of the A-effects and the reparametrization of the variance of $(\alpha\beta)_{ij}$ in $\sigma_{\alpha\beta}^2 \frac{a-1}{a}$, as well as the constraints (6.87), have an effect on the expected mean squares. The expectations of the MS's are now

$$E(MS_A) = \sigma^2 + r\sigma_{\alpha\beta}^2 + \frac{br\sum_{i=1}^a \alpha_i^2}{a-1} \tag{6.92}$$

$$E(MS_B) = \sigma^2 + ar\sigma_\beta^2 \tag{6.93}$$

$$E(MS_{A\times B}) = \sigma^2 + r\sigma_{\alpha\beta}^2 \tag{6.94}$$

$$E(MS_{Error}) = \sigma^2 \ . \tag{6.95}$$

The test statistic for testing H_0: *no A-effect*, i.e., H_0: $\alpha_i = 0$ (for all i), is

$$F_A = F_{a-1,(a-1)(b-1)} = \frac{MS_A}{MS_{A\times B}} \ . \tag{6.96}$$

The test statistic for H_0: $\sigma_\beta^2 = 0$ is

$$F_B = F_{b-1,ab(r-1)} = \frac{MS_B}{MS_{Error}} \ . \tag{6.97}$$

The test statistic for H_0: $\sigma_{\alpha\beta}^2 = 0$ is

$$F_{A\times B} = F_{(a-1)(b-1),ab(r-1)} = \frac{MS_{A\times B}}{MS_{Error}} \ . \tag{6.98}$$

Estimation of the Variance Components

The variance components may be estimated by solving the following system (6.92) – (6.95) in its sample version:

$$
\begin{aligned}
MS_A &= \frac{br}{a-1}\sum\alpha_i^2 &&+ \ r\hat\sigma_{\alpha\beta}^2 \ + \ \hat\sigma^2 \\
MS_B &= &ar\hat\sigma_\beta^2 \quad\quad &&+ \ \hat\sigma^2 \\
MS_{A\times B} &= &&r\hat\sigma_{\alpha\beta}^2 \ + \ \hat\sigma^2 \\
MS_{Error} &= &&\hat\sigma^2
\end{aligned}
$$

$$\implies \hat\sigma^2 = MS_{Error} \tag{6.99}$$

$$\hat\sigma_{\alpha\beta}^2 = \frac{MS_{A\times B} - MS_{Error}}{r} \tag{6.100}$$

$$\hat\sigma_\beta^2 = \frac{MS_B - MS_{Error}}{ar} \ . \tag{6.101}$$

In addition to the standard model with intraclass correlation structure, several other versions of the mixed model exist (cf. Hocking, 1973). An important version is the model with independent interaction effects that assumes

$$(\alpha\beta)_{ij} \overset{i.i.d.}{\sim} N(0, \sigma_{\alpha\beta}^2) \quad \text{(for all } i, j) \ . \tag{6.102}$$

Source	SS	df	$E(MS)$	F
A	SS_A	$a-1$	$\sigma^2 + r\sigma^2_{\alpha\beta} + \frac{br}{a-1}\sum\alpha^2_i$	$F_A = \frac{MS_A}{MS_{A\times B}}$
B	SS_B	$b-1$	$\sigma^2 + ar\sigma^2_\beta$	$F_B = \frac{MS_B}{MS_{Error}}$
$A\times B$	$SS_{A\times B}$	$(a-1)(b-1)$	$\sigma^2 + r\sigma^2_{\alpha\beta}$	$F_{A\times B} = \frac{MS_{A\times B}}{MS_{Error}}$
Error	SS_{Error}	$ab(r-1)$	σ^2	
Total	SS_{Total}	$abr-1$		

Table 6.20: Analysis of variance table in the mixed model (standard model, dependent interaction effects)

Furthermore, independence of the $(\alpha\beta)_{ij}$ from the β_j and the ϵ_{ij} is assumed as in the standard model.

$E(MS_B)$ now changes to

$$E(MS_B) = \sigma^2 + r\sigma^2_{\alpha\beta} + ar\sigma^2_\beta \tag{6.103}$$

and the test statistic for H_0: $\sigma^2_\beta = 0$ changes to

$$F_B = F_{b-1,(a-1)(b-1)} = \frac{MS_B}{MS_{A\times B}} \tag{6.104}$$

Source	SS	df	$E(MS)$	F
A	SS_A	$a-1$	$\sigma^2 + r\sigma^2_{\alpha\beta} + \frac{br}{a-1}\sum\alpha^2_i$	$F_A = \frac{MS_A}{MS_{A\times B}}$
B	SS_B	$b-1$	$\sigma^2 + r\sigma^2_{\alpha\beta} + ar\sigma^2_\beta$	$F_B = \frac{MS_B}{MS_{A\times B}}$
$A\times B$	$SS_{A\times B}$	$(a-1)(b-1)$	$\sigma^2 + r\sigma^2_{\alpha\beta}$	$F_{A\times B} = \frac{MS_{A\times B}}{MS_{Error}}$
Error	SS_{Error}	$ab(r-1)$	σ^2	
Total	SS_{Total}	$abr-1$		

Table 6.21: Analysis of variance table in the mixed model with independent interaction effects

The choice of mixed models should always be dictated by the data. In model (6.83), we have for the covariance within the response values

$$\text{Cov}(y_{i_1 j_1 k_1}, y_{i_2 j_2 k_2}) = \delta_{j_1 j_2}\sigma^2_\beta + \text{Cov}[(\alpha\beta)_{i_1 j_1}, (\alpha\beta)_{i_2 j_2}] + \sigma^2 \quad. \tag{6.105}$$

If factor B represents, e.g., b time intervals (24-hour measure of blood pressure) and if factor A represents the fixed effect placebo/medicament (p/m), then the assumption $\text{Cov}[(\alpha\beta)_{Pj}, (\alpha\beta)_{Mj}] = 0$ would be reasonable which is the opposite of (6.88). Similarly, (6.89) would have to be changed to $\text{Cov}[(\alpha\beta)_{Pj_1}, (\alpha\beta)_{Pj_2}] \neq 0$ or $\text{Cov}[(\alpha\beta)_{Mj_1}, (\alpha\beta)_{Mj_2}] \neq 0$ $(j_1 \neq j_2)$ respectively. These models are described in Chapter 7.

6.7   Three-Factorial Designs

The inclusion of a third factor in the experiment increases the number of parameters to be estimated. At the same time, the interpretation also becomes more difficult.

We denote the three factors (treatments) by A, B, and C and their factor levels by $i = 1, \ldots, a$, $j = 1, \ldots, b$, and $k = 1, \ldots, c$. Furthermore, we assume r replicates each, e.g., the randomized block design with r blocks and abc observations each. The appropriate model is the following additive model

$$
\begin{aligned}
y_{ijkl} = \;& \mu + \alpha_i + \beta_j + \gamma_k + (\alpha\beta)_{ij} + (\alpha\gamma)_{ik} + (\beta\gamma)_{jk} + (\alpha\beta\gamma)_{ijk} \\
& + \tau_l + \epsilon_{ijkl} \quad . \\
& (l = 1, \ldots, r)
\end{aligned}
\tag{6.106}
$$

In addition to the two-way interactions $(\alpha\beta)_{ij}$, $(\beta\gamma)_{jk}$, and $(\alpha\gamma)_{ik}$, we now have the three-way interaction $(\alpha\beta\gamma)_{ijk}$. We assume the usual constraints for the main effects and the two-way interactions. Additionally we assume

$$
\sum_i (\alpha\beta\gamma)_{ijk} = \sum_j (\alpha\beta\gamma)_{ijk} = \sum_k (\alpha\beta\gamma)_{ijk} = 0 \quad .
\tag{6.107}
$$

The test strategy is similar to the two-factorial model, that is the three-way interaction is tested first. If H_0: $(\alpha\beta\gamma)_{ijk} = 0$ is rejected, then all of the two-way interactions and the main effects cannot be interpreted seperately. The test strategy and especially the interpretation of submodels will be discussed in detail in Chapter 10 for models with categorical response. The results of Chapter 10 are valid for models with continuous response analogously.

The total response values are given in Table 6.22. The sums of squares are as follows:

$$
C = \frac{Y_{....}^2}{abcr} \quad \text{(correction term)}
$$

$$
SS_{Total} = \sum\sum\sum\sum y_{ijkl}^2 - C
$$

$$
SS_{Block} = \frac{1}{abc}\sum_{l=1}^{r} Y_{...l}^2 - C
$$

$$
SS_A = \frac{1}{bcr}\sum_i Y_{i...}^2 - C
$$

$$
SS_B = \frac{1}{acr}\sum_j Y_{.j..}^2 - C
$$

$$
SS_{A\times B} = \frac{1}{cr}\sum_i\sum_j Y_{ij..}^2 - C - SS_A - SS_B
$$

$$
SS_C = \frac{1}{abr}\sum_k Y_{..k.}^2 - C
$$

$$SS_{A \times C} = \frac{1}{br} \sum_i \sum_k Y_{i \cdot k \cdot}^2 - C - SS_A - SS_C$$

$$SS_{B \times C} = \frac{1}{ar} \sum_j \sum_k Y_{\cdot j k \cdot}^2 - C - SS_B - SS_C$$

$$SS_{A \times B \times C} = \frac{1}{r} \sum_i \sum_j \sum_k Y_{ijk \cdot}^2 - C$$
$$-SS_A - SS_B - SS_C$$
$$-SS_{A \times B} - SS_{A \times C} - SS_{B \times C}$$

$$SS_{Error} = SS_{Total} - SS_{Block}$$
$$-SS_A - SS_B - SS_C$$
$$-SS_{A \times B} - SS_{A \times C} - SS_{B \times C}$$
$$-SS_{A \times B \times C} \quad .$$

As in the above models with fixed effects, $MS = \frac{SS}{df}$ holds (cf. Table 6.23). The test statistics in general are

$$F_{Effect} = \frac{MS_{Effect}}{MS_{Error}} \quad . \tag{6.108}$$

Factor A	Factor B	Factor C				
		1	2	$\cdots$	c	Sum
1	1	$Y_{111 \cdot}$	$Y_{112 \cdot}$	$\cdots$	$Y_{11c \cdot}$	$Y_{11 \cdot \cdot}$
	2	$Y_{121 \cdot}$	$Y_{122 \cdot}$	$\cdots$	$Y_{12c \cdot}$	$Y_{12 \cdot \cdot}$
	$\vdots$	$\vdots$	$\vdots$		$\vdots$	$\vdots$
	b	$Y_{1b1 \cdot}$	$Y_{1b2 \cdot}$	$\cdots$	$Y_{1bc \cdot}$	$Y_{1b \cdot \cdot}$
	Sum	$Y_{1 \cdot 1 \cdot}$	$Y_{1 \cdot 2 \cdot}$	$\cdots$	$Y_{1 \cdot c \cdot}$	$Y_{1 \cdots}$
$\vdots$	$\vdots$			$\vdots$		$\vdots$
a	1	$Y_{a11 \cdot}$	$Y_{a12 \cdot}$	$\cdots$	$Y_{a1c \cdot}$	$Y_{a1 \cdot \cdot}$
	2	$Y_{a21 \cdot}$	$Y_{a22 \cdot}$	$\cdots$	$Y_{a2c \cdot}$	$Y_{a2 \cdot \cdot}$
	$\vdots$	$\vdots$	$\vdots$		$\vdots$	$\vdots$
	b	$Y_{ab1 \cdot}$	$Y_{ab2 \cdot}$	$\cdots$	$Y_{abc \cdot}$	$Y_{ab \cdot \cdot}$
	Sum	$Y_{a \cdot 1 \cdot}$	$Y_{a \cdot 2 \cdot}$	$\cdots$	$Y_{a \cdot c \cdot}$	$Y_{a \cdots}$
Sum		$Y_{\cdot \cdot 1 \cdot}$	$Y_{\cdot \cdot 2 \cdot}$	$\cdots$	$Y_{\cdot \cdot c \cdot}$	$Y_{\cdots}$

Table 6.22: Total response per block of the (A, B, C)–factor combinations

Example 6.6: The firmness Y of a ceramic material is dependent on the pressure (A), on the temperature (B), and on an additive (C). A three-factorial experiment, that includes all three factors at two levels low/high, is to analyse the influence on the response Y. A randomized block design is chosen with $r = 2$

Source	SS	df	MS	F
Block	SS_{Block}	$r-1$	MS_{Block}	F_{Block}
A	SS_A	$a-1$	MS_A	F_A
B	SS_B	$b-1$	MS_B	F_B
C	SS_C	$c-1$	MS_C	F_C
$A \times B$	$SS_{A\times B}$	$(a-1)(b-1)$	$MS_{A\times B}$	$F_{A\times B}$
$A \times C$	$SS_{A\times C}$	$(a-1)(c-1)$	$MS_{A\times C}$	$F_{A\times C}$
$B \times C$	$SS_{B\times C}$	$(b-1)(c-1)$	$MS_{B\times C}$	$F_{B\times C}$
$A \times B \times C$	$SS_{A\times B\times C}$	$(a-1)(b-1)(c-1)$	$MS_{A\times B\times C}$	$F_{A\times B\times C}$
Error	SS_{Error}	$(r-1)(abc-1)$	MS_{Error}	
Total	SS_{Total}	$abcr-1$		

Table 6.23: Three-factorial analysis of variance table

blocks of workpieces that are homogeneous within the blocks and heterogeneous between the blocks. The results are shown in Table 6.24.

We compute ($N = abcr = 2^4 = 16$)

$$C = \frac{Y_{....}^2}{N} = \frac{249^2}{16} = 3875.06$$

$$SS_{Total} = 5175 - C = 1299.94$$

$$SS_{Block} = \frac{1}{8}(108^2 + 141^2) - C = 3943.13 - C = 68.07$$

$$SS_A = \frac{1}{8}(116^2 + 133^2) - C = 3893.13 - C = 18.07$$

$$SS_B = \frac{1}{8}((42 + 54)^2 + (74 + 79)^2) - C = 4078.13 - C = 203.07$$

$$SS_{A\times B} = \frac{1}{4}(42^2 + 74^2 + 54^2 + 79^2) - C - SS_A - SS_B$$
$$= 4099.25 - C - SS_A - SS_B = 3.05$$

$$SS_C = \frac{1}{8}(105^2 + 144^2) - C = 3970.13 - C = 95.07$$

$$SS_{A\times C} = \frac{1}{4}(48^2 + 68^2 + 57^2 + 76^2) - C - SS_A - SS_C = 0.05$$

$$SS_{B\times C} = \frac{1}{4}((14 + 16 + 18 + 20)^2 + (4 + 8 + 6 + 10)^2$$
$$+ (7 + 11 + 9 + 10)^2 + (24 + 32 + 26 + 34)^2)$$
$$- C - SS_B - SS_C = 885.05$$

$$SS_{A\times B\times C} = \frac{1}{2}((14 + 16)^2 + \cdots + (26 + 34)^2) - C$$
$$- SS_A - SS_B - SS_{A\times B} - SS_C - SS_{A\times C} - SS_{B\times C} = 3.08$$

$$SS_{Error} = 24.43$$

Result: The F-tests with $F_{1,7;0.95} = 5.99$ show significance for the following effects: *block, B, C*, and $B \times C$. The influence of A is significant for none of the effects, hence the analysis can be done in a two-factorial $B \times C$-design. (Table 6.26, $F_{1,11;0.95} = 4.84$). The response Y is maximized for the combination $B_2 \times C_2$.

		Block 1 2	Block 1 2	
		C_1	C_2	Sum
A_1	B_1	14 , 16	4 , 8	42
	B_2	7 , 11	24 , 32	74
		48	68	116
A_2	B_1	18 , 20	6 , 10	54
	B_2	9 , 10	26 , 34	79
		57	76	133
	Sum	105	144	249

$$Y_{...1} = 108 \quad , \quad Y_{...2} = 141$$

Table 6.24: Response values for Example 6.6

	SS	df	MS	F	
Block	68.07	1	68.07	19.50	*
A	18.07	1	18.07	5.18	
B	203.07	1	203.07	58.19	*
C	95.07	1	95.07	27.24	*
$A \times B$	3.05	1	3.05	0.87	
$A \times C$	0.05	1	0.05	0.01	
$B \times C$	885.05	1	885.05	253.60	*
$A \times B \times C$	3.08	1	3.08	0.88	
Error	24.43	7	3.49		
Total	1299.94	15			

Table 6.25: Analysis of variance in the $A \times B \times C$ design for Example 6.6

6.8 Split-Plot Design

In many practical applications of the randomized block design it is not possible to arrange all factor combinations at random within one block. This is the case if the factors require different numbers of experimental units, e.g., because of technical reasons. Consider some examples (cf. Montgomery, 1976, pp. 292–300, Petersen, 1985, pp. 134–145):

	SS	df	MS	F
Block	68.07	1	68.07	15.37 *
B	203.07	1	203.07	45.84 *
C	95.07	1	95.07	21.46 *
$B \times C$	885.05	1	885.05	199.79 *
Error	48.68	11	4.43	
Total	1299.94	15		

Table 6.26: Analysis of variance in the $B \times C$ design for Example 6.6

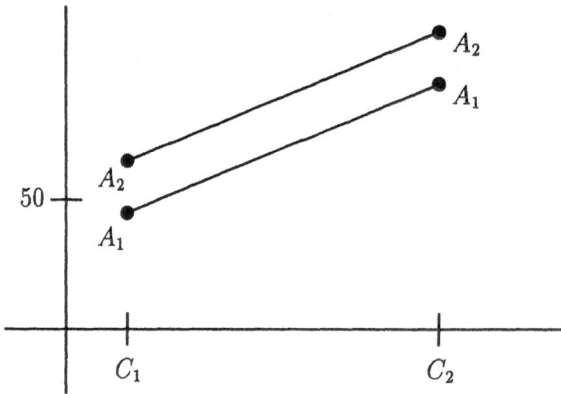

Figure 6.5: $A \times C$ – response

- Employment of various drill machines (factor B, only possible on larger fields) and of various fertilizers (factor C, may be employed on smaller fields as well). In this case factor B is set and only factor C is randomized in the blocks.

- Combination of three different paper pulp preperation methods and of four different temperatures in the paper manufacturing. Each replicate of the experiment requires 12 observations. In a completely randomized design, a factor combination (pulp i, temperature j) would have to be chosen at random within the block. In this example however, this procedure may not be economical. Hence, the three types of pulp are divided in four sample units and the temperature is randomized within these units.

Split-plot designs are used if the possibilities for randomization are restricted. The large units are called *whole-plots* while the smaller units are called *subplots* (or split-plots).

In this design of experiment, the whole-plot factor effects are estimated from the large units while the subplot effects and the interaction whole-plot – subplot is estimated from the small units. This design however leads to two experimental

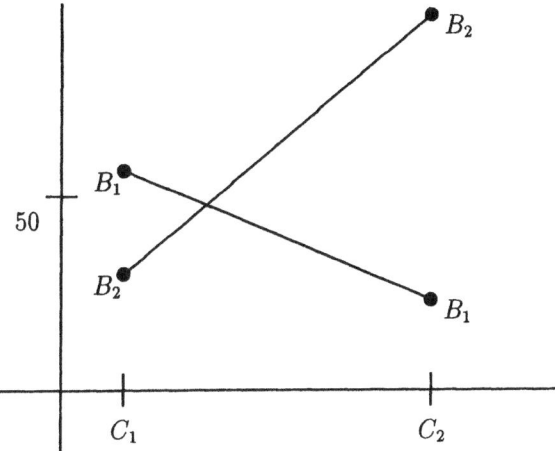

Figure 6.6: $B \times C$ – response

errors. The error associated with the subplot is the smaller one. The reason for this is the larger number of degrees of freedom of the subplot error, as well as the fact that the units in the subplots tend to be positively correlated in the response.

In our examples

- the drill machine is the whole-plot and the fertilizer the subplot,

- the type of pulp is the whole-plot and the temperature is the subplot.

The linear model for the two-factorial split-plot design is (Montgomery, 1976, p. 293)

$$y_{ijk} = \mu + \tau_i + \beta_j + (\tau\beta)_{ij} + \gamma_k + (\tau\gamma)_{ik} + (\beta\gamma)_{jk} + (\tau\beta\gamma)_{ijk} + \epsilon_{ijk} \quad (6.109)$$
$$(i = 1, \ldots, a; \, j = 1, \ldots, b; \, k = 1, \ldots, c).$$

Here the parameters

τ_i : random block effect (factor A)
β_j : whole-plot effect (factor B)
$(\tau\beta)_{ij}$: whole-plot error ($= A \times B$ interaction)

are the whole-plot parameters and the subplot parameters are:

γ_k : treatment effect factor C

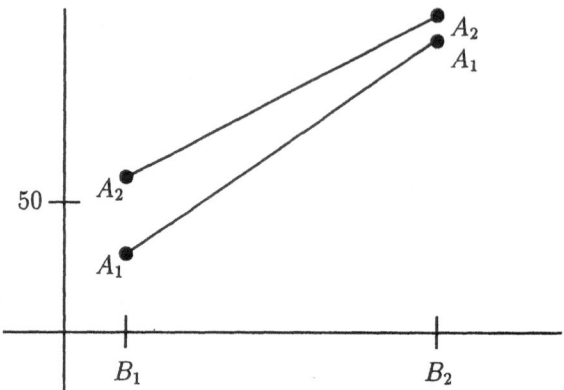

Figure 6.7: $A \times B$ – response

$(\tau\gamma)_{ik}$: $A \times C$ interaction
$(\beta\gamma)_{jk}$: $B \times C$ interaction
$(\tau\beta\gamma)_{ijk}$: subplot error ($= A \times B \times C$ interaction).

The sums of squares are computed as in the three-factorial model without replication (i.e., $r = 1$ in the SS's of the the previous section).

   The test statistics are given in Table 6.27. The effects to be tested are the main effects of factor B and factor C as well as the interaction $B \times C$. The test strategy starts out as in the two-factorial model, that is, with the $B \times C$-interaction.

Source	SS	df	MS	F
Block(A)	SS_A	$a-1$	MS_A	
B	SS_B	$b-1$	MS_B	$F_B = \dfrac{MS_B}{MS_{A\times B}}$
Error $(A \times B)$	$SS_{A\times B}$	$(a-1)(b-1)$	$MS_{A\times B}$	
C	SS_C	$c-1$	MS_C	$F_C = \dfrac{MS_C}{MS_{A\times B\times C}}$
$A \times C$	$SS_{A\times C}$	$(a-1)(c-1)$	$MS_{A\times C}$	
$B \times C$	$SS_{B\times C}$	$(b-1)(c-1)$	$MS_{B\times C}$	$F_{B\times C} = \dfrac{MS_{B\times C}}{MS_{A\times B\times C}}$
Error $(A\times B\times C)$	$SS_{A\times B\times C}$	$a(b-1)(c-1)$	$MS_{A\times B\times C}$	
Totel	SS_{Total}	$abc-1$		

Table 6.27: Analysis of variance in the split-plot design

Example 6.7: A laboratory has two furnaces of which one can only be heated up to 500°C. The hardness of a ceramic having dependence upon two additives and the temperature is to be tested in a split-plot design.

Factor A (block)   : replication on $r = 3$ days
Factor B (whole-plot) : temperature
                          B_1 : 500°C (furnace I),
                          B_2 : 750°C (furnace II)
Factor C (subplot)   : additive
                          C_1 : 10%
                          C_2 : 20%.

Because of $F_{1,2;0.95} = 18.51$, only factor C is significant (Table 6.29). Hence the experiment can be conducted with a single factor additive (Table 6.30).

	I			II			III	
	B_1	B_2		B_1	B_2		B_1	B_2
	C_1	C_2		C_2	C_2		C_2	C_1
	4	6		7	7		9	9
	C_2	C_1		C_1	C_1		C_1	C_2
	7	5		5	6		4	10

Block	B_1	B_2	Sum
I	11	11	22
II	12	13	25
III	13	19	32
Sum	36	43	79

	B_1	B_2	
C_1	13	20	33
C_2	23	23	46
	36	43	79

Table 6.28: Response tables

	SS	df	MS	F
Block (A)	13.17	2	6.58	
B	4.08	1	4.08	$F_B = 1.58$
Error ($A \times B$)	5.17	2	2.58	
C	14.08	1	14.08	$F_C = 24.14$ *
$A \times C$	1.17	2	0.58	
$B \times C$	4.08	1	4.08	$F_{B \times C} = 7.00$
Error ($A \times B \times C$)	1.17	2	0.58	
Total	42.92	11		

Table 6.29: Analysis of variance table for Example 6.7

6.9 2^k-Factorial Design

Especially in the industrial area, factorial designs in the first stage of an analysis are usually conducted with only two factor levels for each of the included factors.

Source	SS	df	MS	F
Factor C	14.08	1	14.08	$F_C = 4.88$
Error	28.83	10	2.88	
Total	42.92	11		

Table 6.30: One-factor analysis of variance table (Example 6.7)

The idea of this procedure is to make the important effects identifiable so that the analysis in the following stages can test factor combinations more specifically and more cost-effective. A complete analysis with k factors, each of two levels, requires 2^k replications for one trial. This fact leads to the nomenclature of the design: 2^k-experiment. The restriction to two levels for all factors makes a minimum of observations possible for a complete factorial experiment with all two-way and higher order interactions. We assume fixed effects and complete randomization. The same linear models and constraints as for the previous two- and three-factorial designs are valid in the 2^k-design too. The advantage of this design is the immediate computation of the sums of squares from special constraints which are linked to the effects.

6.9.1 The 2^2-Design

The 2^2 design has already been introduced in Section 6.1. Two factors A and B are run at two levels each (e.g., low and high). The chosen parametrization is usually

$$\text{low: } 0\ , \qquad \text{high: } 1\ .$$

The high levels of the factors are represented by a, or b respectively, and the low level is denoted by the absence of the corresponding letter. If both factors are at the low level, (1) is used as representation:

$$
\begin{aligned}
(0,0) &\longrightarrow (1) \\
(1,0) &\longrightarrow a \\
(0,1) &\longrightarrow b \\
(1,1) &\longrightarrow ab\ .
\end{aligned}
$$

Here (1), a, b, ab denote the response for all r replicates. The average effect of a factor is defined as the reaction of the response to a change of level of this factor, averaged over the levels of the other factor. The effect of A at the low level of B is $[a - (1)]/r$ and the effect of A at the high level of B is $[ab - b]r$. The average effect of A is then

$$A = \frac{1}{2r}[ab + a - b - (1)]\ . \tag{6.110}$$

The average effect of B is

$$B = \frac{1}{2r}[ab + b - a - (1)]\ . \tag{6.111}$$

The interaction effect AB is defined as the average difference between the effect of A at the high level of B and the effect of A at the low level of B. Thus

$$
\begin{aligned}
AB &= \frac{1}{2r}[(ab - b) - (a - (1))] \\
&= \frac{1}{2r}[ab + (1) - a - b] \quad .
\end{aligned}
$$

(6.112)

Similarly, the effect BA may be defined as the average difference between the effect of B at the high level of A (i.e., $(ab - a)/r$) and the effect of B at the low level of A (i.e., $(b - (1))/r$). We obviously have $AB = BA$. Hence, the average effects A, B, and AB are linear orthogonal contrasts in the total response values $(1), a, b, ab$, except for the factor $\frac{1}{2r}$.

Let $\boldsymbol{Y}_* = ((1), a, b, ab)'$ be the vector of the total response values. Then

$$
\left.
\begin{aligned}
A &= \frac{1}{2r}\boldsymbol{c}'_A \boldsymbol{Y}_*, \quad B = \frac{1}{2r}\boldsymbol{c}'_B \boldsymbol{Y}_* \\
AB &= \frac{1}{2r}\boldsymbol{c}'_{AB} \boldsymbol{Y}_*
\end{aligned}
\right\}
$$

(6.113)

holds where the contrasts $\boldsymbol{c}_A, \boldsymbol{c}_B, \boldsymbol{c}_{AB}$ are taken from Table 6.31.

	(1)	a	b	ab	Contrast
A	-1	$+1$	-1	$+1$	$\boldsymbol{c}'_A$
B	-1	-1	$+1$	$+1$	$\boldsymbol{c}'_B$
AB	$+1$	-1	-1	$+1$	$\boldsymbol{c}'_{AB}$

Table 6.31: Contrasts in the 2^2–design

We have $\boldsymbol{c}'_A \boldsymbol{c}_A = \boldsymbol{c}'_B \boldsymbol{c}_B = \boldsymbol{c}'_{AB} \boldsymbol{c}_{AB} = 4$.
From Section 4.3.2, we find the following sums of squares

$$
SS_A = \frac{(\boldsymbol{c}'_A \boldsymbol{Y}_*)^2}{(r\boldsymbol{c}'_A \boldsymbol{c}_A)} = \frac{(ab + a - b - (1))^2}{4r}
$$

(6.114)

$$
SS_B = \frac{(\boldsymbol{c}'_B \boldsymbol{Y}_*)^2}{(r\boldsymbol{c}'_B \boldsymbol{c}_B)} = \frac{(ab + b - a - (1))^2}{4r}
$$

(6.115)

$$
SS_{AB} = \frac{(\boldsymbol{c}'_{AB} \boldsymbol{Y}_*)^2}{(r\boldsymbol{c}'_{AB} \boldsymbol{c}_{AB})} = \frac{(ab + (1) - a - b)^2}{4r}
$$

(6.116)

The sum of squares SS_{Total} is computed as usual

$$
SS_{Total} = \sum_{i=1}^{2}\sum_{j=1}^{2}\sum_{k=1}^{r} y_{ijk}^2 - \frac{Y_{...}^2}{4r}
$$

(6.117)

and has $(2 \cdot 2 \cdot r) - 1$ degrees of freedom. As usual, we have

$$SS_{Error} = SS_{Total} - SS_A - SS_B - SS_{AB} \quad . \tag{6.118}$$

We now illustrate this procedure with an example.

Example 6.8: We wish to investigate the influence of the factors A (tempera-ture, 0: low, 1: high) and B (catalytic converter, 0: not used, 1: used) on the response Y (hardness of a ceramic material). The response is shown in Table 6.32.

Combination	Replication 1	2	Total response	Coding
(0,0)	86	92	178	(1)
(1,0)	47	39	86	a
(0,1)	104	114	218	b
(1,1)	141	153	294	ab
			$Y_{...} = 776$	

Table 6.32: Response in Example 6.8

From this table, we obtain the average effects

$$A = \frac{1}{4}[294 + 86 - 218 - 178] = -4$$

$$B = \frac{1}{4}[294 + 218 - 86 - 178] = 62$$

$$AB = \frac{1}{4}[294 + 178 - 86 - 218] = 42$$

and from these the sums of squares

$$SS_A = \frac{(4A)^2}{4 \cdot 2} = 32$$

$$SS_B = \frac{(4B)^2}{4 \cdot 2} = 7688$$

$$SS_{AB} = \frac{(4AB)^2}{4 \cdot 2} = 3528 \quad .$$

Furthermore, we have

$$SS_{Total} = (86^2 + \ldots + 153^2) - \frac{776^2}{8} = 86692 - 75272 = 11420 \quad ,$$

$$SS_{Error} = 172 \quad .$$

The analysis of variance table is shown in Table 6.33.

	SS	df	MS	F
A	32	1	32	$F_A = 0.74$
B	7688	1	7688	$F_B = 178.79$ *
AB	3528	1	3528	$F_{AB} = 82.05$ *
Error	172	4	43	
Total	11420	7		

Table 6.33: Analysis of variance for Example 6.8

6.9.2 The 2^3-Design

Suppose that in a complete factorial experiment three binary factors A, B, C are to be studied. The number of combinations is 8 and with r replicates we have $N = 8r$ observations that are to be analysed for their influence on a response.

Assume the total response values are (in the so-called *standard order*)

$$Y_* = [(1), a, b, ab, c, ac, bc, abc]' \quad . \tag{6.119}$$

In the coding 0: low and 1: high, this corresponds to the triples $(0, 0, 0), (1, 0, 0),$ $(0, 1, 0), (1, 1, 0), \ldots, (1, 1, 1)$. The response values can be arranged as a three-dimensional contingency table (cf. Table 6.35). The effects are determined by linear contrasts

$$c'_{Effect} \cdot ((1), a, b, ab, c, ac, bc, abc) = c'_{Effect} \cdot Y_* \tag{6.120}$$

(cf. Table 6.34).

Factorial effect	Factor combination							
	(1)	a	b	ab	c	ac	bc	abc
I	+	+	+	+	+	+	+	+
A	−	+	−	+	−	+	−	+
B	−	−	+	+	−	−	+	+
AB	+	−	−	+	+	−	−	+
C	−	−	−	−	+	+	+	+
AC	+	−	+	−	−	+	−	+
BC	+	+	−	−	−	−	+	+
ABC	−	+	+	−	+	−	−	+

Table 6.34: Algebraic structure for the computation of the effects from the total response values

The first row in Table 6.34 is a basic element. With this element, the total response $Y_{....} = 1'Y_*$ can be computed. If the other rows are multiplied with the first row, they stay unchanged (therefore I for identity). Every other row has the same numbers of + and − signs. [If + is replaced by 1 and − is replaced by −1, we obtain vectors of orthogonal contrasts with the norm 8].

   If each row is multiplied by itself, we obtain I (row 1). The product of any two rows leads to a different row of Table 6.34. For example, we have

$$
\begin{aligned}
A \cdot B &= AB, \\
(AB) \cdot (B) &= A \cdot B^2 = A \\
(AC) \cdot (BC) &= A \cdot C^2 B = AB \quad .
\end{aligned}
$$

The sums of squares in the 2^3-design are

$$
SS_{Effect} = \frac{(\text{Contrast})^2}{8r} \quad . \tag{6.121}
$$

Estimation of the Effects

The algebraic structure of Table 6.34 immediately leads to the estimates of the average effects. For instance, the average effect A is

$$
A = \frac{1}{4r} [a - (1) + ab - b + ac - c + abc - bc] \quad . \tag{6.122}
$$

Explanation: The average effect of A at the low level of B and C is

$$
(1\,0\,0) - (0\,0\,0) \quad : \quad [a - (1)]/r \quad .
$$

The average effect of A at the high level of B and the low level of C is

$$
(1\,1\,0) - (0\,1\,0) \quad : \quad [ab - b]/r \quad .
$$

The average effect of A at the low level of B and the high level of C is

$$
(1\,0\,1) - (0\,0\,1) \quad : \quad [ac - c]/r \quad .
$$

The average effect of A at the high level of B and C is

$$
(1\,1\,1) - (0\,1\,1) \quad : \quad [abc - bc]/r \quad .
$$

Hence for all combinations of B and C the average effect of A is the average of these 4 values, which equals (6.122). Similarly, we obtain the other average effects:

$$
B = \frac{1}{4r} [b + ab + bc + abc - (1) - a - c - ac] \tag{6.123}
$$

$$
C = \frac{1}{4r} [c + ac + bc + abc - (1) - a - b - ab] \tag{6.124}
$$

$$AB = \frac{1}{4r}[(1) + ab + c + abc - a - b - ac - bc] \qquad (6.125)$$

$$AC = \frac{1}{4r}[(1) + b + ac + abc - a - ab - c - bc] \qquad (6.126)$$

$$BC = \frac{1}{4r}[(1) + a + bc + abc - b - ab - c - ac] \qquad (6.127)$$

$$ABC = \frac{1}{4r}[(abc - bc) - (ac - c) - (ab - b) + (a - (1))]$$

$$= \frac{1}{4r}[abc + a + b + c - ab - ac - bc - (1)] \qquad (6.128)$$

Example 6.9: We demonstrate the analysis by means of Table 6.35. We have $r = 2$.

		Factor B		
	0		1	
	Factor C		Factor C	
Factor A	0	1	0	1
	4	7	20	10
0	5	9	14	6
	9=(1)	16=c	34=b	16=bc
	4	2	4	14
1	11	7	6	16
	15=a	9=ac	10=ab	30=abc

Table 6.35: Example for a 2^3-design with $r = 2$ replicates

Average Effects

$$A = \frac{1}{8}[15 - 9 + 10 - 34 + 9 - 16 + 30 - 16] = \frac{1}{8}[64 - 75] = -\frac{11}{8} = -1.375$$

$$B = \frac{1}{8}[34 + 10 + 16 + 30 - (9 + 15 + 16 + 9)] = \frac{1}{8}[90 - 49] = \frac{41}{8} = 5.125$$

$$C = \frac{1}{8}[16 + 9 + 16 + 30 - (9 + 15 + 34 + 10)] = \frac{1}{8}[71 - 68] = \frac{3}{8} = 0.375$$

$$AB = \frac{1}{8}[9 + 10 + 16 + 30 - (15 + 34 + 9 + 16)] = \frac{1}{8}[65 - 74] = -\frac{9}{8} = -1.125$$

$$AC = \frac{1}{8}[9 + 34 + 9 + 30 - (15 + 10 + 16 + 16)] = \frac{1}{8}[82 - 57] = \frac{25}{8} = 3.125$$

$$BC = \frac{1}{8}[9 + 15 + 16 + 30 - (34 + 10 + 16 + 9)] = \frac{1}{8}[70 - 69] = \frac{1}{8} = 0.125$$

$$ABC = \frac{1}{8}[30 + 15 + 34 + 16 - (10 + 9 + 16 + 9)] = \frac{1}{8}[95 - 44] = \frac{51}{8} = 6.375$$

	SS	df	MS	F	
A	7.56	1	7.56	0.87	
B	105.06	1	105.06	12.09	*
AB	5.06	1	5.06	0.58	
C	0.56	1	0.56	0.06	
AC	39.06	1	39.06	4.49	
BC	0.06	1	0.06	0.01	
ABC	162.56	1	162.56	18.71	*
Error	69.52	8	8.69		
Total	389.44	15			

Table 6.36: Analysis of variance for Table 6.35

The sums of squares are (cf. (6.121))

$$SS_A = \frac{11^2}{16} = 7.56, \qquad\qquad SS_{AB} = \frac{9^2}{16} = 5.06$$
$$SS_B = \frac{41^2}{16} = 105.06, \qquad\qquad SS_{AC} = \frac{25^2}{16} = 39.06$$
$$SS_C = \frac{3^2}{16} = 0.56, \qquad\qquad SS_{BC} = \frac{1^2}{16} = 0.06$$
$$SS_{ABC} = \frac{51^2}{16} = 162.56$$
$$SS_{Total} = (4^2 + 5^2 + \ldots + 14^2 + 16^2) - \frac{139^2}{16}$$
$$= 1597 - 1207.56 = 389.44$$
$$SS_{Error} = 69.52$$

The critical value for the F-statistics is $F_{1,8;0.95} = 5.32$ (cf. Table 6.36). Since the ABC-effect is significant, no reduction to a two-factorial model is possible.

6.10   Excercises and Questions

6.10.1   What advantages does a two-factorial experiment (A,B) have, compared to two one-factor experiments (A) and (B)?

6.10.2   Name the score function for parameter estimation in a two-factorial model with interaction.

Name the parameter estimates of the overall mean and of the two main effects.

6.10.3   Fill in the degrees of freedom and the F-statistics (A in a levels, B in b levels, r replicates) in the two-factorial design with fixed effects.

		df	MS	F
A	SS_A			
B	SS_B			
$A \times B$	$SS_{A \times B}$			
Error	SS_{Error}			
Total	SS_{Total}			

6.10.4   At least how many replicates r are needed in order to be able to show interaction?

6.10.5   What is meant by a saturated model and what is meant by the independence model?

6.10.6   How are the following test results to be interpreted (that is, which model corresponds to the two-factorial design with fixed effects)?

a)
$$\begin{array}{ll} F_A & * \\ F_B & * \\ F_{A \times B} & * \end{array}$$

b)
$$\begin{array}{ll} F_A & * \\ F_B & * \\ F_{A \times B} & \end{array}$$

c)
$$\begin{array}{ll} F_A & \\ F_B & \\ F_{A \times B} & * \end{array}$$

d)
$$\begin{array}{ll} F_A & * \\ F_B & \\ F_{A \times B} & * \end{array}$$

e)
$$\begin{array}{ll} F_A & \\ F_B & * \\ F_{A \times B} & \end{array}$$

6.10.7   Of what rank is the design matrix X in the two-factorial model ($A : a$, $B : b$ levels, r replicates)?

6.10.8   Let $a = b = 2$ and $r = 1$. Describe the two-factorial model with interaction in effect coding.

6.10.9 Of what form is the covariance matrix of the OLS estimate in the two-factorial model with fixed effects in effect coding?

$$V\left(\hat{\mu}, \hat{\alpha}, \hat{\beta}, (\hat{\alpha\beta})\right) = \sigma^2 \quad ?$$

In what way do the parameter estimates $\hat{\mu}$, $\hat{\alpha}$, $\hat{\beta}$ change if $F_{A \times B}$ is not significant?

How does the estimate $\hat{\sigma}^2$ change?

In what way do the confidence intervals for $\hat{\alpha}$, $\hat{\beta}$ and the test statistics F_A and F_B change?

Is the test more conservative than in the model with significant interaction?

6.10.10 Carry out the following test in the two-factorial model with fixed effects and define the final model.

	df	MS	F
SS_A	130	1		
SS_B	630	2		
$SS_{A \times B}$	40	2		
SS_{Error}	150	18		
SS_{Total}		23		

6.10.11 Assume the two-factorial experiment with fixed effects to be designed as a randomized block design. Specify the model.

In what way do the parameter estimates and the SS's for the other parameters or effects change, compared to the model without block effects?

Name the SS_{Error}.

What meaning does a significant block effect have?

6.10.12 Analyse the following two-factorial experiment with $a = b = 2$ and $r = 2$ replicates (randomized design, no block design):

	B_1	B_2	
	17	4	
A_1	18	6	
	35	10	45
	6	15	
A_2	4	10	
	10	25	35
	45	35	80

$$C = \frac{?^2}{N}$$

$$SS_{Total} = \sum\sum\sum y_{ijk}^2 - C$$

$$SS_A = \frac{1}{br}\sum_i Y_{i..}^2 - C$$

$$SS_B =$$

$$SS_{Subtotal} = \frac{1}{2}(35^2 + 10^2 + 10^2 + 25^2) - C$$

$$SS_{AxB} = SS_{Subtotal} - SS_A - SS_B$$

$$SS_{Error} =$$

6.10.13 Name the assumptions for μ, α_i, β_j, $(\alpha\beta)_{ij}$ in the two-factorial model with random effects.

Complete the following:

- $Var(y_{ijk}) =$

- $E \begin{pmatrix} \alpha \\ \beta \\ \alpha\beta \\ \epsilon \end{pmatrix} (\alpha, \beta, \alpha\beta, \epsilon)' =$

- Solve the following system

$$\begin{array}{rcccccc} MS_A &=& br\hat{\sigma}_\alpha^2 & & + & r\hat{\sigma}_{\alpha\beta}^2 & + & \hat{\sigma}^2 \\ MS_B &=& & ar\hat{\sigma}_\beta^2 & + & r\hat{\sigma}_{\alpha\beta}^2 & + & \hat{\sigma}^2 \\ MS_{AxB} &=& & & & r\hat{\sigma}_{\alpha\beta}^2 & + & \hat{\sigma}^2 \\ MS_{Error} &=& & & & & & \hat{\sigma}^2 \end{array}$$

- Compute the test statistics

$$F_{AxB} =$$
$$F_A =$$
$$F_B =$$

- Name the test statistics if F_{AxB} is not significant.

6.10.14 The covariance matrix in the mixed two-factorial model (A fixed, B random) has a compound symmetric structure, i.e., $\Sigma = ?$

Therefore, we have a generalized linear regression model. According to which method are the estimates of the fixed effects obtained?

The test statistics in the model with the interactions correlated over the A-levels are:

$$F_{A \times B} = \frac{MS_{A \times B}}{MS_{Error}}$$

$$F_B = \frac{MS_B}{?}$$

$$F_A = \frac{MS_A}{?}$$

and in the model with independent interactions

$$F_B = \frac{MS_B}{?}$$

6.10.15 Name the test statistics for the three-factorial $A \times B \times C$-design with fixed effects

$$F_{Effect} = \frac{\quad}{\quad}$$

(Effect e.g., A, B, C, $A \times B$, $A \times B \times C$) ?

6.10.16 The following table is used in the 2^2-design with fixed effects and r replications

	(1)	a	b	ab
A	−1	+1	−1	+1
B	−1	−1	+1	+1
AB	+1	−1	−1	+1

Here (1) is the total response for (0,0) (A low, B high), (a) for (1,0), (b) for (0,1) and (ab) for (1,1). Hence, the vector of the total response values is $Y_* - ((1), a, b, ab)'$.

Compute the average effects A, B and AB in the following 2^2-design.

	Replications 1	2	Total response
(0,0)	85	93	
(1,0)	46	40	
(0,1)	103	115	
(1,1)	140	154	

How are SS_A, SS_B, and $SS_{A \times B}$ computed (Hint: the contrasts are used)?

7 Repeated Measures Model

7.1 The Fundamental Model for One Population

In contrast to the previous chapters, we now assume that instead of having only one observation per object/subject (e.g., patient) we now have repeated observations. These repeated measurements are collected at previously exactly defined times. The principle idea is that these observations give information about the development of a response Y. This response might for instance be the blood pressure (measured every hour) for a fixed therapy (medicament A), the blood sugar level (measured every day of the week) or the monthly training performance of sprinters for training method A etc., that is variables which change with time (or a different scale of measurement). The aim of a design like this is not so much the description of the average behaviour of a group (with a fixed treatment), rather than the comparison of two or more treatments in their effect across the scale of measurement (e.g., time), that is the treatment or therapy comparison.

First of all, before we deal with this interesting question, let us introduce the model for one treatment, that is, for one sample from one population.

The Model

We index the I elements (e.g., patients) with $i = 1, \ldots, I$ and the measurements with $j = 1, \ldots, p$, so that the response of the jth measurement on the ith element (individual) is denoted by y_{ij}. The general basis for many analyses is the specific modelling approach of a mixed model

$$y_{ij} = \mu_{ij} + \alpha_{ij} + \epsilon_{ij} \tag{7.1}$$

with the three components

(i) μ_{ij} is the average response of y_{ij} over hypothetical repetitions with randomly chosen individuals from the population. Thus, μ_{ij} would stay unchanged if the ith element is substituted by any other element of the sample.

(ii) α_{ij} represents the deviation between y_{ij} and μ_{ij} for the particular individual of the sample that was selected as the ith element. Thus, under hypothetical repetitions, this indiviual would have mean $\mu_{ij} + \alpha_{ij}$.

(iii) ϵ_{ij} describes the random deviation of the ith individual from the hypothetical mean $\mu_{ij} + \alpha_{ij}$.

μ_{ij} is a fixed effect. α_{ij} on the other hand is a random effect that varies over the index i (that is, over the individuals, e.g., patients), hence α_{ij} is a specific characteristic of the individual. *"To be poetic, μ_{ij} is an immutable constant of the universe, α_{ij} is a lasting characteristic of the individual"* (Crowder and Hand, 1990, p. 15). Since μ_{ij} does not vary over the individuals, the index i could be dropped. However, we retain this index in order to be able to identify the individuals.

The vector $\boldsymbol{\mu}_i = (\mu_{i1}, \ldots, \mu_{ip})'$ is called the **μ-profile of the individual**. The following assumptions are made:

(A1) The α_{ij} are random effects that vary over the population for given j according to

$$E(\alpha_{ij}) = 0 \quad \text{(for all } i, j) \tag{7.2}$$
$$\text{Var}(\alpha_{ij}) = \sigma^2_{\alpha_{ij}} \;. \tag{7.3}$$

(A2) The errors ϵ_{ij} vary over the individuals for given j according to

$$E(\epsilon_{ij}) = 0 \quad \text{(for all } i, j) \tag{7.4}$$
$$\text{Var}(\epsilon_{ij}) = \sigma^2_j \;. \tag{7.5}$$

(A3) For different individuals $i \neq i'$ the α-profiles are uncorrelated, i.e.,

$$\text{Cov}(\alpha_{ij}, \alpha_{i'j'}) = 0 \quad (i \neq i') \;. \tag{7.6}$$

However, for different measurements $j \neq j'$, the α-profiles of an individual i are correlated:

$$\text{Cov}(\alpha_{ij}, \alpha_{ij'}) = \sigma^2_{\alpha_{jj'}} \quad (j \neq j') \;. \tag{7.7}$$

This assumption is essential for the repeated measures model, since it models the natural assumption that the response of an element over the j is an individual interdependent characteristic of the individual.

(A4) The random errors are uncorrelated according to

$$E(\epsilon_{ij}\epsilon_{i'j'}) = 0 \quad \text{(for all } i, i', j, j') \;. \tag{7.8}$$

(A5) The random components α_{ij} and ϵ_{ij} are uncorrelated according to

$$E(\alpha_{ij}\epsilon_{i'j'}) = 0 \quad \text{(for all } i, i', j, j') \;. \tag{7.9}$$

(A6) The α_{ij} and ϵ_{ij} are normally distributed.

From these assumptions, it follows that

$$E(y_{ij}) = \mu_{ij} \tag{7.10}$$

and (with δ_{ij}: Kronecker-symbol)

$$
\begin{aligned}
\mathrm{Cov}(y_{ij},\, y_{i'j'}) &= \mathrm{E}\left((\alpha_{ij} + \epsilon_{ij})(\alpha_{i'j'} + \epsilon_{i'j'})\right) \\
&= \mathrm{E}(\alpha_{ij}\alpha_{i'j'} + \alpha_{ij}\epsilon_{i'j'} + \epsilon_{ij}\alpha_{i'j'} + \epsilon_{ij}\epsilon_{i'j'}) \\
&= \delta_{ii'}(\sigma^2_{\alpha_{jj'}} + \delta_{jj'}\sigma^2_j).
\end{aligned}
\tag{7.11}
$$

If homogeneity of the variance over the j is called for, i.e.,

$$\sigma^2_{\alpha_{jj'}} = \sigma^2_\alpha \tag{7.12}$$

and

$$\sigma^2_j = \sigma^2 \quad, \tag{7.13}$$

then the covariance (7.11) simplifies to

$$\mathrm{Cov}(y_{ij},\, y_{i'j'}) = \delta_{ii'}(\sigma^2_\alpha + \delta_{jj'}\sigma^2) \quad. \tag{7.14}$$

Thus, the variance is

$$\mathrm{Var}(y_{ij}) = \sigma^2_\alpha + \sigma^2 \quad. \tag{7.15}$$

The relation (7.14) expresses that two different individuals $i \neq i'$ are uncorrelated, although the observations of an individual i are correlated over the measurements:

$$
\begin{aligned}
\mathrm{Cov}(y_{ij},\, y_{i'j'}) &= 0 \quad & (i \neq i') \tag{7.16} \\
\mathrm{Cov}(y_{ij},\, y_{ij'}) &= \sigma^2_\alpha \quad & (j \neq j') \quad. \tag{7.17}
\end{aligned}
$$

If the intraclass correlation coefficient for one individual over different measurements is taken, then

$$\rho(j, j') = \rho = \frac{\mathrm{Cov}(y_{ij}, y_{ij'})}{\sqrt{\mathrm{Var}(y_{ij})\mathrm{Var}(y_{ij'})}} = \frac{\sigma^2_\alpha}{\sigma^2_\alpha + \sigma^2} \quad. \tag{7.18}$$

The covariance matrix of every individual i $(i = 1, \ldots, I)$ is then of the following form

$$
\mathrm{Var}\begin{pmatrix} y_{i1} \\ \vdots \\ y_{ip} \end{pmatrix} = \mathrm{Var}(\boldsymbol{y}_i)
$$

$$= \Sigma = \sigma^2 \boldsymbol{I}_p + \sigma^2_\alpha \boldsymbol{J}_p \tag{7.19}$$

with $J_p = 1_p 1_p'$ (cf. A.7). This matrix, that we already got acquainted with in Section 3.9, is called compound symmetric.

Remark: The designs of Chapter 4 to 6 always had a covariance structure $\sigma^2 I$, with exception of the mixed model from Section 6.6.2 (cf. (6.91)). Hence, the assumptions of the classical linear regression model (3.51) were valid.

Because of the compound symmetry, we now have a generalized linear regression model and the parameter vector β has to be estimated according to the Gauss-Markov-Aitken Theorem by the generalized least-squares estimate

$$b = (X' \Sigma^{-1} X)^{-1} X' \Sigma^{-1} y.$$

However, according to Theorem 3.17 by McElroy, the ordinary and the generalized LS estimates are identical if and only if Σ has the structure (7.19), under the assumption that the model contains the constant 1. The error structure Σ from (7.19) is ignored if the ordinary LS estimate is applied, i.e., it does not have to be estimated. Hence, more degrees of freedom are available for the residual variance. This explains the preference given to the univariate ANOVA compared to the MANOVA for the comparison of therapies in two groups, if they are treated according to the repeated measures design, and if the assumption of compound symmetry holds for both groups seperately, or rather if an assumption derived from this holds for the difference in response. This will be discussed in detail further on.

7.2 The Repeated Measures Model for Two Populations

We assume that two treatments I and II are to be compared with the repeated measures design. Additionally we assume:

- n_1 individuals receive treatment I

- n_2 individuals receive treatment II

- both groups are homogeneous relating to all essential prognostic factors for a response variable Y of interest

- realization of repeated measurements at the same timings $j = 1, \ldots, p$.

This results in two matrices of sample vectors

$$
Y(I) \;=\;
\begin{array}{c}
\text{occasions} \\[2pt]
\begin{array}{ccc}
1 & \cdots & p
\end{array} \\
\begin{pmatrix}
y_{111} & \cdots & y_{11p} \\
& \cdots & \\
y_{1n_1 1} & \cdots & y_{1n_1 p}
\end{pmatrix}
\end{array}
\quad
\begin{array}{l}
\text{individual } I_1 \\[6pt]
\cdots \\[6pt]
\text{individual } I_{n_1}
\end{array}
$$

$$
\begin{matrix}
 & & \text{occasions} \\
 & & 1 \quad \cdots \quad p
\end{matrix}
$$

$$
\boldsymbol{Y}(II) \;=\; \begin{pmatrix} y_{211} & \cdots & y_{21p} \\ & \cdots & \\ y_{2n_21} & \cdots & y_{2n_2p} \end{pmatrix}
\begin{array}{l}
\text{individual } II_1 \\
\cdots \\
\text{individual } II_{n_2}
\end{array}
$$

$$
\begin{array}{lll}
k = 1 \text{ oder } 2 & : & \text{treatment I or II}
\end{array}
$$
The subscripts of y_{ijk} stand for $\quad i = 1, \ldots, n_i \quad : \quad$ individual
$$
\begin{array}{lll}
j = 1, \ldots, p & : & \text{occasion (time of measurement)} .
\end{array}
$$
The response matrices $\boldsymbol{Y}(I)$ and $\boldsymbol{Y}(II)$ are assumed to be independent. We introduce the fixed factor "treatment" into the model (7.1) and choose the following parametrization

$$
y_{kij} = \mu + \alpha_k + \beta_j + (\alpha\beta)_{kj} + a_{ki} + \epsilon_{kij} \quad . \tag{7.20}
$$

The components have the following meaning

μ: overall mean
α_k: treatment effect
β_j: occasion effect (= time effect)
$(\alpha\beta)_{kj}$:   treatment $\times$ time interaction
a_{ki}:   random effect of the ith individual
        in the kth treatment
ϵ_{kij}:   random error .

The effects α_k, β_j, $(\alpha\beta)_{kj}$ are assumed to be fixed with the usual constraints for fixed effects, i.e., $\sum \alpha_k = 0$, $\sum \beta_j = 0$, $\sum_i(\alpha\beta)_{ij} = \sum_j(\alpha\beta)_{ij} = 0$. The effects α_{ki} and the errors ϵ_{kij} however are random. Hence, (7.20) is a mixed model.
     For the random variables the following assumptions hold:

(i) The vector $\epsilon_k = (\epsilon_{k11}, \ldots, \epsilon_{kn_kp})'$, $k = 1, 2$ is normally distributed according to

$$
\epsilon_k \;\sim\; N(\boldsymbol{0}, \sigma^2 \boldsymbol{I}) \quad . \tag{7.21}
$$

(ii) The vector $\boldsymbol{a}_k = (\alpha_{k1}, \ldots, \alpha_{kn_k})'$, $k = 1, 2$ is normally distributed according to

$$
\boldsymbol{a}_k \;\sim\; N(\boldsymbol{0}, \sigma_\alpha^2 \boldsymbol{I}) \quad . \tag{7.22}
$$

(iii) Both random variables are independent

$$
\mathrm{E}(\epsilon_k \boldsymbol{a}'_{k'}) = \boldsymbol{0} \qquad (k, k' = 1, 2) \quad . \tag{7.23}
$$

With these assumptions, we obtain the expectation of y_{kij}

$$E(y_{kij}) = \mu_{kj} = \mu + \alpha_k + \beta_j + (\alpha\beta)_{kj} \qquad (7.24)$$

and for the expectation vector of the ith individual in the kth treatment, i.e., for $\boldsymbol{y}_{ki} = (y_{ki1}, \ldots, y_{kip})'$, we obtain

$$E(\boldsymbol{y}_{ki}) = \boldsymbol{\mu}_k = (\mu_{k1}, \ldots, \mu_{kp})', \qquad k = 1, 2. \qquad (7.25)$$

The vector $\boldsymbol{\mu}_k$, that represents the mean vector over the p observations of an individual and that is identical for all n_k individuals of a group, is called the $\boldsymbol{\mu}_k$-**profile** of the individuals (Crowder and Hand, 1990, p.26, Morrison, 1984, p.153). The observation vector $\boldsymbol{y}_{ki}$ on the other hand is called the **curve of progress** of the ith individual in the kth treatment group.

With (7.24) and the assumptions (7.21) – (7.23), we have

$$\mathrm{Cov}(y_{kij}, y_{k'i'j'}) = \begin{cases} \sigma_\alpha^2 + \sigma^2 & : \text{if } k = k', i = i', j = j' \\ \sigma_\alpha^2 & : \text{if } k = k', i = i', j \neq j' \\ 0 & : \text{otherwise}. \end{cases} \qquad (7.26)$$

Hence, the $p \times p$-covariance matrix Σ_k, $k = 1, 2$ of the ith observation vector $\boldsymbol{y}_{ki}$, $k = 1, 2$, $i = 1, \ldots n_k$ is of the form

$$\Sigma_k = \sigma^2 \boldsymbol{I}_p + \sigma_\alpha^2 \boldsymbol{J}_p \qquad (7.27)$$

(cf. (7.19)), which is the structure of compound symmetry.

Remark: The reparametrization of (7.1) into (7.20) maintained all assumptions of Section 7.1. Model (7.20) has the advantage that it can adopt the structure of the mixed models, as well as the estimation and interpretation of the parameters. For the correlation between the observations

$$\rho(y_{kij}, y_{k'i'j'}) = \begin{cases} \dfrac{\sigma_\alpha^2}{\sigma_\alpha^2 + \sigma^2} & : & \text{if } k = k', i = i', j \neq j' \\ 1 & : & \text{if } k = k', i = i', j = j' \\ 0 & : & \text{otherwise,} \end{cases} \qquad (7.28)$$

we find

(1) The observations, and hence the observation vectors, of individuals from different groups are uncorrelated. Due to the normal distribution they are independent as well.

(2) Observations, or rather observation vectors, of different individuals of the same group are uncorrelated (independent).

(3) Observations of an individual at different times of measurement are correlated (dependent) with the so-called intraclass correlation

$$\rho = \frac{\sigma_\alpha^2}{\sigma_\alpha^2 + \sigma^2}. \qquad (7.29)$$

7.3 Univariate and Multivariate Analysis

Parametric procedures for analysing continuous data require the assumption of a distribution. Here the normal distribution as an extensive and, after the elimination of outliers or smoothing, an adequate class of distributions is available. Often however, the variables have to be transformed first. The comparison of therapies is part of the complex of general mean comparisons of normally distributed populations. However, therapy comparison requires only the far more weak assumption that the distances (differences) of the populations are normal.

Multivariate procedures for mean comparison of two independent normal distributions are constructed in analogy to univariate procedures. The major principles will be explained in the following section.

7.3.1 The Univariate One-Sample Case

Given a sample $(y_1, \ldots, y_n)$ from $N(\mu, \sigma^2)$ with y_i i.i.d. Then $\bar{y} \sim N(\mu, \sigma^2/n)$ and $s^2(n-1)/\sigma^2 \sim \chi^2_{n-1}$. The t-test for $H_0 : \mu = \mu_0$ is given by $t_{n-1} = \dfrac{(\bar{y} - \mu_0)}{s}\sqrt{n}$.

7.3.2 The Multivariate One-Sample Case

We assume that not only *one* random variable is observed, but a p-dimensional vector of random variables. The sample size is n. The sample is then of the form

$$
\underset{n,p}{Y} = \begin{pmatrix} y'_1 \\ \vdots \\ y'_n \end{pmatrix} = \begin{pmatrix} y_{11}, \ldots, y_{1p} \\ \vdots \\ y_{n1}, \ldots, y_{np} \end{pmatrix}
$$

and we assume for every vector $y_i \overset{i.i.d}{\sim} N_p(\mu, \Sigma)$, with $\mu' = (\mu_1, \ldots, \mu_p)$ and Σ positive definite. Hence

$$
Y \sim N_p \left(\begin{pmatrix} \mu \\ \vdots \\ \mu \end{pmatrix}, \begin{pmatrix} \Sigma & & 0 \\ & \ddots & \\ 0 & & \Sigma \end{pmatrix} \right). \tag{7.30}
$$

The sample mean vector is

$$
y_{..} = (y_{.1}, \ldots, y_{.p})' \tag{7.31}
$$

with

$$
y_{.j} = \frac{1}{n} \sum_{i=1}^{n} y_{ij} \quad (j = 1, \ldots, p) \tag{7.32}
$$

and the sample covariance matrix is

$$
S = (S_{jh}) = \frac{1}{n-1} \sum_{i=1}^{n} (y_i - y_{..})(y_i - y_{..})' \tag{7.33}
$$

with the elements

$$S_{jh} = (n-1)^{-1} \sum_{i=1}^{n} (y_{ij} - y_{j.})(y_{ih} - y_{h.}) . \tag{7.34}$$

Hence

$$\mathbf{y}_{..} \sim N_p(\boldsymbol{\mu}, \boldsymbol{\Sigma}/n) \tag{7.35}$$

with $\boldsymbol{\mu}' = (\mu_1, \ldots, \mu_p)$ and

$$(n-1)\mathbf{S} \sim W_p(\boldsymbol{\Sigma}, n-1) \tag{7.36}$$

distributed independently, where W_p denotes the p-dimensional Wishart distribution with $(n-1)$ degrees of freedom.

Definition 7.1 _Let_ $\mathbf{X} = (\mathbf{x}_1, \ldots, \mathbf{x}_n)'$ _be a_ $(n \times p)$ _data matrix from a_ $N_p(\mathbf{0}, \boldsymbol{\Sigma})$, _where_ $\mathbf{x}_1, \ldots, \mathbf{x}_n$ _are independent and identically_ $N_p(\mathbf{0}, \boldsymbol{\Sigma})$-_distributed. The_ $p \times p$-_matrix_

$$\mathbf{W} = \mathbf{X}'\mathbf{X} = \sum_{i=1}^{n} \mathbf{x}_i \mathbf{x}_i' \sim W_p(\boldsymbol{\Sigma}, n)$$

then has a Wishart distribution with n _degrees of freedom._

For $p = 1$, we have $\mathbf{X}'\mathbf{X} = \sum_{i=1}^{n} x_i^2 = \mathbf{x}'\mathbf{x} \sim W_1(\sigma^2, n)$ so that $W_1(\sigma^2, n) = \sigma^2 \chi_n^2$ holds. Hence, the Wishart distribution is the multivariate analogue of the χ^2-distribution.

Definition 7.2 _A random variable_ u _has a Hotelling_ T^2-_distribution with the parameters_ p _and_ n _if it can be expressed as_

$$u = n\mathbf{x}'\mathbf{W}^{-1}\mathbf{x} \tag{7.37}$$

with

$$\mathbf{x} \sim N_p(\mathbf{0}, \mathbf{I}) \quad and \quad \mathbf{W} \sim W_p(\mathbf{I}, n)$$

being independent. We write

$$u \sim T^2(p, n) . \tag{7.38}$$

Remark: If $\mathbf{x} \sim N_p(\boldsymbol{\mu}, \boldsymbol{\Sigma})$ and $\mathbf{W} \sim W_p(\boldsymbol{\Sigma}, n)$ and $\mathbf{x}$ and $\mathbf{W}$ are independent, then

$$n(\mathbf{x} - \boldsymbol{\mu})'\mathbf{W}^{-1}(\mathbf{x} - \boldsymbol{\mu}) \sim T^2(p, n) . \tag{7.39}$$

The T^2-distribution is equivalent to the F-distribution (Mardia et al. 1979, p. 74):

$$T^2(p, n) \sim \frac{np}{n-p+1} F_{p,n-p+1} . \tag{7.40}$$

The multivariate two-sided hypothesis

$$H_0 : \boldsymbol{\mu} = \boldsymbol{\mu}_0 \quad \text{against} \quad H_1 : \boldsymbol{\mu} \neq \boldsymbol{\mu}_0 \tag{7.41}$$

is tested in analogy to the t-test with the test statistic by Hotelling

$$T^2 = n(\boldsymbol{y}_{..} - \boldsymbol{\mu}_0)' \boldsymbol{S}^{-1} (\boldsymbol{y}_{..} - \boldsymbol{\mu}_0) \quad , \tag{7.42}$$

where $(\boldsymbol{y}_{..} - \boldsymbol{\mu}_0)' \boldsymbol{S}^{-1} (\boldsymbol{y}_{..} - \boldsymbol{\mu}_0)$ is the Mahalanobis-D^2 statistic. If H_0 holds, then the test statistic

$$F = \frac{n - p}{p(n - 1)} T^2 \tag{7.43}$$

has an $F_{p,n-p}$-distribution, according to (7.36) and (7.40) (replace n by $n - 1$). The decision rule is as follows:

do not reject $H_0 : \boldsymbol{\mu} = \boldsymbol{\mu}_0$ if

$$T^2 \leq \frac{p(n - 1)}{n - p} F_{p,n-p;1-\alpha}. \tag{7.44}$$

Idea of proof: This test procedure is dealt with in detail in standard literature for multivariate analysis (cf. e.g., Timm, 1975, pp. 158–166, Morrison, 1984, pp. 128–134). Hence, we only want to give a short outline of the proof.

The decision rule (7.44) is derived by the union-intersection principle that dates back to Roy (1953, 1957). Assume $\boldsymbol{y} \sim N_p(\boldsymbol{\mu}, \boldsymbol{\Sigma})$ and let $\boldsymbol{a} \neq \boldsymbol{0}$ be any $p \times 1$-vector. Hence (cf. A 82)

$$\boldsymbol{a}'\boldsymbol{y} \quad \sim \quad N_1(\boldsymbol{a}'\boldsymbol{\mu}, \boldsymbol{a}'\boldsymbol{\Sigma}\boldsymbol{a}) = N_1(\mu_a, \sigma_a^2) \quad . \tag{7.45}$$

If H_0: $\boldsymbol{\mu} = \boldsymbol{\mu}_0$ (7.41) is true, then $H_0 a$: $\mu_a = \boldsymbol{a}'\boldsymbol{\mu}_0 = \mu_{0a}$ is true for all vectors $\boldsymbol{a}$ as well.
If on the other hand $H_0 a$ is true for every $\boldsymbol{a} \neq \boldsymbol{0}$, H_0 is true as well.

Hence, the multivariate hypothesis H_0: $\boldsymbol{\mu} = \boldsymbol{\mu}_0$ is the intersection of the univariate hypotheses

$$H_0 = \bigcap_{a \neq 0} H_0 a \quad . \tag{7.46}$$

Let $\underset{n,p}{\boldsymbol{Y}}$ be a sample from $N(\boldsymbol{\mu}, \boldsymbol{\Sigma})$ with $\boldsymbol{y}'_{..} = (\boldsymbol{y}_{1.}, \ldots, \boldsymbol{y}_{p.})$ and $\boldsymbol{S}$ from (7.33).

Every univariate hypothesis $H_0 a$: $\boldsymbol{a}'\boldsymbol{\mu} = \boldsymbol{a}'\boldsymbol{\mu}_0$ is tested against its two-sided alternative $H_1 a$: $\boldsymbol{a}'\boldsymbol{\mu} \neq \boldsymbol{a}'\boldsymbol{\mu}_0$ by the t-statistic:

$$t(\boldsymbol{a}) = \frac{\boldsymbol{a}'(\boldsymbol{y}_{..} - \boldsymbol{\mu}_0)}{\sqrt{\boldsymbol{a}'\boldsymbol{S}\boldsymbol{a}}} \sqrt{n} \quad , \tag{7.47}$$

and the acceptance region for H_0 is given by

$$t^2(a) \leq t^2_{n-1,1-\frac{\alpha}{2}} \quad . \tag{7.48}$$

Hence, the multivariate acceptance region is the intersection of all univariate acceptance regions:

$$\bigcap_{a \neq 0} (t^2(a) \leq t^2_{n-1,1-\frac{\alpha}{2}}) \quad . \tag{7.49}$$

Therefore, this area has to contain the largest $t^2(a)$, so that (7.49) is equivalent to

$$\max_a t^2(a) \leq t^2_{n-1,1-\frac{\alpha}{2}} \quad . \tag{7.50}$$

Hence, the multivariate test for H_0: $\mu = \mu_0$ can be based on $t^2(a)$. Since $t^2(a)$ is dimensionless and unaffected by a change of scale of the elements of a, this indeterminacy can be eliminated by a constraint, as for instance

$$a'Sa = 1 \quad . \tag{7.51}$$

The optimization problem $\max_a \{t^2(a)|a'Sa = 1\}$ is now equivalent to

$$\max_a \{a'(y_{..} - \mu_0)(y_{..} - \mu_0)'an - \lambda(a'Sa - 1)\} \quad . \tag{7.52}$$

Differentiation with respect to a and to the Lagrangian multiplier λ (Theorems A 91 – 95) yields the system of normal equations

$$[(y_{..} - \mu_0)(y_{..} - \mu_0)'n - \lambda S]\, a = 0 \tag{7.53}$$

and

$$a'Sa = 1 \quad . \tag{7.54}$$

Premultiplication of (7.53) by a', and taking (7.54) and (7.47) into account, gives

$$\begin{aligned}
\hat{\lambda} &= a'(y_{..} - \mu_0)(y_{..} - \mu_0)'an \\
&= t^2(a|a'Sa = 1) \quad . \tag{7.55}
\end{aligned}$$

On the other hand, (7.53) as a homogeneous system in a has a non-trivial solution $a \neq 0$, as long as the determinant of the matrix equals zero. The matrix $(y_{..} - \mu_0)(y_{..} - \mu_0)'$ is of rank 1. With the determinantal constraint (S is assumed to be regular), (7.53) yields according to

$$\begin{aligned}
0 &= |(y_{..} - \mu_0)(y_{..} - \mu_0)'n - \lambda S| \\
&= |S^{-1/2}(y_{..} - \mu_0)(y_{..} - \mu_0)'S^{-1/2}n - \lambda I_p||S|
\end{aligned}$$

the characteristic equation for the first matrix, which is symmetric and of rank 1 as well.

The only non-trivial eigenvalue of a matrix of rank 1 is the trace of this matrix (Corollary to Theorem A 28):

$$\hat{\lambda} = \mathrm{tr}\{S^{-1/2}(y_{..} - \mu_0)(y_{..} - \mu_0)'S^{-1/2}n\}$$
$$= (y_{..} - \mu_0)'S^{-1}(y_{..} - \mu_0)n \quad . \tag{7.56}$$

Hence $t^2(a|a'Sa = 1)$ equals Hotelling's T^2 from (7.42).

The test statistic derived according to the union-intersection principle is equivalent to the likelihood-ratio statistic. However, this equivalence is not true in general. The advantage of the union-intersection test is that in case of a rejection of H_0, it is possible to test which one of the rejection regions caused this. By choosing $a = e_i$, it can be tested which components of μ are responsible for the rejection of H_0: $\mu = \mu_0$. This is not possible for the likelihood-ratio test. Furthermore, the importance of the union-intersection principle also lies in the fact that simultaneous confidence intervals for μ can be computed (Fahrmeir and Hamerle, 1984, p.81). With

$$\max_{a \neq 0} t^2(a) = n(y_{..} - \mu_0)'S^{-1}(y_{..} - \mu_0)$$
$$= T^2 \tag{7.57}$$

and (cf. (7.43))

$$T^2 = \frac{p(n-1)}{n-p} F_{p,n-p} \tag{7.58}$$

we have for $\mu = \mu_0$

$$P\left\{\frac{n-p}{p(n-1)} T^2 \leq F_{p,n-p,1-\alpha}\right\} = 1 - \alpha \tag{7.59}$$

or equivalently

$$P\left(\bigcap_{a \neq 0} \frac{(n-p)n}{p(n-1)} \frac{a'(y_{..} - \mu)^2}{a'Sa} \leq F_{p,n-p,1-\alpha}\right) = 1 - \alpha \quad . \tag{7.60}$$

These confidence regions are simultaneously true for all $a'\mu$ with $a \in \mathcal{R}^p$. If only a few comparisons are of interest, i.e., only a few a_i, then we have

$$P\left(a_i'y_{..} - c \leq a_i'\mu \leq a_i'y_{..} + c\right) \geq 1 - \alpha \tag{7.61}$$

with

$$c^2 = F_{p,n-p,1-\alpha} \frac{p(n-1)}{(n-p)n} a'Sa \quad . \tag{7.62}$$

In order to assure the confidence coefficient $1 - \alpha$ for the chosen comparisons, that is for $a_1'\mu, \ldots, a_k'\mu$ with $k \leq p$, and to simultaneously shorten the length of the interval, the **Bonferroni method** is applied. Assume E_i $(i = 1, \ldots, k)$ is the event that the ith confidence interval covers the parameter $a_i'\mu$, and assume that $\alpha_i = 1 - P(E_i) = P(\overline{E}_i)$ is the corresponding significance level. Let $\overline{E}_i$ be the appropriate complementary event, then

$$P\left(\bigcap_{i=1}^{k} E_i\right) = 1 - P\left(\bigcup_{i=1}^{k} \overline{E}_i\right) \geq 1 - \sum_{i=1}^{k} P(\overline{E}_i) = 1 - \sum_{i=1}^{k} \alpha_i \quad . \tag{7.63}$$

Hence, $(1 - \sum \alpha_i)$ is a lower limit for the real simultaneous confidence coefficient

$$1 - \delta = P\left(\bigcap_{i=1}^{k} E_i\right) \quad .$$

If $\alpha_i = \frac{\alpha}{k}$ is chosen, then

$$P\left(\bigcap_{i=1}^{k} E_i\right) \geq 1 - \alpha \quad .$$

The corresponding simultaneous confidence intervals are

$$a_i' y_{..} \pm \sqrt{F_{1, n-1, 1-\frac{\alpha}{k}} \frac{a' S a}{n}} \quad . \tag{7.64}$$

7.4 The Univariate Two-Sample Case

Suppose that we are given two independent samples

$$(x_1, \ldots, x_{n_1}) \qquad \text{from} \qquad N(\mu_1, \sigma^2) \tag{7.65}$$

and

$$(y_1, \ldots, y_{n_2}) \qquad \text{from} \qquad N(\mu_2, \sigma^2) \; . \tag{7.66}$$

In case of equal variances, the test statistic for $H_0 : \mu_1 = \mu_2$ is

$$t_{n_1 + n_2 - 2} = \frac{(\overline{x} - \overline{y})}{s\sqrt{\frac{1}{n_1} + \frac{1}{n_2}}} \tag{7.67}$$

with the pooled sample variance

$$s^2 = \frac{(n_1 - 1)s_x^2 + (n_2 - 1)s_y^2}{n_1 + n_2 - 2} \; . \tag{7.68}$$

The assumption of equal variances has to be tested with the F-test. In case of a rejection of $H_0 : \sigma_x^2 = \sigma_y^2$, no exact solution exists. This is called the Behrens-Fisher problem. The comparison of means in case of $\sigma_x \neq \sigma_y$ is done approximately by a t_v-statistic, where the sample variances influence the degrees of freedom v.

7.5 The Multivariate Two-Sample Case

The multivariate analogue of the t-test for testing $H_0 : \boldsymbol{\mu}_x = \boldsymbol{\mu}_y$ ($p \times 1$-vectors each) is defined as Hotelling's two-sample T^2

$$T^2 = (n_1^{-1} + n_2^{-1})^{-1}(\boldsymbol{x}_{..} - \boldsymbol{y}_{..})'S^{-1}(\boldsymbol{x}_{..} - \boldsymbol{y}_{..}) \qquad (7.69)$$

with the pooled sample covariance matrix (within-groups)

$$(n_1 + n_2 - 2)S = (n_1 - 1)S_x + (n_2 - 1)S_y . \qquad (7.70)$$

The statistic T^2 is in fact an estimate of the Mahalanobis distance $D^2 = (\boldsymbol{\mu}_x - \boldsymbol{\mu}_y)'\Sigma^{-1}(\boldsymbol{\mu}_x - \boldsymbol{\mu}_y)$ of both populations. Under $H_0 : \boldsymbol{\mu}_x = \boldsymbol{\mu}_y$, T^2 has the following relationship to the central F-distribution

$$F_{p,v} = \frac{n_1 + n_2 - p - 1}{(n_1 + n_2 - 2)p}T^2 \qquad (7.71)$$

with the degrees of freedom of the denominator

$$v = n_1 + n_2 - p - 1 . \qquad (7.72)$$

The decision rule based on the union-intersection principle (Roy, 1953, 1957) — or equivalently on the likelihood-ratio principle — yields the rejection region for $H_0 : \boldsymbol{\mu}_x = \boldsymbol{\mu}_y$ as

$$T^2 > \frac{(n_1 + n_2 - 2)p}{v}F_{p,v,1-\alpha} . \qquad (7.73)$$

Hotelling's T^2 statistic for the model with fixed effects assumes the equality of the covariance matrices Σ_x and Σ_y, in analogy to the univariate comparison of means. This equality can be tested by various measures.

Remark: If $H_0 : \boldsymbol{\mu}_x = \boldsymbol{\mu}_y$ is replaced by $H_0 : C(\boldsymbol{\mu}_x - \boldsymbol{\mu}_y) = 0$ where C is a contrast matrix for differences, then the statistic F (7.71) has one degree of freedom less in the numerator as well as in the denominator, i.e., p is to be replaced by $p - 1$.

7.6 Testing of H_0: $\Sigma_x = \Sigma_y$

Box (1949) has given the following generalization of Bartlett's test for equality of two univariate variances to $H_0 : \Sigma_x = \Sigma_y$ in the multivariate (p-dimensional case).

Assume that S (7.70) is the pooled sample covariance matrix of the two p-variate normal distributions. The **Box-M statistic** is αM with

$$M = (n_1 - 1)\ln\left(\frac{|S|}{|S_x|}\right) + (n_2 - 1)\ln\left(\frac{|S|}{|S_y|}\right) \qquad (7.74)$$

and

$$\alpha = 1 - \frac{1}{6}(2p^2 + 3p - 1)(p+1)^{-1}\left\{\frac{1}{n_1 - 1} + \frac{1}{n_2 - 1} - \frac{1}{n_1 + n_2 - 2}\right\}. \quad (7.75)$$

Under H_0: $\Sigma_x = \Sigma_y$, we have the following approximate distribution

$$\alpha M \sim \chi^2_{p(p+1)/2}. \quad (7.76)$$

Remark: Box (1949) developed this statistic for the general comparison of $g \geq 2$ normal distributions and gave equivalent representations as F-statistic. For the comparison of g independent normal distributions $N_p(\boldsymbol{\mu}_1, \boldsymbol{\Sigma}_1), \ldots, N_p(\boldsymbol{\mu}_g, \boldsymbol{\Sigma}_g)$, the test problem is

$$H_0 : \ \Sigma_1 = \ldots = \Sigma_g \quad (7.77)$$

against

$$H_1 : \ H_0 \text{ not true.}$$

Let $\boldsymbol{S}_i$ be the unbiased estimates (i.e., the appropriate sample covariance matrices) of Σ_i ($i = 1, \ldots, g$) and let n_i be the corresponding sample size. We assume

$$N = \sum_{i=1}^{g} n_i, \quad v_i = n_i - 1 \quad (7.78)$$

and denote the pooled sample covariance matrix by $\boldsymbol{S}$

$$\boldsymbol{S} = \frac{1}{N-g}\sum_{i=1}^{g} v_i \boldsymbol{S}_i \quad . \quad (7.79)$$

The test statistic is then of the form αM (cf. Timm, 1975, p.252) with

$$M = (N-g)\ln|\boldsymbol{S}| - \sum_{i=1}^{g} v_i \ln|\boldsymbol{S}_i| \quad (7.80)$$

and

$$\alpha = 1 - C, \quad (7.81)$$

$$C = \frac{2p^2 + 3p - 1}{6(p+1)(g-1)}\left(\sum_{i=1}^{g}\frac{1}{v_i} - \frac{1}{N-g}\right). \quad (7.82)$$

The approximate distribution is

$$\alpha M \sim \chi^2_v \quad \text{with } v = \frac{p(p+1)(g-1)}{2} \quad . \quad (7.83)$$

For $g = 2$, we have α specified by (7.75).

7.7 Univariate Analysis of Variance in the Repeated Measures Model

7.7.1 Testing of Hypotheses in Case of Compound Symmetry

Consider the model (7.20) formulated in Section 7.2

$$y_{kij} = \mu + \alpha_k + \beta_j + (\alpha\beta)_{kj} + a_{ki} + \epsilon_{kij} \quad , \tag{7.84}$$

which can be interpreted as a mixed model, that is, as a two-factorial model (fixed factors: treatments $k = 1, 2$ and occasions $j = 1, \ldots, p$) with interaction and one random effect α_{ki} (individual).

The univariate analysis of variance assumes equal covariance matrices of the two subpopulations ($k = 1$ and 2). Furthermore, the structure of compound symmetry (7.19) is required for both covariance matrices. This assumption is sufficient for the validity of the univariate F-tests. Compound symmetry is a special case of a more general covariance structure which ensures the exact F-distribution. This situation, that occurs often in practice, will be discussed in detail in Section 7.7.2.

In the mixed model, the following hypotheses tailored to the situation of the repeated measures model are tested:

(i) The null hypothesis of homogeneous levels of both treatments

$$H_0: \quad \alpha_1 = \alpha_2 \quad . \tag{7.85}$$

(ii) The null hypothesis of homogeneous occasions (cf. Figure 7.2)

$$H_0: \quad \beta_1 = \ldots = \beta_p \quad . \tag{7.86}$$

(iii) The null hypothesis of no interaction between the treatment and time effects (cf. Figure 7.1)

$$H_0: \quad (\alpha\beta)_{ij} = 0 \quad (k = 1, 2, \ j = 1, \ldots, p) \quad . \tag{7.87}$$

We define the correction term once again as

$$C = \frac{Y_{\ldots}^2}{N}$$

with $N = (n_1 + n_2)p = np$. Taking the possibly unbalanced sample sizes ($n_1 \neq n_2$) into consideration, we obtain the following sums of squares (cf. (6.17) – (6.22) and Morrison, 1984, p. 213)

$$
\begin{aligned}
SS_{Total} &= \sum\sum\sum (y_{kij} - y_{\ldots})^2 \\
&= \sum\sum\sum y_{kij}^2 - C
\end{aligned}
\tag{7.88}
$$

Figure 7.1: No interaction (H_0: $(\alpha\beta)_{ij} = 0$ not rejected) and a time effect

Figure 7.2: No interaction and no time effect

$$
\begin{aligned}
SS_A = SS_{Treat} &= \sum\sum\sum (y_{k..} - y_{...})^2 \\
&= \frac{1}{p}\sum_{k=1}^{2}\frac{1}{n_k}Y_{k..}^2 - C && (7.89)
\end{aligned}
$$

$$
\begin{aligned}
SS_B = SS_{Time} &= \sum\sum\sum (y_{..j} - y_{...})^2 \\
&= \frac{1}{n_1 + n_2}\sum_{j=1}^{p}Y_{..j}^2 - C && (7.90)
\end{aligned}
$$

$$
\begin{aligned}
SS_{Subtotal} &= \sum\sum\sum (y_{k.j} - y_{...})^2 \\
&= \sum_k \frac{1}{n_k}\sum_j Y_{k.j}^2 - C && (7.91)
\end{aligned}
$$

$$
\begin{aligned}
SS_{A\times B} &= SS_{Treat\times Time} \\
&= SS_{Subtotal} - SS_{Treat} - SS_{Time} && (7.92)
\end{aligned}
$$

$$
\begin{aligned}
SS_{Ind} &= \sum\sum\sum (y_{.i.} - y_{k..})^2 \\
&= \frac{1}{p}\sum_{k=1}^{2}\sum_{i=1}^{n_k}Y_{.i.}^2 - \frac{1}{p}\sum_{k=1}^{2}\frac{1}{n_k}Y_{k..}^2 && (7.93)
\end{aligned}
$$

$$
SS_{Error} = SS_{Total} - SS_{Subtotal} - SS_{Ind} \quad . \qquad (7.94)
$$

The test statistics are (cf. Greenhouse and Geisser, 1959):

$$
F_{Treat} = \frac{MS_{Treat}}{MS_{Ind}} \qquad (7.95)
$$

$$F_{Time} = \frac{MS_{Time}}{MS_{Error}} \tag{7.96}$$

$$F_{Treat \times Time} = \frac{MS_{Treat \times Time}}{MS_{Error}} . \tag{7.97}$$

Source	SS	df	MS	F-values
Treatment	SS_{Treat}	1	SS_{Treat}	$F_{Treat} = \frac{MS_{Treat}}{MS_{Ind}}$
Occasion	SS_{Time}	$p-1$	$\frac{SS_{Time}}{p-1}$	$F_{Time} = \frac{MS_{Time}}{MS_{Error}}$
treatment × Occasion	$SS_{Treat \times Time}$	$p-1$	$\frac{SS_{Treat \times Time}}{p-1}$	$F_{Treat \times Time} = \frac{MS_{Treat \times Time}}{MS_{Error}}$
Individual	SS_{Ind}	$n-2$	$\frac{SS_{Ind}}{n-2}$	
Error	SS_{Error}	$(p-1)(n-2)$	$\frac{SS_{Error}}{(p-1)(n-2)}$	
Total	SS_{Total}	$np-1$		

Table 7.1: Table of the univariate analysis of variance in the repeated measures model

These F-tests are called **unadjusted univariate F-tests** — as opposed to the **adjusted F-tests** named according to the Greenhouse-Geisser strategy.

Remark: The assumption of a compound symmetric structure is not very realistic in the repeated measures model, since this requirement means that the correlation of the response between two occasions is identical. This assumption, however, can not be expected for all situations. Hence, the question of interest is whether and when univariate tests may be applied in case of a more general covariance structure (sphericity of the contrast covariance matrix) (cf. Girden, 1992).

7.7.2 Testing of Hypotheses in Case of Sphericity

We assume that the two populations have an identical covariance matrix Σ. The comparison of therapies, i.e., the testing of the linear hypotheses (7.85) – (7.87), is done by means of linear contrasts. The comparison of the p means of the p occasions requires a system of $p-1$ orthogonal contrasts. The test statistic follows an F-distribution, if and only if the covariance matrix of the orthogonal contrasts is a scalar multiple of the identity matrix. This condition is called the **circularity** or **sphericity condition**.

This condition can be expressed in a number of alternative ways.

For example, it can be demanded that all the variances of pairwise differences of the response values of an individual are equal. For any random variables x_i and x_j, the following is valid

$$\text{Var}(x_i - x_j) = \text{Var}(x_i) + \text{Var}(x_j) - 2\text{Cov}(x_i, x_j) \quad .$$

If $\text{Var}(x_i) = \text{Var}(x_j)$ and $\text{Cov}(x_i, x_j)$ is constant (for all i, j), then compound symmetry holds. However, more general dependent structures exist under which the condition

$$\text{Var}(x_i - x_j) = \text{const}$$

is valid, from which sphericity of every contrast covariance matrix follows, as long as sphericity is proven for one specific covariance matrix.

The necessary and sufficient condition is known as *Huynh-Feldt condition* (Huynh and Feldt, 1970). It can be expressed in three equivalent (alternative) forms:

Huynh-Feldt Condition (H Pattern)

(i) The common covariance matrix Σ of both populations is $\Sigma = (\sigma_{jj'})$ with

$$\sigma_{jj'} = \begin{cases} \alpha_j + \alpha_{j'} + \lambda & j = j' \\ \alpha_j + \alpha_{j'} & j \neq j' \end{cases} \quad . \tag{7.98}$$

(ii) All possible differences $y_{kij} - y_{kij'}$ of the response variables have equal variance, i.e., $\text{Var}(y_{kij} - y_{kij'}) = 2\lambda$ is valid for every individual i from each of the two groups.

(iii) For the Huynh-Feldt epsilon $\varepsilon_{HF} = 1$ holds, where

$$\varepsilon_{HF} = \frac{p^2(\bar{\sigma}_d - \bar{\sigma}_{..})^2}{(p-1)(\sum\sum \sigma_{rs}^2 - 2p \sum \bar{\sigma}_{r.}^2 + p^2 \bar{\sigma}_{..}^2)} \quad . \tag{7.99}$$

Here $\Sigma = (\sigma_{rs})$ is the population covariance matrix with

$\bar{\sigma}_d$: average of the diagonal elements,
$\bar{\sigma}_{..}$: overall mean of the σ_{rs},
$\bar{\sigma}_{r.}$: average of the rth row.

Testing the Huynh-Feldt Condition

Huynh and Feldt (1970) proved that the necessary and sufficient conditions (i), (ii) or (iii) are valid, if

$$\tilde{C}_H \Sigma \tilde{C}_H' = \lambda I \tag{7.100}$$

holds where $\tilde{C}_H$ is the normalized form of C_H. C_H is the suborthogonal $(p-1) \times p$-submatrix of the orthogonal Helmert matrix:

$$\begin{pmatrix} 1_p'/\sqrt{p} \\ C_H \end{pmatrix} \quad , \tag{7.101}$$

that is formed from the Helmert contrasts. The Helmert matrix C_H in (7.101) contains the following elements:

$$C_H \underset{p-1,p}{=} \begin{pmatrix} c'_1 \\ c'_2 \\ \vdots \\ c'_p \end{pmatrix} = \begin{pmatrix} (p-1) & -1 & -1 & \cdots & -1 & -1 \\ 0 & (p-2) & -1 & \cdots & -1 & -1 \\ & \vdots & & & & \vdots \\ 0 & 0 & 0 & \cdots & 1 & -1 \end{pmatrix}. \qquad (7.102)$$

The vectors c'_s $(s = 1, \ldots, p-1)$ are called *Helmert contrasts*. They are orthogonal

$$c'_{s_1} c_{s_2} = 0 \qquad (s_1 \neq s_2)$$

and $\sum_{j=1}^{p} c_{sj} = 0$, i.e.,

$$c'_s 1_p = 0.$$

However, the c_s are not normed ($c'_s c_s \neq 1$). Therefore, the vector $1'_p$ or its standardized version $1_p/\sqrt{p}$ is included in the contrast matrix as the first row, although strictly speaking this is not a contrast ($1'_p 1_p = p \neq 0$, i.e., the second property of contrasts is not fulfilled).

  Standard software is available that converts the contrasts C_H into orthonormal contrasts $\tilde{C}_H$.

Remark: Based on the standardized Helmert matrix $\tilde{C}_H$, we give a short outline of proof of the equivalence of (ii) and (7.100):
Case $p = 2$:
The Helmert matrix is $C_H = (1, -1)$, hence $\tilde{C}_H = (\frac{1}{\sqrt{2}}, -\frac{1}{\sqrt{2}})$. Thus, (7.100) is

$$\left(\tfrac{1}{\sqrt{2}} \quad -\tfrac{1}{\sqrt{2}} \right) \begin{pmatrix} \sigma_1^2 & \sigma_{12} \\ \sigma_{12} & \sigma_2^2 \end{pmatrix} \begin{pmatrix} \tfrac{1}{\sqrt{2}} \\ -\tfrac{1}{\sqrt{2}} \end{pmatrix} = \lambda$$

$$\Longleftrightarrow \quad \sigma_1^2 + \sigma_2^2 - 2\sigma_{12} = 2\lambda$$

Case $p = 3$:
We obtain $\tilde{C}_H \Sigma \tilde{C}'_H = \lambda I$ as

$$\begin{pmatrix} \tfrac{2}{\sqrt{6}} & -\tfrac{1}{\sqrt{6}} & -\tfrac{1}{\sqrt{6}} \\ 0 & \tfrac{1}{\sqrt{2}} & -\tfrac{1}{\sqrt{2}} \end{pmatrix} \begin{pmatrix} \sigma_1^2 & \sigma_{12} & \sigma_{13} \\ \sigma_{12} & \sigma_2^2 & \sigma_{23} \\ \sigma_{13} & \sigma_{23} & \sigma_3^2 \end{pmatrix} \begin{pmatrix} \tfrac{2}{\sqrt{6}} & 0 \\ -\tfrac{1}{\sqrt{6}} & \tfrac{1}{\sqrt{2}} \\ -\tfrac{1}{\sqrt{6}} & -\tfrac{1}{\sqrt{2}} \end{pmatrix} = \lambda I_2$$

$$\Longleftrightarrow$$

Element $(1,1):$ $\qquad \tfrac{1}{6}[4\sigma_1^2 + \sigma_2^2 + \sigma_3^2 - 4\sigma_{12} - 4\sigma_{13} + 2\sigma_{23}] = \lambda$
Element $(1,2) = (2,1): \sigma_3^2 - \sigma_2^2 + 2\sigma_{12} - 2\sigma_{13} = 0$
$\qquad\qquad\qquad \Longrightarrow \quad \sigma_2^2 = [\sigma_3^2 + 2\sigma_{12} - 2\sigma_{13}]$
Element $(2,2):$ $\qquad \tfrac{1}{2}[\sigma_2^2 + \sigma_3^2 - 2\sigma_{23}] = \lambda$.

Form

$$\sigma_1^2 + \sigma_2^2 - 2\sigma_{12} = \sigma_1^2 + \left[\sigma_3^2 + 2\sigma_{12} - 2\sigma_{13}\right] - 2\sigma_{12}$$
$$= \sigma_1^2 + \sigma_3^2 - 2\sigma_{13} \quad .$$

Equate $(1,1) = (2,2)$ (since the right-hand sides are equal) $\Longrightarrow$

$$(\sigma_1^2 + \sigma_2^2 - 2\sigma_{12}) + (\sigma_1^2 + \sigma_3^2 - 2\sigma_{13}) = 2(\sigma_2^2 + \sigma_3^2 - 2\sigma_{23}) = 4\lambda \quad .$$

Both terms on the left are identical

$$\Longrightarrow \quad \sigma_j^2 + \sigma_{j'}^2 - 2\sigma_{jj'} = 2\lambda \quad (j \neq j').$$

The Condition of Sphericity or Circularity

Compound symmetry is a special case of covariance structures, for which the univariate F-tests are valid. Let us first consider the case of a therapy group measured on p occasions. We can apply $(p-1)$ orthonormal contrasts for testing the differences in the p occasions.

The univariate statistics $(c_j' y_{ki})^2$ follow exact F-distributions if and only if the covariance matrix of the contrasts has equal variances and zero covariances, that is, if it has the form $\sigma^2 I$ (circularity or sphericity). This corresponds to the assumption of the mixed model that the differences in the y_{ki} are caused only by unequal means and not by variance inhomogeneity.

The model of compound symmetry is a special case of the model of sphericity of the orthonormal contrasts. Every term on the main diagonal of the covariance matrix of orthonormal contrasts estimates the denominator in the univariate F-statistic of the corresponding contrast. Thus, **when sphericity holds**, each element estimates the same thing. Hence, a better statistic is the average of these elements. This is called the **averaged F-test**. If sphericity does not hold, the denominators of the F-statistics may become too large or too small so that the test is biased.

Comparison of Two or More Therapy Groups — Test for Sphericity

Similar to the above arguments, univariate F-tests only stay valid if the covariance matrix of orthonormal contrasts within therapy groups are spherical and — additionally — are identical across the therapy groups so that global sphericity holds. This assumption may be weakened, for instance, by demanding sphericity only for the main effects (e.g., j fixed, comparison of two therapies by means of a linear contrast).

For the test of **global sphericity** (7.100), the equality of the covariance matrices in the therapy groups is tested first. This is done by the Box-M statistic

(7.74). If H_0: $\boldsymbol{\Sigma}_1 = \boldsymbol{\Sigma}_2$ is not rejected, then the **test for sphericity by Mauchly** (1940) may be applied. The test statistic is (Morrison, 1984, p. 251):

$$W = \frac{q^q |\boldsymbol{R}|}{(\mathrm{tr}\{\boldsymbol{R}\})^q} \tag{7.103}$$

with $q = p - 1$,

$$\boldsymbol{R} = \tilde{\boldsymbol{C}}_H \boldsymbol{S} \tilde{\boldsymbol{C}}_H' \tag{7.104}$$

and $\tilde{\boldsymbol{C}}_H$ is the $q \times p$-matrix of orthonormal Helmert contrasts. In addition to the exact critical values (cf. Tables in Kres (1983)), the approximate distribution

$$-\left[(N-1) - \frac{2p^2 - 3p + 3}{6(p-1)}\right] \ln W \sim \chi_v^2 \tag{7.105}$$

with

$$v = \frac{1}{2}(p-2)(p+1) = \frac{1}{2}p(p-1) - 1 \tag{7.106}$$

may be used in case of equal sample sizes $n_1 = n_2 = N$.

Tests relating to the covariance structure — especially the Box-M test and the Mauchly test — are sensitive to non-normality in general. Huynh and Mandeville (1979) analysed the robustness of the Mauchly test to such departure by means of simulation studies. The following conclusions are drawn:

(i) the W-test tends to err on the conservative side for light-tailed distributions, the difference between the empirical type I error and the nominal significance level α increases for large samples and for small α,

(ii) for heavy-tailed distributions the reverse is true, i.e., H_0: *sphericity* is rejected earlier, even though H_0 is true.

7.7.3 The Problem of Non-Sphericity

After the pre-tests (univariate F-tests, Box-M test, Mauchly test) are carried out, the following questions have to be settled (cf. Crowder and Hand, 1990, pp. 50–56):

(i) Which effect occurs if the F-test is applied in spite of a rejection of sphericity?

(ii) What is to be done if the assumptions seem unjustifiable altogether?

To (i): If sphericity does not hold, then the actual level of significance $\hat{\alpha}$ of the univariate F-tests will exceed the nominal level α with the effect, that too many true null hypotheses are rejected. For tests with complete systems of orthonormal contrasts, this effect can be analysed by studying the ε correction factor. Rouanet and Lepine (1970), Mitzel and Games (1981), and Boik (1981) discuss the effect of non-sphericity on single contrasts. Boik concludes that the type I error is out of control. Rouanet and Lepine (1970) recommend to use all relevant statistics.

To (ii): What is to be done in case of non-sphericity? The multivariate analysis only assumes the equality of the covariance matrices, but not any specific form of the (common) covariance matrix. If however sphericity holds, then the MANOVA has a relatively low power compared to the univariate approach.

Hence, the direct application of a multivariate analysis, that is without previously testing the possibility of sphericity, is not the best strategy.

7.7.4 Application of Univariate Modified Approaches in Case of Non-Sphericity

Let $\{c\}$ be a set of $(p-1)$ orthonormal contrasts with the covariance matrix Σ_c. The **Greenhouse-Geisser epsilon** is then defined as:

$$\varepsilon_{G-H} = \frac{(\mathrm{tr}\Sigma_c)^2}{(p-1)\,\mathrm{tr}(\Sigma_c^2)} = \frac{(\sum \theta_j)^2}{(p-1)\sum \theta_j^2} \tag{7.107}$$

where θ_j are the eigenvalues of Σ_c. If $\Sigma_c = I$, then all $\theta_j = 1$ and ε is equal to 1. Otherwise we have $\varepsilon_{G-H} < 1$. The overall F-tests for an occasion effect and for interaction in case of two therapy groups with $n = n_1 + n_2$ individuals and p measures involves the $F_{p-1,(p-1)(n-2)}$ distribution (cf. test statistics (7.96) and (7.97)). In the case of non-sphericity, the $F_{\varepsilon_{G-H}(p-1),\varepsilon_{G-H}(p-1)(n-2)}$ distribution is used for testing. Hence, for $\varepsilon_{G-H} < 1$ the critical values increase, i.e., the null hypotheses are rejected less often. This counteracts the previously described effect (answer to (i)).

Since ε_{G-H} will not be known, it will have to be estimated. Hence the question arises, what influence does the estimation error of $\hat{\varepsilon}_{G-H}$ have on the power of the F-test corrected by $\hat{\varepsilon}_{G-H}$.

Greenhouse-Geisser Test Strategy

In order to avoid this problem, Greenhouse and Geisser (1959) suggest a conservative approach. This strategy consists of the following steps:

- standard F-test (unmodified). If H_0 is not rejected, then stop.

- If H_0 is rejected, then the smallest ε-value is chosen (lower bound epsilon)

$$\varepsilon_{min} = 1/(p-1) \tag{7.108}$$

and tested with the modified F-test. If H_0 is rejected by this most conservative test, then the decision is accepted and stop.

If H_0 is not rejected, then ε_{G-H} is estimated (7.107) and the $\hat{\varepsilon}_{G-H}$-F-test is conducted and its decision is accepted.

As a universal answer for the entire problem, we conclude:
If strong prior reasons favour the assumption of sphericity (that is, for the independence of the univariate distributions of the contrasts), then the univariate F-tests should be conducted. Otherwise, either a modified ε-F-test or a multivariate test or a non-parametric approach should be applied. It is obvious that this problem cannot be solved academically, but only on the basis of the data.

Test Procedure in the Two-Sample Case in the Mixed Model

1. Testing for interaction and for occasions effects (H_0 from (7.87) and (7.86))

   (a) $\Sigma_1 = \Sigma_2 \Longrightarrow$ MANOVA

   (b) $\tilde{C}_H(\Sigma_1 - \Sigma_2)\tilde{C}'_H = \lambda I \Longrightarrow$ ANOVA (averaged F-test)

   (c) $\tilde{C}_H(\Sigma_1 - \Sigma_2)\tilde{C}'_H \neq \lambda I \Longrightarrow$ ANOVA (modified) or MANOVA

   Comment: If sphericity holds, then the ANOVA (unmodified) is more powerful than the MANOVA.
   If we have non-sphericity, the power of the ANOVA (modified) compared to the MANOVA depends on the ϵ-values (Huynh-Feldt ϵ or Greenhouse-Geisser ϵ) or rather on the estimation errors in $\hat{\epsilon}$.

2. Testing for the main effect H_0: $\alpha_1 = \alpha_2$ (7.85) under the assumption of H_0: $(\alpha\beta)_{ij} = 0$

$$\Sigma_1 = \Sigma_2 \quad \Longrightarrow \quad \text{univariate } F\text{-test}$$
$$(\text{MANOVA} = \text{unmodified ANOVA})$$
$$\Sigma_1 \neq \Sigma_2 \quad \Longrightarrow \quad \text{non-parametric approach} .$$

7.7.5 Multiple Tests

If a global treatment effect is proven, that is, if H_0: $\boldsymbol{\mu}_1 = \boldsymbol{\mu}_2$ is rejected, then the question of interest is whether regions with a multiple treatment effect exist. Multiple treatment effect means that $\mu_{1j} \neq \mu_{2j}$ for some j.

Of special interest are connected regions with *local multiple treatment effects*, as for example

$$\mu_{1j} \neq \mu_{2j}, \quad j = 1, \ldots, \tilde{p}; \quad \tilde{p} < p , \qquad (7.109)$$

i.e., treatment effects from the first occasion until a specific occasion $\tilde{p}$. For this, a multiple testing procedure is performed that meets the multiple α-level. This is done by defining so-called Holm-adjusted quantiles (cf. Lehmacher, 1987, p. 29), starting out with Bonferroni's inequality.

Holm-Procedure for Local Multiple Treatment Effects

To begin with, the global treatment effect is tested, i.e., H_0: $\boldsymbol{\mu}_1 = \boldsymbol{\mu}_2$ is tested with Hotelling's T^2 (cf. (7.69)). If H_0 is not significant the procedure stops. If however H_0 is not rejected, then the Holm-procedure is conducted, which sorts all p univariate t-statistics of the p single occasions by their size (thus in analogy to the size of the p-values, starting with the smallest p-value). These p-values are compared to the Holm-adjusted sequence

$j = 1$	$j = 2$	$j = 3$	$j = 4$	$\ldots$	$j = p - 1$	$j = p$
$\frac{\alpha}{p-1}$	$\frac{\alpha}{p-1}$	$\frac{\alpha}{p-2}$	$\frac{\alpha}{p-3}$	$\ldots$	$\frac{\alpha}{2}$	α

As soon as one p-value of a t_j lies above its appropriate Holm limit, the procedure is terminated and H_0: $\mu_{1j} = \mu_{2j}$, $j = 1, \ldots, \tilde{p}$ is rejected in favour of H_1 (7.109). *Interpretation:* A local multiple treatment effect exists for all occasions j with a p-value of $t_j \leq j$th Holm limit. This means that all univariate hypotheses $H_0 j$: $\mu_{1j} = \mu_{2j}$, whose test statistics have p-values below the appropriate Holm limit, are rejected in favour of a local multiple treatment effect.

7.7.6 Examples

Example 7.1: Two treatments 1 and 2 over $p = 3$ measures with $n_1 = n_2 = 4$ individuals each are compared (Table 7.2).

Treatment	Occasion A	B	C	$Y_{ki.}$
	10	19	27	56
1	9	13	25	47
	4	10	20	34
	5	6	12	23
	13	16	19	48
2	11	18	28	57
	17	28	25	70
	20	23	29	72

Table 7.2: Repeated measures design for the treatment comparison

Call in SPSS:

```
MANOVA A B C by Treat (1,2)
/ws factors = Time(3)
/contrast(Time) = difference
/ws design
/print = homogeneity(boxm) transform error (cor)
        signig(averf) param(estim)
/design .
```

The steps of the test are:

(i) H_0: $\Sigma_1 = \Sigma_2$

The Box-M statistic is $\alpha M = 3.93638$, i.e., approximately (cf. (7.76))

$$\chi^2_{p(p+1)/2} = \chi^2_6 = 1.80417 \quad \text{(p-value 0.937)} .$$

Hence H_0 is not rejected. After the test procedure, the MANOVA may be performed. Befor doing this however, it should be tested whether sphericity holds for the contrast covariance matrix.

(ii) H_0: $\tilde{C}_H \Sigma \tilde{C}'_H = \lambda I$

$$\text{We have } \tilde{C}_H = \begin{pmatrix} \frac{2}{\sqrt{6}} & -\frac{1}{\sqrt{6}} & -\frac{1}{\sqrt{6}} \\ 0 & \frac{1}{\sqrt{2}} & -\frac{1}{\sqrt{2}} \end{pmatrix} .$$

Test involving 'Time' Within Subject Effect

```
Mauchly sphericity test, W =        .90352
Chi-square approx. =                .50728  with 2 D.F.
Significance =                      .776
Greenhouse-Geisser Epsilon =        .91201
Huynh-Feldt Epsilon =              1.00000
```

Hence H_0: *Sphericity* is not rejected and we may conduct the unadjusted F-tests of the ANOVA.

According to the test strategy in the mixed model, we first test

$$H_0 : \quad (\alpha\beta)_{ij} = 0$$

with (cf. (7.97) and Table 7.1)

$$F_{Treat \times Time} = F_{(p-1);(p-1)(p-2)} = \frac{MS_{Treat}}{MS_{Error}}.$$

From Table 7.2, we get

	A	B	C	
$Y_{1 \cdot j}$	28	48	84	$Y_{1 \cdot \cdot} = 160$
$Y_{2 \cdot j}$	61	85	101	$Y_{2 \cdot \cdot} = 247$
$Y_{\cdot \cdot j}$	89	133	185	$Y_{\cdot \cdot \cdot} = 407$

$$\begin{aligned}
N &= 2 \cdot 3 \cdot 4 = 24 \\
C &= \frac{Y^2_{\cdots}}{N} = \frac{407^2}{24} = 6902.04 \\
SS_{Total} &= 8269 - C = 1366.96 \\
SS_{Treat} &= \frac{1}{12}(160^2 + 247^2) - C
\end{aligned}$$

$$
\begin{aligned}
&= 7217.42 - C = 315.38 \\
SS_{Time} &= \frac{1}{8}(89^2 + 133^2 + 185^2) - C \\
&= 7479.38 - C = 577.33 \\
SS_{Subtotal} &= \frac{1}{4}(28^2 + 48^2 + 84^2 + 61^2 + 85^2 + 101^2) - C \\
&= 7822.75 - C = 920.71 \\
SS_{Treat \times Time} &= SS_{Subtotal} - SS_{Treat} - SS_{Time} \\
&= 920.71 - 315.38 - 577.33 \\
&= 28.00 \\
SS_{Ind} &= \frac{1}{3}(56^2 + 47^2 + \ldots + 70^2 + 72^2) - \frac{1}{12}(160^2 + 247^2) \\
&= 7555.67 - 7217.42 = 338.25 \\
SS_{Error} &= SS_{Total} - SS_{Subtotal} - SS_{Ind} \\
&= 108.00
\end{aligned}
$$

	SS	df	MS	F	p-value
Treat	315.38	1	315.38	5.59	0.056
Time	577.33	2	288.67	32.07	0.000
Treat×Time	28.00	2	14.00	1.56	0.251
Ind	338.25	6	56.38		
Error	108.00	12	9.00		
Total	1366.96	23			

Table 7.3: Analysis of variance table in the model with interaction

We have

$$
F_{Treat \times Time} = \frac{MS_{Treat \times Time}}{MS_{Error}} = 1.56 \quad .
$$

Because of $1.56 < F_{2,12;0.95} = 3.88$, H_0: $(\alpha\beta)_{ij} = 0$ is not rejected. Hence we return to the independence model for testing the main effect "Time". $SS_{Treat \times Time}$ is added to SS_{Error}. The treatment effect (p-value 0.056) is not significant; the time effect is significant. The test statistic of the treatment effect is identical in both tables: $F_{Treat} = \frac{MS_{Treat}}{MS_{Ind}}$.

Example 7.2: Two blood pressure lowering drugs B and a combination (of B and another drug) are to be compared. On three control days, the diastolic blood pressure is measured in intervals of two hours. The last day is then analysed. This results in a repeated measures design with $p = 12$ measures. The sample sizes are $n_1 = 24$ (B) and $n_2 = 27$ (combination). The analysis is done with SPSS.

	SS	df	MS	F	p-value	
Treat	315.38	1	315.38	5.59	0.056	
Time	577.33	2	288.67	29.73	0.000	*
Ind	338.25	6	56.38			
Error	136.00	14	9.71			
Total	1366.96	23				

Table 7.4: Analysis of variance table in the independence model

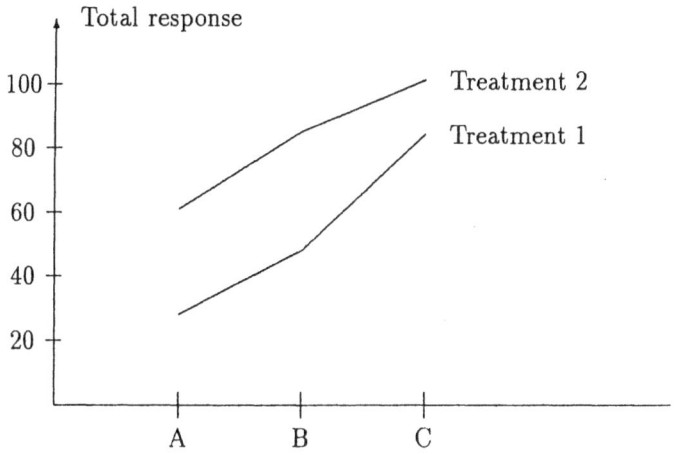

Figure 7.3: Total response treatment 1 and treatment 2 (Example 7.1)

```
MANOVA X1 TO X12 by Treat(1,2)
/wsfactors=Interval(12)
/contrast(Interval)=Difference
/Print=Homogeneity(BoxM)
/Design=Treat .
```

(i) Test of the homogeneity of variance, i.e., $H_0: \Sigma_B = \Sigma_{comb}$.

```
BoxsM =                      109.59084
F with (78,7357)DF =           1.03211  ,  P = .401  (Approx.)
Chi-square with 78 DF =       81.66664  ,  P = .366  (Approx)
```

With $p = 12$, we have $p(p+1)/2 = 78$, so that the Box-M statistic αM follows a χ^2_{78} (cf. (7.76)).

Hence, the null hypothesis $H_0: \Sigma_B = \Sigma_{comb} = \Sigma$ is not rejected.

The univariate unadjusted F-tests require, in addition to the assumption of the homogeneity of variance, the special structure of compound symmetry.

This assumption is included in the sphericity of the contrast covariance matrix as a special case.

(ii) Testing of H_0: $\tilde{C}_H \Sigma \tilde{C}'_H = \lambda I$
The test statistic by Mauchly is (cf. (7.103)) $W \sim \chi^2_v$ with $v = \frac{1}{2}(p-2)(p+1) = \frac{1}{2}(12-2)(12+1) = 65$ degrees of freedom.

```
Mauchly sphericity test, W =        .00478
Chi-square approx.     =     241.17785  with 65 D.F.
Significance                 .00000
```

Hence sphericity (and of course compound symmetry as well) is rejected and the unadjusted (averaged) univariate F-tests may not be applied. However, the adjusted univariate F-tests according to the Greenhouse-Geisser strategy can now be conducted.

(iii) Greenhouse-Geisser strategy
The measures for sphericity/non-sphericity are

Greenhouse-Geisser epsilon (7.107): $\hat{\epsilon}_{G-H} = 0.41$
Huynh-Feldt epsilon (7.99): $\hat{\epsilon} = 0.46$

They are distinctly smaller than 1 and indicate non-sphericity of the contrast covariance matrix $\tilde{C}_H \Sigma \tilde{C}'_H$. The Greenhouse-Geisser strategy now corrects the univariate test statistics according to their degrees of freedom.

Source	SS	df	MS	F	p-value	
Treat	5014.49	1	5014.49	4.24	0.045	*
Time	32414.11	11	2946.74	41.64	0.000	*
Treat×Time	2135.01	11	194.09	2.74	0.002	*
Ind	57996.61	49	1183.60			
Error	38141.34	539	70.76			
Total	135701.56	611				

Table 7.5: Unadjusted univariate averaged F-tests

The null hypothesis H_0: $(\alpha\beta)_{ij} = 0$ is rejected by the unadjusted univariate F-test. The test value of $F_{Treat \times Time} = 2.74$ is now assessed with respect to the $F_{\epsilon(p-1),\epsilon(p-1)(n-2)}$ distribution, where we start with the lower-bound epsilon $\epsilon_{min} = 1/(p-1) = 1/11$. We have $2.74 < F_{1,49;0.95} = 4.04$, hence the interaction is not significant, i.e., H_0:$(\alpha\beta)_{ij} = 0$ is not rejected.

Now the next step of the Greenhouse-Geisser strategy is to be carried out. The value estimated with SPSS is $\hat{\epsilon}_{G-H} = 0.41$, hence the adjusted F-statistic has $11 \cdot 0.41 = 4.5$ degrees of freedom in the numerator and $539 \cdot 0.41 = 221$ degrees of freedom in the denominator. Because of

$$F_{Treat \times Time} = 2.74 > 2.32 = F_{4.5, 221; 0.95}$$

H_0: $(\alpha\beta)_{ij} = 0$ is rejected. This decision is accepted.

Source		p-value
Treat×Time	$F_{11,39} = 1.75$	0.099
Time	$F_{11,39} = 18.01$	0.000 *
Treat	$F_{1,49} = 4.24$	0.045 *

Table 7.6: Results of the MANOVA

Results of the MANOVA and the corrected ANOVA:
At the 5%-level, the model with interaction holds for the ANOVA, and for the MANOVA the independence model holds. Hence, the significant main effects "treatment" and "time" can be interpreted seperately only in case of the MANOVA. If the 10%-level is chosen the independence model holds for the adjusted ANOVA as well.
Multiple Tests:
The overall treatment effect is significant. Hence the multiple test procedure from Section 7.7.5. may be applied.

j	p-values
1	0.006
2	0.003
3	0.002
4	0.061
5	0.329
6	0.374
7	0.424
8	0.893
9	0.536
10	0.117
11	0.582
12	0.024

From the table of the p-values of the univariate comparison of means, we find values in ascending order, which we compare with the adjusted Holm limits:

	$j = 3$	$j = 2$	$j = 1$	$j = 12$	$j = 4$	
p-values	0.002	0.003	0.006	0.024	0.061	...
Holm 5%	$\frac{0.05}{11} = 0.0045$	0.0045	$\frac{0.05}{10} = 0.005$	$\frac{0.05}{9} = 0.0056$	$\frac{0.05}{8} = 0.0063$	...
Holm 10%	0.0091	0.0091	0.010	0.011	0.0125	...

Hence the following local multiple treatment effects are significant:

(i) 5%-level: $j = 2$ and $j = 3$

(ii) 10%-level: $j = 1, 2$ and 3.

7.8 Multivariate Rank Tests in the Repeated Measures Model

In case of continuous but not necessarily normal response values, the same hypotheses as in the previous sections may be tested by statistics that are based on ranks. Starting point is once again a multivariate two-sample problem. Assume the following observation vectors

$$\boldsymbol{y}_{ki} = (y_{ki1}, \ldots, y_{kip})' \quad k = 1, 2, \ i = 1, \ldots, n_k . \tag{7.110}$$

For the observation vectors, we assume that the $\boldsymbol{y}_{ki}$ have independent distributions with a continuous distribution function

$$F_k(\boldsymbol{y}_{ki}) = G(\boldsymbol{y}_{ki} - \boldsymbol{m}_k), \quad k = 1, 2 , \tag{7.111}$$

where $\boldsymbol{m}_k = (m_{k1}, \ldots, m_{kp})'$ is the vector of medians of the kth group for the p measures.

The function G characterizes the type of distribution and $\boldsymbol{m}_k$ represents the localization parameter.

The null hypothesis H_0 : *no treatment effect* means $H_0 : F_1 = F_2$ and implies

$$H_0 : \ \boldsymbol{m}_1 = \boldsymbol{m}_2 , \tag{7.112}$$

so that both distributions are identical. The null hypothesis H_0 : *no time effect* means

$$H_0 : \ m_{k1} = \ldots = m_{kp}, \ k = 1, 2 \tag{7.113}$$

(cf. Koch, 1969). The test procedures are to be carried out considering the fact whether we have a significant interaction *treatment* $\times$ *time* or not. A detailed description of these tests can be found in Koch (1969) (cf. Puri and Sen, 1971). Since these non-parametric tests are quite burdensome and not implemented in standard software, we confine ourselves to a short description of the tests for one treatment effect. In case of continuous but not necessarily normal response, it is more practical to go over to loglinear models by applying categorical coding. The tests may then be conducted according to Chapter 10.

For the construction of a test for H_0 from (7.112), we proceede as follows: Let

$$r_{kij} := [\text{rank of } y_{kij} \text{ in } y_{11j}, \ldots, y_{1n_1j}, y_{21j}, \ldots, y_{2n_2j}] \tag{7.114}$$

with $k = 1, 2$, $i = 1, \ldots, n_k$, $j = 1, \ldots, p$, i.e., for every occasion j, $j = 1, \ldots, p$ the ranks $1, \ldots, N = n_1 + n_2$ are assigned.

If ties occur, then the averaged ranks are used.

Since the distribution is assumed to be continuous, we can assume

$$P\left(y_{kij} = y_{k'i'j}\right) = 0. \tag{7.115}$$

Hence, we disregard the ties in the following.

If the r_{kij} (cf.(7.114)) are combined for each individual, we get the rank observation vector of the ith individual in the kth group:

$$\boldsymbol{r}_{ki} = \left(r_{ki1}, \ldots, r_{kip}\right)', \quad k = 1, 2, \ i = 1, \ldots, n_k. \tag{7.116}$$

This yields N rank vectors that can be summarized by the $p \times N$ rank matrix

$$\boldsymbol{R} = \left(\boldsymbol{r}_{11}, \ldots, \boldsymbol{r}_{1n_1}, \boldsymbol{r}_{21}, \ldots, \boldsymbol{r}_{2n_2}\right). \tag{7.117}$$

Because of the rank assignment (cf.(7.114)), each of the p rows of $\boldsymbol{R}$ is a permutation of the numbers $1, \ldots, N$.

If the columns of $\boldsymbol{R}$ are exchanged in a way that the first row of $\boldsymbol{R}$ contains the ordered ranks, we find the matrix

$$\boldsymbol{R}_{per} = \begin{pmatrix} 1 & \cdots & N \\ r_{21}^{per} & \cdots & r_{2N}^{per} \\ \vdots & & \vdots \\ r_{p1}^{per} & \cdots & r_{pN}^{per} \end{pmatrix} = \left(\boldsymbol{r}_1, \ldots, \boldsymbol{r}_N\right), \tag{7.118}$$

which is a permutation-equivalent to $\boldsymbol{R}$ (cf.(7.117)).

Since the p observations y_{kij}, $j = 1, \ldots, p$ are not independent, the common distribution of the elements of $\boldsymbol{R}$ (or of $\boldsymbol{R}_{per}$) is dependent on the unknown distributions, even if H_0 holds.

Assume $\{\boldsymbol{R}_{per}\}$ is the set of all possible realizations of $\boldsymbol{R}_{per}$. For the size of $\{\boldsymbol{R}_{per}\}$, we have:

$$|\{\boldsymbol{R}_{per}\}| = (N!)^{p-1}. \tag{7.119}$$

In general, the distribution of $\boldsymbol{R}_{per}$ over $\{\boldsymbol{R}_{per}\}$ is dependent on the distributions F_1 and F_2.

If however $H_0 : F_1 = F_2$ holds, then the observation vectors $\boldsymbol{y}_{ki}, k = 1, 2\ i = 1, \ldots, n_k$ are independent and identically distributed. Hence, their common distribution stays invariant in case of a permutation within itself, i.e., it is of no great importance from which treatment group the vectors are derived.

This means however that under H_0, $\boldsymbol{R}$ is uniformly distributed over the set $\{\boldsymbol{R}_{per}\}$ of the $N!$ possible realizations that we get by all possible permutations of the columns of $\boldsymbol{R}_{per}$.

Hence we have

$$P\left(\boldsymbol{R} = \boldsymbol{r}_S \mid \{\boldsymbol{R}_{per}\}, H_0\right) = \frac{1}{N!} \quad \text{for all } \boldsymbol{r}_S \in \{\boldsymbol{R}_{per}\}. \tag{7.120}$$

Denote this (conditional) probability distribution by P_0.

Assume that the N rank observation vectors $\boldsymbol{r}_{ki}$, $k = 1, 2$, $i = 1, \ldots, n_k$ (cf.(7.116)) are known and that these are represented by $\boldsymbol{R}_{per}$, then the following holds (cf. Koch, 1969):

The probability that a rank observation vector $\boldsymbol{r}_{ki}$ takes the value $\boldsymbol{r}$ is

$$P\left(\boldsymbol{r}_{ki} = \boldsymbol{r}\right) = \frac{(N-1)!}{N!} = \frac{1}{N} \quad \text{for } \boldsymbol{r} = \boldsymbol{r}_1, \ldots, \boldsymbol{r}_N \quad . \tag{7.121}$$

Hence for the expectation of $\boldsymbol{r}_{ki}$, $k = 1, 2$, $i = 1, \ldots, n_k$, we have:

$$\begin{aligned} \mathrm{E}\left(\boldsymbol{r}_{ki} \mid \mathrm{H}_0\right) &= \sum_{j=1}^{N} \frac{1}{N} \boldsymbol{r}_j \\ &= \frac{1}{N} \frac{N(N+1)}{2} \boldsymbol{1}_p = \frac{N+1}{2} \boldsymbol{1}_p. \end{aligned} \tag{7.122}$$

For the construction of an appropriate test statistic, we define the rank mean vector of the kth group:

$$\boldsymbol{r}_{k.} = \frac{1}{n_k} \sum_{i=1}^{n_k} \boldsymbol{r}_{ki}, \quad k = 1, 2. \tag{7.123}$$

With (7.122), we obtain

$$\mathrm{E}\left(\boldsymbol{r}_{k.}\right) = \frac{N+1}{2} \boldsymbol{1}_p \quad . \tag{7.124}$$

The hypothesis H_0 can now be tested with the following test statistic (cf. Puri and Sen, 1971, p. 186):

$$L_I = \sum_{k=1}^{2} n_k \left(\boldsymbol{r}_{k.} - \frac{N+1}{2} \boldsymbol{1}_p\right)' \boldsymbol{S}_I^{-1} \left(\boldsymbol{r}_{k.} - \frac{N+1}{2} \boldsymbol{1}_p\right), \tag{7.125}$$

where we assume that the empirical rank covariance matrix $\boldsymbol{S}_I$ is regular

Remark: The matrix $\boldsymbol{S}_I$ measures the interaction treatment$\times$time. If no interaction exists, $\boldsymbol{S}_I$ equals the identity matrix (except for a variance factor) and the multivariate test statistic L_I equals the univariate statistic by Kruskal-Wallis (cf. (4.149)).

We have

$$\boldsymbol{S}_I = \frac{1}{N} \sum_{k=1}^{2} \sum_{i=1}^{n_k} \left(\boldsymbol{r}_{ki} - \frac{N+1}{2} \boldsymbol{1}_p\right)\left(\boldsymbol{r}_{ki} - \frac{N+1}{2} \boldsymbol{1}_p\right)'. \tag{7.126}$$

The test statistic L_I is the multivariate version of the statistic of the Kruskal-Wallis test and is equivalent to a generalized Lawley-Hotelling T^2-statistic.

It can be shown that L_I has an asymptotic χ^2-distribution under H_0 with p degrees of freedom (cf. Puri and Sen, 1971, p. 193).

Based on the construction of the test, large values of L_I indicate a violation of the null hypothesis H_0 from (7.112). Hence, H_0 is rejected if

$$L_I \geq \chi^2_{p;1-\alpha}. \qquad (7.127)$$

Example 7.3: In the following we demonstrate the calculation of the test statistic by means of a simple example. Suppose that we are given the following data set for $p = 3$ repeated measures:

$$
\begin{array}{l}
\text{Group 1} \\[2mm]
\\[2mm]
\text{Group 2}
\end{array}
\left(
\begin{array}{ccc}
2 & 3 & 6 \\
5 & 6 & 4 \\
4 & 5 & 5 \\
\hline
8 & 14 & 10 \\
10 & 12 & 14 \\
12 & 13 & 12
\end{array}
\right)
\underset{\text{ranks}}{\Longrightarrow}
\left(
\begin{array}{ccc}
1 & 1 & 3 \\
3 & 3 & 1 \\
2 & 2 & 2 \\
\hline
4 & 6 & 4 \\
5 & 4 & 6 \\
6 & 5 & 5
\end{array}
\right)
$$

$$
\begin{aligned}
\boldsymbol{R} &= \left(
\begin{array}{cccccc}
1 & 3 & 2 & 4 & 5 & 6 \\
1 & 3 & 2 & 6 & 4 & 5 \\
3 & 1 & 2 & 4 & 6 & 5
\end{array}
\right) \\[2mm]
&= \left(\ \boldsymbol{r}_{11}\ \ \boldsymbol{r}_{12}\ \ \boldsymbol{r}_{13}\ \ \boldsymbol{r}_{21}\ \ \boldsymbol{r}_{22}\ \ \boldsymbol{r}_{23}\ \right).
\end{aligned}
$$

The rank means in the two therapy groups are:

$$
\begin{aligned}
\boldsymbol{r}_{1.} &= \frac{1}{n_k}(\boldsymbol{r}_{11} + \boldsymbol{r}_{12} + \boldsymbol{r}_{13}) \\[2mm]
&= \frac{1}{3}\left[
\left(\begin{array}{c}1\\1\\3\end{array}\right) +
\left(\begin{array}{c}3\\3\\1\end{array}\right) +
\left(\begin{array}{c}2\\2\\2\end{array}\right)
\right] \\[2mm]
&= \frac{1}{3}\left(\begin{array}{c}6\\6\\6\end{array}\right) =
\left(\begin{array}{c}2\\2\\2\end{array}\right).
\end{aligned}
$$

$$
\begin{aligned}
\boldsymbol{r}_{2.} &= \frac{1}{3}\left[
\left(\begin{array}{c}4\\4\\6\end{array}\right) +
\left(\begin{array}{c}5\\4\\6\end{array}\right) +
\left(\begin{array}{c}6\\5\\5\end{array}\right)
\right] \\[2mm]
&= \frac{1}{3}\left(\begin{array}{c}15\\15\\15\end{array}\right) =
\left(\begin{array}{c}5\\5\\5\end{array}\right).
\end{aligned}
$$

From this, we calculate according to (7.125)

$$r_{i\cdot} - \frac{N+1}{2}\mathbf{1}_p = r_{i\cdot} - \frac{6+1}{2}\mathbf{1}_3 \quad (i = 1, 2)$$

$$\begin{aligned}
\left(r_{1\cdot} - \frac{7}{2}\mathbf{1}_3\right) &= -\frac{3}{2}\mathbf{1}_3 \\
\left(r_{2\cdot} - \frac{7}{2}\mathbf{1}_3\right) &= \frac{3}{2}\mathbf{1}_3
\end{aligned} .$$

This yields the covariance matrix S_I from (7.126):

$$S_I = \frac{1}{6 \cdot 4} \begin{pmatrix} 70 & 58 & 50 \\ 58 & 70 & 38 \\ 50 & 38 & 70 \end{pmatrix}$$

and

$$S_I^{-1} = \frac{24}{51840} \begin{pmatrix} 3456 & -2160 & -1296 \\ -2160 & 2400 & 240 \\ -1296 & 240 & 1536 \end{pmatrix} .$$

For L_I from (7.125), we have:

$$L_I = \sum_{k=1}^{2} n_k \left(r_{k\cdot} - \frac{N+1}{2}\mathbf{1}_3\right)' S_I^{-1} \left(r_{k\cdot} - \frac{N+1}{2}\mathbf{1}_3\right) = 6.00 \quad .$$

Hence, the test for H_0: $m_1 = m_2$ (cf. (7.112)) with

$$L_I = 6.00 < 7.81 = \chi^2_{3;0.95}$$

does not lead to a rejection of H_0.

7.9 Categorical Regression for the Analysis of Repeated Binary Response Data

7.9.1 Logit Models for Repeated Binary Response for the Comparison of Therapies

Unlike the previous sections of this chapter, we now assume categorical response. In order to explain the problems, we start with binary response $y_{ijk} = 1$ or $y_{ijk} = 0$. These categories can stand for a reaction above/below an average. In an example, the blood pressure of each patient above/below the median blood pressure of a control group is measured in this way.

Remark: In Chapter 10, the methods of the loglinear and the logistic models are discussed in detail. Because of didactic reasons, we assume the knowledge of Chapter 10.

Let $I = 2$ (response categories) and assume two therapies (P: placebo and M: medicament) to be compared. We define the logit for the response distribution of the kth subpopulation (therapy P or M, i.e., $k = 1$ or $k = 2$) for occasion j $(j = 1, \ldots, m)$ as

$$L(j; k) = \ln \left[P_1(j; k)/P_2(j; k) \right] . \tag{7.128}$$

The independence model in effect coding

$$L(j; k) = \mu + \lambda_1^P + \lambda_j^V \quad (j = 1, \ldots, m-1) \tag{7.129}$$

contains the main effects

λ_1^P : placebo effect,

$\lambda_j^V \ (j = 1, \ldots, m-1)$: occasions effect,

where the constraints of effect coding (cf. Chapter 6) hold

$$\lambda_2^M = -\lambda_1^P \qquad \text{(medicament effect)} \tag{7.130}$$

$$\lambda_m^V = -\sum_{j=1}^{m-1} \lambda_j^V . \tag{7.131}$$

The inclusion of interaction effects λ_{1j}^{PV} is possible (saturated model).

The ML estimation of the parameters of the model (7.129) is quite complicated since marginal probabilities, that are to be estimated from the marginal frequencies, are used for the odds. These marginal frequencies however have no independent multinomial distributions. The ML estimation has to be achieved by maximizing the likelihood under the constraint that the marginal distributions satisfy the model (7.129) of the null hypothesis. For this, iterative procedures (e.g., Koch et al. (1977), Aitchison and Silvey (1958)) have to be applied. These procedures replace the necessary non-linear optimization under linear constraints by stepwise weighted ordinary least squares estimates, and the iterated ML estimates are again used to form the standard χ^2 or G^2-goodness-of-fit statistics.

7.9.2   First-Order Markov Chain Models

A markov chain of the lth order $\{X_t\}$ is a stochastic process with a "memory" of length l, i.e., in case of $l = 1$, we have for a given occasion t

$$P(X_{t+1}|X_0, \ldots X_t) = P(X_{t+1}|X_t) . \tag{7.132}$$

Hence, the conditional probability for a future value X_{t+1} is only dependent on the preceding value X_t and not on the past $X_0, \ldots X_{t-1}$. The common density of $(X_0, \ldots X_m)$ is then of the form

$$f(x_0, \ldots, x_m) = f(x_0) \cdot f(x_1|x_0) \cdot \cdots \cdot f(x_m|x_{m-1}) . \tag{7.133}$$

Hence the common distribution is only dependent on the starting distribution $f(x_0)$ and on the conditional transition probabilities $f(x_i|x_{i-1})$. This corresponds to a loglinear model with the effects

$$(X_0, (X_0, X_1), (X_1, X_2), \ldots, (X_{m-1}, X_m)) .\qquad (7.134)$$

Remark: The transformation of the first-order markov-chain into categorical time-dependent response is the non-parametric counterpart of modelling the process as a time series with first-order auto-correlated errors.

Applied to our problem of binary response $\{X_j\}$ at occasions t_j ($j = 1, \ldots, m$) in the comparison of two therapies (P and M), the probabilities

$$P_{\alpha,\beta}(j - 1, j) \qquad \alpha, \beta = 1, 2 \quad \text{(response)} \qquad (7.135)$$

specify the common distribution of X_{j-1} and X_j.

The conditional probability that the process is in state $\alpha = i$ at occasion j, under the condition that it was in state $\alpha = k$ ($i, k = 1, 2$) at occasion $j - 1$, equals

$$\pi_{i/k}(j) = P(X_j = i|X_{j-1} = k) = \frac{P_{i,k}(j - 1, j)}{\sum\limits_{k=1}^{2} P_{i,k}(j - 1, j)} . \qquad (7.136)$$

Hence, the modelling of this process is equivalent to the loglinear model (7.137). We find the estimates of the $\pi_{i/k}(j)$ by constructing a contingency table and counting the frequencies of possible events. By means of observations in the subpopulations of the prognostic factor (placebo/medicament), we get the estimates $\hat{\pi}_{i/k}^{P}(j)$ and $\hat{\pi}_{i/k}^{M}(j)$ for both subpopulations.

Example 7.4: Binary response X_j, binary prognostic factor (placebo, medicament). Assume

$$X_j^M \text{ and } X_j^P = \begin{cases} 1 & \text{if blood pressure of the patient lies above} \\ & \text{the median of the placebo group at the } j \text{ th} \\ & \text{occasion} \\ 0 & \text{below.} \end{cases}$$

We choose the following fictitious numbers for a therapy group, in order to illustrate the calculation of the estimates of $\pi_{i/k}(j)$.

$j = 1$	
1	80
0	20
	100

$j = 2$	
1	60
0	40
	100

Assume the following counts of transitions for each patient:

$j = 1$		$j = 2$	Number of transitions
1		1	50
1	$\Longrightarrow$	0	30
0		1	10
0		0	10
			100

This yields

$$P_{1,1}(1,2) = \frac{50}{100} = 0.5 \, ,$$

$$P_{1,0}(1,2) = \frac{30}{100} = 0.3 \, ,$$

$$P_{0,1}(1,2) = \frac{10}{100} = 0.1 \, ,$$

$$P_{0,0}(1,2) = \frac{10}{100} = 0.1 \, .$$

Hence the estimated conditional transition probabilities are

$$\left. \begin{array}{l} \hat{\pi}_{1/1}(2) = \dfrac{0.5}{0.8} = 0.625 \\[2mm] \hat{\pi}_{0/1}(2) = \dfrac{0.3}{0.8} = 0.375 \end{array} \right\} \sum = 1$$

$$\left. \begin{array}{l} \hat{\pi}_{1/0}(2) = \dfrac{0.1}{0.2} = 0.5 \\[2mm] \hat{\pi}_{0/0}(2) = \dfrac{0.1}{0.2} = 0.5 \end{array} \right\} \sum = 1 \, .$$

Remark: The seperate modelling for each therapy group by a loglinear model

$$\ln(\hat{\pi}_{i/k}(j)) = \mu + \lambda_1^{X_0 X_1} + \cdots + \lambda_m^{X_{m-1} X_m} \tag{7.137}$$

gives an insight into significant transitions and filters out the best model according to the G^2-criterion.

If both therapy groups are included in one joint model, that is, if the indicator placebo/therapy is chosen as a third dimension, then local statements within the scope of the discrete markov chain models of the following form can be tested: H_0: The effects of the medicament $\lambda_{0,j}^M = -\lambda_{1,j}^P$ on the transition probabilities $\hat{\pi}_{1/0}(j)$ are significant (or significant at some occasions of the day's rythm of blood pressure).

The actual aim – a global measure (overall superiority) or a global test for H_0: "placebo=medicament" – cannot be achieved directly with this model, but only via an additional consideration.

7.9.3 Multinomial Sampling and Loglinear Models for Global Comparison of Therapies

We assume the response of a patient to therapy A or B to be a categorical response (e.g., binary response) over m occasions. Thus, for each therapy, we have m dependent (correlated) response values. If the response is observed in I categories, then the possible response values for the m occasions can be represented in an I^m-dimensional contingency table. Table 7.7 corresponds to $I = 2$ and $m = 4$.

Example 7.5:

$I = 2$ (binary response)

$m = 4$ occasions

Coding of the response: 1

Coding of the non-response: 0

Response	i	Occasion 1	2	3	4	Number
4times	1	1	1	1	1	n_1
3times	2	1	1	1	0	n_2
	3	1	1	0	1	n_3
	4	0	1	1	1	n_4
	5	1	0	1	1	n_5
2times	6	1	1	0	0	n_6
	7	1	0	1	0	n_7
	8	1	0	0	1	n_8
	9	0	1	1	0	n_9
	10	0	1	0	1	n_{10}
	11	0	0	1	1	n_{11}
1time	12	1	0	0	0	n_{12}
	13	0	1	0	0	n_{13}
	14	0	0	1	0	n_{14}
	15	0	0	0	1	n_{15}
0times	16	0	0	0	0	n_{16}
						n

Table 7.7: 2^4-table

Denote by $i = (i_1, \ldots, i_m)$ the cell in the table corresponding to response $i_j = 1$ or $i_j = 0$ $(j = 1, \ldots, m)$ at the occasions $t_1, \ldots, t_m$ and by π_i the probability for this cell. We then have

$$\sum_1^{I^m} \pi_i = 1 \, . \tag{7.138}$$

Let $m_i = n\pi_i$ be the expected cell count of the ith cell. Let the I categories be indexed by h ($h = 1, \ldots I$) and let $P_h(j)$ be the probability of response h at occasion j. The $\{P_h(j), h = 1, \ldots I\}$ for given j are then the jth marginal distribution of the contingency table.

We now consider Table 7.7 with $m = 4$ occasions. For each therapy group (P or M), we count seperately the completely crossed experimental design for the binary response (e.g., 1: above the median blood pressure of the placebo group at occasion j, 0: below), that is, the 2^4-table. We now classify the response according to the independent multinomial scheme $M(n; \pi_1, \ldots, \pi_5)$:

Class 1: 4-times response 1,
           0-times non-response 0
           $\Longrightarrow$ row 1 of Table 7.7
Class 2: 3-times response 1,
           1-time non-response 0
           $\Longrightarrow$ rows 2–5
Class 3: 2-times response 1,
           2-times non-response 0
           $\Longrightarrow$ rows 6–11
Class 4: 1-time response 1,
           3-times non-response 0
           $\Longrightarrow$ rows 12–15
Class 5: 0-times response 1,
           4-times non-response 0
           $\Longrightarrow$ row 16 .

If both therapies (P/M) are included, we receive a 5×2-table. The **disjoint categories of the rows** are often called **profiles**.

cumulated number of response 1	P	M
0	n_{11}	n_{12}
1	n_{21}	n_{22}
2	n_{31}	n_{32}
3	n_{41}	n_{42}
4	n_{51}	n_{52}
	n_{+1}	n_{+2}

Since P and M are independent and since the columns follow the model of the independent multinomial scheme $M(n_{+1}; \boldsymbol{\pi}_P)$, or $M(n_{+2}; \boldsymbol{\pi}_M)$ respectively, the null hypothesis H_0: "independent decomposition according to cumulated response and therapy" can equivalently be formulated by a loglinear model (m_{ij}: under H_0 expected cell frequencies)

$$\ln(m_{ij}) = \mu + \lambda_i^R + \lambda_1^P + \lambda_{i1}^{RP} , \tag{7.139}$$

where

μ : total mean ,

λ_i^R : effect of the ith cumulated response category (ith profile),

λ_1^P : effect of the placebo,

λ_{i1}^{RP} : interaction ith response category - placebo.

If effect coding is chosen, the effect of the medicament is $\lambda_1^M = -\lambda_1^P$.

Example 7.6: We illustrate the global test on a 13-hour blood pressure data set. The data set consists of measures of $n_1 = 63$ and $n_2 = 64$ patients of the therapy groups P (placebo) and M (medicament) over a stretch of $m = 13$ hours (start: $j = 0$, then 12 measures taken in 1-hour intervals). For each patient, it is recorded to which cumulated response category i ($i = 0, \ldots, 13$) he belongs, with i: number of hourly blood pressures above the median of the jth hourly measurement of the placebo group ($j = 0, \ldots, 12$).

The results are shown in Table 7.8. Table 7.9 shows these results summarized according to groups $(0,1), (2,3), \ldots, (12,13)$ (in order to overcome zero-counts in the cells). The parameter estimates and the standardized parameter estimates (∗: significance at the two-sided level of 5%, i.e., comparison with $u_{0.95\,(two-sided)} = 1.96$) are shown in Table 7.10.

i	P	M	Σ
0	5	30	35
1	7	7	14
2	3	6	9
3	4	6	10
4	3	5	8
5	3	3	6
6	5	2	7
7	6	0	6
8	3	2	5
9	9	0	9
10	5	0	5
11	2	2	4
12	2	1	3
13	6	0	6
Σ	63	64	127

Table 7.8: Classification of the 12-hour measures at the end point according to "i-times blood pressure values above the respective hourly median of the placebo group"

Remark: The calculations have been done with the newly developed software

	P	M	Σ
0,1	12	37	49
2,3	8	12	20
4,5	6	8	14
6,7	10	2	12
8,9	13	2	15
10,11	7	2	9
12,13	7	1	8
Σ	63	64	127

Table 7.9: Summary of the classes in Table 7.8

LOGGY 1.0 (cf. Heumann and Jacobsen, 1993), the standard software PCS, as well as additional programs.

Interpretation

(i) Saturated model

$$\ln(m_{ij}) = \mu + \lambda_i^R + \lambda_1^P + \lambda_{i1}^{RP} \ . \tag{7.140}$$

The test statistic for H_0: "saturated model valid" is $G^2 = 0$ (perfect fit) as usual.

The placebo effect $\hat{\lambda}_1^P = 0.35$ (2.57 standardized) is significant. Since code 1 symbolizes high blood pressure (above the respective hourly median of the placebo group), a positive λ_1^P stands for an effect towards higher blood pressure. Hence $(\lambda_1^M = -0.35)$, the medicament significantly lowers the blood pressure.

The significant response effects λ_1^R (categories 0- and 1-time above the median) and λ_2^R (2- and 3-times above the median) are positive and λ_7^R (10- and 11-times above the median) is negative. These two results once again speak (in a qualitive way) for the blood pressure lowering effect of the medicament.

The interactions are hard to interpret seperately.

The analysis of the submodels of the hierarchy lead to the following results:

(ii) Independence model

$$H_0 : \ \ln(m_{ij}) = \mu + \lambda_i^R + \lambda_1^P \ . \tag{7.141}$$

The test value $G^2 = 37$ (p-value 0.000002) is significant, hence H_0 (7.141) is rejected.

parameter	parameter estimate	significant	standardized
μ	1.81	*	13.42
λ_1^P	0.35	*	2.57
λ_1^R	1.24	*	6.35
λ_2^R	0.47	*	2.00
λ_3^R	0.12		0.47
λ_4^R	−0.31		−0.89
λ_5^R	−0.18		−0.53
λ_6^R	−0.49		−1.35
λ_7^R	−0.84	*	−1.98
λ_{11}^{RP}	−0.91	*	−4.67
λ_{21}^{RP}	−0.55	*	−2.34
λ_{31}^{RP}	−0.49		−1.85
λ_{41}^{RP}	0.46		1.29
λ_{51}^{RP}	0.59		1.69
λ_{61}^{RP}	0.28		0.77
λ_{71}^{RP}	0.63		1.33

Table 7.10: Parameter estimates and standardized values for the saturated model $\ln(m_{ij}) = \mu + \lambda_i^R + \lambda_1^P + \lambda_{i1}^{RP}$

(iii) Model for isolated profile effects

$$H_0 : \ln(m_{ij}) = \mu + \lambda_i^R . \tag{7.142}$$

The test value is $G^2 = 37$ (7 df) is significant as well (H_0: (7.142) is rejected).

(iv) Model for isolated treatment effect

$$H_0 : \ln(m_{ij}) = \mu + \lambda_1^P \tag{7.143}$$

The test value is $G^2 = 90$ (12 df) and hence significant.

As a result, it can be stated that the saturated model is the only possible statistical model for the observed profiles of the two subpopulations placebo and medicament. This model indicates

- a blood pressure lowering effect of the medicament,

- profile effects,

and gives evidence for

- significant interactions.

As an interesting result, it can be stated that the therapy effect is not isolated (i.e., it is not an orthogonal component), but has a mutual effect with the time after taking the medicament.

This analysis is confirmed by the following crude-rate analysis for which the profiles 0–6 and 7–13 were combined.

	P	M	Σ
0–6	32	59	91
7–12	31	5	36
Σ	63	64	127

The saturated model

$$\ln(m_{ij}) = \mu + \lambda_1^R + \lambda_1^P + \lambda_{11}^{RP} \tag{7.144}$$

yields the significant parameter estimates

	$\hat{\mu}$	$\hat{\lambda}_1^R$	$\hat{\lambda}_1^P$	$\hat{\lambda}_{11}^{RP}$
	3.15	0.63	0.30	−0.61
standardized	23.77 *	4.72 *	2.69 *	−4.60 *

In the saturated model, we have for the odds ratio

$$\theta = \exp(4\lambda_{11}^{RP}) ,$$
$$\text{i.e.,} \quad \hat{\theta} = 0.0036 ,$$
$$\ln \hat{\theta} = -2.44 \quad \text{(negative interaction)}.$$

The crude model of the 2×2-table is regarded as a robust indicator of interactions in general, that can be broken down by finer structures. The advantage of the 2×2-table is the estimation of a crude interaction over all levels of the categories of the rows.

Remark: The model calculations assume a Poisson sampling scheme for the contingency table – that is unrestricted random sampling, i.e., especially a random total sample size.

The sampling scheme is restricted to independent multinomial sampling in case of the model of therapy comparison. Birch (1963) has proved that the ML estimates are identical for independent multinomial sampling and Poisson sampling, as long as the model contains a term (parameter) for the marginal distribution given by the experimental design. For our case of therapy comparison, this means that the marginal sums n_{+1} and n_{+2} (= number of patients in the placebo group and the medicated group) have to appear as sufficient statistics in the parameter estimates. This is the case in the

(i) saturated model (7.140)

(ii) independence model (7.141)

(iii) model for isolated profile effects (7.142),

but not in the

(iv) model for the isolated treatment effect (7.143).

As our model calculations show, model (7.143) is of no interest, since a treatment effect cannot be detected isolated but only in interaction with the profiles.

Remark: Table 7.8 and Table 7.9 differ slightly due to patients whose blood pressure coincide with the hourly median.

Trend of the Profiles of the Medicated Group

As another non-parametric indicator for the blood pressure lowering effect of the medicament, we now model the crude binary risk

7–12–times over the respective placebo hourly median /
0–6–times over the median

over three observation days (i.e., $i = 1, 2, 3$) by a logistic regression. The results are shown in Table 7.11.

i	7–12	0–6	Logit
1	34	32	0.06
2	12	51	−1.45
3	5	59	−2.47

Table 7.11: Crude profile of the medicated group for the 3 observation days

From this we calculate the model

$$\ln\left(\frac{n_{i1}}{n_{i2}}\right) = \hat{\alpha} + \hat{\beta}i \quad (i = 1, 2, 3)$$
$$= 1.243 - 1.265 \cdot i \quad (7.145)$$

with the correlation coefficient $r = 0.9938$ (p-value 0.0354, one-sided) and the residual variance $\hat{\sigma}^2 = 0.2^2$.

Hence, the negative trend to fall into the unfavourable profile group "7–12" is significant for this model (3 observations, 2 parameters !). However, this result can only be regarded as a crude indicator. Results that are more reliable are

i	0–1	2–3	4–5	6–7	8–9	10–11	12–13
1	4	10	10	13	8	13	8
2	29	14	7	4	4	2	1
3	37	12	8	2	2	2	1

Table 7.12: Fine profiles of the medicated group for the 3 observation days

achieved with Table 7.12, which is subdivided into 7 groups instead of only 2 profiles.

The G^2-analysis in Table 7.12 for testing H_0: "cell counts over the profiles and days are independent " yields a significant value of $G^2_{14} = 70.50$ ($> 23.7 = \chi^2_{14;0.95}$) so that H_0 is rejected.

7.10   Excercises and Questions

7.10.1 How is the correlation of an individual over the occasions defined?
In which way are two individuals correlated?
Name the intraclass-correlation coefficient of an individual over two different
occasions.

7.10.2 What structure does the compound symmetric covariance matrix have?
Name the best linear unbiased estimate of β in the model $y = X\beta + \epsilon$,
$\epsilon \sim (0, \sigma^2 \Sigma)$ with Σ of compound symmetric structure.

7.10.3 Why is the ordinary LS estimate chosen instead of the Aitken estimate in
case of compound symmetry?

7.10.4 Name the repeated measures model for two independent populations.
Why can it be interpreted as a mixed model and as a split-plot design?

7.10.5 What is meant by the μ_k-profile of an individual?

7.10.6 How is the Wishart-distribution defined?

7.10.7 How is H_0: $\mu = \mu_0$ (one-sample problem) tested univariate for $x_1, \ldots, x_n$
i.i.d. $\sim N_p(\mu, \Sigma)$?

7.10.8 How is H_0: $\mu_x = \mu_y$ (two-sample problem) tested multivariate for
$x_1, \ldots, x_{n_1} \sim N_p(\mu_x, \Sigma_x)$ and $y_1, \ldots, y_{n_2} \sim N_p(\mu_y, \Sigma_y)$?
Which conditions have to hold true?

7.10.9 Describe the test strategy (univariate/multivariate) dependent on the ful-
fillment of the sphericity condition.

8 Cross-Over Design

8.1 Introduction

Clinical trials form an important part of the examination of new drugs or medical treatments. The drugs are usually assessed by comparing their effects on human subjects. From an ethical point of view, the risks which patients might be exposed to must be reduced to a minimum and also the number of individuals should be as small as statistically required. Cross-over trials follow the latter, treating each patient successively with two or more treatments. For that purpose, the individuals are divided into randomized groups in which the treatments are given in certain orders. In a 2×2 design, each subject receives two treatments, conventionally labelled as A and B. Half of the subjects receive A first and then *cross over* to B while the remaining subjects receive B first and then cross over to A. Between two treatments a suitable period of time is chosen, where no treatment is applied. This *washout period* is used to avoid the persistence of a treatment applied in one period to a subsequent period of treatment.

The aim of cross-over designs is to estimate most of the main effects using within subject differences (or contrasts). Since it is often the case that there is considerably more variation between subjects than within subjects, this strategy leads to more powerful tests than simply comparing two independent groups using between subject information. As each subject acts as his own control, between subject variation is eliminated as a source of error.

If the washout periods are not chosen long enough, then a treatment may persist in a subsequent period of treatment. This *carry-over effect* will make it more difficult, or nearly impossible, to estimate *direct treatment effects*.

To avoid psychological effects, subjects are treated in a double blinded manner so that neither patients nor doctors know which of the treatments is actually applied.

8.2 Linear Model and Notations

We assume that there are s groups of subjects. Each group receives the M treatments in a different order. It is favourable to use all of the $M!$ orderings of treatments, i.e., to use the orderings AB and BA for comparison of $M = 2$ treatments and $ABC, BCA, CAB, ACB, CBA, BAC$ for $M = 3$ treatments so that $s = M!$

We generally assume that the trial lasts p periods (i.e., $p = M$ periods if all possible orderings are used). Let y_{ijk} denote the response observed on the kth subject ($k = 1, \ldots, n_i$) of group i ($i = 1, \ldots, s$) in period j ($j = 1, \ldots, p$). We first consider the following linear model (cf. Jones and Kenward, 1989, p. 9) which Ratkowsky et al. (1993, pp. 81–84) label as parametrization 1:

$$y_{ijk} = \mu + s_{ik} + \pi_j + \tau_{[i,j]} + \lambda_{[i,j-1]} + \epsilon_{ijk}. \tag{8.1}$$

The terms are:

y_{ijk}	:	the response of the kth subject of group i in period j,
μ	:	the overall mean,
s_{ik}	:	the effect of subject k in group i, $i = 1, \ldots s,\ k = 1, \ldots, n_i$,
π_j	:	the effect of period j, $j = 1, \ldots, p$,
$\tau_{[i,j]}$	:	the direct effect of the treatment administered in period j of group i (treatment effect),
$\lambda_{[i,j-1]}$	:	the carry-over effect (effect of the treatment administered in period $j-1$ of group i) that still persists in period j, where $\lambda_{[i,0]} = 0$,
ϵ_{ijk}	:	random error.

The subject effects s_{ik} are taken to be random. Sample totals will be denoted by capital letters, sample means by small letters. A dot ($\cdot$) will replace a subscript to indicate that the data have been summed over that subscript. For example,

$$\left.\begin{array}{ll} \text{total response:} & Y_{ij\cdot} = \sum_{k=1}^{n_i} y_{ijk}, \quad Y_{i\cdot\cdot} = \sum_{j=1}^{p} Y_{ij\cdot}, \quad Y_{\cdots} = \sum_{i=1}^{s} Y_{i\cdot\cdot} \\ \text{means:} & y_{ij\cdot} = \frac{Y_{ij\cdot}}{n_i}, \qquad\quad y_{i\cdot\cdot} = \frac{Y_{i\cdot\cdot}}{pn_i}, \qquad\quad y_{\cdots} = \frac{Y_{\cdots}}{p\sum_{i=1}^{s} n_i} \end{array}\right\} \tag{8.2}$$

To begin with, we assume that the response has been recorded on a continuous scale.

Remark. Model (8.1) may be called *classical approach* and has been explored intensively since the sixties (Grizzle, 1965). This parametrization, however, shows some inconsistencies concerning the effect caused by the order in which the treatments are given. This so-called sequence effect becomes important, especially regarding higher order designs. For example, using the following plan in a cross-over design trial

Period

	1	2	3	4
	A	B	C	D
	B	D	A	C
Sequence	C	A	D	B
	D	C	B	A

,

the actual sequence (group) might have a fixed effect on the response. Then the between-subject effect s_{ik} would also be stratified by sequences (groups). This effect would have to be considered as an additional parameter γ_i $(i = 1, \ldots, s)$ in model (8.1). Applying the classical approach (8.1) without this sequence effect leads to the sequence effect being confounded with other effects. We will discuss this fact later in this chapter.

8.3 2× 2 Cross-Over (Classical Approach)

We now consider the common comparison of $M = 2$ treatments A and B (cf. Figure 8.1) using a 2×2 cross-over trial with $p = 2$ periods.

	Period 1	Period 2
Group 1	A	B
Group 2	B	A

Figure 8.1: 2 × 2 Cross-over design with 2 treatments

As there are only four sample means $y_{11.}, y_{12.}, y_{21.}$, and $y_{22.}$ available from the 2 × 2 cross-over design, we can only use three degrees of freedom to estimate period, treatment, and carry-over effects. Thus, we have to omit the direct treatment×period interaction which now has to be estimated as an aliased effect confounded with the carry-over effect. Therefore, the 2 × 2 cross-over design has the special parametrization:

$$\tau_1 = \tau_A \quad \text{and} \quad \tau_2 = \tau_B \quad . \tag{8.3}$$

The carry-over effects are simplified as

$$\left. \begin{array}{l} \lambda_1 = \lambda_{[1,1]} = \lambda_{[A,1]} \\ \lambda_2 = \lambda_{[2,1]} = \lambda_{[B,1]} \end{array} \right\} \tag{8.4}$$

Group	Period 1	Period 2
1 (AB)	$\mu + \pi_1 + \tau_1 + s_{1k} + \epsilon_{11k}$	$\mu + \pi_2 + \tau_2 + \lambda_1 + s_{1k} + \epsilon_{12k}$
2 (BA)	$\mu + \pi_1 + \tau_2 + s_{2k} + e_{21k}$	$\mu + \pi_2 + \tau_1 + \lambda_2 + s_{2k} + \epsilon_{22k}$

Table 8.1: The effects in the 2×2 cross-over model

Then λ_1 and λ_2 denote the carry-over effect of treatment A resp. B applied in the first period so that the effects in the full model are as shown in Table 8.1. The subject effects s_{ik} are regarded as random.

The random effects are assumed to be distributed as follows:

$$\left.\begin{array}{rcl} s_{ik} & \overset{i.i.d.}{\sim} & N(0,\sigma_s^2) \\ \epsilon_{ijk} & \overset{i.i.d.}{\sim} & N(0,\sigma^2) \\ E\left(\epsilon_{ijk}s_{ik}\right) & = & 0 \quad (\text{all } i,j,k) \end{array}\right\} \tag{8.5}$$

8.3.1 Analysis Using t-Tests

The analysis of data from a 2×2 cross-over trial using t-tests was first suggested by Hills and Armitage (1979). Jones and Kenward (1989) note that these are valid, whatever the covariance structure of the two measurements y_A and y_B taken on each subject during the active treatment periods.

Testing Carry-Over Effects, i.e., H_0: $\lambda_1 = \lambda_2$

The first test we consider is the test on equality of the carry-over effects λ_1 and λ_2. Only if equality is not rejected, the following tests on main effects are valid, since the difference of carry-over effects $\lambda_d = \lambda_1 - \lambda_2$ is the aliased effect of the treatment$\times$period interaction.

We note that the subject total $Y_{1 \cdot k}$ of the kth subject in group 1

$$Y_{1 \cdot k} = y_{11k} + y_{12k} \tag{8.6}$$

has the expectation (cf. Table 8.1)

$$\begin{aligned} E\left(Y_{1 \cdot k}\right) & = E\left(y_{11k}\right) + E\left(y_{12k}\right) \\ & = \left(\mu + \pi_1 + \tau_1\right) + \left(\mu + \pi_2 + \tau_2 + \lambda_1\right) \\ & = 2\mu + \pi_1 + \pi_2 + \tau_1 + \tau_2 + \lambda_1. \end{aligned} \tag{8.7}$$

In group 2 (BA) we get

$$Y_{2 \cdot k} = y_{21k} + y_{22k} \tag{8.8}$$

and

$$E\left(Y_{2 \cdot k}\right) = 2\mu + \pi_1 + \pi_2 + \tau_1 + \tau_2 + \lambda_2. \tag{8.9}$$

Under the null hypothesis

$$H_0 : \lambda_1 = \lambda_2 \tag{8.10}$$

these two expectations are equal

$$E\left(Y_{1 \cdot k}\right) = E\left(Y_{2 \cdot k}\right) \qquad \text{for all } k. \tag{8.11}$$

Now we can apply the two-sample t-test to the subject totals and define

$$\lambda_d = \lambda_1 - \lambda_2 \qquad . \tag{8.12}$$

Then

$$\hat{\lambda}_d = \frac{Y_{1 \cdot \cdot}}{n_1} - \frac{Y_{2 \cdot \cdot}}{n_2} = 2(y_{1 \cdot \cdot} - y_{2 \cdot \cdot}) \tag{8.13}$$

is an unbiased estimator for λ_d, i.e.,

$$E\left(\hat{\lambda}_d\right) = \lambda_d \quad . \tag{8.14}$$

Using

$$Y_{i\cdot k} - E\left(Y_{i\cdot k}\right) = 2s_{ik} + \epsilon_{i1k} + \epsilon_{i2k}$$

and

$$\text{Var}(Y_{i\cdot k}) = 4\sigma_s^2 + 2\sigma^2$$

we get

$$\text{Var}\left(\frac{Y_{i\cdot\cdot}}{n_i}\right) = \frac{1}{n_i^2}\sum_{k=1}^{n_i}\text{Var}(Y_{i\cdot k}) = \frac{4\sigma_s^2 + 2\sigma^2}{n_i}$$

$(i = 1, 2)$. Therefore we have

$$\begin{aligned}\text{Var}\left(\hat{\lambda}_d\right) &= 2(2\sigma_s^2 + \sigma^2)\left(\frac{1}{n_1} + \frac{1}{n_2}\right) \\ &= \sigma_d^2\left(\frac{n_1 + n_2}{n_1 n_2}\right)\end{aligned} \tag{8.15}$$

where

$$\sigma_d^2 = 2(2\sigma_s^2 + \sigma^2) \quad . \tag{8.16}$$

To estimate σ_d^2 we use the pooled sample variance

$$s^2 = \frac{(n_1 - 1)s_1^2 + (n_2 - 1)s_2^2}{n_1 + n_2 - 2} \tag{8.17}$$

which has $(n_1 + n_2 - 2)$ degrees of freedom, with s_1^2 and s_2^2 denoting the sample variances of the response totals within groups, where

$$s_i^2 = \frac{1}{n_i - 1}\sum_{k=1}^{n_i}(Y_{i\cdot k} - \frac{Y_{i\cdot\cdot}}{n_i})^2 = \frac{1}{n_i - 1}\left(\sum_{k=1}^{n_i}Y_{i\cdot k}^2 - \frac{Y_{i\cdot\cdot}^2}{n_i}\right) \quad (i = 1, 2). \tag{8.18}$$

We construct the test statistic

$$T_\lambda = \frac{\hat{\lambda}_d}{s}\sqrt{\frac{n_1 n_2}{n_1 + n_2}} \tag{8.19}$$

that follows a Student's t-distribution with $(n_1 + n_2 - 2)$ degrees of freedom under H_0 (8.10).

According to Jones and Kenward (1989), it is usual practice to follow Grizzle (1965) to run this test at the $\alpha = 0.1$ level. If this test does not reject H_1, we can proceed to test the main effects.

Testing Treatment Effects (Given $\lambda_1 = \lambda_2 = \lambda$)

If we can assume that $\lambda_1 = \lambda_2 = \lambda$, then the period differences

$$
\begin{array}{ll}
d_{1k} = y_{11k} - y_{12k} & \text{(group 1, i.e., A–B)} \\
d_{2k} = y_{21k} - y_{22k} & \text{(group 2, i.e., B–A)}
\end{array}
\tag{8.20}
$$

have expectations

$$
\begin{array}{ll}
E(d_{1k}) = \pi_1 - \pi_2 + \tau_1 - \tau_2 - \lambda , \\
E(d_{2k}) = \pi_1 - \pi_2 + \tau_2 - \tau_1 - \lambda .
\end{array}
\tag{8.21}
$$

Under the null hypothesis H_0: no treatment effect, i.e.,

$$
H_0 : \tau_1 = \tau_2
\tag{8.22}
$$

these two expectations coincide. The difference of the treatment effects

$$
\tau_d = \tau_1 - \tau_2
\tag{8.23}
$$

is estimated by

$$
\hat{\tau}_d = \frac{1}{2}(d_{1\cdot} - d_{2\cdot})
\tag{8.24}
$$

which is unbiased

$$
E(\hat{\tau}_d) = \tau_d ,
\tag{8.25}
$$

and has variance

$$
\begin{aligned}
\mathrm{Var}(\hat{\tau}_d) &= \frac{2\sigma^2}{4}\left(\frac{1}{n_1} + \frac{1}{n_2}\right) \\
&= \frac{\sigma_D^2}{4}\left(\frac{1}{n_1} + \frac{1}{n_2}\right)
\end{aligned}
\tag{8.26}
$$

where

$$
\sigma_D^2 = 2\sigma^2 .
\tag{8.27}
$$

The pooled estimate of σ_D^2 according to (8.17), replacing s_i^2 by

$$
s_{iD}^2 = \frac{1}{n_i - 1}\sum_{k=1}^{n_i}(d_{ik} - d_{i\cdot})^2
$$

becomes

$$
s_D^2 = \frac{(n_1 - 1)s_{1D}^2 + (n_2 - 1)s_{2D}^2}{n_1 + n_2 - 2} .
\tag{8.28}
$$

Under the null hypothesis H_0: $\tau_d = 0$, the statistic

$$
T_\tau = \frac{\hat{\tau}_d}{\frac{1}{2}s_D}\sqrt{\frac{n_1 n_2}{n_1 + n_2}} ,
\tag{8.29}
$$

follows a t-distribution with $(n_1 + n_2 - 2)$ degrees of freedom.

Testing Period Effects (Given $\lambda_1 + \lambda_2 = 0$)

Finally we test for period effects using the null hypothesis

$$H_0 : \pi_1 = \pi_2 \quad . \tag{8.30}$$

The 'cross-over' differences

$$\begin{aligned} c_{1k} &= d_{1k} \ , \\ c_{2k} &= -d_{2k} \end{aligned} \tag{8.31}$$

have expectations

$$\begin{aligned} E\left(c_{1k}\right) &= \pi_1 - \pi_2 + \tau_1 - \tau_2 - \lambda_1 \\ E\left(c_{2k}\right) &= \pi_2 - \pi_1 + \tau_1 - \tau_2 + \lambda_2 \quad . \end{aligned} \tag{8.32}$$

Under the null hypothesis H_0: $\pi_1 = \pi_2$ and the familiar reparametrization $\lambda_1 + \lambda_2 = 0$, these expectations coincide, i.e., $E\left(c_{1k}\right) = E\left(c_{2k}\right)$. An unbiased estimater for the difference of the period effects $\pi_d = \pi_1 - \pi_2$ is given by

$$\hat{\pi}_d = \frac{1}{2}(c_{1.} - c_{2.}) \tag{8.33}$$

and we get the test statistic with s_D from (8.28)

$$T_\pi = \frac{\hat{\pi}_d}{\frac{1}{2} s_D} \sqrt{\frac{n_1 n_2}{n_1 + n_2}} , \tag{8.34}$$

which again follows a t-distribution with $(n_1 + n_2 - 2)$ degrees of freedom.

Unequal Carry-Over Effects

If the hypothesis $\lambda_1 = \lambda_2$ is rejected, the above procedure for testing $\tau_1 = \tau_2$ should not be used since it is based on biased estimators. Given $\lambda_d = \lambda_1 - \lambda_2 \neq 0$, we get

$$E\left(\hat{\tau}_d\right) = E\left(\frac{d_{1.} - d_{2.}}{2}\right) = \tau_d - \frac{\lambda_d}{2} \quad . \tag{8.35}$$

With

$$\hat{\lambda}_d = y_{11.} + y_{12.} - y_{21.} - y_{22.} \tag{8.36}$$

and

$$\hat{\tau}_d = \frac{1}{2}(y_{11.} - y_{12.} - y_{21.} + y_{22.}) \tag{8.37}$$

an unbiased estimator $\hat{\tau}_{d|\lambda_d}$ of τ_d is given by

$$\begin{aligned} \hat{\tau}_{d|\lambda_d} &= \frac{1}{2}(y_{11.} - y_{12.} - y_{21.} + y_{22.}) + \frac{1}{2}(y_{11.} + y_{12.} - y_{21.} - y_{22.}) \\ &= y_{11.} - y_{21.} \end{aligned} \tag{8.38}$$

The unbiased estimator of τ_d for $\lambda_d \neq 0$ is identical to the estimator of a parallel group study. The estimator is based on between subject information of the first period and the measurements. Testing for $H_0: \tau_d = 0$ is done following a two-sample t-test, but using the measurements of the first period only, to estimate the variance. Thus, the sample size might become too small to get significant results for the treatment effect.

Regarding the reparametrization

$$\lambda_1 + \lambda_2 = 0 \quad , \tag{8.39}$$

we see that the estimator $\hat{\pi}_d$ is still unbiased:

$$
\begin{aligned}
E(\hat{\pi}_d) &= E\left(\frac{c_1. - c_2.}{2}\right) \\
&= \frac{1}{2}E\left(\frac{1}{n_1}\sum_{k=1}^{n_1} c_{1k} - \frac{1}{n_2}\sum_{k=1}^{n_2} c_{2k}\right) \\
&= \frac{1}{2}\left(\frac{1}{n_1}\sum_{k=1}^{n_1} E(c_{1k}) - \frac{1}{n_2}\sum_{k=1}^{n_2} E(c_{2k})\right) \\
&= \frac{1}{2}(2\pi_1 - 2\pi_2 - (\lambda_1 + \lambda_2)) \quad &\text{[cf. (8.32)]} \\
&= \pi_d \quad . \quad &\text{[cf. (8.39)]}
\end{aligned}
$$

and thus $\hat{\pi}_d$ is unbiased even if $\lambda_d = \lambda_1 - \lambda_2 \neq 0$ but $\lambda_1 + \lambda_2 = 0$.

8.3.2 Analysis of Variance

Considering higher order cross-over designs, it is useful to test the effects using F-tests obtained from an analysis of variance table. Such a table was presented by Grizzle (1965) for the special case $n_1 = n_2$. The first general table was given by Hills and Armitage (1979). The sums of squares may be derived for the 2 × 2 cross-over design as a simple example of a split-plot design. The subjects form the main plots while the periods are treated as the subplots at which repeated measurements are taken (cf. Section 6.8). With this in mind, we get

$$SS_{Total} = \sum_{i=1}^{2}\sum_{j=1}^{2}\sum_{k=1}^{n_i} y_{ijk}^2 - \frac{Y_{...}^2}{2(n_1 + n_2)}$$

between-subjects:

$$SS_{Carry-over} = \frac{2n_1 n_2}{(n_1 + n_2)}(y_{1..} - y_{2..})^2$$

$$SS_{b-s\ Residual} = \sum_{i=1}^{2}\sum_{k=1}^{n_i}\frac{Y_{i \cdot k}^2}{2} - \sum_{i=1}^{2}\frac{Y_{i..}^2}{2n_i}$$

within-subjects:

$$SS_{Treat} = \frac{n_1 n_2}{2(n_1 + n_2)}(y_{11.} - y_{12.} - y_{21.} + y_{22.})^2$$

$$SS_{Period} = \frac{n_1 n_2}{2(n_1 + n_2)}(y_{11.} - y_{12.} + y_{21.} - y_{22.})^2$$

$$SS_{w-s\ Residual} = \sum_{i=1}^{2}\sum_{j=1}^{2}\sum_{k=1}^{n_i} y_{ijk}^2 - \sum_{i=1}^{2}\sum_{j=1}^{2} \frac{Y_{ij\cdot}^2}{n_i} - SS_{b-s\ Residual}$$

Source	SS	df	MS	F
Between subjects				
Carry-over	SS_{c-o}	1	MS_{c-o}	F_{c-o}
Residual (between subjects)	$SS_{Residual(b-s)}$	$n_1 + n_2 - 2$	$MS_{Residual(b-s)}$	
Within subjects				
Direct treatment effect	SS_{Treat}	1	MS_{Treat}	F_{Treat}
Period effect	SS_{Period}	1	MS_{Period}	F_{Period}
Residual (within subjects)	$SS_{Residual(w-s)}$	$n_1 + n_2 - 2$	$MS_{Residual(w-s)}$	
Total	SS_{Total}	$2(n_1 + n_2) - 1$		

Table 8.2: Analysis of variance table for 2 × 2 cross-over designs (Jones and Kenward, 1989, p. 31, Hills and Armitage, 1979)

MS	E(MS)
MS_{c-o}	$\frac{2n_1 n_2}{n_1+n_2}(\lambda_1 - \lambda_2)^2 + (2\sigma_s^2 + \sigma^2)$
$MS_{Residual(b-s)}$	$(2\sigma_s^2 + \sigma^2)$
MS_{Treat}	$\frac{2n_1 n_2}{(n_1+n_2)}[(\tau_1 - \tau_2) - \frac{(\lambda_1-\lambda_2)}{2}]^2 + \sigma^2$
MS_{Period}	$\frac{2n_1 n_2}{(n_1+n_2)}(\pi_1 - \pi_2)^2 + \sigma^2$
$MS_{Residual(w-s)}$	σ^2

Table 8.3: E(MS)

The F-statistics are built according to Table 8.3.

Under H_0: $\lambda_1 = \lambda_2$, the expressions MS_{c-o} and $MS_{Residual(b-s)}$ have the same expectations and we use the statistic $F_{c-o} = \frac{MS_{c-o}}{MS_{Residual(b-s)}}$.

Assuming $\lambda_1 = \lambda_2$ and H_0: $\tau_1 = \tau_2$, MS_{Treat} and $MS_{Residual(w-s)}$ have equal expectations σ^2. Therefore, we get $F_{Treat} = \frac{MS_{Treat}}{MS_{Residual(w-s)}}$.

Testing for period effects does not depend upon the assumption that $\lambda_1 = \lambda_2$ holds. Since MS_{Period} and $MS_{Residual(w-s)}$ have expectation σ^2 considering H_0: $\pi_1 = \pi_2$, the statistic $F_{Period|H_0} = \frac{MS_{Period}}{MS_{Residual(w-s)}}$ follows a central F-distribution.

Example 8.1: A clinical trial is used to compare the effect of two soporifics A and B. Response is the prolongation of sleep (in minutes).

| Group 1 | | Patient | | | | $Y_{1j\cdot}$ | $y_{1j\cdot}$ |
Period	Treatment	1	2	3	4		
1	A	20	40	30	20	110	27.5
2	B	30	50	40	40	160	40.0
	$Y_{1\cdot k}$	50	90	70	60	$Y_{1\cdot\cdot} = 270$	
						$Y_{1\cdot\cdot}/4 = 67.50$	
						$y_{1\cdot\cdot} = 33.75$	
Differences d_{1k}		-10	-10	-10	-20	$d_{1\cdot} = -12.5$	

| Group 2 | | Patient | | | | $Y_{2j\cdot}$ | $y_{2j\cdot}$ |
Period	Treatment	1	2	3	4		
1	B	30	40	20	30	120	30.0
2	A	20	50	10	10	90	22.5
	$Y_{2\cdot k}$	50	90	30	40	$Y_{2\cdot\cdot} = 210$	
						$Y_{2\cdot\cdot}/4 = 52.50$	
						$y_{2\cdot\cdot} = 26.25$	
Differences d_{2k}		10	-10	10	20	$d_{2\cdot} = 7.5$	

t-Tests

H_0: $\lambda_1 = \lambda_2$ (no carry-over effect):

$$(8.13): \quad \hat{\lambda}_d = \frac{Y_{1\cdot\cdot}}{4} - \frac{Y_{2\cdot\cdot}}{4} = \frac{270}{4} - \frac{210}{4} = 15$$

$$(8.18): \quad 3s_1^2 = \sum_{k=1}^{4}(Y_{1\cdot k} - \frac{Y_{1\cdot\cdot}}{n_i})^2$$

$$= (50 - 67.5)^2 + \cdots + (60 - 67.5)^2 = 875$$

$$(8.18): \quad 3s_2^2 = (50 - 52.5)^2 + \cdots + (40 - 52.5)^2 = 2075$$

$$(8.17): \quad s^2 = \frac{2950}{6} = 491.67 = 22.17^2$$

$$(8.19): \quad T_\lambda = \frac{15}{22.17}\sqrt{\frac{16}{8}} = 0.96$$

Decision: $T_\lambda = 0.96 < 1.94 = t_{6;0.90}(\text{two-sided}) \Rightarrow H_0$: $\lambda_1 = \lambda_2$ is not rejected. Therefore, we can go on testing the main effects.

H_0: $\tau_1 = \tau_2$ (no treatment effect):

We compute

$$d_{1\cdot} = \frac{-10 - 10 - 10 - 20}{4} = -12.5$$

$$d_{2\cdot} = \frac{10 - 10 + 10 + 20}{4} = 7.5 \quad,$$

$$(8.24): \quad \hat{\tau}_d = \frac{1}{2}(d_{1\cdot} - d_{2\cdot}) = -10$$

$$3s_{1D}^2 = \sum(d_{1k} - d_{1\cdot})^2$$

$$= (-10 + 12.5)^2 + \cdots + (-20 + 12.5)^2 = 75$$

$$3s_{2D}^2 = (10 - 7.5)^2 + \cdots + (20 - 7.5)^2 = 475$$

(8.28): $\quad s_D^2 = \dfrac{75 + 475}{6} = 9.57^2$

(8.29): $\quad T_\tau = \dfrac{-10}{9.57/2}\sqrt{\dfrac{4\cdot 4}{4 + 4}} = -2.96 \quad .$

Decision: With $t_{6;0.95}(\text{two-sided}) = 2.45$ and $t_{6;0.95}(\text{one-sided}) = 1.94$ the hypothesis $H_0: \tau_1 = \tau_2$ is rejected one-sided as well as two-sided, which means a significant treatment effect.

$H_0: \pi_1 = \pi_2$ (no period effect):

We calculate

(8.33): $\quad \hat{\pi}_d = \dfrac{1}{2}(c_1. - c_2.) = \dfrac{1}{2}(d_1. + d_2.)$

$$= \dfrac{1}{2}(-12.5 + 7.5) = -2.5$$

(8.34): $\quad T_\pi = \dfrac{-2.5}{9.57/2}\sqrt{2} = -0.74 \quad .$

$H_0: \pi_1 = \pi_2$ can not be rejected (one and two-sided) .
From the analysis of variance we get the same $F_{1,6} = t_6^2$-statistics.

	SS	df	MS	F	
Carry-over	225	1	225.00	$0.92 = 0.96^2$	
Residual (b-s)	1475	6	245.83		
Treatment	400	1	400.00	$8.73 = 2.96^2$	*
Period	25	1	25.00	$0.55 = 0.74^2$	
Residual (w-s)	275	6	45.83		
Total	2400	15			

$$SS_{Total} = 16800 - \dfrac{480^2}{2\cdot 8} = 2400$$

$$SS_{c-o} = \dfrac{2\cdot 4\cdot 4}{4 + 4}(33.75 - 26.25)^2 = 225$$

$$SS_{Residual(b-s)} = \dfrac{1}{2}(50^2 + 90^2 + \cdots + 40^2) - (\dfrac{270^2}{8} - \dfrac{210^2}{8})$$

$$= \dfrac{32200}{2} - \dfrac{117000}{8}$$

$$= 16100 - 14625 = 1475$$

$$SS_{Treat} = \dfrac{4\cdot 4}{2(4 + 4)}(27.5 - 40.0 - 30.0 + 22.5)^2$$

$$= (-20)^2 = 400$$

$$SS_{Period} = (27.5 - 40.0 + 30.0 - 22.5)^2$$
$$= (-5)^2 = 25$$
$$SS_{Residual(w-s)} = 16800 - \frac{1}{4}(110^2 + 160^2 + 120^2 + 90^2) - 1475$$
$$= 16800 - 15050 - 1475 = 275 \quad .$$

8.3.3 Residual Analysis and Plotting the Data

In addition to t- and F-tests, it is often desirable to represent the data using plots. We will now describe three methods of plotting the data which will allow us to detect patients being conspicuous by their response (outliers) and interactions such as carry-over effects.

Subject profile plots are produced for each group by plotting each subject's reponse against the period label. To summarize the data, we choose a **groups-by-periods** plot in which the group-by-period means are plotted against the period labels and points which refer to the same treatment are connected. Using Example 8.1 we get the following plots.

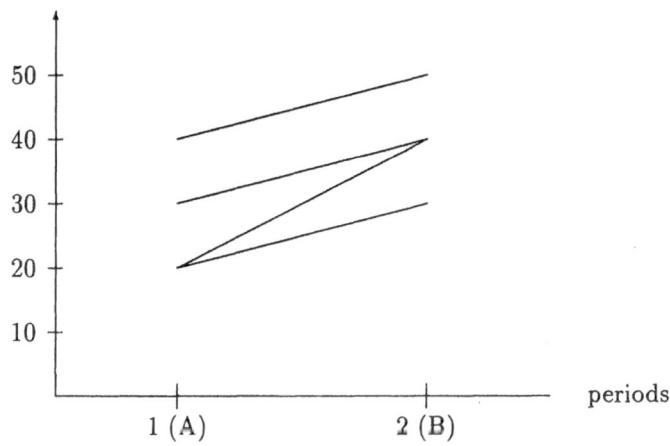

Figure 8.2: Individual profiles (group 1)

All patients in group 1 show increasing response when they cross-over from treatment A to B. In group 2, the profile of patient 2 (uppermost line) exhibits a decreasing response while the other three profiles show an increasing tendency.

Figure 8.4 shows that in both periods treatment B leads to higher response than treatment A (difference of means $B - A$: $30 - 27.5 = 2.5$ for period 1, $40 - 22.5 = 17.5$ for period 2 so that $\hat{\tau}_d(B-A) = \frac{1}{2}(17.5+2.5) = 10 = -\hat{\tau}_d(A-B)$). It would also be possible to say that treatment A shows a slight carry-over effect that strengthens B (or B has a carry-over effect that reduces A). This difference in the treatment effects is not statistically significant according to the results

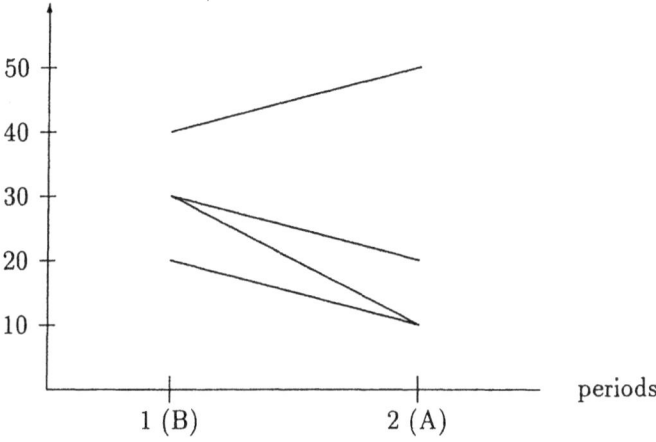

Figure 8.3: Individual profiles (group 2)

we obtained from testing treatment×period interactions (= carry-over effect). Without doubt, we can say that A has lower response than B in period 1 and this effect is even more pronounced in period 2. Another interesting view is given by the **differences-by-totals** plot where the subjects' differences d_{ik} are plotted against the total responses $Y_{i\cdot k}$. Plotting the pairs $(d_{ik}, Y_{i\cdot k})$ and connecting the outermost points of each group by a convex hull, we get a clear impression of carry-over and treatment effects. Since the statistic for carry-over is based on $\hat{\lambda}_d = (Y_{1\cdot\cdot}/n_1 - Y_{2\cdot\cdot}/n_2)$, the two hulls will be separated horizontally if $\lambda_d \neq 0$. In the same way the treatment effect based on $\hat{\tau}_d = \frac{1}{2}(d_{1\cdot} - d_{2\cdot})$ will manifest if the two hulls being vertically separated.

Figure 8.5 shows vertically separated hulls indicating a treatment effect (which we already know is significant according to our tests). On the other hand, the hulls are not separated horizontally and indicate no carry-over effect.

Analysis of Residuals
The components $\hat{\epsilon}_{ijk}$ of $\hat{\epsilon} = (\boldsymbol{y} - \boldsymbol{X}\hat{\boldsymbol{\beta}})$ are the estimated residuals which are used to check the model assumptions on the errors ϵ_{ijk}. Using appropriate plots, we can check for outliers and revise our assumptions on normal distribution and independency. The response values corresponding to unusually large standardized residuals are called outliers. A standardized residual is given by

$$r_{ijk} = \frac{\hat{\epsilon}_{ijk}}{\sqrt{\text{Var}(\hat{\epsilon}_{ijk})}} \quad , \tag{8.40}$$

with the variance factor σ^2 being estimated with $MS_{Residual(w-s)}$.

From the 2×2 cross-over, we get

$$\hat{y}_{ijk} = y_{i\cdot k} + y_{ij\cdot} - y_{i\cdot\cdot} \tag{8.41}$$

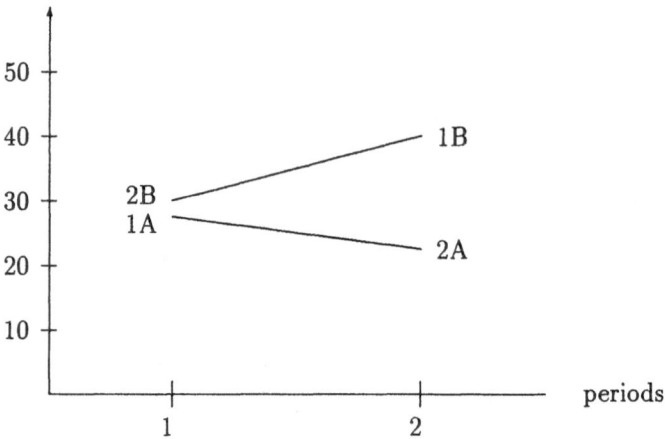

Figure 8.4: Group-period-plots

and

$$\text{Var}(\hat{\epsilon}_{ijk}) = \text{Var}(y_{ijk} - \hat{y}_{ijk}) = \frac{(n_i - 1)}{2n_i}\sigma^2 \quad . \tag{8.42}$$

Then

$$r_{ijk} = \frac{\hat{\epsilon}_{ijk}}{\sqrt{MS_{Residual(w-s)}\frac{(n_i-1)}{2n_i}}} \quad . \tag{8.43}$$

This is the internally studentized residual and follows a beta-distribution. We, however, regard r_{ijk} as $N(0,1)$-distributed and choose the two-sided quantile 2.00 (instead of $u_{0.975} = 1.96$) to test for y_{ijk} being an outlier.

Remark: If a more exact analysis is required, externally studentized residuals should be used, since they follow the F-distribution (and can therefore be tested directly) and additionally are more sensitive to outliers (cf. Beckmann and Trussel, 1974, Toutenburg, 1992, pp. 185–187).

| | Group 1 (AB) | | | | | Group 2 (BA) | | | |
Patient	y_{ijk}	$\hat{y}_{ijk}$	$\hat{\epsilon}_{ijk}$	r_{ijk}	Patient	y_{ijk}	$\hat{y}_{ijk}$	$\hat{\epsilon}_{ijk}$	r_{ijk}
1	20	18.75	1.25	0.30	1	30	28.75	1.25	0.30
2	40	38.75	1.25	0.30	2	40	48.75	−8.75	−2.10 *
3	30	28.75	1.25	0.30	3	20	18.75	1.25	0.30
4	20	23.75	−3.75	−0.90	4	30	23.75	6.25	1.51

Hence, patient 2 in group 2 is an outlier.

Remark: If $\epsilon_{ijk} \sim N(0,\sigma^2)$ is not tenable, the response values are substituted by their ranks and the hypothesises are tested with the Wilcoxon-Mann-Whitney test (cf. Section 2.5) instead of using t-tests.

A detailed discussion of the various approaches for the 2×2 cross-over and

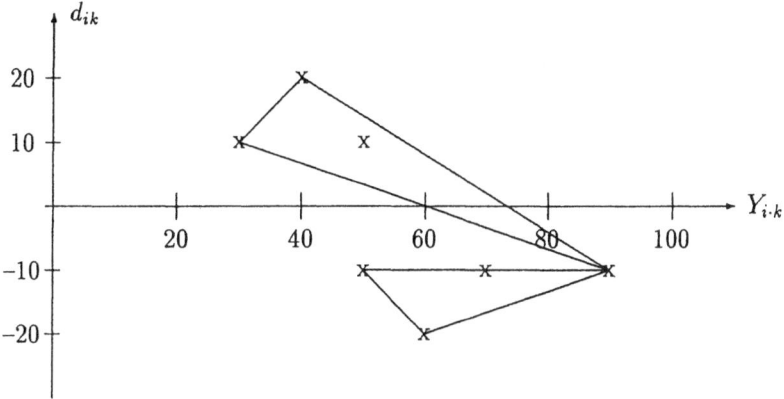

Figure 8.5: Difference-response-total-plot to Example 8.1

especially their interpretations may be found in Jones and Kenward (1989, Chapter 2) and Ratkowsky, Evans and Alldredge (1993).

Comment on the Procedure of Testing

Grizzle (1965) suggested to test carry-over effects on a quite high level of significance ($\alpha = 0.1$) first. If this leads to a significant result then the test for treatment effects is to be based on the data of the first period only. If it is not significant, then the treatment effects are tested using the differences between the periods. This procedure has certain disadvantages. For example, Brown (1980) showed that this pretest is of minor efficiency in case of real carry-over effects.

The hypothesis of no carry-over effect is very likely to be rejected even if there is a true carry-over effect. Hence, the biased test (8.29) (biased, because the carry-over was not recognized) is used to test for treatment differences. This test is conservative in case of a true positive carry-over effect and therefore is insensitive to potential differences in treatments. On the other hand, this test will exceed the level of significance if there is a true negative carry-over effect (not very likely in practice, since this refers to a withdrawal-effect).

If there is no true carry-over effect, the null hypothesis is very likely to be rejected erroneously ($\alpha = 0.1$) and the less efficient test using first period data only is performed.

Brown (1980) concluded that this method is not very useful to test treatment effects as it depends upon the outcome of the pretest.

Remark: Further comments are given in Section 8.3.4.

8.3.4 Alternative Parametrizations in 2×2 Cross-Over

Model (8.1) was introduced as the classical approach and is labeled parametrization No. 1 using the notation of Ratkowsky et al. (1993). A more general parametrization of the 2×2 cross-over design, that includes a sequence effect γ_i, is given by

$$y_{ijk} = \mu + \gamma_i + s_{ik} + \pi_j + \tau_t + \lambda_r + \epsilon_{ijk} \quad , \tag{8.44}$$

with $i, j, t, r = 1, 2$ und $k = 1, \ldots, n_i$. The data are summarized in a table containing the cell means $y_{ij\cdot}$, i.e.

$$
\begin{array}{ccc}
 & & \text{Period} \\
 & & \begin{array}{cc} 1 & \quad 2 \end{array} \\
\text{Sequence} \begin{array}{c} 1 \\ 2 \end{array} &
\begin{array}{|cc|}
\hline
y_{11\cdot} & y_{12\cdot} \\
y_{21\cdot} & y_{22\cdot} \\
\hline
\end{array}
\end{array}
$$

Here sequence 1 indicates that the treatments are given in the order (AB) and sequence 2 has the (BA) order. Using the common restrictions

$$\gamma_2 = -\gamma_1 \,, \; \pi_2 = -\pi_1 \,, \; \tau_2 = -\tau_1 \,, \; \lambda_2 = -\lambda_1 \tag{8.45}$$

and writing $\gamma_1 = \gamma$, $\pi_1 = \pi$, $\tau_1 = \tau$, $\lambda_1 = \lambda$ for brevity, we get the following equations representing the four expectations:

$$
\begin{aligned}
\mu_{11} &= \mu + \gamma + \pi + \tau \\
\mu_{12} &= \mu + \gamma - \pi - \tau + \lambda \\
\mu_{21} &= \mu - \gamma + \pi - \tau \\
\mu_{22} &= \mu - \gamma - \pi + \tau - \lambda \quad .
\end{aligned}
$$

In matrix notation this is equivalent to

$$
\begin{pmatrix} \mu_{11} \\ \mu_{12} \\ \mu_{21} \\ \mu_{22} \end{pmatrix} = \boldsymbol{X}\boldsymbol{\beta} =
\begin{pmatrix}
1 & 1 & 1 & 1 & 0 \\
1 & 1 & -1 & -1 & 1 \\
1 & -1 & 1 & -1 & 0 \\
1 & -1 & -1 & 1 & -1
\end{pmatrix}
\begin{pmatrix} \mu \\ \gamma \\ \pi \\ \tau \\ \lambda \end{pmatrix} \quad . \tag{8.46}
$$

This $4{\times}5$-matrix $\boldsymbol{X}$ has rank 4, so that $\boldsymbol{\beta}$ is only estimable if one of the parameters is removed. Various parametrizations are possible depending on which of the five parameters is removed and then confounded with the remaining ones.

Parametrization No. 1

The classical approach ignores the sequence parameter. Its expectations may therefore be represented as a submodel of (8.46) by dropping the second column of $\boldsymbol{X}$

$$X_1\beta_1 = \begin{pmatrix} 1 & 1 & 1 & 0 \\ 1 & -1 & -1 & 1 \\ 1 & 1 & -1 & 0 \\ 1 & -1 & 1 & -1 \end{pmatrix} \begin{pmatrix} \mu \\ \pi \\ \tau \\ \lambda \end{pmatrix} . \tag{8.47}$$

From this we get

$$X_1'X_1 = \begin{pmatrix} E & 0 \\ 0 & H \end{pmatrix}$$

where

$$E = 4I_2$$

$$H = \begin{pmatrix} 4 & -2 \\ -2 & 2 \end{pmatrix} , \quad |X_1'X_1| = 64 ,$$

$$(X_1'X_1)^{-1} = \begin{pmatrix} E^{-1} & 0 \\ 0 & H^{-1} \end{pmatrix} \quad \text{[cf. Theorem A 19]}$$

with $E^{-1} = \frac{1}{4}I_2$, $H^{-1} = \begin{pmatrix} 1/2 & 1/2 \\ 1/2 & 1 \end{pmatrix}$. The least squares estimate of β_1 is

$$\hat{\beta}_1 = \begin{pmatrix} \hat{\mu} \\ \hat{\pi} \\ \hat{\tau} \\ \hat{\lambda} \end{pmatrix} = (X_1'X_1)^{-1}X_1' \begin{pmatrix} y_{11.} \\ y_{12.} \\ y_{21.} \\ y_{22.} \end{pmatrix} . \tag{8.48}$$

We calculate

$$X_1' \begin{pmatrix} y_{11.} \\ y_{12.} \\ y_{21.} \\ y_{22.} \end{pmatrix} = \begin{pmatrix} 1 & 1 & 1 & 1 \\ 1 & -1 & 1 & -1 \\ 1 & -1 & -1 & 1 \\ 0 & 1 & 0 & -1 \end{pmatrix} \begin{pmatrix} y_{11.} \\ y_{12.} \\ y_{21.} \\ y_{22.} \end{pmatrix}$$

$$= \begin{pmatrix} y_{11.} + y_{12.} + y_{21.} + y_{22.} \\ y_{11.} - y_{12.} + y_{21.} - y_{22.} \\ y_{11.} - y_{12.} - y_{21.} + y_{22.} \\ y_{12.} - y_{22.} \end{pmatrix} . \tag{8.49}$$

Therefore, the least squares estimation gives

$$\hat{\beta}_1 = \begin{pmatrix} \hat{\mu} \\ \hat{\pi} \\ \hat{\tau} \\ \hat{\lambda} \end{pmatrix} = (X_1'X_1)^{-1}X_1' \begin{pmatrix} y_{11.} \\ y_{12.} \\ y_{21.} \\ y_{22.} \end{pmatrix} \tag{8.50}$$

$$= \begin{pmatrix} (y_{11.} + y_{12.} + y_{21.} + y_{22.})/4 \\ (y_{11.} - y_{12.} + y_{21.} - y_{22.})/4 \\ (y_{11.} - y_{21.})/2 \\ (y_{11.} + y_{12.} - y_{21.} - y_{22.})/2 \end{pmatrix} , \tag{8.51}$$

from which we get the following results:

$$\hat{\mu} = y_{\cdots} \;, \tag{8.52}$$

$$\hat{\pi} = (y_{\cdot 1\cdot} - y_{\cdot 2\cdot})/2 = (c_{1\cdot} - c_{2\cdot})/4 = \frac{\hat{\pi}_d}{2} \quad [\text{cf. (8.33)}] \;, \tag{8.53}$$

$$\hat{\tau} = (y_{11\cdot} - y_{21\cdot})/2 = \frac{\hat{\tau}_{d/\lambda_d}}{2} \quad [\text{cf. (8.38)}] \;, \tag{8.54}$$

$$\hat{\lambda} = y_{1\cdots} - y_{2\cdots} = \hat{\lambda}_d/2 \quad [\text{cf. (8.13)}] \;. \tag{8.55}$$

The estimators $\hat{\tau}$ and $\hat{\lambda}$ are correlated:

$$V(\hat{\tau}, \hat{\lambda}) = \sigma^2 H^{-1} = \sigma^2 \begin{pmatrix} 1/2 & 1/2 \\ 1/2 & 1 \end{pmatrix} \;,$$

with $\rho(\hat{\tau}, \hat{\lambda}) = \frac{1}{2}/(\frac{1}{2} \cdot 1)^{1/2} = 0.707$. The estimation of $\hat{\tau}$ is always twice as accurate as the estimation of $\hat{\lambda}$, although $\hat{\tau}$ uses data of the first period only and is confounded with the difference between the two groups (sequences).

Remark: In fact, parametrization No. 1 is a three-factorial design with the main effects π, τ, and λ and with τ and λ being correlated. On the other hand, the classical approach uses the split-plot model in addition to parametrization (8.1). So it is obvious that we will get different results depending on which parametrization we use. We will demonstrate this in Example 8.2, where the four different parametrizations are applied to our data set of Example 8.1.

Parametrization No. 1a

If the test for no carry-over effect does not reject H_0: $\lambda = 0$ against H_1: $\lambda \neq 0$ using the test statistic $F_{1,df} = \frac{\hat{\lambda}_d^2}{Var(\hat{\lambda}_d)}$ (cf. (8.19)), our model can be reduced to the following

$$\tilde{X}_1\tilde{\beta}_1 = \begin{pmatrix} 1 & 1 & 1 \\ 1 & -1 & -1 \\ 1 & 1 & -1 \\ 1 & -1 & 1 \end{pmatrix} \begin{pmatrix} \mu \\ \pi \\ \tau \end{pmatrix} \tag{8.56}$$

and we get the same estimators $\hat{\mu}$ (8.52) and $\hat{\pi}$ (8.53) as before, but now the estimator $\hat{\tau}$ is based on both periods' data,

$$\hat{\tau} = (y_{11\cdot} - y_{12\cdot} - y_{21\cdot} + y_{22\cdot})/4$$
$$= (d_{1\cdot} - d_{2\cdot})/4$$
$$= \hat{\tau}_d/2 \quad [\text{cf. (8.24)}] \;. \tag{8.57}$$

The results of parametrizations No. 1 and No. 1a are the same as the classical univariate results we obtained in Section 8.3.1 (except for a factor of $1/2$ in $\hat{\pi}$, $\hat{\tau}$ and $\hat{\lambda}$). But in addition, the dependency in estimating the treatment effect τ and the carry-over effect λ is explained.

Parametrization No. 2

In the first parametrization, the interaction treatment×period was aliased with the carry-over effect λ. We now want to parametrize this interaction directly. Dropping the sequence effect, the model of expectations is as follows:

$$\mathrm{E}\left(y_{ijk}\right) = \mu_{ij} = \mu + \pi_j + \tau_t + (\tau\pi)_{tj} \quad . \tag{8.58}$$

Using effect coding, the codings of the interaction effects are just the products of the involved main effects. Therefore, we get

$$\begin{pmatrix} \mu_{11} \\ \mu_{12} \\ \mu_{21} \\ \mu_{22} \end{pmatrix} = \boldsymbol{X}_2\boldsymbol{\beta}_2 = \begin{pmatrix} 1 & 1 & 1 & 1 \\ 1 & -1 & -1 & 1 \\ 1 & 1 & -1 & -1 \\ 1 & -1 & 1 & -1 \end{pmatrix} \begin{pmatrix} \mu \\ \pi \\ \tau \\ (\pi\tau) \end{pmatrix} \quad . \tag{8.59}$$

Since the column vectors are orthogonal, we easily get $(\boldsymbol{X}_2'\boldsymbol{X}_2) = 4\boldsymbol{I}_4$ and therefore the parameter estimations are independent (cf. Section 6.3). The estimators are

$$\hat{\boldsymbol{\beta}}_2 = \begin{pmatrix} \hat{\mu} \\ \hat{\pi} \\ \hat{\tau} \\ \widehat{(\pi\tau)} \end{pmatrix} = \begin{pmatrix} y_{\cdots} \\ \hat{\pi}_d/2 \\ (y_{11\cdot} - y_{12\cdot} - y_{21\cdot} + y_{22\cdot})/4 \\ (y_{11\cdot} + y_{12\cdot} - y_{21\cdot} - y_{22\cdot})/4 \end{pmatrix} \quad . \tag{8.60}$$

Note that $\hat{\mu}$ and $\hat{\pi}$ are as in the first parametrization. The estimator $\hat{\tau}$ in (8.60) and the estimator $\hat{\tau}$ (8.57) in the reduced model (8.56) coincide. The estimator $\widehat{(\pi\tau)}$ may be written as (cf. (8.55))

$$\widehat{(\pi\tau)} = (y_{1\cdots} - y_{2\cdots})/2 = \hat{\lambda}_d/4 = \hat{\lambda}/2 \quad , \tag{8.61}$$

and coincides – except for a factor of $1/2$ – with the estimation of the carry-over effect (8.55) in model (8.47). So it is obvious that there is an intrinsic aliasing between the two parametes λ and $(\pi\tau)$.

Parametrization No. 3

Supposing that a carry-over effect λ or alternatively an interaction effect $(\pi\tau)$ may be excluded from analysis, the model now contains only main effects. We already discussed model (8.56). Now we want to introduce the sequence effect γ as an additional main effect. With $\gamma_2 = -\gamma_1 = \gamma$, we get

$$\begin{pmatrix} \mu_{11} \\ \mu_{12} \\ \mu_{21} \\ \mu_{22} \end{pmatrix} = \boldsymbol{X}_3\boldsymbol{\beta}_3 = \begin{pmatrix} 1 & 1 & 1 & 1 \\ 1 & 1 & -1 & -1 \\ 1 & -1 & 1 & -1 \\ 1 & -1 & -1 & 1 \end{pmatrix} \begin{pmatrix} \mu \\ \gamma \\ \pi \\ \tau \end{pmatrix} \quad , \tag{8.62}$$

$$(\boldsymbol{X}_3'\boldsymbol{X}_3) = 4\boldsymbol{I}_4 \quad ,$$

$$\hat{\beta}_3 = \begin{pmatrix} \hat{\mu} \\ \hat{\gamma} \\ \hat{\pi} \\ \hat{\tau} \end{pmatrix} = \frac{1}{4} X_3' \begin{pmatrix} y_{11.} \\ y_{12.} \\ y_{21.} \\ y_{22.} \end{pmatrix}$$

$$= \begin{pmatrix} y_{...} \\ (y_{11.} + y_{12.} - y_{21.} - y_{22.})/4 \\ (y_{11.} - y_{12.} + y_{21.} - y_{22.})/4 \\ (y_{11.} - y_{12.} - y_{21.} + y_{22.})/4 \end{pmatrix} \qquad (8.63)$$

$$= \begin{pmatrix} y_{...} \\ (y_{1..} - y_{2..})/2 \\ (y_{.1.} - y_{.2.})/2 \\ \hat{\tau}_d/2 \end{pmatrix} . \qquad (8.64)$$

The sequence effect γ is estimated using the contrast in the total response of both groups (AB) and (BA) and we see the equivalence $\hat{\gamma} = \widehat{(\pi\tau)} = \hat{\lambda}_d/4$. The period effect π is estimated using the contrast in the total response of both periods and coincides with $\hat{\pi}$ in parametrizations No. 1 (cf. (8.53)) and No. 2 (cf. (8.60)) The estimation of $\hat{\tau}$ is the same as $\hat{\tau}$ (8.57) in the reduced model (8.56) and $\hat{\tau}$ (cf. (8.60)) in parametrization No. 2. Furthermore, the estimates in $\hat{\beta}_3$ are independent, so that, e.g., H_0: $\tau = 0$ can be tested not depending on $\gamma = \lambda_d = 0$ (in contrast to parametrization No. 1).

Parametrization No. 4

Here, the main effects treatment and sequence and their interaction are represented in a two-factorial model (cf. Milliken and Johnson, 1984)

$$E(y_{ijk}) = \mu_{ij} = \mu + \gamma_i + \tau_t + (\gamma\tau)_{it} \;, \qquad (8.65)$$

i.e.,

$$\begin{pmatrix} \mu_{11} \\ \mu_{12} \\ \mu_{21} \\ \mu_{22} \end{pmatrix} = X_4 \beta_4 = \begin{pmatrix} 1 & 1 & 1 & 1 \\ 1 & 1 & -1 & -1 \\ 1 & -1 & -1 & 1 \\ 1 & -1 & 1 & -1 \end{pmatrix} \begin{pmatrix} \mu \\ \gamma \\ \tau \\ (\gamma\tau) \end{pmatrix} . \qquad (8.66)$$

Since $X_4' X_4 = 4I_4$, the components of β_4 can be estimated independently as

$$\hat{\beta}_4 = \begin{pmatrix} \hat{\mu} \\ \hat{\gamma} \\ \hat{\tau} \\ \widehat{(\gamma\tau)} \end{pmatrix} = \begin{pmatrix} y_{...} \\ (y_{1..} - y_{2..})/2 \\ \hat{\tau}_d/2 \\ (y_{.1.} - y_{.2.})/2 \end{pmatrix} . \qquad (8.67)$$

Values of $\hat{\gamma}$ in parametrizations 3 and 4 are the same. Analogously, the values of $\hat{\tau}$ coincide in parametrizations 2, 3 and 4 whereas the interaction effect sequence×treatment $\widehat{(\gamma\tau)}$ refers to the period effect π in parametrizations 1, 2 and 3.

Remark: From the various parametrizations we get the following results:

Parametrization

	Classical	No. 1	No. 1a	No. 2	No. 3	No. 4
$\hat\mu$	$y...$	$y...$	$y...$	$y...$	$y...$	$y...$
$\hat\gamma$	—	—	—	—	$\hat\lambda_d/4$	$\hat\lambda_d/4$
$\hat\pi$	$\hat\pi_d = \frac{1}{2}(d_1. + d_2.)$	$\hat\pi_d/2$	$\hat\pi_d/2$	$\hat\pi_d/2$	$\hat\pi_d/2$	—
$\hat\tau$	$\hat\tau_{d/\lambda_d} = y_{11}. - y_{21}.$	$\hat\tau_{d/\lambda_d}/2$	$\hat\tau_d/2$	$\hat\tau_d/2$	$\hat\tau_d/2$	$\hat\tau_d/2$
$\hat\lambda$	$\hat\lambda_d = 2(y_1.. - y_2..)$	$\hat\lambda_d/2$	—	—	—	—
$\widehat{(\tau\pi)}$	—	—	—	$\hat\lambda_d/4$	—	—
$\widehat{(\gamma\tau)}$	—	—	—	—	—	$\hat\pi_d/2$

Table 8.4: Estimators using six different parametrizations

(i) In parametrization No. 1, the estimators of τ and λ are correlated. In contrast to the arguments of Ratkowsky et al. (1993, pp.89–90), the values of $E(MS)$ given in Table 8.3 are valid. $E(MS_{Treat})$ depends on $(\lambda_1 - \lambda_2) = 2\lambda$ so that testing for H_0: $\tau = 0$ may either be done using a central t-test if $\lambda = 0$ or using a non-central t-test if λ is known. A difficulty in the argumentation certainly is that τ and λ are correlated but not represented in the two-factorial hierarchy "main effect A, main effect B, interaction A×B".

(ii) In parametrization No. 2, the carry-over effect is indirectly represented as the alias effect of the interaction $(\pi\tau)$. We can use the common hierarchical test procedure as in a two-factorial model with interaction, since the design is orthogonal. If the interaction is not significant the estimators of the main effects remain the same (in contrast to parametrization No. 1).

(iii) Analysis of data of a 2×2 cross-over design is done in two steps. In the first step, we test for carry-over using one of the parametrizations in which the carry-over effect is separable from the main effects, e.g., parametrization No. 3 – and it is not surprising that the result will be the same as if we had used the sequence effect.

We consider the following experiment: We take two groups of subjects and apply the treatments in both groups in the same order (AB). If there is an interaction effect (maybe a significant carry-over effect in the classical approach of Grizzle or a significant sequence effect in parametrization No. 3 of Ratkowsky et al.), then we conclude that the two groups must consist of two different classes of subjects. There is either a difference per se between the subjects of the two groups or treatment A shows different persistencies in the two groups. Since the latter is not very likely, it is clear that the subjects of both groups are different in their reactions. And therefore it is a sequence effect but not a carry-over effect. We try to avoid this confusion by randomizing the subjects.

Regarding the classical (AB) / (BA) design, there are two ways to interpret a significant interaction effect:

(a) Either it is a true sequence effect as a result of insufficient randomization

(b) or it is a true carry-over effect. This will be the case if there is no doubt about the randomization process.

Since the actual set of data may hardly be used to decide whether the randomization succeeded or failed, it is necessary to make a distinction before we analyse our data.

If the subjects have not been randomized, the possibility of a sequence effect should attract our attention. The F-statistics given for parametrization No. 3 are valid and do not depend upon whether the sequence effect is significant or not, because there is no natural link between a sequence effect and a treatment or a period effect.

Given the case that we did randomize our subjects, then there is no need to consider a sequence effect and therefore the interaction effect is to be regarded as a result of carry-over.

The carry-over effect was introduced as the persisting effect of a treatment during the subsequent period of treatment and it is represented as an additive component in our model. Therefore, it is evident that the F-statistics for treatment and period effects, derived from parametrization No. 3 or from the classical approach, are not longer valid if the carry-over effect is significant.

To continue our examination, we choose one of the following alternatives:

(a) We try to test treatment effects using the data of the first period only. This might be difficult because the sample size is likely to be too small for a parallel group design. Of course we then omit the sequence effect from our analysis (because we just have this first period).

(b) A significant carry-over effect may also be regarded as a suffcient indicator that the two treatments differ in their effects. At least we can state that the two treatments have different persistencies and therefore they are not equal.

It can be assumed that Ratkowsky et al. regarded the analysis of variance tables as to be read simultaneously and that the given F-statistics for carry-over, treatment, and period effects are always valid. But they are not. This is only the case if the carry-over effect was proven to be non-significant. Only with a non-significant carry-over effect the expressions for treatment and period effect are valid. If the label *carry-over* is replaced by the label *sequence effect*, then the ordering of tests is not important and the table is no longer misleading to readers who only just glance at the literature. The interpretation of the results must reflect this relabeling too. Then, of course, we do not know anything about the carry-over effect which mostly is of more importance than a sequence effect. Using the classical approach, the analysis of variance table is valid.

Period

		1	2	3	4	5
Sequence	1	Baseline	A	Wash-out	B	Wash-out
	2	Baseline	B	Wash-out	A	Wash-out

Figure 8.6: Extended 2×2 Cross-over design

(iv) From a theoretical point of view, it is interesting to extend the 2×2-design by three additional periods: a baseline period and two wash-out periods (Figure 8.6). This approach was suggested by Ratkowsky et al. (1993, Chapter 3.6), but it is rarely applied because of the larger efforts.

The linear model then contains two additional period effects and carry-over effects of first and second order. The main advantages are that all parameters are estimable, there is no dependence between treatment and carry-over effects and we get reduced variance.

(v) Possible modifications of the 2×2 cross-over are $2 \times n$-designs like

Period

		1	2	3
Sequence	1	A	B	B
	2	B	A	A

Period

		1	2	3
Sequence	1	A	B	A
	2	B	A	B

or $n \times 2$-designs like

Period

		1	2
	1	A	B
Sequence	2	B	A
	3	A	A
	4	B	B

Adding baseline and wash-out periods may further improve these designs. A comprehensive treatment of this subject matter is given by Ratkowsky et. al. (1993, Chapter 4).

Example 8.2: (Continuation of Example 8.1). The data of Example 8.1 are now analysed with parametrizations 2, 3 and 4 using the SAS procedure GLM. In the split-plot model (classical approach) the following analysis of variance table was obtained for the data of Example 8.1 (cf. Section 8.3.2)

Source	SS	df	MS	F
Carry-over	225	1	225.00	0.92
Residual (b–s)	1475	6	245.83	
Treatment	400	1	400.00	8.73 *
Period	25	1	25.00	0.55
Residual (w–s)	275	6	45.83	
Total	2400	15		

The treatment effect was found to be significant.

Parametrization No. 1 does not take the split-plot character of the design (limited randomization) into account. Therefore, the two sums of squares SS (b–s) and SS (w–s) are added for $SS_{Residual} = 1750$. Table 8.5 shows this result in the upper part (SS Type I). The lower part (SS Type II) gives the result using first period data only, because the model contains the carry-over effect. All other parametrizations do not contain carry-over effects and the important sums of squares are found in the lower part (SS Type II) of the table. We note that the following F-values coincide

Carry-over (resp. Sequence)	$F = 0.92$ (classical, No. 3, No. 4)
Treatment	$F = 8.73$ (classical, No. 3, No. 4)
Period	$F = 0.55$ (classical, No. 3) .

The different parametrizations were calculated using the following small SAS programs.

```
proc glm;
class seq subj period treat carry;
model y = period treat carry /solution ss1 ss2;
title "Parametrization 1";
run;

proc glm;
class seq subj period treat carry;
model y = period treat treat(period) /solution ss1 ss2;
title "Parametrization 2";
run;

proc glm;
class seq subj period treat carry;
model y = seq subj(seq) period treat /solution ss1 ss2;
random subj(seq);
title "Parametrization 3";
run;
```

```
proc glm;
class seq subj period treat carry;
model y = seq subj(seq) treat seq(treat) /solution ss1 ss2;
random subj(seq);
title "Parametrization 4";
run;

data Example 8.2;
input subj seq period treat $ carry $ y @@;

cards;
1 1 1 a 0 20   1 1 2 b a 30
2 1 1 a 0 40   2 1 2 b a 50
3 1 1 a 0 30   3 1 2 b a 40
4 1 1 a 0 20   4 1 2 b a 40
1 2 1 b 0 30   1 2 2 a b 20
2 2 1 b 0 40   2 2 2 a b 50
3 2 1 b 0 20   3 2 2 a b 10
4 2 1 b 0 30   4 2 2 a b 10
run;
```

Parametrization No. 1				
source	**df**	**SS type I**	**MS**	**F**
periods	1	25.00	25.00	0.17
treatments	1	400.00	400.00	2.74
carry-over	1	225.00	225.00	1.54
residual	12	1750.00	145.83	
	df	**SS type I**	**MS**	**F**
treatments	1	12.50	12.50	0.09
carry-over	1	225.00	225.00	1.54
residual	12	1750.00	145.83	

Parametrization No. 2				
source	**df**	**SS type I**	**MS**	**F**
periods (P)	1	25.00	25.00	0.17
treatments (T)	1	400.00	400.00	2.74
P×T	1	225.00	225.00	1.54
residual	12	1750.00	145.83	
	df	**SS type I**	**MS**	**F**
treatments	1	400.00	400.00	2.74
P×T	1	225.00	225.00	1.54
residual	12	1750.00	145.83	

Parametrization No. 3				
source	**df**	**SS type I**	**MS**	**F**
b–s				
sequence	1	225.00	225.00	0.92
residual	6	1475.00	245.83	
	df	**SS type I**	**MS**	**F**
w–s				
periods	1	25.00	25.00	0.55
treatments	1	400.00	400.00	8.73
residual	6	275.00	45.83	

Parametrization No. 4				
source	**df**	**SS type I**	**MS**	**F**
b–s				
sequence	1	225.00	225.00	0.92
residual	6	1475.00	245.83	
	df	**SS type I**	**MS**	**F**
w–s				
treatments	1	400.00	400.00	8.73
seq×treat.	1	25.00	25.00	0.55
residual	6	275.00	45.83	

Table 8.5: GLM results of the four parametrizations

8.3.5 Cross-Over Analysis Using Rank Tests

Known rank tests from other designs with two independent groups offer a non-parametric approach to analyse a cross-over trial. These tests are based on the model given in Table 8.1. However, the random effects may now follow any continuous distribution with expectation zero. The advantage of using nonparametric methods is that there is no need to assume a normal distribution. According to the difficulties mentioned above, we now assume either that there are no carry-over effects or that they are at least ignorable.

Rank Test on Treatment Differences

The null hypothesis that there are no differences between the two treatments implies that the period differences follow the same distribution.

$$H_0: \quad F_{d1}(d_{1k}) = F_{d2}(d_{2k}), \quad k = 1, \ldots, n_i \quad . \tag{8.68}$$

Here F_{d1} and F_{d2} are continuous distributions with identical variances. Then, the null hypothesis of no treatment effects may be tested using the statistic of Wilcoxon, Mann and Whitney (cf. Section 2.5 and Koch, 1972).

We calculate the period differences d_{1k} and d_{2k} (cf. (8.20)). These $N = (n_1 + n_2)$ differences then get ranks from 1 to N. Let

$$r_{ik}^{\phi} = [\text{rank of } d_{ik} \text{ in} \{d_{11}, \ldots, d_{1n_1}, d_{21}, \ldots, d_{2n_2}\}], \tag{8.69}$$

with $i = 1, 2$, $k = 1, \ldots, n_i$. In case of ties we use mean ranks. For both groups (AB) and (BA), we get the sum of ranks R_1 resp. R_2 which are used to build the test statistics U_1 resp. U_2 [(2.34) resp. (2.35)].

Rank Tests on Period Differences

The null hypothesis of no period differences is:

$$H_0: \quad F_{c1}(c_{1k}) = F_{c2}(c_{2k}), \quad k = 1, \ldots, n_i, \tag{8.70}$$

and so the distribution of the difference $c_{1k} = (y_{11k} - y_{12k})$ equals the distribution of the difference $c_{2k} = (y_{22k} - y_{21k})$. Again, F_{ci} $(i = 1, 2)$ are continuous distributions with equal variances.

The null hypothesis H_0 is then tested in the same way as H_1 in (8.68) using the test of Wilcoxon, Mann and Whitney.

8.4 2×2 Cross-Over and Categorical (Binary) Response

8.4.1 Introduction

In many applications, the response is categorical. This is the case in pretests when only a rough overview of possible relations is needed. Often a continuous response is not available. For example, recovering from a mental illness can not be measured on a continuous scale, categories like "worse, constant, better" would be sufficient.

Example: Patients suffering from depression participate in two treatments A and B. Their response to each treatment is coded binary with 1 for improvement and 0: no change. The profile of each subject is then one of the pairs $(0,0), (0,1), (1,0)$ und $(1,1)$. To summarize the data we count how often each pair occurs.

Group	(0,0)	(0,1)	(1,0)	(1,1)	Total
1 (AB)	n_{11}	n_{12}	n_{13}	n_{14}	$n_{1.}$
2 (BA)	n_{21}	n_{22}	n_{23}	n_{24}	$n_{2.}$
Total	$n_{.1}$	$n_{.2}$	$n_{.3}$	$n_{.4}$	$n_{..}$

Table 8.6: 2×2 Cross-over with binary response

Contingency Tables and Odds Ratio

The two columns in the middle of this 2×4-contingency table may indicate a treatment effect. Assuming no period effect and under the null hypothesis H_0: "no treatment effect", the two responses $n_A = (n_{13} + n_{22})$ for A and $n_B = (n_{12} + n_{23})$ for B have equal probabilities and follow the same binomial distribution n_A resp. $n_B \sim B(n_{.2} + n_{.3}; \frac{1}{2})$.

The odds ratio

$$\widehat{OR} = \frac{n_{12}n_{23}}{n_{22}n_{13}} \tag{8.71}$$

may also indicate a treatment effect.

Testing for carry-over effects is done – similar to the test statistic T_λ (8.19), which is based mainly on $\hat{\lambda} = Y_{1..}/n_1 - Y_{2..}/n_2$ – by comparing the differences in the total response values for the profiles $(0,0)$ and $(1,1)$. Instead of differences, we choose the odds ratio

$$\widehat{OR} = \frac{n_{11}n_{24}}{n_{14}n_{21}} \tag{8.72}$$

which should equal 1 under H_0: "no treatment×period-effect". Using the 2 × 2-table $\begin{array}{|c|c|} \hline A & B \\ \hline C & D \\ \hline \end{array}$, the odds ratio is $\widehat{OR} = \frac{AD}{BC}$ with the following asymptotic distribution

$$\frac{(\ln(\widehat{OR}))^2}{\hat{\sigma}^2_{\ln(\widehat{OR})}} \sim \chi^2_1 \quad , \tag{8.73}$$

where

$$\hat{\sigma}^2_{\ln(\widehat{OR})} = \left(\frac{1}{A} + \frac{1}{B} + \frac{1}{C} + \frac{1}{D} \right) \tag{8.74}$$

(cf. Agresti, 1990). We can now test the significance of the two odds ratios (8.71) and (8.72).

McNemar's Test

Application of this test assumes no period effects. Only values of subjects are considered, who show a preference of one of the treatments. These subjects either have a $(0,1)$ or $(1,0)$ response profile.

There are $n_P = (n_{.2} + n_{.3})$ subjects who show a preference of one of the treatments. $n_A = (n_{13} + n_{22})$ prefer treatment A and $n_B = (n_{12} + n_{23})$ prefer B. Under the null hypothesis of no treatment effects, n_A resp. n_B are binomial distributed $B(n_P; \frac{1}{2})$. The hypothesis is tested using the following statistic (cf. Jones and Kenward, 1989, p.93)

$$\chi^2_{MN} = \frac{(n_A - n_B)^2}{n_P},$$ (8.75)

where χ^2_{MN} is asymptotically χ^2-distributed with one degree of freedom under the null hypothesis.

Mainland-Gart Test

Based on a logistic model, Gart (1969) proposed a test for treatment differences which is equivalent to Fisher's exact test using the following 2×2 contingency table:

Group	(0,1)	(1,0)	Total
1 (AB)	n_{12}	n_{13}	$n_{12} + n_{13} = m_1$
2 (BA)	n_{22}	n_{23}	$n_{22} + n_{23} = m_2$
Total	$n_{.2}$	$n_{.3}$	$m_.$

This test is described in Jones and Kenward (1989, p.113). Asymptotically, the hypothesis of no treatment differences may be tested using one of the common tests for 2×2 contingency tables, e.g., the χ^2-statistic

$$\chi^2 = \frac{m_.(n_{12}n_{23} - n_{13}n_{22})^2}{m_1 m_2 n_{.2} n_{.3}}.$$ (8.76)

This statistic follows a χ^2_1 distribution under the null hypothesis. This test and the test with $\ln(\widehat{OR})$ (cf. (8.73)) coincide.

Prescott Test

The above tests have one thing in common: subjects showing no preference of one of the treatments are discarded from the analysis. Prescott (1981) includes these subjects in his test, by means of the marginal sums $n_{1.}$ und $n_{2.}$. The following 2×3 table will be used:

Group	(0,1)	(0,0) or (1,1)	(1,0)	Total
1 (AB)	n_{12}	$n_{11} + n_{14}$	n_{13}	$n_{1.}$
2 (BA)	n_{22}	$n_{21} + n_{24}$	n_{23}	$n_{2.}$
Total	$n_{.2}$	$n_{.1} + n_{.4}$	$n_{.3}$	$n_{..}$

We first consider the difference between the first and the second response. Depending on the response profile $(1,0),(0,0),(1,1)$ or $(0,1)$, this difference takes the values $+1$, 0 or -1.

Assuming that treatment A is better, we expect the first group (AB) to have a higher mean difference than the second group (BA). The mean difference of the response values in group 1 (AB) is

$$\frac{1}{n_{1\cdot}} \sum_{k=1}^{n_{1\cdot}} (y_{12k} - y_{11k}) = \frac{n_{12} - n_{13}}{n_{1\cdot}} = -d_{1\cdot}. \tag{8.77}$$

and in group 2 (BA)

$$\frac{1}{n_{2\cdot}} \sum_{k=1}^{n_{2\cdot}} (y_{22k} - y_{21k}) = \frac{n_{22} - n_{23}}{n_{2\cdot}} = -d_{2\cdot}. \tag{8.78}$$

Prescott's test statistic (cf. Jones and Kenward, 1989, p.100) under the null hypothesis H_0: *no direct treatment effect* (i.e., $E(d_{1\cdot} - d_{2\cdot}) = 0$) is

$$\chi^2(P) = [(n_{12} - n_{13})n_{\cdot\cdot} - (n_{\cdot 2} - n_{\cdot 3})n_{1\cdot}]^2/V \tag{8.79}$$

with

$$V = n_{1\cdot}n_{2\cdot}[(n_{\cdot 2} + n_{\cdot 3})n_{\cdot\cdot} - (n_{\cdot 2} - n_{\cdot 3})^2]/n_{\cdot\cdot}. \tag{8.80}$$

Asymptotically, $\chi^2(P)$ follows the χ_1^2-distribution under H_0.

This test, however, has the disadvantage, that only the hypothesis of no treatment differences can be tested. As a uniform approach for testing all important hypotheses one could choose the approach of Grizzle, Starmer, and Koch (1969).

Remark: Another and often more efficent method of analysis is given by log-linear models, especially models with uncorrelated two-dimensional binary response. These were examined thoroughly over the last years (cf. Chapter 10).

Example 8.3: A comparison between a placebo A and a new drug B for treating depression might have shown the following results (1: improvement, 0: no improvement)

Group	(0,0)	(0,1)	(1,0)	(1,1)	Total
1(AB)	5	14	3	6	28
2(BA)	10	7	18	10	45
Total	15	21	21	16	73

We check for H_0: "treatment×period-effect $= 0$" (i.e., no carry-over effect) using the odds ratio (8.72)

$$\widehat{OR} = \frac{5 \cdot 10}{6 \cdot 10} = 0.83 \quad \text{and} \quad \ln(\widehat{OR}) = -0.1823.$$

We get

$$\hat{\sigma}^2_{\ln\widehat{OR}} = \frac{1}{5} + \frac{1}{10} + \frac{1}{6} + \frac{1}{10} = 0.5667$$

and

$$\frac{(\ln(\widehat{OR}))^2}{\hat{\sigma}^2_{\ln\widehat{OR}}} = 0.06 < 3.84 = \chi^2_{1;0.95} \quad ,$$

so that H_0 cannot be rejected. In the same way, we get for the odds ratio (8.71)

$$\widehat{OR} = \frac{14 \cdot 18}{7 \cdot 3} = 12 \quad , \quad \ln(\widehat{OR}) = 2.48 \quad ,$$

$$\hat{\sigma}^2_{\ln\widehat{OR}} = \left(\frac{1}{14} + \frac{1}{18} + \frac{1}{7} + \frac{1}{3}\right) = 0.60 \quad ,$$

$$\frac{(\ln(\widehat{OR}))^2}{\hat{\sigma}^2_{\ln OR}} = 10.24 > 3.84 \quad ,$$

and this test rejects H_0: no treatment effect. Since there is no carry-over effect, we can use McNemar's test

$$\chi^2_{MN} = \frac{((3+7) - (14+18))^2}{21+21}$$

$$= \frac{22^2}{42} = 11.53 > 3.84 \quad ,$$

which gives the same result. For Prescott's test we get

$$V = 28 \cdot 45[(21+21) \cdot 73]/73$$
$$= 28 \cdot 45 \cdot 42 = 52920 \quad ,$$
$$\chi^2(P) = [(14-3) \cdot 73 - (21-21) \cdot 28]^2/V$$
$$= (11 \cdot 73)^2/V = 12.28 > 3.84 \quad ,$$

and H_0: no treatment effect is also rejected.

8.4.2 Loglinear and Logit Models

In Table 8.6, we see that group 1 (AB) and group 2 (BA) are represented with four distinct categorical response profiles $(0,0)$, $(0,1)$, $(1,0)$ and $(1,1)$. We assume that each row (and therefore each variable) is an independent observation from a multinomial distribution $M(n_{i.}; \pi_{i1}, \pi_{i2}, \pi_{i3}, \pi_{i4})$ $(i = 1, 2)$. Using appropriate parametrizations and logit or loglinear models, we try to define a bivariate binary variable (Y_1, Y_2), which represents the four profiles and their probabilities according the model of the 2×2 cross-over design. There are various approaches available for handling this.

Bivariate Logistic Model

Generally, Y_1 and Y_2 denote a pair of correlated binary variables. We first want to follow the approach of Jones and Kenward (1989, p.106) who use the following bivariate logistic model according to Cox (1970) and McCullagh and Nelder (1989):

$$P(Y_1 = y_1, Y_2 = y_2) = \exp(\beta_0 + \beta_1 y_1 + \beta_2 y_2 + \beta_{12} y_1 y_2) \quad , \qquad (8.81)$$

with the binary response being coded with $+1$ and -1 in contrast to the former coding. This coding relates to the transformation $Z_i = (2Y_i - 1)$ $(i = 1, 2)$, which was used by Cox (1972). The parameter β_0 is a scaling constant to assure that the four probabilities sum to 1. It depends upon the other three parameters. The parameter β_{12} measures the correlation between the two variables. β_1 and β_2 depict the main effects.

The four possible observations are now put into (8.81) in order to get the joint distribution

$$
\begin{aligned}
\ln P(Y_1 = 1, Y_2 = 1) &= \beta_0 + \beta_1 + \beta_2 + \beta_{12} \quad, \\
\ln P(Y_1 = 1, Y_2 = -1) &= \beta_0 + \beta_1 - \beta_2 - \beta_{12} \quad, \\
\ln P(Y_1 = -1, Y_2 = 1) &= \beta_0 - \beta_1 + \beta_2 - \beta_{12} \quad, \\
\ln P(Y_1 = -1, Y_2 = -1) &= \beta_0 - \beta_1 - \beta_2 + \beta_{12} \quad.
\end{aligned}
$$

Bayes' theorem gives

$$
\begin{aligned}
\frac{P(Y_1 = 1|Y_2 = 1)}{P(Y_1 = -1|Y_2 = 1)} &= \frac{P(Y_1 = 1, Y_2 = 1)/P(Y_2 = 1)}{P(Y_1 = -1, Y_2 = 1)/P(Y_2 = 1)} \\
&= \frac{\exp(\beta_0 + \beta_1 + \beta_2 + \beta_{12})}{\exp(\beta_0 - \beta_1 + \beta_2 - \beta_{12})} \\
&= \exp 2(\beta_1 + \beta_{12}) \quad.
\end{aligned}
$$

We now get the logits

$$
\begin{aligned}
\text{Logit}[P(Y_1 = 1|Y_2 = 1)] &= \ln \frac{P(Y_1 = 1|Y_2 = 1)}{P(Y_1 = -1|Y_2 = 1)} = 2(\beta_1 + \beta_{12}) \quad, \\
\text{Logit}[P(Y_1 = 1|Y_2 = -1)] &= \ln \frac{P(Y_1 = 1|Y_2 = -1)}{P(Y_1 = -1|Y_2 = -1)} = 2(\beta_1 - \beta_{12})
\end{aligned}
$$

and the conditional log-odds ratio

$$\text{Logit}[P(Y_1 = 1|Y_2 = 1)] - \text{Logit}[P(Y_1 = 1|Y_2 = -1)] = 4\beta_{12} \quad, \qquad (8.82)$$

i.e.,

$$\frac{P(Y_1 = 1|Y_2 = 1)P(Y_1 = -1|Y_2 = -1)}{P(Y_1 = -1|Y_2 = 1)P(Y_1 = 1|Y_2 = -1)} = \exp(4\beta_{12}) \quad. \qquad (8.83)$$

This refers to the relation

$$\frac{m_{11}m_{22}}{m_{12}m_{21}} = \exp(4\lambda_{11}^{XY})$$

between odds-ratio and interaction parameter in the loglinear model (cf. Chapter 10). In the same way we get for $i, j = 1, 2$ $(i \neq j)$

$$\text{Logit}[P(Y_i = 1|Y_j = y_j)] = 2(\beta_i + y_j\beta_{12}) \quad . \tag{8.84}$$

For a specific subject of one of the groups $(AB$ or $BA)$, a treatment effect exists if the response is either $(1,-1)$ or $(-1,1)$. From the log-odds ratio for this combination we get

$$\text{Logit}[P(Y_1 = 1|Y_2 = -1)] - \text{Logit}[P(Y_2 = 1|Y_1 = -1)] = 2(\beta_1 - \beta_2) \quad . \tag{8.85}$$

This is an indicator for a treatment effect within a group.

Assuming the same parameter β_{12} for both groups AB and BA, the following expression is an indicator for a period effect:

$$\text{Logit}[P(Y_i^{AB} = 1|Y_j^{AB} = y_j)] - \text{Logit}[P(Y_i^{BA} = 1|Y_j^{BA} = y_j)] = 2(\beta_i^{AB} - \beta_i^{BA}) \quad . \tag{8.86}$$

This relation is directly derived from (8.84) with an additional indexing for the two groups AB and BA. The assumption $\beta_{12}^{AB} = \beta_{12}^{BA}$ is important, i.e., identical interaction in both groups.

Logit Model of Jones and Kenward for the Classical Approach

Let y_{ijk} denote the binary response of subject k of group i in period j $(i = 1, 2,\ j = 1, 2,\ k = 1, \ldots, n_i)$. Again we choose the coding as in Table 8.6 with $y_{ijk} = 1$ denoting success and $y_{ijk} = 0$ for failure. Using logit-links we want to reparametrize the model according to Table 8.1 for bivariate binary response (y_{i1k}, y_{i2k}).

$$\text{Logit}(\pi_{ij}) = \ln\left(\frac{\pi_{ij}}{1 - \pi_{ij}}\right) = X\beta \quad , \tag{8.87}$$

where X denotes the design-matrix using effect coding for the two groups and the two periods (cf. (8.47))

$$X = \begin{pmatrix} 1 & 1 & 1 & 0 \\ 1 & -1 & -1 & 1 \\ 1 & 1 & -1 & 0 \\ 1 & -1 & 1 & -1 \end{pmatrix} \tag{8.88}$$

and $\beta = (\mu\ \pi\ \tau\ \lambda)'$ is the parameter vector using the reparametrization conditions

$$\pi = -\pi_1 = \pi_2, \quad \tau = -\tau_1 = \tau_2, \quad \lambda = -\lambda_1 = \lambda_2 \quad . \tag{8.89}$$

(i) For both of the two groups and the two periods of the 2×2 cross-over with binary response, the logits show the following relation to the model in Table 8.1:

$$\text{Logit}\,P(y_{11k} = 1) = \ln\left(\frac{P(y_{11k} = 1)}{P(y_{11k} = 0)}\right) = \ln\left(\frac{P(y_{11k} = 1)}{1 - P(y_{11k} = 1)}\right)$$

$$= \mu - \pi - \tau$$

$$\text{Logit}\,P(y_{12k} = 1) = \mu + \pi + \tau - \lambda$$

$$\text{Logit}\,P(y_{21k} = 1) = \mu - \pi + \tau$$

$$\text{Logit}\,P(y_{22k} = 1) = \mu + \pi - \tau + \lambda \quad.$$

We get, for example,

$$P(y_{11k} = 1) = \frac{\exp(\mu - \pi - \tau)}{1 + \exp(\mu - \pi - \tau)} \quad,$$

and

$$P(y_{11k} = 0) = \frac{1}{1 + \exp(\mu - \pi - \tau)} \quad.$$

(ii) To start with, we assume that the two observations of each subject in period 1 and 2 are independent. The joint probabilities π_{ij}

Group	(0,0)	(0,1)	(1,0)	(1,1)
1 (AB)	π_{11}	π_{12}	π_{13}	π_{14}
2 (BA)	π_{21}	π_{22}	π_{23}	π_{24}

are the product of the probabilities defined above. We introduce a normalizing constant for the case of non-response $(0,0)$ to adjust the other probabilities. The constant c_1 is chosen so that the four probabilities sum to 1 (in group 2 this constant is c_2):

$$\left.\begin{array}{l} \pi_{11} = P(y_{11k} = 0,\, y_{12k} = 0) = \exp(c_1) \\ \pi_{12} = P(y_{11k} = 0,\, y_{12k} = 1) = \exp(c_1 + \mu + \pi + \tau - \lambda) \\ \pi_{13} = P(y_{11k} = 1,\, y_{12k} = 0) = \exp(c_1 + \mu - \pi - \tau) \\ \pi_{14} = P(y_{11k} = 1,\, y_{12k} = 1) = \exp(c_1 + 2\mu - \lambda) \end{array}\right\} \quad. \tag{8.90}$$

Then

$$\exp(c_1)[1 + \exp(\mu + \pi + \tau - \lambda) + \exp(\mu - \pi - \tau) + \exp(2\mu - \lambda)] = 1 \quad,$$

will give $\exp(c_1)$.

	Group 1	Group 2
(0,0)	$\ln \pi_{11} = c_1 + \sigma + \phi$	$\ln \pi_{21} = c_2 + \sigma - \phi$
(0,1)	$\ln \pi_{12} = c_1 + \mu + \pi + \tau - \lambda - \sigma - \phi$	$\ln \pi_{22} = c_2 + \mu + \pi - \tau + \lambda - \sigma + \phi$
(1,0)	$\ln \pi_{13} = c_1 + \mu - \pi - \tau - \sigma - \phi$	$\ln \pi_{23} = c_2 + \mu - \pi + \tau - \sigma + \phi$
(1,1)	$\ln \pi_{14} = c_1 + 2\mu - \lambda + \sigma + \phi$	$\ln \pi_{24} = c_2 + 2\mu + \lambda + \sigma - \phi$

Table 8.7: Logit model of Jones and Kenward

(iii) Jones and Kenward (1989, p.109) choose the following parametrization to represent the interaction referring to β_{12}. They introduce a new parameter σ to denote the mean interaction of both the groups (i.e., $\sigma = (\beta_{12}^{AB} + \beta_{12}^{BA})/2$) and another parameter ϕ that measures the interaction difference ($\phi = (\beta_{12}^{AB} - \beta_{12}^{BA})/2$). In the logarithms of the probabilities, the model for the two groups is as follows (Table 8.7).

The values of c_i and μ are somewhat difficult to interpret. The nuisance parameters σ and ϕ represent the dependency in the structure of the subjects of the two groups.

From Table 8.7 we obtain the following relations among the parameters π, τ and λ and the odds ratios:

$$\pi = \frac{1}{4}(\ln \pi_{12} + \ln \pi_{22} - \ln \pi_{13} - \ln \pi_{23})$$

$$= \frac{1}{4} \ln \left(\frac{\pi_{12}\pi_{22}}{\pi_{13}\pi_{23}} \right) \quad , \tag{8.91}$$

$$\lambda = \frac{1}{2} \ln \left(\frac{\pi_{11}\pi_{24}}{\pi_{14}\pi_{21}} \right) \tag{8.92}$$

(cf. (8.72)),

$$\tau = \frac{1}{4} \ln \left(\frac{\pi_{12}\pi_{23}}{\pi_{13}\pi_{22}} \right) \tag{8.93}$$

(cf. (8.71)).

The null hypotheses $H_0: \pi = 0$, $H_0: \tau = 0$, $H_0: \lambda = 0$ can be tested using likelihood ratio tests in the appropriate 2×2 table:

for π:

$\hat{m}_{12}$	$\hat{m}_{13}$
$\hat{m}_{23}$	$\hat{m}_{22}$

(second and third column of Table 8.6, where the second row BA is reversed to get the same order AB as the first row)

for λ:

$\hat{m}_{11}$	$\hat{m}_{14}$
$\hat{m}_{21}$	$\hat{m}_{24}$

(first and last column of Table 8.6)

for τ:

$\hat{m}_{12}$	$\hat{m}_{13}$
$\hat{m}_{22}$	$\hat{m}_{23}$

(second and third column of Table 8.6)

The estimators $\hat{m}_{ij}$ are taken from the appropriate loglinear model, corresponding to the hypothesis.

Remark: The modelling (8.90) of the probabilities π_{1j} of the first group (and analogously for the second group) is based on the assumption that the response of each subject is independent over the two periods. Since this assumption cannot be justified in a cross-over design, this within-subject dependency has to be introduced afterwards using the parameters σ and ϕ. This guarantees the formal independency of $\ln(\hat{\pi}_{ij})$ and therefore the applicability of loglinear models. This approach however is critically examined by Ratkowsky et al. (1993, p.300), who suggest the following alternative approach.

Sequence	$(1,1)$	$(1,0)$	$(0,1)$	$(0,0)$
1 (A B)	$m_{11} =$	$m_{12} =$	$m_{13} =$	$m_{14} =$
	$n_1.P_A P_{B|A}$	$n_1.P_A(1 - P_{B|A})$	$n_1.(1 - P_A)P_{B|\bar{A}}$	$n_1.(1 - P_A)(1 - P_{B|\bar{A}})$
2 (B A)	$m_{21} =$	$m_{22} =$	$m_{23} =$	$m_{24} =$
	$n_2.P_B P_{A|B}$	$n_2.P_B(1 - P_{A|B})$	$n_2.(1 - P_B)P_{A|\bar{B}}$	$n_2.(1 - P_B)(1 - P_{A|\bar{B}})$

Table 8.8: Expectations m_{ij} of the 2×4-contingency table

Logit Model of Ratkowsky, Evans and Alldredge

The cross-over experiment aims to analyse the relationship between the transitions $(0,1)$ and $(1,0)$ and the constant response profiles $(0,0)$ and $(1,1)$. We define the following probabilities:

(i) unconditional:

$$P_A \ : \ P(\text{success of } A)$$
$$P_B \ : \ P(\text{success of } B)$$

(ii) conditional: (conditioned on the preceeding treatment)

$$P_{A|B} \ : \ P(\text{success of } A|\text{success of } B)$$
$$P_{A|\bar{B}} \ : \ P(\text{success of } B|\text{no success of } B)$$

and analogously $P_{B|A}$ and $P_{B|\bar{A}}$. The contingency tables of the two groups then have the following expectations m_{ij} of cell counts (Table 8.8). The proper table of observed response values is as follows (Table 8.6 transformed and using N_{ij} instead of n_{ij}):

$(1,1)$	$(1,0)$	$(0,1)$	$(0,0)$	
N_{11}	N_{12}	N_{13}	N_{14}	$n_1.$
N_{21}	N_{22}	N_{23}	N_{24}	$n_2.$

The loglinear model for sequence i (group, $i = 1, 2$) can then be written as follows

$$
\begin{pmatrix}
\ln(N_{i1}) \\
\ln(N_{i2}) \\
\ln(N_{i3}) \\
\ln(N_{i4})
\end{pmatrix}
= X\beta_i + \epsilon_i \quad ,
\tag{8.94}
$$

where the vector of errors ϵ_i is such that $p \lim \epsilon_i = 0$. From Table 8.8, we get the design matrix for the two groups

$$
X =
\begin{pmatrix}
1 & 1 & 0 & 1 & 0 & 0 & 0 \\
1 & 1 & 0 & 0 & 1 & 0 & 0 \\
1 & 0 & 1 & 0 & 0 & 1 & 0 \\
1 & 0 & 1 & 0 & 0 & 0 & 1
\end{pmatrix}
\tag{8.95}
$$

and the vectors of the parameters

$$
\beta_1 =
\begin{pmatrix}
\ln(n_1.) \\
\ln(P_A) \\
\ln(1 - P_A) \\
\ln(P_{B|A}) \\
\ln(1 - P_{B|A}) \\
\ln(P_{B|\bar{A}}) \\
\ln(1 - P_{B|\bar{A}})
\end{pmatrix}
\quad , \quad
\beta_2 =
\begin{pmatrix}
\ln(n_2.) \\
\ln(P_B) \\
\ln(1 - P_B) \\
\ln(P_{A|B}) \\
\ln(1 - P_{A|B}) \\
\ln(P_{A|\bar{B}}) \\
\ln(1 - P_{A|\bar{B}})
\end{pmatrix}
\quad .
\tag{8.96}
$$

Under the usual assumption of independent multinomial distributions $M(n_i., \pi_{i1}, \pi_{i2}, \pi_{i3}, \pi_{i4})$, we get the estimators of the parameters $\hat{\beta}_i$ by solving iteratively the likelihood equations using the Newton-Raphson procedure. An algorithm to solve this problem is given in Ratkowsky et al. (1993, Appendix 7.A). The authors mention that the implementation is quite difficult.

Taking advantage of the structure of Table 8.8, this difficulty can be avoided by transforming the problem (equivalently reducing it) to a standard problem that can be solved with standard software.

From Table 8.8, we get the following relations

$$
\left.
\begin{aligned}
(m_{11} + m_{12})/n_1. &= P_A P_{B|A} + P_A(1 - P_{B|A}) = P_A \\
(m_{13} + m_{14})/n_1. &= (1 - P_A)
\end{aligned}
\right\}
\tag{8.97}
$$

$\Rightarrow$

$$
\begin{aligned}
\ln(m_{11} + m_{12}) - \ln(m_{13} + m_{14}) &= \ln(P_A) - \ln(1 - P_A) \\
&= \mathrm{Logit}(P_A) \tag{8.98} \\
\ln(m_{11}) - \ln(m_{12}) &= \mathrm{Logit}(P_{B|A}) \tag{8.99} \\
\ln(m_{13}) - \ln(m_{14}) &= \mathrm{Logit}(P_{B|\bar{A}}) \tag{8.100}
\end{aligned}
$$

and analogously

$$\ln(m_{21} + m_{22}) - \ln(m_{23} + m_{24}) = \text{Logit}(P_B) \qquad (8.101)$$
$$\ln(m_{21}) - \ln(m_{22}) = \text{Logit}(P_{A|B}) \qquad (8.102)$$
$$\ln(m_{23}) - \ln(m_{24}) = \text{Logit}(P_{A|\bar{B}}) \quad . \qquad (8.103)$$

The logits, as a measure for the various effects in the 2×2 cross-over, are developed using one of the four parametrizations given in Section 8.3.4 for the main effects and the additional effects for the within-subject correlation. To avoid over-parametrization, we drop the carry-over effect λ which is represented as an alias effect anyhow, using the other interaction effects (cf. Section 8.3.4). The model of Ratkowsky, Evans, and Alldredge (REA-model) has the following structure

REA-Model

$$\text{Logit}(P_A) = \mu + \gamma_1 + \pi_1 + \tau_1$$
$$\text{Logit}(P_{B|A}) = \mu + \gamma_1 + \pi_2 + \tau_2 + \alpha_{11}$$
$$\text{Logit}(P_{B|\bar{A}}) = \mu + \gamma_1 + \pi_2 + \tau_2 + \alpha_{10}$$
$$\text{Logit}(P_B) = \mu + \gamma_2 + \pi_1 + \tau_2$$
$$\text{Logit}(P_{A|B}) = \mu + \gamma_2 + \pi_2 + \tau_1 + \alpha_{21}$$
$$\text{Logit}(P_{A|\bar{B}}) = \mu + \gamma_2 + \pi_2 + \tau_1 + \alpha_{20}$$

μ, γ_i, π_i and τ_i denote the usual parameters for the four main effects overall-mean, sequence, period and treatment. The new parameters have the following meaning

α_{i1} : association effect averaged over subjects of sequence i
   if period 1 treatment was a success
α_{i0} : analogously for failure.

Using the sum-to-zero conventions

$\gamma = \gamma_1 = -\gamma_2$ sequence effect
$\pi - \pi_1 = \pi_2$ period effect
$\tau = \tau_1 = -\tau_2$ treatment effect
and
$\alpha_{i0} = -\alpha_{i1}$ association effect

for the within-subject effects, we can represent the REA-model for the two sequences as follows

$$
\begin{pmatrix} \text{Logit}(P_A) \\ \text{Logit}(P_{B|A}) \\ \text{Logit}(P_{B|\bar{A}}) \\ \text{Logit}(P_B) \\ \text{Logit}(P_{A|B}) \\ \text{Logit}(P_{A|\bar{B}}) \end{pmatrix} = \begin{pmatrix} 1 & 1 & 1 & 1 & 0 & 0 \\ 1 & 1 & -1 & -1 & 1 & 0 \\ 1 & 1 & -1 & -1 & -1 & 0 \\ 1 & -1 & 1 & -1 & 0 & 0 \\ 1 & -1 & -1 & 1 & 0 & 1 \\ 1 & -1 & -1 & 1 & 0 & -1 \end{pmatrix} \begin{pmatrix} \mu \\ \gamma \\ \pi \\ \tau \\ \alpha_{11} \\ \alpha_{21} \end{pmatrix} \quad ,
$$

$$\textbf{Logit} \; = \; \boldsymbol{X}_s \boldsymbol{\beta}_s \; . \tag{8.104}$$

Replacing the estimators of the logits on the left side by the relations (8.98) – (8.103) and replacing the expected counts m_{ij} by the obseverd counts N_{ij}, we get the following solutions:

$$\hat{\boldsymbol{\beta}}_s = \boldsymbol{X}_s^{-1} \widehat{\textbf{Logit}} \; , \tag{8.105}$$

i.e.,

$$\begin{pmatrix} \hat{\mu} \\ \hat{\gamma} \\ \hat{\pi} \\ \hat{\tau} \\ \hat{\alpha}_{11} \\ \hat{\alpha}_{21} \end{pmatrix} = \frac{1}{8} \begin{pmatrix} 2 & 1 & 1 & 2 & 1 & 1 \\ 2 & 1 & 1 & -2 & -1 & -1 \\ 2 & -1 & -1 & 2 & -1 & -1 \\ 2 & -1 & -1 & -2 & 1 & 1 \\ 0 & 4 & -4 & 0 & 0 & 0 \\ 0 & 0 & 0 & 0 & 4 & -4 \end{pmatrix} \begin{pmatrix} \widehat{\text{Logit}}(P_A) \\ \widehat{\text{Logit}}(P_{B|A}) \\ \widehat{\text{Logit}}(P_{B|\bar{A}}) \\ \widehat{\text{Logit}}(P_B) \\ \widehat{\text{Logit}}(P_{A|B}) \\ \widehat{\text{Logit}}(P_{A|\bar{B}}) \end{pmatrix} . \tag{8.106}$$

With (8.98) – (8.103) (m_{ij} replaced by N_{ij}) we get

$$\widehat{\text{Logit}}(P_A) \; = \; \ln\left(\frac{N_{11} + N_{12}}{N_{13} + N_{14}}\right) \; , \tag{8.107}$$

$$\widehat{\text{Logit}}(P_{B|A}) \; = \; \ln\left(\frac{N_{11}}{N_{12}}\right) \; , \tag{8.108}$$

$$\widehat{\text{Logit}}(P_{B|\bar{A}}) \; = \; \ln\left(\frac{N_{13}}{N_{14}}\right) \; , \tag{8.109}$$

$$\widehat{\text{Logit}}(P_B) \; = \; \ln\left(\frac{N_{21} + N_{22}}{N_{23} + N_{24}}\right) \; , \tag{8.110}$$

$$\widehat{\text{Logit}}(P_{A|B}) \; = \; \ln\left(\frac{N_{21}}{N_{22}}\right) \; , \tag{8.111}$$

$$\widehat{\text{Logit}}(P_{A|\bar{B}}) \; = \; \ln\left(\frac{N_{23}}{N_{24}}\right) \; . \tag{8.112}$$

In the saturated model (8.104), rank($\boldsymbol{X}_s$) = 6, so that the parameter estimates $\hat{\boldsymbol{\beta}}_s$ can directly be derived from the estimated logits from (8.105). The parameter estimates in the saturated model (8.104) are:

$$\hat{\alpha}_{11} \; = \; \frac{1}{2}[\widehat{\text{Logit}}(P_{B|A}) - \widehat{\text{Logit}}(P_{B|\bar{A}})]$$

$$= \; \frac{1}{2} \ln(\frac{N_{11} N_{14}}{N_{12} N_{13}}) \; , \tag{8.113}$$

$$\hat{\alpha}_{21} \; = \; \frac{1}{2} \ln(\frac{N_{21} N_{24}}{N_{22} N_{23}}) \; . \tag{8.114}$$

Then $\exp(2\hat{\alpha}_{11})$, e.g., is the odds ratio in the 2 × 2-table of the AB-sequence

$$\begin{array}{c|cc}
 & 1 & 0 \\
\hline
1 & N_{11} & N_{12} \\
0 & N_{13} & N_{14}
\end{array} \quad .$$

$$
\begin{aligned}
8\hat{\mu} &= \ln\left(\frac{N_{11}+N_{12}}{N_{13}+N_{14}}\right)^2\left(\frac{N_{11}N_{13}}{N_{12}N_{14}}\right) \\
&\quad + \ln\left(\frac{N_{21}+N_{22}}{N_{23}+N_{24}}\right)^2\left(\frac{N_{21}N_{23}}{N_{22}N_{24}}\right) \\
&= a_1 + a_2 \ ,
\end{aligned}
$$
(8.115)

$$8\hat{\gamma} = a_1 - a_2 \ ,$$
(8.116)

$$
\begin{aligned}
8\hat{\pi} &= \ln\left(\frac{N_{11}+N_{12}}{N_{13}+N_{14}}\right)^2\left(\frac{N_{12}N_{14}}{N_{11}N_{13}}\right) \\
&\quad + \ln\left(\frac{N_{21}+N_{22}}{N_{23}+N_{24}}\right)^2\left(\frac{N_{22}N_{24}}{N_{21}N_{23}}\right) \\
&= a_3 + a_4 \ ,
\end{aligned}
$$
(8.117)

$$8\hat{\tau} = a_3 - a_4 \ .$$
(8.118)

The covariance matrix of $\hat{\boldsymbol{\beta}}_s$ is derived considering the covariance structure of the logits in the weighted least-squares estimation (cf. Chapter 10). For the saturated model or sub-models (after dropping non-significant parameters), the parameter estimates are given by standard software.

Ratkowsky et al. (1993, p.310) give an example of the application of the procedure SAS PROC CATMOD. The file has to be organized according to (8.107)–(8.112) and Table 8.9 ($Y = 1$: success, $Y = 2$: failure).

Count	Y		Count in Example 8.3
$N_{11}+N_{12}$	1	$\widehat{\text{Logit}}(P_A)$	16
$N_{13}+N_{14}$	2		14
N_{11}	1	$\widehat{\text{Logit}}(\Gamma_{B|A})$	14
N_{12}	2		2
N_{13}	1	$\widehat{\text{Logit}}(P_{B|\bar{A}})$	15
N_{14}	2		9
$N_{21}+N_{22}$	1	$\widehat{\text{Logit}}(P_B)$	23
$N_{23}+N_{24}$	2		15
N_{21}	1	$\widehat{\text{Logit}}(P_{A|B})$	18
N_{22}	2		5
N_{23}	1	$\widehat{\text{Logit}}(P_{A|\bar{B}})$	4
N_{24}	2		11

Table 8.9: Data organization in SAS PROC CATMOD (saturated model)

Example 8.4: The efficiency of a treatment (B) compared to a placebo (A) for a mental illness is examined using a 2 × 2 cross-over experiment (Table 8.10). Coding is 1 : improvement und 0 : no improvement.

Group	(0,0)	(0,1)	(1,0)	(1,1)	Total
1 (AB)	9	5	2	14	30
2 (BA)	11	4	5	18	38
Total	20	9	7	32	68

Table 8.10: Response profiles in a 2×2 cross-over with binary response

We first check for H_0: "treatment×period effect $= 0$" using the odds ratio (8.72)

$$\widehat{OR} = \frac{9 \cdot 18}{14 \cdot 11} = 1.05 \quad ,$$

$$\ln(\widehat{OR}) = 0.05 \quad ,$$

$$\hat{\sigma}^2_{\ln \widehat{OR}} = \frac{1}{9} + \frac{1}{18} + \frac{1}{14} + \frac{1}{11} = 0.33 \quad ,$$

$$\frac{(\ln(\widehat{OR}))^2}{\hat{\sigma}^2_{\ln \widehat{OR}}} = 0.01 < 3.84 \quad ,$$

so that H_0 is not rejected. Now we can run the tests for treatment-effects.

The Mainland-Gart test uses the following 2×2-table:

Group	(0,1)	(1,0)	Total
1 (AB)	5	2	7
2 (BA)	4	5	9
Total	9	7	16

Pearson's χ_1^2-statistic with

$$\chi^2 = \frac{16(5 \cdot 5 - 2 \cdot 4)^2}{9 \cdot 7 \cdot 7 \cdot 9} = 1.17 < 3.84 = \chi^2_{1;0.95}$$

does not indicate a treatment effect (p-value : 0.2804).

The Mainland-Gart test and Fisher's exact test are equivalent. Fisher's exact test (cf. Section 2.6.2) gives for the three tables

2	5
5	4

1	6
6	3

0	7
7	2

the following probabilities

$$P_1 = \frac{7!9!7!9!}{16!} \cdot \frac{1}{5!2!4!5!} = 0.2317$$

$$P_2 = \frac{2 \cdot 4}{6 \cdot 6} P_1 = 0.0515$$

$$P_3 = \frac{1 \cdot 3}{7 \cdot 7} P_2 = 0.0032 \quad ,$$

with $P = P_1 + P_2 + P_3 = 0.2364$, so that $H_0 : P((AB)) = P((BA))$ is not rejected. Prescott's test uses the following 2×3-table:

Group	(0,1)	(0,0) or (1,1)	(1,0)	Total
(AB)	5	9 + 14	2	30
(BA)	4	11 + 18	6	38
Total	9	52	7	68

$$
\begin{aligned}
V &= 30 \cdot 38[(9+7) \cdot 68 - (9-7)^2]/68 \\
  &= \frac{30 \cdot 38}{68}[16 \cdot 68 - 4] = 18172.94 \\
\chi^2(P) &= [(5-2) \cdot 68 - (9-7) \cdot 30]^2/V \\
  &= \frac{144^2}{V} = 1.14 < 3.84 \quad .
\end{aligned}
$$

H_0: treatment effect $= 0$ is not rejected.

Saturated REA-Model

The analysis of the REA-model using SAS gives the following table:
Calling procedure in SAS:

```
PROC CATMOD DATA = BEISPIEL 8.4;
WEIGHT COUNT;
DIRECT SEQUENCE PERIOD TREAT
ASSOC_AB ASSOC_BA;
MODEL Y = SEQUENCE PERIOD TREAT
  ASSOC_AB ASSOC_BA /
  NOGLS ML;
RUN;
```

Effect	Estimate	S.E.	Chi-Square	P-value
INTERCEPT	0.3437	0.1959	3.08	0.0793
SEQUENCE	0.0626	0.1959	0.10	0.7429
PERIOD	-0.0623	0.1959	0.10	0.7470
TREAT	-0.2096	0.1959	1.14	0.2846
ASSOC_AB	1.2668	0.4697	7.27	0.0070 *
ASSOC_BA	1.1463	0.3862	8.81	0.0030 *

None of the main effects is significant.

Remark: The parameter estimates may be directly checked using (8.113)–(8.118):

$$
\hat{\mu} = \frac{1}{8}\ln\left[\left(\frac{14+2}{9+5}\right)^2\frac{14 \cdot 5}{9 \cdot 2}\right] + \frac{1}{8}\ln\left[\left(\frac{18+5}{11+4}\right)^2\frac{18 \cdot 4}{11 \cdot 5}\right]
$$

$$
\begin{aligned}
&= \; 0.2031 + 0.1406 = 0.3437 \\
\hat{\gamma} &= \; 0.2031 - 0.1406 = 0.0625 \\
\hat{\pi} &= \; \frac{1}{8}\ln\left[(\frac{16}{14})^2\frac{18}{70}\right] + \frac{1}{8}\ln\left[(\frac{23}{15})^2\frac{55}{72}\right] \\
&= \; -0.1364 + 0.0732 = -0.0632 \\
\hat{\tau} &= \; -0.1364 - 0.0732 = -0.2096 \\
\hat{\alpha}_{11} &= \; \frac{1}{2}\ln(\frac{9\cdot14}{5\cdot2}) = 1.2668 \\
\hat{\alpha}_{21} &= \; \frac{1}{2}\ln(\frac{11\cdot18}{4\cdot5}) = 1.1463 \quad .
\end{aligned}
$$

Analysis via GEE1 (cf. Chapter 10)

The analysis of the data set using the GEE1-procedure of Heumann (1993) gives
the following results for parametrization No. 2 (model (8.58))

Effect	Estimates	Naive S.E.	Robust S.E.	P-Robust
INTERCEPT	0.1335	0.3569	0.3569	0.7154
TREATMENT	0.2939	0.4940	0.4940	0.5521
PERIOD	0.1849	0.4918	0.4918	0.7071
TREAT x PERIOD	-0.0658	0.7040	0.8693	0.9397

The working correlation is 0.5220. All effects are not significant.

8.5 Excercises and Questions

8.5.1 Give a description of the linear model of cross-over designs.
What is its relationship to repeated measures and split plot designs?
Which are the main effects and the interaction effect?

8.5.2 Review the test strategy in the 2 × 2 cross-over.
Assuming the carry-over effect to be significant, which effect is still testable?
Is this test useful?

8.5.3 What is the difference between the classical approach and the four alternative parametrizations?
Describe the relationship between randomization vs. carry-over effect and parallel groups vs. sequence-effect.

8.5.4 Consider the following 2 × 2 cross-over with binary response:

Group	(0,0)	(0,1)	(1,0)	(1,1)	Total
1 (AB)	n_{11}	n_{12}	n_{13}	n_{14}	$n_{1.}$
2 (BA)	n_{21}	n_{22}	n_{23}	n_{24}	$n_{2.}$

Which contingency tables and corresponding odds ratios are indicators for treatment effect or treatment×period effect?

8.5.5 Review the tests of McNemar, Mainland-Gart, and Prescott (assumptions, objectives).

9 Statistical Analysis of Incomplete Data

9.1 Introduction

A basic problem in the statistical analysis of data sets is the loss of single observations, of variables, or of single values. Rubin (1976) can be regarded as the pioneer of modern theory of *Nonresponse in Sample Surveys*. Little and Rubin (1987) and Rubin (1987) have discussed fundamental concepts for handling missing data based on decision theory and models for the mechanism of nonresponse.

Standard statistical methods have been developed to analyse rectangular data sets, i.e., to analyse a matrix

$$X = \begin{pmatrix} x_{11} & \cdots & \cdots & x_{1m} \\ \vdots & * & & \vdots \\ & & & * \\ \vdots & & * & \vdots \\ x_{n1} & \cdots & \cdots & x_{nm} \end{pmatrix}.$$

The columns of the matrix X represent variables observed for each unit, and the rows of X represent units (cases, observations) of the variables. Here, data on all scales can be observed

- interval-scaled data

- ordinal-scaled data

- nominal-scaled data.

In practice, some of the observations may be missing. This fact is indicated by the symbol '*'.

Examples:

- Persons not always give answers to all items of a questionnaire. Answers may be missing at random (a question was overlooked) or not missing at random (individuals are not willing to give detailed information concerning private items like drinking behaviour, income, sexual behaviour, etc.).

- Mechanical experiments in industry (e.g., quality control by pressure) sometimes destroy the object and the response is missing. If there is a strong causal relationship between the object of the experiment and the loss of response, then it may be expected that the response is not missing at random.

- In clinical long-time studies, some individuals may not cooperate or do not participate over the whole period and drop out. In the analysis of lifetime data, these individuals are called censored. Censoring is a mechanism causing nonrandomly missing data.

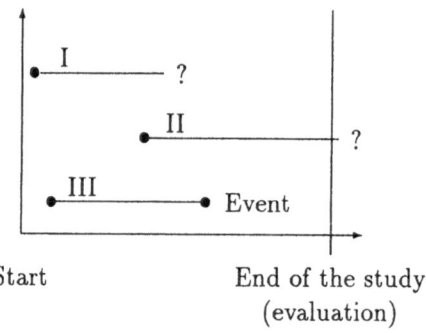

Figure 9.1: Censored individuals (I: drop-out and II: censored by the endpoint) and an individual with response (event) (III)

Statistical Methods with Missing Data

There are mainly three general approaches to handle the missing data problem in statistical analysis.

(i) Complete Case Analysis

Analyses using only complete cases confine attention to those cases (rows of the matrix X) where all m variables are observed. Let X be rearranged according to

$$X = \begin{pmatrix} X_c \\ {}_{n_1,m} \\ X_* \\ {}_{n_2,m} \end{pmatrix}$$

where X_c (c: complete) is fully observed. The statistical analysis makes use of the data in X_c only. The complete case analysis tends to become inefficient if the percentage $(n_2/n) \cdot 100$ is increasing and if there are blocks in the pattern of missing data. The selection of complete cases can lead to a selectivity bias in the estimates if selection is heterogeneous with respect to the covariates. Hence, the crucial concern is whether the complete cases constitute a random subsample of X or not.

Example 9.1: Suppose that age under 60 and age over 60 are the two levels of the binary variable X (age of individuals). Assume the following situation in a lifetime data analysis:

	Start	End
< 60	100	60
> 60	100	40

The drop-out percentage is 40 % and 60 %, respectively. Hence, one has to test if there is a selectivity bias in estimating survivorship models and, if the tests are significant, one has to correct the estimations by adjustment methods (see, e.g., Walther and Toutenburg, 1991).

(ii) Filling-In the Missing Values (Imputation for Nonresponse)

Imputation is a general and flexible alternative to the complete case analysis. The missing cells in the submatrix X_* are replaced by guesses or correlation-based predictors transforming X_* to $\hat{X}_*$. However, this method can lead to severe biases in statistical analysis, as the imputed values in general are different from the true but missing data. We will discuss this problem in detail in case of regression. Sometimes, the statistician has no other choice than to fill-up the matrix X_*, especially if the percentage of complete units is too small. There are several approaches for imputation. Popular among them are the following:

- *hot deck imputation:* Recorded units of the sample are substituted for missing data.

- *cold deck imputation:* A missing value is replaced by a constant value, as for example a unit from external (or previous) samples.

- *mean imputation:* Based on the sample of the responding units, means are substituted for the missing cells.

- *regression (correlation) imputation:* Based on the correlative structure of the matrix X_c, missing values are replaced by predicted values from a regression of the missing item on items observed for the unit.

(iii) Model-Based Procedures

Modelling techniques are generated by factorization of the likelihood according to the observation pattern and the missing pattern. Parameters can be estimated by iterative maximum likelihood procedures starting with the complete cases. These methods are discussed in full detail by Little and Rubin (1987).

Multiple Imputation

The idea of multiple imputation (Rubin, 1987) is to achieve a variability of the estimate by repeated imputation and analysis of each of the so completed data sets. The final estimate can then be calculated, for example, by taking the means.

Missing Data Mechanisms

Ignorable nonresponse: Knowledge of the mechanism for nonresponse is a central element in choosing an appropriate statistical analysis. If the mechanism is under control of the statistician and if it generates a random subsample of the whole sample, then it may be called ignorable.

Example: Assume $Y \sim N(\mu, \sigma^2)$ to be a univariate normally distributed variable and denote by $(y_1, \ldots, y_m, y_{m+1}, \ldots, y_n)$ the planned whole sample. Suppose that indeed only a subsample $\boldsymbol{y}_{obs} = (y_1, \ldots, y_m)'$ of responses is observed and the remaining responses $\boldsymbol{y}_{mis} = (y_{m+1}, \ldots, y_n)'$ are missing. If the values are missing at random (MAR), then the vector $(y_1, \ldots, y_m)'$ is a random subsample. The only disadvantage is a loss of sample size and, hence a loss of efficiency of the unbiased estimators $\bar{y}$ and s_y^2.

Nonignorable nonresponse occurs if the probability $P(y_i \text{ observed})$ is a function of the value y_i itself, as it happens for example in case of censoring. In general, estimators based on nonrandom subsamples are biased.

MAR, OAR and MCAR

Let us assume a bivariate sample of (X, Y) such that X is completely observed but that some values of Y are missing. This structure is a special case of a so-called monotone pattern of missing data.

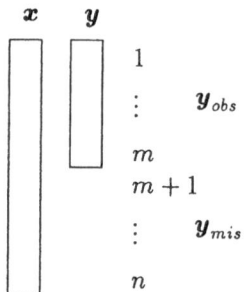

Figure 9.2: Monotone pattern in the bivariate case

This situation is typical for longitudinal studies or questionnaires, when one variable is known for all elements of the sample, but the other variable is unknown for some of them.

Examples:

X	Y
age	income
placebo	blood pressure after 28 days
cancer	life span

The probability of the response of Y can be dependent on X and Y in the following manner

(i) dependent on X and Y

(ii) dependent on X but independent of Y

(iii) independent of X and Y

In case (iii) the missing data is said to be missing at random (MAR) and the observed data is said to be observed at random (OAR). Thus the missing data is said to be missing completely at random (MCAR). As a consequence, the data y_{obs} constitutes a random subsample of $y = (y_{obs}, y_{mis})'$.

In case (ii) the missing data is MAR but the observed values are not necessarily a random subsample of y. However, within fixed X-levels the y-values y_{obs} are OAR.

In case (i) the data is neither MAR nor OAR and hence, the missing data mechanism is not ignorable. In cases (ii) and (iii) the missing data mechanisms are ignorable for methods using the likelihood. In case (iii) this is true for methods based on the sample as well.

If the conditional distribution of $Y|X$ has to be investigated, then MAR is sufficient to have efficient estimators. On the other hand, if the marginal distribution of Y is of interest (e.g., estimation of μ by $\bar{y}$ based on the m complete observations) then MCAR is a necessary assumption to avoid a bias. Suppose that the joint density function of X and Y is factorized as

$$f(X, Y) = f(X)f(Y|X)$$

where $f(X)$ is the marginal density of X and $f(Y|X)$ is the conditional density of $Y|X$. It is obvious, that analysis of $f(Y|X)$ has to be based on the m jointly observed data points. Estimating y_{mis} coincides with the classical prediction.

Example: Suppose that X is a categorical covariate with two categories $X = 1$ (age > 60 years) and $X = 0$ (age ≤ 60 years). Let Y be the lifetime of a dentures. It may happen that the younger group of patients participates less often in the follow-ups compared to the other group. Therefore, one may expect that $P(y_{obs}|X = 1) > P(y_{obs}|X = 0)$.

9.2 Missing Data in the Response

In controlled experiments such as clinical trials, the design matrix X is fixed
and the response is observed for the different factor levels of X. The analysis
is done by means of analysis of variance or the common linear model and the
associated test procedures (cf. Chapter 3). In this situation, it is realistic to
assume that missing values occur in the response y and not in the design matrix
X. This results in unbalanced response. Even if we can assume that MCAR
holds, sometimes it may be more advantageous to fill-up the vector y than to
confine the analysis to the complete cases. This is the fact, for example, in
factorial (cross-classified) designs with few replications.

9.2.1 Least-Squares Analysis for Complete Data

Let Y be the response variable, X the (T, K)-matrix of design and assume the
linear model

$$y = X\beta + \epsilon, \quad \epsilon \sim N(0, \sigma^2 I). \tag{9.1}$$

The OLSE of β is given by $b = (X'X)^{-1}X'y$ and the unbiased estimator of σ^2
is given by

$$
\begin{aligned}
s^2 &= (y - Xb)'(y - Xb)(T - K)^{-1} \\
&= \frac{\sum_{t=1}^{T}(y_t - \hat{y}_t)^2}{T - K}.
\end{aligned}
\tag{9.2}
$$

To test linear hypotheses of the type $R\beta = 0$ (R a $J \times K$-matrix of rank J), we
use the test statistic

$$F_{J,T-K} = \frac{(Rb)'(R(X'X)^{-1}R')^{-1}(Rb)}{Js^2} \tag{9.3}$$

(cf. Sections 3.7 and 3.8).

9.2.2 Least-Squares Analysis for Filled-Up Data

The following method was proposed by Yates (1933). Assume that $(T - m)$
responses in y are missing. Reorganize the data matrices according to

$$\begin{pmatrix} y_{obs} \\ y_{mis} \end{pmatrix} = \begin{pmatrix} X_c \\ X_* \end{pmatrix} \beta + \begin{pmatrix} \epsilon_c \\ \epsilon_* \end{pmatrix} \tag{9.4}$$

(c: complete). The complete case estimator of β is then given by

$$b_c = (X_c'X_c)^{-1}X_c'y_{obs} \tag{9.5}$$

$(X_c : m \times K)$ and the classical predictor of the $(T - m)$-vector y_{mis} is given by

$$\hat{y}_{mis} = X_* b_c. \tag{9.6}$$

Inserting this estimator into (9.4) for $\boldsymbol{y}_{mis}$ and estimating $\boldsymbol{\beta}$ in the filled-up model is equivalent to minimizing the following function with respect to $\boldsymbol{\beta}$ (cf. (3.6))

$$
\begin{aligned}
S(\boldsymbol{\beta}) &= \left\{ \begin{pmatrix} \boldsymbol{y}_{obs} \\ \hat{\boldsymbol{y}}_{mis} \end{pmatrix} - \begin{pmatrix} \boldsymbol{X}_c \\ \boldsymbol{X}_* \end{pmatrix} \boldsymbol{\beta} \right\}' \left\{ \begin{pmatrix} \boldsymbol{y}_{obs} \\ \hat{\boldsymbol{y}}_{mis} \end{pmatrix} - \begin{pmatrix} \boldsymbol{X}_c \\ \boldsymbol{X}_* \end{pmatrix} \boldsymbol{\beta} \right\} \\
&= \sum_{t=1}^{m} (y_t - \boldsymbol{x}_t'\boldsymbol{\beta})^2 + \sum_{t=m+1}^{T} (\hat{y}_t - \boldsymbol{x}_t'\boldsymbol{\beta})^2 \longrightarrow \min_{\boldsymbol{\beta}} ! \qquad (9.7)
\end{aligned}
$$

The first sum is minimized by $\boldsymbol{b}_c$ (9.5). Replacing $\boldsymbol{\beta}$ in the second sum by $\boldsymbol{b}_c$ equates this sum to zero (cf. (9.6)), i.e., to its absolute minimum. Therefore, the estimator $\boldsymbol{b}_c$ minimizes the error-sum-of-squares $S(\boldsymbol{\beta})$ (9.7) and $\boldsymbol{b}_c$ is seen to be the OLSE of $\boldsymbol{\beta}$ in the filled-up model.

Estimating σ^2

(i) If the data is complete, then $s^2 = \sum_{t=1}^{T}(y_t - \hat{y}_t)^2/(T - K)$ is the correct estimator of σ^2.

(ii) If $(T - m)$ values are missing (i.e., $\boldsymbol{y}_{mis}$ in (9.4)), then

$$
\hat{\sigma}^2_{mis} = \sum_{t=1}^{m}(y_t - \hat{y}_t)^2/(m - K) \qquad (9.8)
$$

would be the appropriate estimator of σ^2.

(iii) On the other hand, if the missing data are filled-up according to the method of Yates, we automatically receive the estimator

$$
\begin{aligned}
\hat{\sigma}^2_{\text{Yates}} &= \left\{ \sum_{t=1}^{m}(y_t - \hat{y}_t)^2 + \sum_{t=m+1}^{T} (\hat{y}_t - \hat{y}_t)^2 \right\} /(T - K) \\
&= \sum_{t=1}^{m}(y_t - \hat{y}_t)^2 /(T - K) . \qquad (9.9)
\end{aligned}
$$

Therefore we get the relationship

$$
\hat{\sigma}^2_{\text{Yates}} = \hat{\sigma}^2_{mis} \cdot \frac{m - K}{T - K} < \hat{\sigma}^2_{mis}, \qquad (9.10)
$$

and, hence the method of Yates underestimates the variance. As a consequence of this, the confidence intervals (cf. (3.220), (3.221) and (3.240)) turn out to be too small and the test statistics (cf. (9.3)) become too large, implying that null hypotheses can be rejected more often. To ensure correct tests, the estimate of the variance and all the following statistics would have to be corrected by the factor $\left(\frac{T-K}{m-K}\right)$.

9.2.3   Analysis of Covariance – Bartlett's Method

Bartlett (1937) suggested an improvement of Yates' ANOVA, which is known as
Bartlett's ANCOVA (analysis of covariance). This procedure is as follows:

(i) each missing value is replaced by an arbitrary estimate (guess):
$\boldsymbol{y}_{mis} \Rightarrow \hat{\boldsymbol{y}}_{mis}$,

(ii) define an indicator matrix $\underset{T,(T-m)}{\boldsymbol{Z}}$ as a covariable according to

$$
\boldsymbol{Z} = \begin{pmatrix}
0 & 0 & 0 & \cdots & 0 \\
0 & 0 & 0 & \cdots & 0 \\
\vdots & \vdots & \vdots & & \vdots \\
0 & 0 & 0 & \cdots & 0 \\
1 & 0 & 0 & \cdots & 0 \\
0 & 1 & 0 & \cdots & 0 \\
\vdots & \vdots & \vdots & & \vdots \\
0 & 0 & 0 & \cdots & 1
\end{pmatrix}.
\tag{9.11}
$$

The m null vectors indicate the observed cases and the $(T - m)$ vectors
e'_i indicate the missing values. This covariable $\boldsymbol{Z}$ leads to an additional
parameter $\underset{(T-m),1}{\boldsymbol{\gamma}}$ in the model that has to be estimated:

$$
\begin{pmatrix} \boldsymbol{y}_{obs} \\ \hat{\boldsymbol{y}}_{mis} \end{pmatrix} = \boldsymbol{X}\boldsymbol{\beta} + \boldsymbol{Z}\boldsymbol{\gamma} + \boldsymbol{\epsilon}
$$

$$
= (\boldsymbol{X}, \boldsymbol{Z}) \begin{pmatrix} \boldsymbol{\beta} \\ \boldsymbol{\gamma} \end{pmatrix} + \boldsymbol{\epsilon}.
\tag{9.12}
$$

The OLSE of the parameter vector $\begin{pmatrix} \boldsymbol{\beta} \\ \boldsymbol{\gamma} \end{pmatrix}$ is found by minimizing the error-
sum-of-squares

$$
S(\boldsymbol{\beta}, \boldsymbol{\gamma}) = \sum_{t=1}^{m}(y_t - \boldsymbol{x}'_t\boldsymbol{\beta} - \boldsymbol{0}'\boldsymbol{\gamma})^2 + \sum_{t=m+1}^{T}(\hat{y}_t - \boldsymbol{x}'_t\boldsymbol{\beta} - \boldsymbol{e}'_t\boldsymbol{\gamma})^2.
\tag{9.13}
$$

The first term is minimal for $\hat{\boldsymbol{\beta}} = \boldsymbol{b}_c$ (9.5), whereas the second term becomes
minimal (equating to zero) for $\hat{\boldsymbol{\gamma}} = \hat{\boldsymbol{y}}_{mis} - \boldsymbol{X}_*\boldsymbol{b}_c$. Hence, the sum total is minimal
for $(\boldsymbol{b}_c, \hat{\boldsymbol{\gamma}})'$, and so

$$
\begin{pmatrix} \boldsymbol{b}_c \\ \hat{\boldsymbol{y}}_{mis} - \boldsymbol{X}_*\boldsymbol{b}_c \end{pmatrix}
\tag{9.14}
$$

is the OLSE of $\begin{pmatrix} \boldsymbol{\beta} \\ \boldsymbol{\gamma} \end{pmatrix}$ in the model (9.12). Choosing the guess $\hat{\boldsymbol{y}}_{mis} = \boldsymbol{X}_*\boldsymbol{b}_c$ (as
in Yates' method), we get $\hat{\boldsymbol{\gamma}} = 0$. Both methods lead to the complete case OLSE

b_c as estimate of β. Introducing the additional parameter γ (which is not of any statistical interest) has one advantage: the degree of freedom in estimating σ^2 in model (9.12) is now T minus the number of estimated parameters, i.e., $T - K - (T - m) = m - K$, and hence correct. Therefore Bartlett's ANCOVA leads to $\hat{\sigma}^2 = \hat{\sigma}^2_{mis}$ (cf. (9.8)), an unbiased estimator of σ^2.

9.3 Missing Values in the X-Matrix

In econometric models, other than in experimental design in biology or pharmacy, the matrix X is not fixed but contains observations of exogeneous variables. Hence X may be a matrix of random variables and missing observations can occur. In general, we may assume the following structure of the data

$$\begin{pmatrix} \boldsymbol{y}_{obs} \\ \boldsymbol{y}_{mis} \\ \boldsymbol{y}^*_{obs} \end{pmatrix} = \begin{pmatrix} \boldsymbol{X}_{obs} \\ \boldsymbol{X}^*_{obs} \\ \boldsymbol{X}_{mis} \end{pmatrix} \boldsymbol{\beta} + \boldsymbol{\epsilon}. \tag{9.15}$$

Estimation of $\boldsymbol{y}_{mis}$ corresponds to the prediction problem. The classical prediction is equivalent to the method of Yates. Based on these arguments, we may confine ourselves to the substructure

$$\begin{pmatrix} \boldsymbol{y}_{obs} \\ \boldsymbol{y}^*_{obs} \end{pmatrix} = \begin{pmatrix} \boldsymbol{X}_{obs} \\ \boldsymbol{X}_{mis} \end{pmatrix} \boldsymbol{\beta} + \boldsymbol{\epsilon} \tag{9.16}$$

of (9.15) and change the notation as follows

$$\begin{pmatrix} \boldsymbol{y}_c \\ \boldsymbol{y}_* \end{pmatrix} = \begin{pmatrix} \boldsymbol{X}_c \\ \boldsymbol{X}_* \end{pmatrix} \boldsymbol{\beta} + \begin{pmatrix} \boldsymbol{\epsilon}_c \\ \boldsymbol{\epsilon}_* \end{pmatrix}, \quad \begin{pmatrix} \boldsymbol{\epsilon}_c \\ \boldsymbol{\epsilon}_* \end{pmatrix} \sim (\boldsymbol{0}, \sigma^2 \mathbf{I}). \tag{9.17}$$

The submodel

$$\boldsymbol{y}_c = \boldsymbol{X}_c \boldsymbol{\beta} + \boldsymbol{\epsilon}_c \tag{9.18}$$

stands for the completely observed data (c: complete), and we have $\boldsymbol{y}_c : m \times 1$, $\boldsymbol{X}_c : m \times K$ and rank $(\boldsymbol{X}_c) = K$. Assume that $\boldsymbol{X}$ is nonstochastic. If not, we would use conditional expectations.

The other submodel

$$\boldsymbol{y}_* = \boldsymbol{X}_* \boldsymbol{\beta} + \boldsymbol{\epsilon}_* \tag{9.19}$$

is of the dimension $(T - m) = J$. The vector $\boldsymbol{y}_*$ is observed completely. In the matrix $\boldsymbol{X}_*$ some observations are missing. The notation $\boldsymbol{X}_*$ shall underline that $\boldsymbol{X}_*$ is partially incomplete, in contrast to the matrix $\boldsymbol{X}_{mis}$, which is completely missing. Combining both of the submodels in model (9.17) corresponds to the so-called mixed model (cf. Toutenburg, 1992). Therefore, it seems to be natural to use the method of mixed estimation

The optimal estimator of $\boldsymbol{\beta}$ in model (9.17) is given by the mixed estimator (cf. Toutenburg, 1992, Chapter 5)

$$\begin{aligned} \hat{\boldsymbol{\beta}}(\boldsymbol{X}_*) &= (\boldsymbol{X}'_c \boldsymbol{X}_c + \boldsymbol{X}'_* \boldsymbol{X}_*)^{-1}(\boldsymbol{X}'_c \boldsymbol{y}_c + \boldsymbol{X}'_* \boldsymbol{y}_*) \\ &= \boldsymbol{b}_c + \boldsymbol{S}^{-1}_c \boldsymbol{X}'_*(\mathbf{I}_J + \boldsymbol{X}_* \boldsymbol{S}^{-1}_c \boldsymbol{X}'_*)^{-1}(\boldsymbol{y}_* - \boldsymbol{X}_* \boldsymbol{b}_c) \end{aligned} \tag{9.20}$$

where

$$b_c = (X'_c X_c)^{-1} X'_c y_c \tag{9.21}$$

is the OLSE in the complete case submodel (9.18) and

$$S_c = X'_c X_c \tag{9.22}$$

The covariance matrix of $\hat{\beta}(X_*)$ is

$$V(\hat{\beta}(X_*)) = \sigma^2 (S_c + S_*)^{-1} \tag{9.23}$$

with

$$S_* = X'_* X_*. \tag{9.24}$$

The mixed estimator (9.20) is not operational though, due to the fact that X_* is partially unknown.

9.3.1   Missing Values and Loss of Efficiency

Before we discuss the different methods for estimating missing values, let us study the consequences of confining the analysis to the complete case model (9.18). Our measure to compare $\hat{\beta}_c$ and $\hat{\beta}(X_*)$ is the scalar risk

$$R(\hat{\beta}, \beta, S_c) = \operatorname{tr}\{S_c V(\hat{\beta})\}, \tag{9.25}$$

which coincides with the MSE-III-risk. From Theorem A 18 (iii) we have the identity

$$(S_c + X'_* X_*)^{-1} = S_c^{-1} - S_c^{-1} X'_*(I_J + X_* S^{-1} X'_*)^{-1} X_* S_c^{-1}. \tag{9.26}$$

Applying this we get the risk of $\hat{\beta}(X_*)$ as

$$\begin{aligned}
\sigma^{-2} R(\hat{\beta}(X_*), \beta, S_c) &= \operatorname{tr}\{S_c (S_c + S_*)^{-1}\} \\
&= K - \operatorname{tr}\{(I_J + B'B)^{-1} B'B\}, \tag{9.27}
\end{aligned}$$

where $B = S_c^{-1/2} X'_*$.

The $J \times J$-Matrix $B'B$ is nonnegative definite with rank $(B'B) = J^*$. If rank $(X_*) = J < K$ holds, then $J^* = J$ and hence $B'B > 0$.

Let $\lambda_1 \geq \ldots \geq \lambda_J \geq 0$ denote the eigenvalues of B, $\Lambda = \operatorname{diag}(\lambda_1, \ldots, \lambda_J)$ and P the matrix of orthogonal eigenvectors. Then we have (Theorem A 30) $B'B = P\Lambda P'$ and

$$\begin{aligned}
\operatorname{tr}\{(I_J + B'B)^{-1} B'B\} &= \operatorname{tr}\{P(I_J + \Lambda)^{-1} P' P\Lambda P'\} \\
&= \operatorname{tr}\{(I_J + \Lambda)^{-1}\Lambda\} \\
&= \sum_{i=1}^{J} \frac{\lambda_i}{1 + \lambda_i}. \tag{9.28}
\end{aligned}$$

The MSE-III-risk of b_c is

$$\sigma^{-2} R(b_c, \beta, S_c) = \text{tr}\{S_c S_c^{-1}\} = K. \qquad (9.29)$$

Using the MSE-III-criterion, we may conclude that

$$R(b_c, \beta, S_c) - R(\hat{\beta}(X_*), \beta, S_c) = \sum \frac{\lambda_i}{1 + \lambda_i} \geq 0, \qquad (9.30)$$

and hence, that $\hat{\beta}(X_*)$ is superior to b_c. We want to continue the comparison according to a different criterion, which compares the size of the risks instead of their differences.

Definition 9.1 *The relative efficiency of an estimator $\hat{\beta}_1$ compared to another estimator $\hat{\beta}_2$ is defined as the following ratio*

$$\text{eff}(\hat{\beta}_1, \hat{\beta}_2, A) = \frac{R(\hat{\beta}_2, \beta, A)}{R(\hat{\beta}_1, \beta, A)}. \qquad (9.31)$$

$\hat{\beta}_1$ *is said to be less efficient than* $\hat{\beta}_2$ *if*

$$\text{eff}(\hat{\beta}_1, \hat{\beta}_2, A) \leq 1.$$

Using (9.27) – (9.29) we find

$$\text{eff}(b_c, \hat{\beta}(X_*), S_c) = 1 - \frac{1}{K} \sum \frac{\lambda_i}{1 + \lambda_i} \leq 1. \qquad (9.32)$$

The relative efficiency of the complete case estimator b_c compared to the mixed estimator in the full model (9.17) is smaller or equal to one:

$$\max\left[0, 1 - \frac{J}{K} \frac{\lambda_1}{1 + \lambda_1}\right] \leq \text{eff}(b_c, \hat{\beta}(X_*), S_c) \leq 1 - \frac{J}{K} \frac{\lambda_J}{1 + \lambda_J} \leq 1. \qquad (9.33)$$

Examples:

(i) Let $X_* = X_c$, so that in the full model the design matrix X_c is used twice. Then $B'B = X_c S_c^{-1} X_c'$ is idempotent of rank $J = K$. Therefore, we have $\lambda_i = 1$ (Theorem A 61 (i)) and hence

$$\text{eff}(b_c, \hat{\beta}(X_c), S_c) = \frac{1}{2}. \qquad (9.34)$$

(ii) $J = 1$ (one row of X is incomplete). Then $X_* = x_*'$ becomes a $1 \times K$-vector and $B'B = x_*' S_c^{-1} x_*$ a scalar. Let $\mu_1 \geq \ldots \geq \mu_K > 0$ be the eigenvalues of S_c and $\Gamma = (\gamma_1, \ldots, \gamma_K)$ the matrix of the corresponding orthogonal eigenvectors.

Therefore, we may write $\hat{\beta}(\boldsymbol{x}_*)$ as

$$\hat{\beta}(\boldsymbol{x}_*) = (\boldsymbol{S}_c + \boldsymbol{x}_*\boldsymbol{x}_*')^{-1}(\boldsymbol{x}_c'\boldsymbol{y}_c + \boldsymbol{x}_*y_*) \tag{9.35}$$

and observe that

$$\mu_1^{-1}\boldsymbol{x}_*'\boldsymbol{x}_* \le \boldsymbol{x}_*'\boldsymbol{S}_c^{-1}\boldsymbol{x}_* = \sum \mu_j^{-1}(\boldsymbol{x}_*'\boldsymbol{\gamma}_j)^2 \le \mu_K^{-1}\boldsymbol{x}_*'\boldsymbol{x}_*. \tag{9.36}$$

According to (9.32), the relative efficiency becomes

$$\text{eff}(\boldsymbol{b}_c, \hat{\beta}(\boldsymbol{x}_*), \boldsymbol{S}_c) = 1 - \frac{1}{K}\frac{\boldsymbol{x}_*'\boldsymbol{S}_c^{-1}\boldsymbol{x}_*}{1 + \boldsymbol{x}_*'\boldsymbol{S}_c^{-1}\boldsymbol{x}_*} = 1 - \frac{1}{K}\frac{\sum \mu_j^{-1}(\boldsymbol{x}_*'\boldsymbol{\gamma}_j)^2}{1 + \sum \mu_j^{-1}(\boldsymbol{x}_*'\boldsymbol{\gamma}_j)^2} \le 1 \tag{9.37}$$

and hence

$$1 - \frac{\mu_1\mu_K^{-1}\boldsymbol{x}_*'\boldsymbol{x}_*}{K(\mu_1 + \boldsymbol{x}_*'\boldsymbol{x}_*)} \le \text{eff}(\boldsymbol{b}_c, \hat{\beta}(\boldsymbol{x}_*), \boldsymbol{S}_c) \le 1 - \frac{\boldsymbol{x}_*'\boldsymbol{x}_*}{K(\mu_1\mu_K^{-1})(\mu_K + \boldsymbol{x}_*'\boldsymbol{x}_*)}. \tag{9.38}$$

The relative efficiency of $\boldsymbol{b}_c$ in comparison to $\hat{\beta}(\boldsymbol{x}_*)$ is dependent on the vector $\boldsymbol{x}_*$ (or rather its quadratic norm $\boldsymbol{x}_*'\boldsymbol{x}_*$), as well as on the eigenvalues of the matrix $\boldsymbol{S}_c$, especially on the so-called condition number μ_1/μ_K and the span $(\mu_1 - \mu_K)$ between the largest and the smallest eigenvalue.

Let $\boldsymbol{x}_* = g\boldsymbol{\gamma}_i$ $(i = 1, \ldots, K)$, where g is a scalar and $\boldsymbol{M} = \text{diag}(\mu_1, \ldots, \mu_K)$. For these $\boldsymbol{x}_*$-vectors, which are parallel to the eigenvectors of $\boldsymbol{S}_c$, the quadratic risk of the estimators $\hat{\beta}(g\boldsymbol{\gamma}_i)$ becomes

$$\sigma^{-2}R(\hat{\beta}(g\boldsymbol{\gamma}_i), \beta, \boldsymbol{S}_c) = \text{tr}\{\boldsymbol{\Gamma}\boldsymbol{M}\boldsymbol{\Gamma}'(\boldsymbol{\Gamma}\boldsymbol{M}\boldsymbol{\Gamma}' + g^2\boldsymbol{\gamma}_i\boldsymbol{\gamma}_i')^{-1}\} = K - 1 + \frac{\mu_i}{\mu_i + g^2}. \tag{9.39}$$

Hence, the relative efficiency of $\boldsymbol{b}_c$ reaches its maximum if $\boldsymbol{x}_*$ is parallel to $\boldsymbol{\gamma}_1$ (eigenvector corresponding to the maximum eigenvalue μ_1). Therefore, the loss in efficiency by removing one row $\boldsymbol{x}_*$ is minimal for $\boldsymbol{x}_* = g\boldsymbol{\gamma}_1$ and maximum for $\boldsymbol{x}_* = g\boldsymbol{\gamma}_K$. This fact corresponds to the result of Silvey (1969), namely that the goodness-of-fit of the OLSE can be improved, if additional observations are taken in the direction which was most imprecise. This is just the direction of the eigenvector corresponding to the minimal eigenvalue μ_K of $\boldsymbol{S}_c$.

9.3.2 Standard Methods for Incomplete X-Matrices

(i) Complete case analysis

The idea of the first method is to confine the analysis to the completely observed submodel (9.18). The corresponding estimator of β is $\boldsymbol{b}_c = \boldsymbol{S}_c^{-1}\boldsymbol{X}_c'\boldsymbol{y}_c$ (9.21), which is unbiased and has the covariance matrix $\text{V}(\boldsymbol{b}_c) = \sigma^2\boldsymbol{S}_c^{-1}$. Using the estimator $\boldsymbol{b}_c$ is only feasible for a small percentage of missing or incomplete rows in $\boldsymbol{X}_*$, that is for $\left(\frac{T-m}{T}\right) \cdot 100\%$ at the most, and assumes that MAR holds. The assumption of MAR may not be tenable if, for instance, too many rows in $\boldsymbol{X}_*$ are parallel to the eigenvector $\boldsymbol{\gamma}_K$ corresponding to the eigenvalue μ_K of $\boldsymbol{S}_c$.

(ii) Zero-order regression (ZOR)

This method by Wilks (1932), also called the method of sample means, replaces a missing value x_{ij} of the jth regressor X_j by the sample mean of the observed values of X_j. Denote the index sets of the missing values of X_j by

$$\Phi_j = \{i : x_{ij} \text{ missing}\}, \quad j = 1, \ldots, K \tag{9.40}$$

and let M_j be the number of elements in Φ_j. Then for j fixed, any missing value x_{ij} in $\boldsymbol{X}_*$ is replaced by

$$\hat{x}_{ij} = \bar{x}_j = \frac{1}{T - M_j} \sum_{i \notin \Phi_j} x_{ij}. \tag{9.41}$$

This method may be recommended, as long as the sample mean is a good estimator for the mean of the jth column. If somehow the data in the jth column is trended or follows a growth curve, then $\bar{x}_j$ is not a good estimator and hence replacing missing values by $\bar{x}_j$ may cause a bias. If all the missing values x_{ij} are replaced by the corresponding column means $\bar{x}_j$ $(j = 1, \ldots, K)$, then the matrix $\boldsymbol{X}_*$ results in a – now completely known – matrix $\boldsymbol{X}_{(1)}$. Hence, an operationalized version of the mixed model (9.17) is

$$\begin{pmatrix} \boldsymbol{y}_c \\ \boldsymbol{y}_* \end{pmatrix} = \begin{pmatrix} \boldsymbol{X}_c \\ \boldsymbol{X}_{(1)} \end{pmatrix} \boldsymbol{\beta} + \begin{pmatrix} \boldsymbol{\epsilon} \\ \boldsymbol{\epsilon}_{(1)} \end{pmatrix}. \tag{9.42}$$

For the vector of errors $\boldsymbol{\epsilon}_{(1)}$, we have

$$\boldsymbol{\epsilon}_{(1)} = (\boldsymbol{X}_* - \boldsymbol{X}_{(1)})\boldsymbol{\beta} + \boldsymbol{\epsilon}_* \tag{9.43}$$

with

$$\boldsymbol{\epsilon}_{(1)} \sim \{(\boldsymbol{X}_* - \boldsymbol{X}_{(1)})\boldsymbol{\beta}, \sigma^2 \boldsymbol{I}_J\} \tag{9.44}$$

and $J = (T - m)$.

In general, replacing missing values can result in a biased mixed model, since $(\boldsymbol{X}_* - \boldsymbol{X}_{(1)}) \neq \boldsymbol{0}$ holds. If $\boldsymbol{X}$ is a matrix of stochastic regressor variables, then at the most one may expect that $\mathrm{E}(\boldsymbol{X}_* - \boldsymbol{X}_{(1)}) = \boldsymbol{0}$ holds.

(iii) First-order regression (FOR)

This term comprises a set of methods, which make use of the structure of the matrix $\boldsymbol{X}$ by setting up additional regressions. Based on the index sets Φ_j in (9.40), the dependence of each column $\boldsymbol{x}_j$ $(j = 1, \ldots, K, j$ fixed) on the other columns is modelled according to the following relationship:

$$x_{ij} = \theta_{0j} + \sum_{\substack{\mu=1 \\ \mu \neq j}}^{K} x_{i\mu} \theta_{\mu j} + u_{ij}, \quad i \notin \Phi = \bigcup_{j=1}^{K} \Phi_j. \tag{9.45}$$

The missing values x_{ij} in $\boldsymbol{X}_*$ are estimated and replaced by

$$\hat{x}_{ij} = \hat{\theta}_{0j} + \sum_{\substack{\mu=1 \\ \mu \neq j}}^{K} x_{i\mu} \hat{\theta}_{\mu j} \quad (i \in \Phi_j). \tag{9.46}$$

(iv) Correlation methods for stochastic X

In case of stochastic regressors $X_1, \ldots, X_K$ (or $X_2, \ldots, X_K$, if $X_1 = 1$), the vector β is estimated by solving the normal equations

$$\mathrm{Cov}(x_i, x_j)\hat{\beta} = \mathrm{Cov}(x_i, y) \quad (i, j = 1, \ldots, K), \tag{9.47}$$

where $\mathrm{Cov}(x_i, x_j)$ is the $K \times K$-sample covariance matrix. The (i, j)th element of $\mathrm{Cov}(x_i, x_j)$ is calculated from the pairwise observed elements of the variables X_i and X_j. Similarily, $\mathrm{Cov}(x_i, y)$ makes use of pairwise observed elements of x_i and y. Since this method frequently leads to unsatisfactory results, we will not deal with this method any further. Based on simulation studies, Haitovsky (1968) concludes that in most situations the complete case estimator b_c is superior to the correlation method.

Maximum-Likelihood Estimates of Missing Values

Suppose that the errors are normally distributed, i.e., $\epsilon \sim N(0, \sigma^2 I_T)$. Moreover, assume a so-called monotone pattern of missing values, which enables a factorization of the likelihood (cf. Little and Rubin, 1987). We confine ourselves to the most simple case and assume that the matrix X_* is completely unobserved. This requires a model which contains no constant. Then X_* in the mixed model (9.17) may be treated as an unknown parameter. The log-likelihood corresponding to the estimators of the unknown parameters β, σ^2 and the "parameter" X_* may be written as

$$\ln L(\beta, \sigma^2, X_*) = -\frac{n}{2}\ln(2\pi) - \frac{n}{2}\ln(\sigma^2)$$
$$-\frac{1}{2\sigma^2}(y_c - X_c\beta, y_* - X_*\beta)' \begin{pmatrix} y_c - X_c\beta \\ y_* - X_*\beta \end{pmatrix}. \tag{9.48}$$

Differentiating with respect to β, σ^2, and X_* leads to the following normal equations

$$\frac{\partial \ln L}{\partial \beta} = \frac{1}{2\sigma^2}\{X_c'(y_c - X_c\beta) + X_*'(y_* - X_*\beta)\} = 0, \tag{9.49}$$

$$\frac{\partial \ln L}{\partial \sigma^2} = \frac{1}{2\sigma^2}\{-n + \frac{1}{\sigma^2}(y_c - X_c\beta)'(y_c - X_c\beta)$$
$$+ \frac{1}{\sigma^2}(y_* - X_*\beta)'(y_* - X_*\beta)\} = 0 \tag{9.50}$$

and

$$\frac{\partial \ln L}{\partial X_*} = \frac{1}{2\sigma^2}(y_* - X_*\beta)\beta' = 0. \tag{9.51}$$

This results in the ML estimators for β and σ^2

$$\hat{\beta} = b_c = S_c^{-1}X_c'y_c \;, \tag{9.52}$$

$$\hat{\sigma}^2 = \frac{1}{m}(y_c - X_c b_c)'(y_c - X_c b_c), \tag{9.53}$$

which are only based on the complete submodel (9.18). Hence, the ML estimator $\hat{X}_*$ is solution (cf. (9.36) with $\hat{\beta} = b_c$) of

$$y_* = \hat{X}_* b_c. \tag{9.54}$$

Only if $K = 1$, the solution is unique

$$\hat{x}_* = \frac{y_*}{b_c} \tag{9.55}$$

where $b_c = (x'_c x_c)^{-1} x'_c y_c$ (cf. Kmenta, 1971). For $K > 1$, a $J \times (K-1)$-fold set of solutions $\hat{X}_*$ exists. If any solution $\hat{X}_*$ of (9.39) is substituted for X_* in the mixed model, i.e.,

$$\begin{pmatrix} y_c \\ y_* \end{pmatrix} = \begin{pmatrix} X_c \\ \hat{X}_* \end{pmatrix} \beta + \begin{pmatrix} \epsilon_c \\ \epsilon_* \end{pmatrix}, \tag{9.56}$$

then the following identity holds

$$\begin{aligned} \hat{\beta}(\hat{X}_*) &= (S_c + \hat{X}'_* \hat{X}_*)^{-1}(X'_c y_c + \hat{X}'_* y_*) \\ &= (S_c + \hat{X}'_* \hat{X}_*)^{-1}(S_c \beta + X'_c \epsilon_c + \hat{X}'_* \hat{X}_* \beta + \hat{X}'_* \hat{X}_* S_c^{-1} X'_c \epsilon_c) \\ &= \beta + (S_c + \hat{X}'_* \hat{X}_*)^{-1}(S_c + \hat{X}'_* \hat{X}_*) S_c^{-1} X'_c \epsilon_c \\ &= \beta + S_c^{-1} X'_c \epsilon_c \\ &= b_c. \end{aligned} \tag{9.57}$$

Remark: The OLSE $\hat{\beta}(\hat{X}_*)$ in the model filled up with the ML estimator $\hat{X}_*$ equals the OLSE b_c in the submodel with the incomplete observations. This is true for other monotone patterns as well.

On the other hand, if the pattern is not monotone, then the ML-equations have to be solved by iterative procedures as for example the EM-algorithm by Dempster et al. (1977) (cf. algorithms by Oberhofer and Kmenta, 1974).

Further discussions of the problem of estimating missing values can be found in Little and Rubin (1987), Weisberg (1980) and Toutenburg (1992, Chapter 8). Toutenburg et al. (1995) propose a unique solution of the normal equation (9.49) according to

$$\min_{\hat{X}_*, \lambda} \{|S_c + \hat{X}'_* \hat{X}_*|^{-1} - 2\lambda'(y_* - \hat{X}_* b_c)\}. \tag{9.58}$$

The solution is

$$\hat{X}_* = \frac{y_* y'_c X_c}{y'_c x_X S_c^{-1} X'_c y_c}. \tag{9.59}$$

9.4   Adjusting for Missing Data in a 2×2 Cross-Over Design

In Chapter 8, procedures for testing a 2×2 cross-over design were introduced for continuous response. In practice, small sample sizes are an important factor for the employment of the cross-over design. Hence, for studies of this kind, it is especially important to use all available information and to include the data of incomplete observations in the analysis as well.

9.4.1   Notation

We assume that data is only missing for the second period of treatment. Moreover, we assume that the response (y_{i1k}, y_{i2k}) of group i is ordered, so that the first m_i pairs represent the complete data sets. The last $(n_i - m_i)$ pairs are then the incomplete pairs of response. The first m_i values of the response of period j, which belong to complete observation pairs of group i, are now stacked in the vector

$$\boldsymbol{y}_{ij}' = (y_{ij1}, \ldots, y_{ijm_i}). \tag{9.60}$$

Those observations of the first period which are assigned to incomplete response pairs are denoted by

$$\boldsymbol{y}_{i1}^{*\prime} = (y_{i1(m_i+1)}, \ldots, y_{i1n_i}) \tag{9.61}$$

for group i. The $m \times 2$ data matrix $\boldsymbol{Y}$ of the complete data and the $(n - m) \times 1$ vector $\boldsymbol{y}_1^*$ of the incomplete data can now be written as

$$\boldsymbol{Y} = \begin{pmatrix} \boldsymbol{y}_{11} & \boldsymbol{y}_{12} \\ \boldsymbol{y}_{21} & \boldsymbol{y}_{22} \end{pmatrix}, \qquad \boldsymbol{y}_1^* = \begin{pmatrix} \boldsymbol{y}_{11}^* \\ \boldsymbol{y}_{21}^* \end{pmatrix}, \tag{9.62}$$

with $m = m_1 + m_2$ and $n = n_1 + n_2$. Additionally, we assume that

$$(y_{i1k}, y_{i2k}) \overset{i.i.d.}{\sim} N\left((\mu_{i1}, \mu_{i2}), \Sigma\right) \quad \text{for } k = 1, \ldots, m_i \quad,$$
$$y_{i1k} \overset{iid}{\sim} N(\mu_{i1}, \sigma_{11}^2) \qquad \text{for } k = m_i + 1, \ldots, n_i \quad. \tag{9.63}$$

Here Σ denotes the covariance matrix

$$\Sigma = \begin{pmatrix} \sigma_{11} & \sigma_{12} \\ \sigma_{21} & \sigma_{22} \end{pmatrix} \tag{9.64}$$

with

$$\sigma_{jj'} = \text{Cov}\,(y_{ijk}, y_{ij'k}) \tag{9.65}$$

and hence $\sigma_{11} = \text{Var}\,(y_{i1k})$ and $\sigma_{22} = \text{Var}\,(y_{i2k})$. The correlation coefficient ρ can now be written as

$$\rho = \frac{\sigma_{12}}{\sqrt{\sigma_{11}\sigma_{22}}}. \tag{9.66}$$

Additionally, we assume that the rows of the matrix Y are independent of the rows of the vector y_1^*. The entire sample can now be described by the two vectors $u' = (y_{11}', y_{21}', y_1^{*\prime})$ and $v' = (y_{12}', y_{22}')$. Hence, the $n \times 1$ vector u represents the observations of the first period and the $m \times 1$ vector v those of the second period. Since we interpret the observed response pairs as independent realizations of a random sample of a bivariate normal distribution, we can express the density function of (u, v) as the product of the marginal density of u and the conditional density of v given u. The density function of u is

$$f_u = \left(\frac{1}{\sqrt{2\pi\sigma_{11}}}\right)^n \exp\left(-\frac{1}{2\sigma_{11}} \sum_{i=1}^{2} \sum_{k=1}^{n_i} (y_{i1k} - \mu_{i1})^2\right) \tag{9.67}$$

and the conditional density of v given u is

$$f_{v|u} = \left(\frac{1}{\sqrt{2\pi\sigma_{22}(1-\rho^2)}}\right)^m \exp\left(-\frac{1}{2\sigma_{22}(1-\rho^2)} \sum_{i=1}^{2} \sum_{k=1}^{m_i} \left(y_{i2k} - \mu_{i2} - \rho\sqrt{\tfrac{\sigma_{22}}{\sigma_{11}}}(y_{i1k} - \mu_{i1})\right)^2\right). \tag{9.68}$$

The joint density function $f_{u,v}$ of (u, v) is now

$$f_{u,v} = f_u f_{v|u}. \tag{9.69}$$

9.4.2 Maximum-Likelihood Estimator (Patel, 1985)

We now estimate the unknown parameters $\mu_{11}, \mu_{21}, \mu_{12}$ and μ_{22}, as well as the unknown components $\sigma_{jj'}$ of the covariance matrix Σ. The log-likelihood is $\ln L = \ln f_u + \ln f_{v|u}$ with

$$\ln f_u = -\frac{n}{2}\ln(2\pi\sigma_{11}) - \frac{1}{2\sigma_{11}} \sum_{i=1}^{2} \sum_{k=1}^{n_i} (y_{i1k} - \mu_{i1})^2 \tag{9.70}$$

and

$$\ln f_{v|u} = -\frac{m}{2}\ln(2\pi\sigma_{22}(1-\rho^2)) - \frac{1}{2\sigma_{22}(1-\rho^2)} \sum_{i=1}^{2} \sum_{k=1}^{m_i} \left(y_{i2k} - \mu_{i2} - \rho\sqrt{\tfrac{\sigma_{22}}{\sigma_{11}}}(y_{i1k} - \mu_{i1})\right)^2 \tag{9.71}$$

Let us introduce the following notation

$$\sigma^* = \sigma_{22}(1 - \rho^2) \ , \tag{9.72}$$

$$\beta = \rho\sqrt{\frac{\sigma_{22}}{\sigma_{11}}} \ , \tag{9.73}$$

$$\mu_{i2}^* = \mu_{i2} - \beta\mu_{i1} \ . \tag{9.74}$$

(9.71) can now be transformed, and we get

$$\ln f_{v|u} = -\frac{m}{2}\ln(2\pi\sigma^*) - \frac{1}{2\sigma^*} \sum_{i=1}^{2} \sum_{k=1}^{m_i} \left(y_{i2k} - \mu_{i2}^* - \beta y_{i1k}\right)^2. \tag{9.75}$$

This leads to a factorization of the log-likelihood into two terms (9.70) and (9.75), where no two of the unknown parameters $\mu_{11}, \mu_{21}, \mu_{12}^*, \mu_{22}^*, \sigma_{11}, \sigma^*$ and β show up in one summand at the same time. Hence maximization of the log-likelihood can be done independently for the unknown parameters and we find the maximum-likelihood estimates

$$\left.\begin{aligned}
\hat{\mu}_{i1} &= y_{i1\cdot}^{(n_i)} \ , \\
\hat{\mu}_{i2} &= y_{i2\cdot}^{(m_i)} + \hat{\beta}\left(\hat{\mu}_{i1} - y_{i1\cdot}^{(m_i)}\right) \ , \\
\hat{\beta} &= \frac{s_{12}}{s_{11}} \ , \\
\hat{\sigma}_{11} &= \frac{1}{n}\sum_{i=1}^{2}\sum_{k=1}^{n_i}\left(y_{i1k} - \hat{\mu}_{i1}\right)^2 \ , \\
\hat{\sigma}_{22} &= s_{22} + \hat{\beta}^2\left(\hat{\sigma}_{11} - s_{11}\right) \ , \\
\hat{\sigma}_{12} &= \hat{\beta}\hat{\sigma}_{11} \ .
\end{aligned}\right\} \qquad (9.76)$$

If we write

$$y_{ij\cdot}^{(c)} = \frac{1}{a}\sum_{k=1}^{a} y_{ijk},$$

$$s_{jj'} = \frac{1}{m_1 + m_2}\sum_{i=1}^{2}\sum_{k=1}^{m_i}\left(y_{ijk} - y_{ij\cdot}^{(m_i)}\right)\left(y_{ij'k} - y_{ij'\cdot}^{(m_i)}\right), \qquad (9.77)$$

then $\hat{\beta}$ and $\hat{y}_{ij\cdot}^{(c)}$ are independent for $a = n_i, m_i$. Consequently, the covariance matrix $\mathbf{\Gamma}_i = ((\gamma_{i,uv}))$ of $(\hat{\mu}_{i1}, \hat{\mu}_{i2})$ is

$$\mathbf{\Gamma}_i = \begin{pmatrix} \dfrac{\sigma_{11}}{n_i} & \dfrac{\sigma_{12}}{n_i} \\ \dfrac{\sigma_{12}}{n_i} & \dfrac{\sigma_{22} + \left(1 - \frac{m_i}{n_i}\right)\sigma_{11}\left(Var(\hat{\beta}) - \beta^2\right)}{m_i} \end{pmatrix} \qquad (9.78)$$

with

$$Var\left(\hat{\beta}\right) = E\left(Var\left(\hat{\beta}|y_1\right)\right) = \frac{\sigma_{22}(1 - \hat{\rho}^2)}{\sigma_{11}(m - 4)}, \qquad (9.79)$$

$$\hat{\rho} = \hat{\beta}\sqrt{\frac{\hat{\sigma}_{11}}{\hat{\sigma}_{22}}}. \qquad (9.80)$$

9.4.3 Test Procedures

We now develop test procedures for large and small sample sizes and formulate the hypotheses $H_0^{(1)}$: no interaction, $H_0^{(2)}$: no treatment effect and $H_0^{(3)}$: no effect of the period:

$$H_0^{(1)}: \quad \theta_1 = \mu_{11} + \mu_{12} - \mu_{21} - \mu_{22} = 0 \ , \qquad (9.81)$$

$$H_0^{(2)}: \quad \theta_2 = \mu_{11} - \mu_{12} - \mu_{21} + \mu_{22} = 0 \ , \qquad (9.82)$$

$$H_0^{(3)}: \quad \theta_2 = \mu_{11} - \mu_{12} + \mu_{21} - \mu_{22} = 0 \ . \qquad (9.83)$$

Large Samples

The estimates (9.76) lead to the maximum-likelihood estimate $\hat{\theta}_1$ of θ_1. For large sample sizes m_1 and m_2, the distribution of Z_1 defined by

$$Z_1 = \frac{\hat{\theta}_1}{\sqrt{\sum\limits_{i=1}^{2} (\tilde{\gamma}_{i,11} + 2\tilde{\gamma}_{i,12} + \tilde{\gamma}_{i,22})}}, \tag{9.84}$$

can be approximated by the $N(0,1)$-distribution if $H_0^{(1)}$ holds. Here $\tilde{\gamma}_{i,uv}$ denote the estimates of the elements of the covariance matrix Γ_i. These are found by replacing $\hat{\sigma}_{11}$ (9.76) and $s_{jj'}$ (9.77) by their unbiased estimates

$$\tilde{\sigma}_{11} = \frac{n}{n-2}\hat{\sigma}_{11}, \tag{9.85}$$

$$\tilde{s}_{jj'} = \frac{m}{m-2}s_{jj'}. \tag{9.86}$$

The maximum-likelihood estimate $\hat{\theta}_2$ for θ_2 is derived from the estimates in (9.76). The test statistic Z_2 given by

$$Z_2 = \frac{\hat{\theta}_2}{\sqrt{\sum\limits_{i=1}^{2} (\tilde{\gamma}_{i,11} - 2\tilde{\gamma}_{i,12} + \tilde{\gamma}_{i,22})}} \tag{9.87}$$

is approximatively $N(0,1)$ distributed for large samples m_1 and m_2 under $H_0^{(2)}$. Analogously, we find the distribution of the test statistic Z_3

$$Z_3 = \frac{\hat{\theta}_3}{\sqrt{\sum\limits_{i=1}^{2} (\tilde{\gamma}_{i,11} - 2\tilde{\gamma}_{i,12} + \tilde{\gamma}_{i,22})}} \tag{9.88}$$

and construct the maximum-likelihood estimate $\hat{\theta}_3$ for θ_3.

Small Samples

For small sample sizes m_1 and m_2, Patel (1985) suggests to approximate the distribution of Z_1 by a t-distribution with $v_1 = \frac{1}{2}(n + m - 5)$ degrees of freedom. The choice of v_1 degrees of freedom is explained by Patel as follows: The estimates of the variances σ_{11} and σ^* ($\hat{\sigma}^* = s_{22} - \hat{\beta}s_{12}$) are based on $(n-2)$ and $(n-3)$ degrees of freedom, and their mean is $v_1 = \frac{1}{2}(n + m - 5)$. If there are no missing values in the second period ($n = m$) then a t-distribution with $(n - 2)$ degrees of freedom should be chosen. This test then corresponds to the previously introduced test based on T_λ (8.19).

Patel chooses a t-distribution with $v_2 = (m - 2)$ degrees of freedom for the approximation of the distribution of Z_2 and Z_3. Here, Patel refers to the results of Morrison (1973). Morrison constructs a test for a comparison of the means of a bivariate normal distribution for missing values in one variable at the most. Morrison derives the test statistic from the maximum-likelihood estimate and specifies its distribution as a t-distribution, where the degrees of freedom are only dependent on the number of completely observed response pairs. These tests are equivalent to the tests in Section 8.1.3 if no data is missing.

Example 9.2: In Example 8.1, patient 2 in group 2 was identified as an outlier. We now want to check, to what extent the estimates of the effects vary when the observation of this patient in the second period is excluded from the analysis. We reorganize the data so that patient 2 in group 2 comes last

Group 1		Group 2	
A	B	B	A
20	30	30	20
40	50	20	10
30	40	30	10
20	40	40	—

Summarizing in matrix notation (cf. (9.62)) we have

$$Y = \begin{pmatrix} 20 & 30 \\ 40 & 50 \\ 30 & 40 \\ 20 & 40 \\ \hline 30 & 20 \\ 20 & 10 \\ 30 & 10 \end{pmatrix}, \qquad y_1^* = (40). \tag{9.89}$$

The unbiased estimates are calculated with $n_1 = 4, n_2 = 4, m_1 = 4$ and $m_2 = 3$ by inserting (9.85) and (9.86) in (9.76). We calculate:

$$y_{11.}^{(n_1)} = \frac{1}{4}(20 + 40 + 30 + 20) = 27.50,$$

$$y_{11.}^{(m_1)} = \frac{1}{4}(20 + 40 + 30 + 20) = 27.50,$$

$$y_{12.}^{(m_1)} = \frac{1}{4}(30 + 50 + 40 + 40) = 40.00,$$

$$y_{21.}^{(n_2)} = \frac{1}{4}(30 + 20 + 30 + 40) = 30.00,$$

$$y_{21.}^{(m_2)} = \frac{1}{3}(30 + 20 + 30) = 26.67,$$

$$y_{22.}^{(m_1)} = \frac{1}{3}(20 + 10 + 10) = 13.33$$

and

$$\tilde{s}_{11} = \frac{1}{7-2}[(20-27.50)^2 + \cdots + (20-27.50)^2$$
$$+ (30-26.67)^2 + \cdots + (30-26.67)^2] = 68.33,$$

$$\tilde{s}_{22} = \frac{1}{7-2}[(30-40.00)^2 + \cdots + (40-40.00)^2 +$$
$$+ (20-13.33)^2 + (10-13.33)^2 + (10-13.33)^2] = 53.33,$$

$$\tilde{s}_{12} = \frac{1}{7-2}[(20-27.50)(30-40) + \cdots + (20-27.50)(40-40) +$$
$$+ (30-26.67)(20-13.33) + \cdots + (30-26.67)(10-13.33)]$$
$$= 46.67,$$

$$\tilde{s}_{21} = \tilde{s}_{12}.$$

With

$$\hat{\beta} = \frac{\tilde{s}_{12}}{\tilde{s}_{11}} = \frac{53.33}{68.33} = 0.68$$

we find

$$\hat{\mu}_{11} = y_{11\cdot}^{(n_1)} = 27.50,$$
$$\hat{\mu}_{21} = y_{21\cdot}^{(n_2)} = 30.00,$$
$$\hat{\mu}_{12} = 40.00 + 0.68 \cdot (27.50 - 27.50) = 40.00,$$
$$\hat{\mu}_{22} = 13.33 + 0.68 \cdot (30.00 - 26.67) = 15.61$$

and with

$$\tilde{\sigma}_{11} = \frac{1}{8-2}[(20-27.50)^2 + \cdots + (20-27.50)^2 +$$
$$+ (30-30)^2 + \cdots + (30-30)^2] = 79.17,$$

$$\tilde{\sigma}_{22} = 53.33 + 0.68^2 \cdot (79.17 - 68.33) = 58.39,$$
$$\tilde{\sigma}_{12} = 0.68 \cdot 79.17 = 54.07,$$
$$\tilde{\sigma}_{21} = \tilde{\sigma}_{12}$$

we get

$$\hat{\rho} = 0.68 \cdot \sqrt{\frac{79.17}{58.39}} = 0.80 \qquad [\text{cf. (9.80)}],$$

$$\widehat{\text{Var}}(\hat{\beta}) = \frac{58.39 \cdot (1 - 0.80^2)}{79.17 \cdot (7-4)} = 0.09 \qquad [\text{cf. (9.79)}].$$

We now determine the two covariance matrices (9.78)

$$\Gamma_1 = \begin{pmatrix} \frac{79.17}{4} & \frac{54.07}{4} \\ \frac{54.07}{4} & \frac{58.39 + \left(1 - \frac{4}{4}\right) \cdot 79.17 \cdot \left(0.09 - 0.68^2\right)}{4} \end{pmatrix}$$

$$= \begin{pmatrix} 19.79 & 13.52 \\ 13.52 & 14.60 \end{pmatrix},$$

$$\Gamma_2 = \begin{pmatrix} 19.79 & 13.52 \\ 13.52 & 16.98 \end{pmatrix}.$$

Finally, our test statistics are

$$\text{interaction} \quad Z_1 = \frac{21.89}{11.19} = 1.96 \qquad [\text{5 degrees of freedom}]$$

$$\text{treatment} \quad Z_2 = \frac{-26.89}{4.13} = -6.50 \quad [\text{5 degrees of freedom}]$$

$$\text{period} \quad Z_3 = \frac{1.89}{4.13} = 0.46 \qquad [\text{5 degrees of freedom}]$$

The following table shows a comparison with the results of the analysis of the complete data set:

	Complete			Incomplete		
	t	df	p-value	t	df	p-value
Carry-over	0.96	6	0.376	1.96	5	0.108
Treatment	-2.96	6	0.026	-6.50	5	0.001
Period	0.74	6	0.488	0.46	5	0.667

An interesting result is that by excluding the second observation of patient 2, the treatment effect achieves an even higher level of significance of $p = 0.001$ (compared to $p = 0.026$ before). However, the carry-over effect of $p = 0.108$ is now very close to the limit of significance of $p = 0.100$ proposed by Grizzle. This is easily seen in the difference-response-total-plot (Figure 9.3), which shows a clear separation of the covering in horizontal as well as vertical direction (cf. Figure 8.5).

9.5 Missing Categorical Data

The procedures which were introduced so far are all based on the linear regression model (9.1) with one continuous endogeneous variable Y. In many applications however, this assumption does not hold. Often Y is defined as a binary response variable and hence has a binomial distribution. Because of this, statistical analysis of incompletely observed categorical data demands different procedures than those previously described. For a clear and understandable representation of the different procedures, a three-dimensional contingency table is chosen where only one of the three categorical variables is assumed to be observed incompletely.

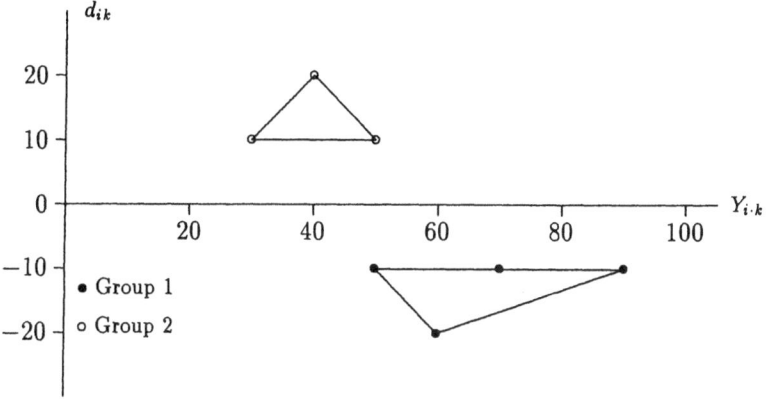

Figure 9.3: Difference-response-total-plot of the incomplete data set

9.5.1 Introduction

Let Y be a binary outcome variable and X_1, X_2 two covariates with J and K categories. The contingency table is thus of the dimension $2 \times J \times K$. We assume that only X_2 is observed incompletely. The response of the covariate X_2 is indicated by an additional variable

$$R_2 = \begin{cases} 1 & \text{if } X_2 \text{ is not missing} \\ 0 & \text{if } X_2 \text{ is missing .} \end{cases} \tag{9.90}$$

This leads to a new random variable

$$Z_2 = \begin{cases} X_2 & \text{if } R_2 = 1 \\ K+1 & \text{if } R_2 = 0 . \end{cases} \tag{9.91}$$

Assume that Y is related to X_1 and X_2 by the logistic model, a generalized linear model with logit link. This model assesses the effects of the covariates X_1 and X_2 on the outcome variable Y.

Let $\mu_{i|jk} = P(Y = i \mid X_1 = j, X_2 = k)$ be the conditional distribution of the binary variable Y, given the values of the covariates X_1 and X_2. The logistic model without interaction is

$$\ln\left(\frac{\mu_{1|jk}}{1 - \mu_{1|jk}}\right) = \beta_0 + \beta_{1j} + \beta_{2k} \tag{9.92}$$

or

$$\mu_{1|jk} = \frac{\exp(\beta_0 + \beta_{1j} + \beta_{2k})}{1 + \exp(\beta_0 + \beta_{1j} + \beta_{2k})} . \tag{9.93}$$

The parameters β_{1j} and β_{2k} describe the effect of the jth category of X_1 and the kth category of X_2 on the outcome variable Y. The parameter vector $\beta' = (\beta_0, \beta_{11}, \ldots, \beta_{1J}, \beta_{21}, \ldots, \beta_{2K})$ is estimated by the maximum-likelihood approach.

9.5.2 Maximum-Likelihood Estimation in the Complete Data Case

Let $\pi_{ijk}^* = P(Y = i, X_1 = j, X_2 = k)$ be the joint distribution of the three variables for the complete data case and define

$$
\begin{aligned}
\gamma_{k|j} &= P(X_2 = k \mid X_1 = j) \\
\tau_j &= P(X_1 = j).
\end{aligned}
\tag{9.94}
$$

This parametrization allows a factorization of the joint distribution of Y, X_1 and X_2:

$$
\begin{aligned}
\pi_{ijk}^* &= \mu_{i|jk}\, \gamma_{k|j}\, \tau_j \\
&= \left(\mu_{1|jk}\right)^i \left(1 - \mu_{1|jk}\right)^{1-i} \gamma_{k|j}\, \tau_j .
\end{aligned}
\tag{9.95}
$$

The contribution of a single observation with the values $Y = i, X_1 = j, X_2 = k$ to the log-likelihood is

$$
\ln\left(\left(\mu_{1|jk}\right)^i \left(1 - \mu_{1|jk}\right)^{1-i}\right) + \ln \gamma_{k|j} + \ln \tau_j .
\tag{9.96}
$$

Hence, the log-likelihood is additive in the parameters and can be maximized independently for β, γ and τ. The maximum-likelihood estimate of β results from maximizing the log-likelihood of the entire sample

$$
l_n^*(\beta) = \sum_{i=0}^{1} \sum_{j=1}^{J} \sum_{k=1}^{K} n_{ijk}^*\, l^*(\beta; i, j, k)
\tag{9.97}
$$

with

$$
l^*(\beta; , i, j, k) = \ln\left(\left(\mu_{1|jk}\right)^i \left(1 - \mu_{1|jk}\right)^{1-i}\right),
$$

where n_{ijk}^* is the number of elements with $Y = i, X_1 = j$ and $X_2 = k$. However, these equations are nonlinear in β, and hence the maximization task involves an iterative method. A standard procedure for nonlinear optimization is the Newton-Raphson method or one of its variants, like the Fisher-scoring method.

9.5.3 Ad-hoc Methods

Complete Case Analysis

Similar to the previously described situation with continuous variables, the complete case analysis is a standard approach for incomplete categorical data as well: the incompletely observed cases are eliminated from the data set. This reduced sample can now be analysed by the maximum-likelihood approach for completely observed contingency tables (cf. Section 9.5.2).

Filling the Contingency Table

Unlike imputation methods that fill up the gaps in the data set (cf. Section 9.1), the filling method by Vach and Blettner (1991) fills up the cells of the contingency table. This is done by distributing the elements with a missing value of X_2, that is, with the value $Z_2 = K+1$, to the other cells, dependent on the (known) values of Y and X_1.

Let n_{ijk} be the number of elements with the values $Y = i$, $X_1 = j$ and $Z_2 = k$, that is the cell counts of the $2 \times J \times (K+1)$ contingency table. The filled up contingency table is then

$$n_{ijk}^{FILL} = n_{ijk} + n_{ijK+1} \frac{n_{ijk}}{\sum_{k=1}^{K} n_{ijk}}. \tag{9.98}$$

To this new $2 \times J \times K$-table, the maximum-likelihood procedure for completely observed contingency tables is applied, according to Section 9.5.2.

9.5.4   Model-Based Methods

Maximum-Likelihood Estimation in the Incomplete Data Case

Let $\pi_{ijk} = P(Y = i, X_1 = j, Z_2 = k)$ be the joint distribution of the variables Y, X_1 and Z_2, and define

$$q_{ijk} = P(R_2 = 1 \mid Y = i, X_1 = j, X_2 = k). \tag{9.99}$$

The parametrization (9.94) and (9.99) enables a decomposition of the joint distribution (cf. Vach and Schumacher, 1992, p.7). However, we have to distinguish between the case that the value of X_2 is known:

$$\pi_{ijk} = P(Y = i, X_1 = j, Z_2 = k)$$

$$= P(Y = i, X_1 = j, X_2 = k, R_2 = 1)$$

$$= P(R_2 = 1 \mid Y = i, X_1 = j, X_2 = k)\, P(Y = i \mid X_1 = j, X_2 = k)$$
$$\times P(X_2 = k \mid X_1 = j)\, P(X_1 = j)$$

$$= q_{ijk}\, (\mu_{1|jk})^i (1 - \mu_{1|jk})^{1-i}\, \gamma_{k|j}\, \tau_j. \tag{9.100}$$

and the case that the value of X_2 is missing, i.e., $k = K+1$:

$$\pi_{ijK+1} = P(Y = i, X_1 = j, Z_2 = K+1)$$

$$= P(Y = i, X_1 = j, R_2 = 0)$$

$$= P(R_2 = 0 \mid Y = i, X_1 = j)\, P(Y = i \mid X_1 = j)\, P(X_1 = j)$$

$$= \left(\sum_{k=1}^{K} P(R_2 = 0 \mid Y = i, X_1 = j, X_2 = k)\, P(Y = i \mid X_1 = j, X_2 = k) \right.$$
$$\left. \times P(X_2 = k \mid X_1 = j) \right) P(X_1 = j)$$

$$= \left(\sum_{k=1}^{K} (1 - q_{ijk})\, (\mu_{1|jk})^i (1 - \mu_{1|jk})^{1-i}\, \gamma_{k|j} \right) \tau_j. \tag{9.101}$$

Note that this distribution, unlike to the complete data case, is dependent on the parameter q. Furthermore, the log-likelihood is not additive in the parameters β, γ, τ and q, and hence cannot be maximized seperately for the parameters.

If the missing values are missing at random (MAR), then the missing probability is independent of the true value k of X_2, i.e.,

$$P(R_2 = 1 \mid Y = i, X_1 = j, X_2 = k) \equiv P(R_2 = 1 \mid Y = i, X_1 = j) \qquad (9.102)$$

and thus $q_{ijk} \equiv q_{ij}$. For the joint distribution of Y, X_1 and Z_2 (cf. (9.100) and (9.101)) this leads to

$$\pi_{ijk} = q_{ij} \left(\mu_{1|jk}\right)^{i}(1 - \mu_{1|jk})^{1-i} \gamma_{k|j} \tau_j \qquad (9.103)$$

for $k = 1, \ldots, K$ and to

$$\pi_{ijK+1} = (1 - q_{ij}) \left(\sum_{k=1}^{K}(\mu_{1|jk})^{i}(1 - \mu_{1|jk})^{1-i} \gamma_{k|j}\right) \tau_j \qquad (9.104)$$

for $k = K+1$.

The contribution of a single element to the log-likelihood under the MAR assumption is now

$$\ln q_{ij} + \ln \left((\mu_{1|jk})^{i}(1 - \mu_{1|jk})^{1-i}\right) + \ln \gamma_{k|j} + \ln \tau_j \qquad (9.105)$$

for $k = 1, \ldots, K$ and

$$\ln (1 - q_{ij}) + \ln \left(\sum_{k=1}^{K}(\mu_{1|jk})^{i}(1 - \mu_{1|jk})^{1-i} \gamma_{k|j}\right) + \ln \tau_j \qquad (9.106)$$

for $k = K+1$.

The log-likelihood disintegrates into three summands; hence maximizing the log-likelihood for β can now be done independent of q. If the value of X_2 is missing, it is impossible to split the second summand depending on β and γ any further. Hence maximum-likelihood estimation of β requires joint maximization of the following log-likelihood for (β, γ), where γ is regarded as a nuisance parameter:

$$l_n^{ML}(\beta, \gamma) = \sum_{i=0}^{1}\sum_{j=1}^{J}\sum_{k=1}^{K+1} n_{ijk} \, l^{ML}(\beta, \gamma; i, j, k) \qquad (9.107)$$

with

$$l^{ML}(\beta, \gamma; i, j, k) = \begin{cases} \ln \left((\mu_{1|jk})^{i}(1 - \mu_{1|jk})^{1-i}\right) + \ln \gamma_{k|j} & \text{for } k = 1, \ldots, K \\ \ln \left(\sum_{k=1}^{K}(\mu_{1|jk})^{i}(1 - \mu_{1|jk})^{1-i} \gamma_{k|j}\right) & \text{for } k = K+1, \end{cases}$$

where n_{ijk} is the number of elements with $Y = i$, $X_1 = j$ and $Z_2 = k$.

Analogously to the complete data case, the computation of the estimates of β and γ requires an iterative procedure such as the Fisher-scoring method. Let $\theta = (\beta, \gamma)$. The iteration step of the Fisher-scoring method is

$$\theta^{(t+1)} = \theta^{(t)} + (I_{\theta\theta}^{ML}(\theta^{(t)}, \hat{\tau}^n, \hat{q}^n))^{-1} S_n^{ML}(\theta^{(t)}), \tag{9.108}$$

with the score function

$$S_n^{ML}(\theta) = \frac{1}{n} \frac{\partial}{\partial \theta} l_n^{ML}(\theta) \tag{9.109}$$

and the information matrix

$$I_{\theta\theta}^{ML}(\theta, \tau, q) = -E_{\theta,\tau,q} \left(\frac{\partial^2}{\partial\theta\partial\theta'} l^{ML}(\beta; Y, X_1, Z_2) \right). \tag{9.110}$$

Pseudo-Maximum-Likelihood Estimation (PML)

In order to simplify the computation of the maximum-likelihood estimate of the regression parameter β, the nuisance parameter γ may be estimated from the observed values of X_1 and Z_2 and inserted into the log-likelihood, instead of joint iterative estimation along with β. A possible estimate (cf. Pepe and Fleming, 1991) is

$$\hat{\gamma}_{k|j} = \frac{n_{+jk}}{\sum_{k=1}^{K} n_{+jk}}. \tag{9.111}$$

This estimate is only consistent for γ under very strict assumptions for the missing mechanism. Vach and Schumacher (1992) suggest to apply this estimate to the filled up contingency table of the filling method (cf. Section 9.5.3)

$$\tilde{\gamma}_{k|j} = \frac{n_{+jk}^{FILL}}{\sum_{k=1}^{K} n_{+jk}^{FILL}} = \frac{n_{0jk} \frac{\sum_{k=1}^{K+1} n_{0jk}}{\sum_{k=1}^{K} n_{0jk}} + n_{1jk} \frac{\sum_{k=1}^{K+1} n_{1jk}}{\sum_{k=1}^{K} n_{1jk}}}{\sum_{k=1}^{K+1} n_{+jk}}. \tag{9.112}$$

This estimate is consistent for γ if the MAR assumption holds. PML estimation of β is now achieved by iterative maximization of the following log-likelihood

$$l_n^{PML}(\beta) = \sum_{i=0}^{1} \sum_{j=1}^{J} \sum_{k=1}^{K+1} n_{ijk} \, l^{PML}(\beta, \tilde{\gamma}; i, j, k) \tag{9.113}$$

with

$$l^{PML}(\beta, \tilde{\gamma}; i, j, k) = \begin{cases} \ln \left((\mu_{1|jk})^i (1 - \mu_{1|jk})^{1-i} \right) & \text{for } k = 1, \ldots, K \\ \ln \left((\sum_{k=1}^{K} \mu_{1|jk} \tilde{\gamma}_{k|j})^i (1 - \sum_{k=1}^{K} \mu_{1|jk} \tilde{\gamma}_{k|j})^{1-i} \right) & \text{for } k = K+1. \end{cases}$$

9.6   Exercises and Questions

9.6.1 What is a selectivity bias and what is meant by drop-out in longterm studies?

9.6.2 Name the essential methods for imputation and describe them.

9.6.3 Explain the missing data mechanisms MAR, OAR and MCAR by means of a bivariate sample.

9.6.4 Describe the OLS-methods by Yates and Bartlett. Where is the difference?

9.6.5 Assume that in a regression model values in the matrix X are missing and are to be replaced. Which methods may be used? Explain the effect on the unbiasedness of the final estimator $\hat{\beta}$.

10 Models for Categorical Response

10.1 Generalized Linear Models

10.1.1 Extension of the Regression Model

Generalized linear models are a generalization of the classical linear models of the regression analysis and analysis of variance, which model the relationship between the expectation of a response variable and unknown predictors according to

$$
\begin{aligned}
\mathrm{E}\,(y_i) &= x_{i1}\beta_1 + \ldots + x_{ip}\beta_p \\
&= \boldsymbol{x}_i'\boldsymbol{\beta}
\end{aligned}
\tag{10.1}
$$

or

$$
\mathrm{E}\,(y_{ij}) = \mu + \sum \alpha_i = \tilde{\boldsymbol{x}}_i'\tilde{\boldsymbol{\beta}}.
\tag{10.2}
$$

The parameters are estimated according to the principle of least squares and are optimal according to the qualities of orthogonal projection, or in case of a normal distribution, are optimal according to the ML theory (cf. Chapter 3).

Assuming an additive random error ϵ_i, the density function can be written as

$$
f(y_i) = f_{\epsilon_i}(y_i - \boldsymbol{x}_i'\boldsymbol{\beta}),
\tag{10.3}
$$

where $\eta_i = \boldsymbol{x}_i'\boldsymbol{\beta}$ is the linear predictor. Hence, for continuous normally distributed data, we have the following distribution and mean structure

$$
y_i \sim N(\mu_i, \sigma^2), \quad \mathrm{E}\,(y_i) = \mu_i, \quad \mu_i = \eta_i = \boldsymbol{x}_i'\boldsymbol{\beta}.
\tag{10.4}
$$

In analysing categorical response variables, three major distributions may arise: the binomial, multinomial, and Poisson distribution, which belong to the natural exponential family (along with the normal distribution).

In analogy to the normal distribution, the effect of covariates on the expectation of the response variables may be modelled by linear predictors for these distributions as well:

Binomial distribution

Assume that I predictors $\eta_i = \boldsymbol{x}_i'\boldsymbol{\beta}$ $(i = 1, \ldots, I)$ and N_i realizations respectively are given, and furthermore, assume that the response has a binomial distribution

$$y_i \sim B(N_i; \pi_i) \quad \text{with} \quad \mathrm{E}\,(y_i) = N_i\pi_i = \mu_i \quad .$$

Let $g(\pi_i) = \mathrm{Logit}(\pi_i)$ be the chosen link function between μ_i and η_i:

$$
\begin{aligned}
\mathrm{Logit}(\pi_i) &= \ln\left(\frac{\pi_i}{1 - \pi_i}\right) \\
&= \ln\left(\frac{N_i\pi_i}{N_i - N_i\pi_i}\right) = \boldsymbol{x}_i'\boldsymbol{\beta} \quad .
\end{aligned}
\tag{10.5}
$$

With the inverse function $g^{-1}(\boldsymbol{x}_i'\boldsymbol{\beta})$ we then have

$$N_i\pi_i = \mu_i = N_i\frac{\exp(\boldsymbol{x}_i'\boldsymbol{\beta})}{1 + \exp(\boldsymbol{x}_i'\boldsymbol{\beta})} = g^{-1}(\eta_i) \quad .\tag{10.6}$$

Poisson distribution

Let y_i $(i = 1, \ldots, I)$ have a Poisson distribution with $\mathrm{E}\,(y_i) = \mu_i$

$$P(y_i) = \frac{e^{-\mu_i}\mu_i^{y_i}}{y_i!} \quad \text{for} \quad y_i = 0, 1, 2, \ldots \quad .\tag{10.7}$$

The link function is then chosen as $\ln(\mu_i) = \boldsymbol{x}_i'\boldsymbol{\beta}$.

Contingency tables

The cell frequencies y_{ij} of an $I \times J$ contingency table of two treatments can have a Poisson, multinomial, or binomial distribution (depending on the sampling design). By choosing appropriate design vectors $\boldsymbol{x}_{ij}$, the expected cell frequencies can be described by a loglinear model

$$
\begin{aligned}
\ln(m_{ij}) &= \mu + \alpha_i^A + \beta_j^B + (\alpha\beta)_{ij}^{AB} \\
&= \boldsymbol{x}_{ij}'\boldsymbol{\beta}
\end{aligned}
\tag{10.8}
$$

and hence we have

$$\mu_{ij} = m_{ij} = \exp(\boldsymbol{x}_{ij}'\boldsymbol{\beta}) = \exp(\eta_{ij}) \quad .\tag{10.9}$$

In contrast to the classical model of the regression analysis, where $\mathrm{E}\,(y)$ is linear in the parameter vector $\boldsymbol{\beta}$, so that $\mu = \eta = \boldsymbol{x}'\boldsymbol{\beta}$ holds, the generalized models are of the following form

$$\mu = g^{-1}(\boldsymbol{x}'\boldsymbol{\beta}) \quad ,\tag{10.10}$$

where g^{-1} is the inverse function of the link function. Furthermore, the additivity of the random error is no longer a necessary assumption so that in general

$$f(y) = f(y; \boldsymbol{x}'\boldsymbol{\beta})\tag{10.11}$$

is assumed, instead of (10.3).

10.1.2 Structure of the Generalized Linear Models (GLM)

The **generalized linear model (GLM)** (cf. Nelder and Wedderburn, 1972) is defined as follows. A GLM consists of three components

- the *random component*, which specifies the probability distribution of the response variable,

- the *systematic component*, which specifies a linear function of the explanatory variables,

- the *link function*, which describes a functional relationship between the systematic component and the expectation of the random component.

The three components are specified as follows:

1. The random component Y consists of N independent observations $y' = (y_1, y_2, \ldots, y_N)$ of a distribution belonging to the natural exponential family (cf. Agresti 1990, p.80). Hence, each observation y_i has the following probability density function:

$$f(y_i; \theta_i) = a(\theta_i) b(y_i) \exp(y_i Q(\theta_i)) \quad . \tag{10.12}$$

Remark: The parameter θ_i can vary over $i = 1, 2, \ldots, N$, depending on the value of the explanatory variable, which influences y_i through the systematic component.

Special distributions of particular importance in this family are the Poisson, the binomial, and the multinomial distribution. $Q(\theta_i)$ is called the *natural parameter* of the distribution. Likewise, if the y_i are independent, the common distribution is a member of the exponential family.

Remark: A more general parametrization allows to include scaling or nuisance variables.

An alternative parametrization with an additional scaling parameter ϕ (the so-called dispersion parameter) is given by

$$f(y_i | \theta_i, \phi) = \exp\left\{ \frac{y_i \theta_i - b(\theta_i)}{a(\phi)} + c(y_i; \phi) \right\} \quad , \tag{10.13}$$

where θ_i is called the natural parameter. If ϕ is known, (10.13) represents a linear exponential family. If on the other hand ϕ is unknown, then (10.13) is called *exponential dispersion model*. With ϕ and θ_i, (10.13) is a two-parametrical distribution for $i = 1, \ldots, N$, which is used for normal or gamma distributions for instance.

2. The systematic component relates a vector $\boldsymbol{\eta} = (\eta_1, \eta_2, \ldots, \eta_N)$ to a set of explanatory variables through a linear model

$$\boldsymbol{\eta} = \boldsymbol{X}\boldsymbol{\beta} . \tag{10.14}$$

Here $\boldsymbol{\eta}$ is called the linear predictor, $\boldsymbol{X}$ $(N \times p)$ is the matrix of observations on the explanatory variables and $\boldsymbol{\beta}$ is the $(p \times 1)$-vector of parameters.

3. The link function connects the systematic component with the expectation of the random component. Let $\mu_i = \mathrm{E}\,(y_i)$, then μ_i is linked to η_i by $\eta_i = g(\mu_i)$. Here g is a monotonic and differentiable function:

$$g(\mu_i) = \sum_{j=1}^{p} \beta_j x_{ij} \qquad i = 1, 2, \ldots, N . \tag{10.15}$$

Special cases:

(i) $g(\mu) = \mu$ is called the *identity link* $\Rightarrow \eta_i = \mu_i$.

(ii) $g(\mu) = Q(\theta_i)$ is called the *canonical (natural) link*

$\Rightarrow Q(\theta_i) = \sum_{j=1}^{p} \beta_j x_{ij}.$

Properties of the Density Function (10.13)

Let

$$l_i = l(\theta_i, \phi; y_i) = \ln f(y_i; \theta_i, \phi) \tag{10.16}$$

be the contribution of the ith observation y_i to the log-likelihood. Then

$$l_i = [y_i \theta_i - b(\theta_i)]/a(\phi) + c(y_i; \phi) \tag{10.17}$$

holds and we get the following derivatives with respect to θ_i

$$\frac{\partial l_i}{\partial \theta_i} = [y_i - b'(\theta_i)]/a(\phi) \quad , \tag{10.18}$$

$$\frac{\partial^2 l_i}{\partial \theta_i^2} = -b''(\theta_i)/a(\phi) \quad , \tag{10.19}$$

where $b'(\theta_i) = \partial b(\theta_i)/\partial \theta_i$ and $b''(\theta_i) = \partial^2 b(\theta_i)/\partial \theta_i^2$ are the first and second derivatives of the function $b(\theta_i)$, assumed to be known. By equating (10.18) to zero, it becomes obvious that the solution of the likelihood equations is independent of $a(\phi)$. Since our interest belongs to the estimation of θ and $\boldsymbol{\beta}$ in $\eta = \boldsymbol{x}'\boldsymbol{\beta}$, we could assume $a(\phi) = 1$ without any loss of generality (this corresponds to assuming $\sigma^2 = 1$ in case of a normal distribution). For the present however, we retain $a(\phi)$.

Under certain assumptions of regularity, the order of integration und differentiation may be interchangeable, so that

$$\mathrm{E}\left(\frac{\partial l_i}{\partial \theta_i}\right) = 0 \qquad (10.20)$$

$$-\mathrm{E}\left(\frac{\partial^2 l_i}{\partial \theta_i^2}\right) = \mathrm{E}\left(\frac{\partial l_i}{\partial \theta_i}\right)^2 . \qquad (10.21)$$

Hence we have from (10.18) and (10.20)

$$\mathrm{E}\left(y_i\right) = \mu_i = b'(\theta_i) \quad . \qquad (10.22)$$

Similarly, from (10.19) and (10.21), we find

$$
\begin{aligned}
b''(\theta_i)/a(\phi) &= \mathrm{E}\left\{[y_i - b'(\theta_i)]^2/a^2(\phi)\right\} \\
&= \mathrm{Var}\left(y_i\right)/a^2(\phi) \quad ,
\end{aligned}
\qquad (10.23)
$$

since $\mathrm{E}\left[y_i - b'(\theta_i)\right] = 0$, and hence

$$V(\mu_i) = \mathrm{Var}\left(y_i\right) = b''(\theta_i)a(\phi) \quad . \qquad (10.24)$$

Under the assumption that the y_i $(i = 1, \ldots, N)$ are independent, the log-likelihood of $\boldsymbol{y}' = (y_1, \ldots, y_N)$ equals the sum of $l_i(\theta_i, \phi; y_i)$. Let $\boldsymbol{\theta}' = (\theta_1, \ldots, \theta_N)$,

$$\boldsymbol{\mu}' = (\mu_1, \ldots, \mu_N), \ \boldsymbol{X} = \begin{pmatrix} \boldsymbol{x}_1' \\ \vdots \\ \boldsymbol{x}_N' \end{pmatrix} \ \text{and} \ \boldsymbol{\eta} = (\eta_1, \ldots, \eta_N)' = \boldsymbol{X\beta}. \ \text{We then have}$$

from (10.22)

$$\boldsymbol{\mu} = \frac{\partial b(\boldsymbol{\theta})}{\partial \boldsymbol{\theta}} = \left(\frac{\partial b(\theta_1)}{\partial \theta_1}, \ldots, \frac{\partial b(\theta_1)}{\partial \theta_N}\right)' \qquad (10.25)$$

and in analogy to (10.24) for the covariance matrix of $\boldsymbol{y}' = (y_1, \ldots, y_N)$

$$\mathrm{Cov}\left(\boldsymbol{y}\right) = V(\boldsymbol{\mu}) = \frac{\partial^2 b(\boldsymbol{\theta})}{\partial \boldsymbol{\theta} \partial \boldsymbol{\theta}'} = a(\phi)\mathrm{diag}(b''(\theta_1), \ldots, b''(\theta_N)) \quad . \qquad (10.26)$$

These relations hold in general, as we will show in the following.

10.1.3 Score Function and Information Matrix

The likelihood of the random sample is the product of the density functions:

$$L(\boldsymbol{\theta}, \phi; \boldsymbol{y}) = \prod_{i=1}^{N} f(y_i; \theta_i, \phi) \quad . \qquad (10.27)$$

The log-likelihood $\ln L(\boldsymbol{\theta}, \phi; \boldsymbol{y})$ for the sample $\boldsymbol{y}$ of independent y_i $(i = 1, \ldots, N)$ is of the form

$$l = l(\boldsymbol{\theta}, \phi; \boldsymbol{y}) = \sum_{i=1}^{N} l_i = \sum_{i=1}^{N} \{(y_i\theta_i - b(\theta_i))/a(\phi) + c(y_i; \phi)\} \quad . \qquad (10.28)$$

The vector of first derivatives of l with respect to θ_i is needed for determining the ML estimates. This vector is called the *score function*. For now, we neglect the parametrization with ϕ in the representation of l and L and thus get the score function as

$$s(\boldsymbol{\theta}; \boldsymbol{y}) = \frac{\partial}{\partial\boldsymbol{\theta}} l(\boldsymbol{\theta}; \boldsymbol{y}) = \frac{1}{L(\boldsymbol{\theta}; \boldsymbol{y})} \frac{\partial}{\partial\boldsymbol{\theta}} L(\boldsymbol{\theta}; \boldsymbol{y}) \quad . \tag{10.29}$$

Let

$$\frac{\partial^2 l}{\partial\boldsymbol{\theta}\partial\boldsymbol{\theta}'} = \left(\frac{\partial^2 l}{\partial\theta_i\partial\theta_j}\right)_{\substack{i=1,\ldots,N \\ j=1,\ldots,N}}$$

be the matrix of the second derivatives of the log-likelihood. Then

$$F_{(N)}(\boldsymbol{\theta}) = \mathrm{E}\left(\frac{-\partial^2 l(\boldsymbol{\theta}; \boldsymbol{y})}{\partial\boldsymbol{\theta}\partial\boldsymbol{\theta}'}\right) \tag{10.30}$$

is called the *Fisher-information matrix* of the sample $\boldsymbol{y}' = (y_1, \ldots, y_N)$, where the expectation is to be taken with respect to the following density function

$$f(y_1, \ldots, y_N | \theta_i) = \prod f(y_i | \theta_i) = L(\boldsymbol{\theta}; \boldsymbol{y}).$$

In case of regular likelihood functions (regular: exchange of integration and differentiation is possible), to which the exponential families belong, we have

$$\mathrm{E}\left(s(\boldsymbol{\theta}; \boldsymbol{y})\right) = \mathbf{0} \tag{10.31}$$

and

$$F_{(N)}(\boldsymbol{\theta}) = \mathrm{E}\left(s(\boldsymbol{\theta}; \boldsymbol{y})s'(\boldsymbol{\theta}; \boldsymbol{y})\right) = \mathrm{Cov}\left(s(\boldsymbol{\theta}; \boldsymbol{y})\right) \quad , \tag{10.32}$$

The relation (10.31) follows from

$$\int f(y_1, \ldots, y_N | \boldsymbol{\theta}) dy_1 \cdots dy_N = \int L(\boldsymbol{\theta}; \boldsymbol{y}) d\boldsymbol{y} = 1 \quad , \tag{10.33}$$

by differentiating with respect to $\boldsymbol{\theta}$:

$$\begin{aligned}
\int \frac{\partial L(\boldsymbol{\theta}; \boldsymbol{y})}{\partial\boldsymbol{\theta}} d\boldsymbol{y} &= \int \frac{\partial l(\boldsymbol{\theta}; \boldsymbol{y})}{\partial\boldsymbol{\theta}} L(\boldsymbol{\theta}; \boldsymbol{y}) d\boldsymbol{y} \\
&= \mathrm{E}\left(s(\boldsymbol{\theta}; \boldsymbol{y})\right) = \mathbf{0} \quad .
\end{aligned} \tag{10.34}$$

Remark: We have

$$\begin{aligned}
L(\boldsymbol{\theta}; \boldsymbol{y}) &= \exp\ln L(\boldsymbol{\theta}; \boldsymbol{y}) \\
&= \exp l(\boldsymbol{\theta}; \boldsymbol{y}) \quad , \\
\frac{\partial L(\boldsymbol{\theta}; \boldsymbol{y})}{\partial\boldsymbol{\theta}} &= \frac{\partial l(\boldsymbol{\theta}; \boldsymbol{y})}{\partial\boldsymbol{\theta}} \exp l(\boldsymbol{\theta}; \boldsymbol{y}) = \frac{\partial l(\boldsymbol{\theta}; \boldsymbol{y})}{\partial\boldsymbol{\theta}} L(\boldsymbol{\theta}; \boldsymbol{y}) \quad .
\end{aligned}$$

If (10.34) is differentiated with respect to θ', we get

$$
\begin{aligned}
0 &= \int \frac{\partial^2 l(\boldsymbol{\theta}; \boldsymbol{y})}{\partial \boldsymbol{\theta} \partial \boldsymbol{\theta}'} L(\boldsymbol{\theta}; \boldsymbol{y}) dy \\
&+ \int \frac{\partial l(\boldsymbol{\theta}; \boldsymbol{y})}{\partial \boldsymbol{\theta}} \frac{\partial l(\boldsymbol{\theta}; \boldsymbol{y})}{\partial \boldsymbol{\theta}'} L(\boldsymbol{\theta}; \boldsymbol{y}) dy \\
&= -F_{(N)}(\boldsymbol{\theta}) + \mathrm{E}\left(s(\boldsymbol{\theta}; \boldsymbol{y}) s'(\boldsymbol{\theta}; \boldsymbol{y})\right) \quad,
\end{aligned}
$$

and hence (10.32), because of $\mathrm{E}\left(s(\boldsymbol{\theta}; \boldsymbol{y})\right) = \mathbf{0}$.

10.1.4   Maximum-Likelihood Estimation of the Predictors

Let $\eta_i = \boldsymbol{x}_i' \boldsymbol{\beta} = \sum_{j=1}^{p} x_{ij} \beta_j$ be the predictor of the ith observation of the response variable $(i = 1, \dots, N)$ or – in matrix representation –

$$
\boldsymbol{\eta} = \begin{pmatrix} \eta_1 \\ \vdots \\ \eta_N \end{pmatrix} = \begin{pmatrix} \boldsymbol{x}_1' \boldsymbol{\beta} \\ \vdots \\ \boldsymbol{x}_N' \boldsymbol{\beta} \end{pmatrix} = \boldsymbol{X} \boldsymbol{\beta} \quad. \tag{10.35}
$$

Assume, that the predictors are linked to $\mathrm{E}(\boldsymbol{y}) = \boldsymbol{\mu}$ by a monotonic differentiable function $g(\cdot)$:

$$
g(\mu_i) = \eta_i \quad (i = 1 \dots, N) \quad, \tag{10.36}
$$

or in matrix representation

$$
g(\boldsymbol{\mu}) = \begin{pmatrix} g(\mu_1) \\ \vdots \\ g(\mu_N) \end{pmatrix} = \boldsymbol{\eta} \quad. \tag{10.37}
$$

The parameters θ_i and $\boldsymbol{\beta}$ are then linked by the relation (10.22), that is $\mu_i = b'(\theta_i)$ and with $g(\mu_i) = \boldsymbol{x}_i' \boldsymbol{\beta}$. Hence we have $\theta_i = \theta_i(\boldsymbol{\beta})$. Since we are only interested in estimating $\boldsymbol{\beta}$, we will write the log-likelihood (10.28) as a function of $\boldsymbol{\beta}$

$$
l(\boldsymbol{\beta}) = \sum l_i(\boldsymbol{\beta}) \quad. \tag{10.38}
$$

We can find the derivatives $\partial l_i(\boldsymbol{\beta}) / \partial \beta_j$ according to standard rules:

$$
\frac{\partial l_i(\boldsymbol{\beta})}{\partial \beta_j} = \frac{\partial l_i}{\partial \theta_i} \frac{\partial \theta_i}{\partial \mu_i} \frac{\partial \mu_i}{\partial \eta_i} \frac{\partial \eta_i}{\partial \beta_j} \quad. \tag{10.39}
$$

The partial results are as follows:

$$
\frac{\partial l_i}{\partial \theta_i} = [y_i - b'(\theta_i)] / a(\phi)
$$

[cf. (10.18)]

$$
= [y_i - \mu_i] / a(\phi) \tag{10.40}
$$

[cf. (10.22)],

$$\mu_i = b'(\theta_i) \ ,$$

$$\frac{\partial \mu_i}{\partial \theta_i} = b''(\theta_i) = \text{Var}\,(y_i)/a(\phi) \tag{10.41}$$

[cf. (10.24)],

$$\frac{\partial \eta_i}{\partial \beta_j} = \frac{\partial \sum_{k=1}^{p} x_{ik}\beta_k}{\partial \beta_j} = x_{ij} \ . \tag{10.42}$$

Because of $\eta_i = g(\mu_i)$, the derivative $\partial \mu_i/\partial \eta_i$ is dependent on the link function $g(\cdot)$, or rather its inverse $g^{-1}(\cdot)$. Hence, it cannot be specified until the link is defined.

Summarizing, we now have

$$\frac{\partial l_i}{\partial \beta_j} = \frac{(y_i - \mu_i)x_{ij}}{\text{Var}\,(y_i)} \frac{\partial \mu_i}{\partial \eta_i} \ , \tag{10.43}$$

using the rule

$$\frac{\partial \theta_i}{\partial \mu_i} = \left(\frac{\partial \mu_i}{\partial \theta_i}\right)^{-1}$$

for inverse functions $(\mu_i = b'(\theta_i)\ , \ \theta_i = (b')^{-1}(\mu_i))$. The estimating equations for finding the components β_j are now

$$\sum_{i=1}^{N} \frac{(y_i - \mu_i)x_{ij}}{\text{Var}\,(y_i)} \frac{\partial \mu_i}{\partial \eta_i} = 0 \ , \quad j = 1 \ldots, p. \tag{10.44}$$

The log-likelihood is nonlinear in $\boldsymbol{\beta}$. Hence, the solution of (10.44) has to be found iteratively. For the second derivative with respect to components of $\boldsymbol{\beta}$, we have in analogy to (10.21), with (10.43),

$$\begin{aligned}
\text{E}\left(\frac{\partial^2 l_i}{\partial \beta_j \partial \beta_h}\right) &= -\text{E}\left(\frac{\partial l_i}{\partial \beta_j}\right)\left(\frac{\partial l_i}{\partial \beta_h}\right) \\
&= -\text{E}\left[\frac{(y_i - \mu_i)(y_i - \mu_i)x_{ij}x_{ih}}{(\text{Var}\,(y_i))^2}\left(\frac{\partial \mu_i}{\partial \eta_i}\right)^2\right] \\
&= -\frac{x_{ij}x_{ih}}{\text{Var}\,(y_i)}\left(\frac{\partial \mu_i}{\partial \eta_i}\right)^2 \ , \tag{10.45}
\end{aligned}$$

and hence

$$\text{E}\left(-\frac{\partial^2 l(\boldsymbol{\beta})}{\partial \beta_j \partial \beta_h}\right) = \sum_{i=1}^{N} \frac{x_{ij}x_{ih}}{\text{Var}\,(y_i)}\left(\frac{\partial \mu_i}{\partial \eta_i}\right)^2 \tag{10.46}$$

and in matrix representation for all (j, h)-combinations

$$F_{(N)}(\boldsymbol{\beta}) = \text{E}\left(-\frac{\partial^2 l(\boldsymbol{\beta})}{\partial \boldsymbol{\beta} \partial \boldsymbol{\beta}'}\right) = \boldsymbol{X}'\boldsymbol{W}\boldsymbol{X} \tag{10.47}$$

with

$$W = \text{diag}(w_1 \ldots, w_N) \tag{10.48}$$

and the weights

$$w_i = \left(\frac{\partial \mu_i}{\partial \eta_i}\right)^2 / \text{Var}(y_i) \quad . \tag{10.49}$$

Fisher-Scoring Algorithm

For the iterative determination of the ML estimate of $\boldsymbol{\beta}$, the method of iterative reweighted least squares is used. Let $\boldsymbol{\beta}^{(k)}$ be the kth approximation of the ML estimate $\hat{\boldsymbol{\beta}}$. Furthermore, let $\boldsymbol{q}^{(k)}(\boldsymbol{\beta}) = \partial l(\boldsymbol{\beta})/\partial \boldsymbol{\beta}$ be the vector of the first derivatives at $\boldsymbol{\beta}^{(k)}$ (cf. (10.43)). Analogously, we define $\boldsymbol{W}^{(k)}$. The formula of the Fisher-scoring algorithm is then

$$(\boldsymbol{X}'\boldsymbol{W}^{(k)}\boldsymbol{X})\boldsymbol{\beta}^{(k+1)} = (\boldsymbol{X}'\boldsymbol{W}^{(k)}\boldsymbol{X})\boldsymbol{\beta}^{(k)} + \boldsymbol{q}^{(k)} \quad . \tag{10.50}$$

The vector on the right side of (10.50) has the components (cf. (10.46) and (10.43))

$$\sum_h \left[\sum_i \frac{x_{ij}x_{ih}}{\text{Var}(y_i)} \left(\frac{\partial \mu_i}{\partial \eta_i}\right)^2 \beta_h^{(k)}\right] + \sum_i \frac{(y_i - \mu_i^{(k)})x_{ij}}{\text{Var}(y_i)} \left(\frac{\partial \mu_i}{\partial \eta_i}\right) \quad . \tag{10.51}$$

$$(j = 1, \ldots, p)$$

The entire vector (10.51) can now be written as

$$\boldsymbol{X}'\boldsymbol{W}^{(k)}\boldsymbol{z}^{(k)} \quad , \tag{10.52}$$

where the vector $\boldsymbol{z}^{(k)}$ has the jth element as follows

$$
\begin{aligned}
z_j^{(k)} &= \sum_i x_{ij}\beta_i^{(k)} + (y_j - \mu_j^{(k)})\left(\frac{\partial \eta_j^{(k)}}{\partial \mu_j^{(k)}}\right) \\
&= \eta_j^{(k)} + (y_j - \mu_j^{(k)})\left(\frac{\partial \eta_j^{(k)}}{\partial \mu_j^{(k)}}\right) \quad .
\end{aligned} \tag{10.53}
$$

Hence, the equation (10.50) of the Fisher-scoring algorithm can now be written as

$$(\boldsymbol{X}'\boldsymbol{W}^{(k)}\boldsymbol{X})\boldsymbol{\beta}^{(k+1)} = \boldsymbol{X}'\boldsymbol{W}^{(k)}\boldsymbol{z}^{(k)} \quad . \tag{10.54}$$

This is the estimating equation of a generalized linear model with the response vector $\boldsymbol{z}^{(k)}$ and the random error covariance matrix $(\boldsymbol{W}^{(k)})^{-1}$. If $\text{rank}(\boldsymbol{X}) = p$ holds, we obtain the ML estimate $\hat{\boldsymbol{\beta}}$ as the limit of

$$\hat{\boldsymbol{\beta}}^{(k+1)} = (\boldsymbol{X}'\boldsymbol{W}^{(k)}\boldsymbol{X})^{-1}\boldsymbol{X}'\boldsymbol{W}^{(k)}\boldsymbol{z}^{(k)} \tag{10.55}$$

for $k \to \infty$, with the asymptotic covariance matrix

$$V(\hat{\beta}) = (X'\hat{W}X)^{-1} = F_{(N)}^{-1}(\hat{\beta}) \quad , \tag{10.56}$$

where $\hat{W}$ is determined at $\hat{\beta}$. Once a solution is found, then $\hat{\beta}$ is consistent for β, asymptotically normal, and asymptotically efficient (Fahrmeir and Kaufmann, 1985, cf. Wedderburn, 1976, for existence and uniqueness of the solutions). Hence we have $\hat{\beta} \overset{as.}{\sim} N(\beta, V(\hat{\beta}))$.

Remark: In case of a canonical link function, that is for $g(\mu_i) = \theta_i$, the ML equations simplify and the Fisher-scoring algorithm is identical to the Newton-Raphson algorithm (cf. Agresti, 1990, p. 451). If the values $a(\phi)$ are identical for all observations, then the ML equations are

$$\sum_i x_{ij} y_i = \sum_i x_{ij} \mu_i \quad . \tag{10.57}$$

If on the other hand $a(\phi) = a_i \phi$ $(i = 1, \ldots, N)$ holds, then the ML equations are

$$\sum_i \frac{x_{ij} y_i}{a_i} = \sum_i \frac{x_{ij} \mu_i}{a_i} \quad . \tag{10.58}$$

As starting values for the Fisher-scoring algorithm the estimates $\hat{\beta}^{(0)} = (X'X)^{-1}X'y$ or $\hat{\beta}^{(0)} = (X'X)^{-1}X'g(y)$ may be used.

10.1.5 Goodness of Fit and Testing of Hypotheses

A generalized linear model $g(\mu_i) = x_i'\beta$ is determined by the link function $g(\cdot)$ and the explanatory variables $X_1, \ldots, X_p$, as well as their number p, which determines the number of parameters $\beta_1, \ldots, \beta_p$ to be estimated. If $g(\cdot)$ is chosen, then the model is defined by the design matrix X. Let X_1 and X_2 be two design matrices (models), and assume that the hierarchical order $X_1 \subset X_2$ holds, that is, we have $X_2 = (X_1, X_3)$ with some matrix X_3 and hence $\mathcal{R}(X_1) \subset \mathcal{R}(X_2)$. Let $g(\hat{\mu}_1) = \hat{\eta}_1 = X_1\hat{\beta}_1$ and $g(\hat{\mu}_2) = \hat{\eta}_2 = X_2\hat{\beta}_2$ and $\text{rank}(X_1) = r_1$, $\text{rank}(X_2) = r_2$ and $(r_2 - r_1) = r = df$. The likelihood ratio statistic, which compares a larger model X_2 with a (smaller) submodel X_1, is then defined as follows (L: likelihood function)

$$\Lambda = \frac{\max_{\beta_1} L(X_1)}{\max_{\beta_2} L(X_2)} \tag{10.59}$$

or

$$\begin{aligned} G^2(X_1|X_2) &= -2(\ln \max L(X_1) - \ln \max L(X_2)) \\ &= 2[l(X_2) - l(X_1)] \end{aligned} \tag{10.60}$$

(l: log-likelihood, that is, $l(\cdot) = \ln(\max L(\cdot))$). If X_2 holds, then the test statistic $G^2(X_1|X_2)$ under model X_1 is approximately χ_r^2-distributed. If this statistic is significant, then the additional parameters (belonging to X_3) are significantly different from zero.

Applied to our previous notation and with the assumption $a(\phi) = a_i\phi$, we obtain the statistic in the following form

$$
\begin{aligned}
G^2(X_1|X_2) &= 2\{l(\hat{\mu}_2, \phi; y) - l(\hat{\mu}_1, \phi; y)\} \\
&= 2\sum_{i=1}^{N} \frac{1}{a_i}\{y_i(\hat{\theta}_{2i} - \hat{\theta}_{1i}) - b(\hat{\theta}_{2i}) + b(\hat{\theta}_{1i})\}/\phi \quad . \quad (10.61)
\end{aligned}
$$

Let X be the design matrix of the saturated model that contains the same number of parameters as observations. Denote by $\tilde{\theta}$ the estimate of θ that belongs to the estimates $\tilde{\mu}_i = y_i$ $(i = 1, \ldots, N)$ in the saturated model. For every submodel X_j that is not saturated, we then have

$$
\begin{aligned}
G^2(X_j|X) &= 2\sum \frac{1}{a_i}\{y_i(\tilde{\theta}_i - \hat{\theta}_i) - b(\tilde{\theta}_i) + b(\hat{\theta}_i)\}/\phi \\
&= D(y; \hat{\mu}_j)/\phi \quad\quad\quad (10.62)
\end{aligned}
$$

as a measure for the loss in goodness of fit of the model X_j compared to the perfect fit achieved by the saturated model. The statistic $D(y; \hat{\mu}_j)$ is called *deviance* of the model X_j. We then have

$$
G^2(X_1|X_2) = G^2(X_1|X) - G^2(X_2|X) = [D(y; \hat{\mu}_1) - D(y; \hat{\mu}_2)]/\phi \quad , \quad (10.63)
$$

that is, the test statistic for comparing the model X_1 with the larger model X_2 equals the difference of the goodness-of-fit statistics of the two models, weighted with $1/\phi$.

10.1.6 Overdispersion

In samples of a Poisson or multinomial distribution, it may occur that the elements show a larger variance than that given by the distribution. This may be due to a violation of the assumption of independence, as for example a positive correlation in the sample elements. A frequent cause for this is the cluster-structure of the sample. Examples are

- the behaviour of families of insects in case of the influence of insecticides (Agresti, 1990, p. 42), where the family (cluster, batch) shows a *collective* (correlated) survivorship (many survive or most of them die) rather than an independent survivorship, due to dependence on cluster-specific covariables as for instance the temperature,

- the survivorship of dental implants when two or more implants are incorporated for each patient

- the developement of diseases or social behaviour of the members of a family.

The existence of a larger variation (inhomogeneity) in the sample than in the sample model is called *overdispersion*. Overdispersion is modelled by multiplying the variance with a constant $\phi > 1$, where ϕ is either known (e.g., $\phi = \sigma^2$ for a normal distribution), or has to be estimated from the sample (cf. Fahrmeir and Tutz, 1994, p. 19 for alternative approaches).

Example: (McCullagh and Nelder, 1989, p. 125)
Let N individuals be divided into N/k clusters of equal cluster size k. Assume that the individual response is binary with $P(Y_i = 1) = \pi_i$, so that the total response

$$Y = Z_1 + Z_2 + \cdots + Z_{N/k}$$

equals the sum of independent $B(k; \pi_i)$-distributed binomial variables Z_i $(i = 1, \ldots, N/k)$. The π_i's vary across the clusters and assume that $\mathrm{E}(\pi_i) = \pi$ and $\mathrm{Var}(\pi_i) = \tau^2\pi(1 - \pi)$ with $0 \leq \tau^2 \leq 1$. We then have

$$
\begin{aligned}
\mathrm{E}(Y) &= N\pi \\
\mathrm{Var}(Y) &= N\pi(1 - \pi)\{1 + (k - 1)\tau^2\} \\
&= \phi N\pi(1 - \pi) \quad.
\end{aligned}
\tag{10.64}
$$

The dispersion parameter $\phi = 1 + (k-1)\tau^2$ is dependent on the cluster size k and on the variability of the π_i, but not on the sample size N. This fact is essential for interpreting the variable Y as the sum of binomial variables Z_i and for estimating the dispersion parameter ϕ from the residuals. Because of $0 \leq \tau^2 \leq 1$, we have

$$1 \leq \phi \leq k \leq N \quad.
\tag{10.65}$$

The relationship (10.64) means that

$$\frac{\mathrm{Var}(Y)}{N\pi(1 - \pi)} = 1 + (k - 1)\tau^2 = \phi
\tag{10.66}$$

is constant. An alternative model – the beta-binomial distribution – has the property that the quotient in (10.66), i.e. ϕ, is a linear function of the sample size N. By plotting the residuals against N, it is easy to recognize which of the two models is more likely. Rosner (1984) uses the the beta-binomial distribution for estimation in clusters of the size $k = 2$.

10.1.7 Quasi-Log-Likelihood

The generalized models assume a distribution of the natural exponential family for the data as the random component (cf. (10.12)). If this assumption does not hold, an alternative approach can be used to specify the functional relationship between the mean and the variance. For exponential families, the relationship (10.24) between variance and expectation holds. Assume the general approach

$$\mathrm{Var}(Y) = \phi V(\mu) \quad,
\tag{10.67}$$

where $V(\cdot)$ is an appropriately chosen function.

In the quasi-likelihood approach (Wedderburn, 1974), only assumptions about the first and second moments of the random variables are made. It is not necessary for the distribution itself to be specified. The starting point in estimating the influence of covariables is the score function (10.29) or rather the system of ML equations (10.44). If the general specification (10.67) is inserted into (10.44), we get the system of *estimating equations* for β

$$\sum_{i=1}^{N} \frac{(y_i - \mu_i)}{V(\mu_i)} x_{ij} \frac{\partial \mu_i}{\partial \eta_i} = 0 \quad (j = 1, \ldots, p) \quad , \tag{10.68}$$

which is of the same form as as the likelihood equations (10.44) for GLM's. However, the system (10.68) is a ML equation system only if the y_i's have a distribution of the natural exponential family.

In case of independent response, the modelling of the influence of the covariables X on the mean response $\mathrm{E}(\boldsymbol{y}) = \boldsymbol{\mu}$ is done according to McCullagh and Nelder (1989, p. 324) as follows. Assume that for the response vector we have

$$\boldsymbol{y} \sim (\boldsymbol{\mu}, \phi \boldsymbol{V}(\boldsymbol{\mu})) \tag{10.69}$$

where $\phi > 0$ is an unknown dispersion parameter and $\boldsymbol{V}(\boldsymbol{\mu})$ is a matrix of known functions. The expression $\phi \boldsymbol{V}(\boldsymbol{\mu})$ is called *working variance*.

If the components of $\boldsymbol{y}$ are assumed to be independent, the covariance matrix $\phi \boldsymbol{V}(\boldsymbol{\mu})$ has to be diagonal, that is,

$$\boldsymbol{V}(\boldsymbol{\mu}) = \mathrm{diag}(V_1(\boldsymbol{\mu}), \ldots, V_N(\boldsymbol{\mu})) \quad . \tag{10.70}$$

Here it is realistic to assume that the variance of each random variable y_i is only dependent on the ith component μ_i of $\boldsymbol{\mu}$ meaning thereby

$$\boldsymbol{V}(\boldsymbol{\mu}) = \mathrm{diag}(V_1(\mu_1), \ldots, V_N(\mu_N)). \tag{10.71}$$

A dependency on all components of $\boldsymbol{\mu}$ according to (10.70) is hard to interpret in practice, if independence of the y_i is demanded as well. (Nevertheless, situations as in (10.70) are possible). In many applications it is reasonable to assume, in addition to functional independency (10.71), that the functions V_i's are identical, so that

$$\boldsymbol{V}(\boldsymbol{\mu}) = \mathrm{diag}(v(\mu_1), \ldots, v(\mu_N)) \tag{10.72}$$

holds, with $V_i = v(\cdot)$.

Under the above assumptions, the following function for a component y_i of $\boldsymbol{y}$

$$U = u(\mu_i, y_i) = \frac{y_i - \mu_i}{\phi v(\mu_i)} \tag{10.73}$$

has the properties

$$E(U) = 0 \quad , \tag{10.74}$$

$$\mathrm{Var}(U) = \frac{1}{\phi v(\mu_i)} \quad , \tag{10.75}$$

$$\frac{\partial U}{\partial \mu_i} = \frac{-\phi v(\mu_i) - (y_i - \mu_i)\phi\frac{\partial v(\mu_i)}{\partial \mu_i}}{\phi^2 v^2(\mu_i)}$$

$$-E\left(\frac{\partial U}{\partial \mu_i}\right) = \frac{1}{\phi v(\mu_i)} \quad . \tag{10.76}$$

Hence U has the same properties as the derivative of a log-likelihood, which of course is the score function (10.29). The property (10.74) corresponds to (10.31), while the property (10.76) in combination with (10.75) corresponds to (10.32). Therefore,

$$Q(\boldsymbol{\mu}; \boldsymbol{y}) = \sum_{i=1}^{N} Q_i(\mu_i; y_i) \tag{10.77}$$

with

$$Q_i(\mu_i; y_i) = \int_{y_i}^{\mu_i} \frac{\mu_i - t}{\phi v(t)} dt \tag{10.78}$$

(cf. McCullagh and Nelder, 1989, p. 325) is the analogue of the log-likelihood function. $Q(\boldsymbol{\mu}; \boldsymbol{y})$ is called *quasi-log-likelihood*. Hence, the *quasi-score function*, which is obtained by differentiating $Q(\boldsymbol{\mu}; \boldsymbol{y})$, equals

$$U(\boldsymbol{\beta}) = \phi^{-1} \boldsymbol{D}' \boldsymbol{V}^{-1} (\boldsymbol{y} - \boldsymbol{\mu}) \quad , \tag{10.79}$$

with $\boldsymbol{D} = (\partial\mu_i/\partial\beta_j)$ $(i = 1,\ldots,N,\ j = 1,\ldots,p)$ and $\boldsymbol{V} = \mathrm{diag}(v_1,\ldots,v_N)$. The quasi likelihood estimate $\hat{\boldsymbol{\beta}}$ is the solution of $U(\hat{\boldsymbol{\beta}}) = 0$. It has the asymptotic covariance matrix

$$\mathrm{Cov}(\hat{\boldsymbol{\beta}}) = \phi(\boldsymbol{D}'\boldsymbol{V}^{-1}\boldsymbol{D})^{-1} \quad . \tag{10.80}$$

The dispersion parameter ϕ is estimated by

$$\hat{\phi} = \frac{1}{N-p} \frac{\sum(y_i - \hat{\mu}_i)^2}{v(\hat{\mu}_i)} = \frac{X^2}{N-p}, \tag{10.81}$$

where X^2 is the so-called Pearson statistic. In case of overdispersion (or assumed overdispersion), the influence of covariables (that is, of the vector $\boldsymbol{\beta}$) is to be estimated by a quasi-log-likelihood approach (10.68) rather than by a log-likelihood approach.

10.2 Loglinear Models for Categorical Response

10.2.1 Binary Response

Let Y be a binary random variable, that is, there are only two categories (for instance success/failure or case/control). Hence the response variable Y can always be coded as $(Y = 0, Y = 1)$. Y has a binomial distribution with $P(Y = 1) = \pi$ and $P(Y = 0) = (1 - \pi)$. If Y is realized for N patients, we get N different independent distributions $B(1; \pi_i)$ with $P(Y_i = 1) = \pi_i$ and $P(Y_i = 0) = 1 - \pi_i$ for $i = 1, 2, \ldots, N$. Here we have $\mu_i = \pi_i$ and $\text{Var}(y_i) = V(\mu_i) = \pi_i(1 - \pi_i)$. The density function in the form (10.12) is now

$$
\begin{aligned}
f(y_i; \pi_i) &= \pi_i^{y_i}(1 - \pi_i)^{1 - y_i} \\
&= (1 - \pi_i)\left(\frac{\pi_i}{1 - \pi_i}\right)^{y_i} \\
&= (1 - \pi_i)\exp\left(y_i \ln\left(\frac{\pi_i}{1 - \pi_i}\right)\right).
\end{aligned} \tag{10.82}
$$

The natural parameter $Q(\pi_i) = \ln\left(\frac{\pi_i}{1 - \pi_i}\right)$ is the log-odds of response 1 and is called the logit of π_i. A GLM with the *logit link* is called logit model. The generalized model with the canonical link is given by:

$$
\ln\left(\frac{\pi_i}{1 - \pi_i}\right) = \boldsymbol{x}_i'\boldsymbol{\beta}. \tag{10.83}
$$

The patients are grouped by their X-values. Thus, given $\boldsymbol{x}_j$, n_j of the N_j patients are observed to have response $(Y = 1)$ and $(N_j - n_j)$ are observed to have nonresponse $(Y = 0)$. Now $\hat{\pi}_j = \frac{n_j}{N_j}$ is an estimate of the probability of response for the jth category of the vector of prognostic factors.

Assume the random variables $Y_1, \ldots, Y_N$ to have independent binomial distributions: $Y_i \sim B(N_i, \pi_i)$, and hence $\text{E}(Y_i) = \mu_i = N_i\pi_i$ and $V(\mu_i) = N_i\pi_i(1 - \pi_i)$ holds. Apart from the binomial coefficients $\begin{pmatrix} N_i \\ y_i \end{pmatrix}$, which are independent of π_i, the log-likelihood function (cf. (10.28)) is given by

$$
l = l(\pi_i; y_i) = \sum_{i=1}^{N}\left[y_i \ln\left(\frac{\pi_i}{1 - \pi_i}\right) + N_i \ln(1 - \pi_i)\right]. \tag{10.84}
$$

With $g(\mu_i) = g(N_i\pi_i) = \eta_i = \ln(\pi_i/(1 - \pi_i))$, $\partial\mu_i/\partial\eta_i = N_i\partial\pi_i/\partial\eta_i$ and $\partial\pi_i/\partial\eta_i = (\partial\eta_i/\partial\pi_i)^{-1} = \pi_i(1 - \pi_i)$, the estimating equation (10.44) has the following form

$$
\sum_{i=1}^{N}\frac{(y_i - N_i\pi_i)x_{ij}}{N_i\pi_i(1 - \pi_i)}N_i\pi_i(1 - \pi_i) = \sum_{i=1}^{N}(y_i - N_i\pi_i)x_{ij} = 0 \quad (j = 1, \ldots, p). \tag{10.85}
$$

10.2.2 Loglinear Models for Poisson Distributions

In a contingency table with N cells, we observe the cell entries n_i with $i = 1, 2, \ldots, N$. Here the n_i are random variables with $n_i \geq 0$ and $E(n_i) = m_i$. Thus n_i are the observed and m_i are the expected cell frequencies. Consider the Poisson distribution:

$$P(n_i; \mu_i) = \frac{\exp(-\mu_i)\mu_i^{n_i}}{n_i!} \tag{10.86}$$

with $\mathrm{Var}(n_i) = E(n_i) = \mu_i$. The independent sample $(n_1, n_2, \ldots, n_N)$ with $n = \sum_{i=1}^{N} n_i$ has a Poisson distribution as well, with $E(n) = \sum_{i=1}^{N} m_i$. The probability function of the Poisson variables n_i may be written according to (10.12) ($n_i \hat{=} y_i$ and $m_i \hat{=} \theta_i$):

$$f(n_i; \mu_i) = \frac{\exp(-\mu_i)\,\mu_i^{n_i}}{n_i!} = \underbrace{\exp(-\mu_i)}_{A(\theta_i)} \underbrace{\left(\frac{1}{n_i!}\right)}_{B(y_i)} \underbrace{\exp(n_i \ln(\mu_i))}_{\exp(y_i Q(\theta_i))} . \tag{10.87}$$

The natural parameter is $Q(\theta_i) = \ln(m_i)$ and hence the canonical link is $\eta_i = \ln(m_i) \Rightarrow \ln(m_i) = \sum_{j=1}^{p} \beta_j x_{ij}$ for $i = 1, 2, \ldots, N$. For the parametrization (10.13), we get with $n_i = y_i$

$$\begin{aligned} f(y_i; \mu_i) &= \exp(y_i \ln(\mu_i) - \mu_i - \ln(y_i!)) \\ &= \exp(y_i \theta_i - \exp(\theta_i) - \ln(y_i!)) \quad , \end{aligned} \tag{10.88}$$

where $\theta_i = \ln(\mu_i)$ is the natural parameter. Comparing (10.88) and (10.13), we identify $b(\theta) = \exp(\theta_i)$, $a(\phi) = 1$, $c(y_i; \phi) = -\ln(y_i!)$ as well as $E(y_i) = \mu_i = b'(\theta_i) = \exp(\theta_i)$ and $\mathrm{Var}(y_i) = V(\mu_i) = b''(\theta_i) = \exp(\theta_i) = \mu_i$. Since $\theta_i = \ln(\mu_i)$ is the natural parameter, the canonical link is now $\eta_i = \ln(\mu_i)$. This leads to the loglinear model

$$\ln(\boldsymbol{\mu}) = \boldsymbol{X}\boldsymbol{\beta}. \tag{10.89}$$

With $\mu_i = \exp(\eta_i)$, we have

$$\frac{\partial \mu_i}{\partial \eta_i} = \exp(\eta_i) = \mu_i \quad , \tag{10.90}$$

so that the estimating equations (10.44) simplify to

$$\sum_{i=1}^{N} \frac{(y_i - \mu_i)x_{ij}}{\mu_i}\mu_i = \sum_{i=1}^{N}(y_i - \mu_i)x_{ij} = 0 \quad (j = 1, \ldots, p). \tag{10.91}$$

For the solution $\hat{\boldsymbol{\mu}}$, this means in matrix representation

$$\boldsymbol{X}'\boldsymbol{y} = \boldsymbol{X}'\hat{\boldsymbol{\mu}} \quad . \tag{10.92}$$

Remark: The existence and uniqueness of the solutions was proven by Birch (1963) for $\mathrm{rank}(\boldsymbol{X}) = p$ and $y_i > 0$.

The covariance matrix (10.56) with the elements w_i (10.49) is of the following form

$$V(\hat{\boldsymbol{\beta}}) = (\boldsymbol{X}'\hat{\boldsymbol{W}}\boldsymbol{X})^{-1}, \quad \hat{\boldsymbol{W}} = \mathrm{diag}(\hat{w}_i) = \mathrm{diag}(\hat{\mu}_i) \tag{10.93}$$

because of $w_i = \left(\frac{\partial \mu_i}{\partial \eta_i}\right)^2 / \mathrm{Var}(y_i) = \mu_i$ (cf. (10.49) and (10.90)) and $\hat{w}_i = \hat{\mu}_i$ as solution of (10.92). For a GLM (10.89) in Poisson variables with $\hat{\theta}_i = \ln(\hat{\mu}_i)$ and $b(\hat{\theta}_i) = \exp(\hat{\theta}_i) = \hat{\mu}_i$ in the model (10.89) and $\tilde{\theta}_i = \ln(y_i)$, $b(\tilde{\theta}_i) = y_i$ in the saturated model, the deviance (10.62) is

$$D(\boldsymbol{y}; \hat{\boldsymbol{\mu}}) = 2 \sum_{i=1}^{N} (y_i \ln(y_i/\hat{\mu}_i) - y_i + \hat{\mu}_i) \quad . \tag{10.94}$$

If the matrix $\boldsymbol{X}$ in (10.89) contains the vector $\mathbf{1}$ as a column (that is the matrix is adjusted to an overall mean), then $\sum y_i = \sum \hat{\mu}_i$ and hence

$$D(\boldsymbol{y}; \hat{\boldsymbol{\mu}}) = 2 \sum y_i \ln(y_i/\hat{\mu}_i) \tag{10.95}$$

is the usual G^2-statistic for loglinear models.

The test statistic (10.63) for testing H_0: $\boldsymbol{\beta}_3 = \mathbf{0}$ in the model

$$\ln(\boldsymbol{\mu}) = (\boldsymbol{X}_1, \ \boldsymbol{X}_3) \begin{pmatrix} \boldsymbol{\beta}_1 \\ \boldsymbol{\beta}_3 \end{pmatrix} = \boldsymbol{X}_2 \boldsymbol{\beta}_2 \tag{10.96}$$

simplifies to

$$G^2(\boldsymbol{X}_1|\boldsymbol{X}_2) = 2 \sum_i \left[y_i \ln\left(\frac{\hat{\mu}_{i(2)}}{\hat{\mu}_{i(1)}}\right) - (\hat{\mu}_{i(2)} - \hat{\mu}_{i(1)}) \right] \tag{10.97}$$

and asymptotically we have

$$G^2(\boldsymbol{X}_1|\boldsymbol{X}_2)_{|H_0} \sim \chi^2_{df} \tag{10.98}$$

with $df = \mathrm{rank}(\boldsymbol{X}_3)$.

10.2.3   Loglinear Models for Multinomial Distributions

For Poisson sampling, the total sample size $n = \sum_{i=1}^{N} n_i$ is random. If the condition n *fixed* is introduced into the Poisson model, then the vector $\{n_1, \ldots, n_N\}$ no longer has an independent Poisson distribution, since one variable n_i restricts the range of the other variables.

For $n = \sum n_i$ fixed, we get the conditional distribution for $\{n_1, \ldots, n_N\}$ (cf. Agresti, 1990, p.38)

$$P(n_i \text{ observations in cell } i, i = 1, \ldots, N | \textstyle\sum n_i = n)$$

$$= \frac{P(n_i \text{ observations in cell } i, i = 1, \ldots, N)}{P(\sum n_i = n)}$$

$$= \frac{\prod_{i=1}^{N} \exp(-\mu_i)\mu_i^{n_i}/n_i!}{\exp(-\sum \mu_i)(\sum \mu_i)^n/n!}$$

$$= \frac{n!}{\prod_i n_i!} \prod \pi_i^{n_i} \qquad (10.99)$$

with

$$\pi_i = \mu_i/(\sum_j \mu_i) \quad . \qquad (10.100)$$

This is the multinomial distribution $M(n; \pi_1, \ldots, \pi_N)$. The marginal distributions for n_i are binomial $B(n_i; \pi_i)$ with $\mathrm{E}(n_i) = n\pi_i$ and $\mathrm{Var}(n_i) = n\pi_i(1 - \pi_i)$. Alternatively, the multinomial distribution for the vector $\{n_1, \ldots, n_N\}$ developes from n independent observations with N categories and the vector $\boldsymbol{\pi}' = (\pi_1, \ldots, \pi_N)$ of category probabilities $(\sum \pi_i = 1)$. This is called *multinomial sampling*.

Product-Multinomial Sampling

The following consideration leads to an extension of the multinomial sampling. We assume that observations of a categorical response variable Y (J categories) for different levels of an explanatory variable X are given. n_{ij} is the frequency in cell $(X = i, Y = j)$. Assume that the $n_{i+} = \sum_{j=1}^{J} n_{ij}$ observations of Y for the ith category of X are independent with the distribution $\{\pi_{1|i}, \ldots, \pi_{J|i}\}$. The cell counts $\{n_{ij}, j = 1, \ldots, J\}$ in the ith category of X then have a multinomial distribution according to

$$\frac{n_{i+}!}{\prod_{j=1}^{J} n_{ij}!} \prod_{j=1}^{J} \pi_{j|i}^{n_{ij}} \quad (i = 1, \ldots, I) \quad . \qquad (10.101)$$

If furthermore the samples are independent for the i, then the joint distribution of the n_{ij} over the $I \times J$ cells equals the product of the I multinomial distributions from (10.101). We call this **product-multinomial sampling** or **independent multinomial sample**.

Regarding the multinomial sample and its distribution (10.99), we have for the ith observation (cf. (10.13)) the following representation

$$f(n_i|\pi_i) = \frac{(n!)^{1/N}}{n_i!} \pi_i^{n_i}$$

$$= \exp(n_i \ln(\pi_i) + \ln[(n_i!)^{1/N}/n_i!] \quad , \qquad (10.102)$$

so that $\ln(\pi_i)$ is the natural parameter.

The n_i are not independent, since $\sum_{i=1}^{N} \pi_i = 1$. With $\sum_{i=1}^{N} n_i = n$, we have

$$\mathrm{E}(n_i) = n\pi_i, \qquad (10.103)$$
$$\mathrm{Var}(n_i) = n\pi_i(1 - \pi_i) \qquad (10.104)$$

and

$$\mathrm{Cov}(n_i, n_j) = -n\pi_i\pi_j \quad (i, j = 1, \ldots, N - 1, i \neq j), \qquad (10.105)$$

if we classify the Nth category as *redundant*. In matrix notation, we have for $n'_* = (n_1, \ldots, n_{N-1})$ the following covariance matrix

$$V(n_*) = n\left(\text{diag}(\pi_*) - \pi_* \pi'_*\right) \tag{10.106}$$

with $\pi'_* = (\pi_1, \ldots, \pi_{N-1})$. Hence

$$\pi' = (\pi_1, \ldots, \pi_N) = (\pi'_*, \pi_N) \quad . \tag{10.107}$$

Let $n' = (n_1, \ldots, n_N) = (n'_*, n_N)$ be the vector of all random cell counts. The vector of expected cell frequencies is then

$$E(n) = \mu = n\pi \quad . \tag{10.108}$$

From (10.102) it follows that $\ln(\pi_i)$ is the natural parameter. The canonical link has to take the condition $\sum \pi_i = 1$ into consideration.

For the probabilities of the model, this means

$$\pi_i = \pi_i(\beta) = \frac{\exp(x'_i \beta)}{\sum_{i=1}^{N} \exp(x'_i \beta)} \tag{10.109}$$

or in matrix notation

$$\pi = \frac{\exp(X\beta)}{1' \exp(X\beta)} \quad . \tag{10.110}$$

This leads to the following loglinear model for the vector μ:

$$
\begin{aligned}
\ln(\mu) &= \ln(n\pi) \\
&= X\beta + [\ln n - \ln(1' \exp(X\beta))]1 \\
&= X\beta + 1\mu_0
\end{aligned}
\tag{10.111}
$$

with

$$\mu_0 = \ln n - \ln(\sum_{i=1}^{N} \exp(x'_i \beta)) \quad . \tag{10.112}$$

Hence, the parameter μ_0 ensures the adjustment onto total sample size n for multinomial sampling.

ML Estimation of β

Since the conditional distribution, given the total sum $n = \sum n_i$, of a Poisson sampling scheme $(n_1, \ldots, n_N)$ is multinomial $M(n; \pi_1, \ldots, \pi_N)$ with $\pi_i = \mu_i/(\sum \mu_j)$ (cf. (10.99)), the ML estimate of β can be derived by the following consideration (cf. Agresti, 1990, p. 455 based on results by Birch, 1963 and McCullagh and Nelder, 1989).

The loglinear model (10.89) for the Poisson distribution is specified as an inhomogeneous approach by separation of a constant α as follows

$$\ln(\mu_i) = \alpha + \boldsymbol{x}_i'\boldsymbol{\beta} \quad . \tag{10.113}$$

Using (10.113), the Poisson log-likelihood results from (10.88) (y_i replaced by n_i)

$$
\begin{aligned}
l(\alpha, \boldsymbol{\beta}) &= \sum_{i=1}^{N} n_i \ln(\mu_i) - \sum_{i=1}^{N} \mu_i \\
&= \sum_i n_i(\alpha + \boldsymbol{x}_i'\boldsymbol{\beta}) - \sum_i \exp(\alpha + \boldsymbol{x}_i'\boldsymbol{\beta}) \\
&= n\alpha + \sum_i n_i \boldsymbol{x}_i'\boldsymbol{\beta} - \tau \tag{10.114}
\end{aligned}
$$

with

$$\tau = \sum_{i=1}^{N} \mu_i = \sum_{i=1}^{N} \exp(\alpha + \boldsymbol{x}_i'\boldsymbol{\beta}) \quad . \tag{10.115}$$

Using

$$\ln(\tau) = \alpha + \ln(\sum_{i=1}^{N} \exp(\boldsymbol{x}_i'\boldsymbol{\beta})) \tag{10.116}$$

we obtain the log-likelihood $l(\alpha, \boldsymbol{\beta})$ for the loglinear Poisson model in the following representation

$$
\begin{aligned}
l(\tau, \boldsymbol{\beta}) &= \left[\sum_{i=1}^{N} n_i \boldsymbol{x}_i'\boldsymbol{\beta} - n \ln \left(\sum \exp(\boldsymbol{x}_i'\boldsymbol{\beta}) \right) \right] + [n \ln(\tau) - \tau] \tag{10.117} \\
&= f_1(\boldsymbol{\beta}) + f_2(\tau) \quad . \tag{10.118}
\end{aligned}
$$

The probabilities π_i of the multinomial model are linked to the expected frequencies μ_i of the Poisson model by the relationship $\pi_i = \mu_i/(\sum \mu_j)$. Thus, we have

$$
\begin{aligned}
\pi_i &= \frac{\mu_i}{\sum_{j=1}^{N} \mu_j} = \frac{\exp(\alpha + \boldsymbol{x}_i'\boldsymbol{\beta})}{\sum \exp(\alpha + \boldsymbol{x}_i'\boldsymbol{\beta})} \\
&= \frac{\exp(\alpha) \exp(\boldsymbol{x}_i'\boldsymbol{\beta})}{\exp(\alpha) \sum \exp(\boldsymbol{x}_i'\boldsymbol{\beta})} \quad , \\
\sum n_i \ln(\pi_i) &= \sum n_i \boldsymbol{x}_i'\boldsymbol{\beta} - \sum n_i \ln \left[\sum \exp(\boldsymbol{x}_i'\boldsymbol{\beta}) \right] \\
&= \sum n_i \boldsymbol{x}_i'\boldsymbol{\beta} - n \ln[\sum \exp(\boldsymbol{x}_i'\boldsymbol{\beta})] \\
&= f_1(\boldsymbol{\beta}) \quad . \tag{10.119}
\end{aligned}
$$

According to (10.102), $\sum n_i \ln(\pi_i)$ is the kernel of the log-likelihood of multinomial sampling under the condition $n = \sum_{i=1}^{N} n_i$.

On the other hand if the condition is omitted, then $n = \sum n_i$ in the Poisson model has a Poisson distribution with $E(n) = \sum \mu_i = \tau$, so that the second term $f_2(\tau)$ in (10.118) is the log-likelihood for τ. As β appears in the first term $f_1(\beta)$ only, the ML estimates $\hat{\beta}$ of β for the Poisson log-likelihood $l(\alpha, \beta)$ and the multinomial log-likelihood $f_1(\beta)$ are identical. The loglinear Poisson model requires the definition of the additional parameter α.

Result: The ML parameter estimates $\hat{\beta}$ in the loglinear model for the multinomial sampling and in the model for Poisson sampling are the same.

Remark: This result holds for the product-multinomial sampling as well, as long as the fixed marginal sums are taken into account by the corresponding loglinear models (that is, as long as the estimated expected marginal sums equal the observed marginal sums).

Asymptotic Covariance Matrix

The derivation of the asymptotic distribution of $\hat{\pi}$ and $\hat{\beta}$ is based on the multivariate delta-method and the multivariate Central Limit Theorem (Rao, 1973, p. 128). Here we omit the proof and refer to Chapter 6 and 12 in Agresti (1990).

The unbiased estimate of the asymptotic covariance matrix of $\hat{\beta}$ is (with $\hat{\mu} = n\hat{\pi}$)

$$
\begin{aligned}
\hat{V}(\hat{\beta}) &= \{X'[\mathrm{diag}(\hat{\pi}) - \hat{\pi}\hat{\pi}']X\}^{-1}/n \\
&= \{X'[\mathrm{diag}(\hat{\mu}) - \hat{\mu}\hat{\mu}'/n]X\}^{-1} \quad .
\end{aligned}
\tag{10.120}
$$

Analogously, we have

$$
\hat{V}(\hat{\pi}) = [\mathrm{diag}(\hat{\pi}) - \hat{\pi}\hat{\pi}']X\,V(\hat{\beta})X'[\mathrm{diag}(\hat{\pi}) - \hat{\pi}\hat{\pi}'] \quad .
\tag{10.121}
$$

Remark: In the literature, the notation m is often used for the expected frequencies instead of μ, especially for contingency tables.

10.3 Linear Models for Two-Dimensional Relationships – ANOVA

After these more theoretical preliminary remarks, the models that correspond to the linear models of the two-way ANOVA with continuous response y_{ijk} can now be introduced. We now have the cell counts n_{ij} as response.

Essentially, we will confine ourselves to the sample scheme of a multinomial or product-multinomial distribution. To begin with, we explain the relationship between the two-way analysis of variance and the loglinear model of a two-dimensional contingency table. In the model (6.1) of the two-way analysis of variance, two factors A and B have an effect on a continuous response variable Y:

$$
\begin{aligned}
y_{ijk} = \mu + \alpha_i + \beta_j + (\alpha\beta)_{ij} + \epsilon_{ijk} \\
i = 1, \ldots, I \, , \; j = 1, \ldots, J \, , \; k = 1, \ldots, K \quad .
\end{aligned}
\tag{10.122}
$$

Here we assume that the y_{ijk} are independently and normally distributed:

$$y_{ijk} \sim N(m_{ij}, \sigma^2) \quad \text{with}$$
$$m_{ij} = \mu + \alpha_i + \beta_j + (\alpha\beta)_{ij} \quad . \tag{10.123}$$

The main aim is to examine the structure of the m_{ij}. For this purpose, the OLS estimates (which correspond to the ML estimates)

$$\begin{aligned} \hat{m}_{ij} &= \hat{\mu} + \hat{\alpha}_i + \hat{\beta}_j + \widehat{(\alpha\beta)}_{ij} \\ &= (y_{...}) + (y_{i..} - y_{...}) + (y_{.j.} - y_{...}) + (y_{ij.} - y_{i..} - y_{.j.} + y_{...}) \end{aligned} \tag{10.124}$$

(cf. $(6.12) - (6.16)$) are analysed. The $\hat{m}_{ij}$ are independently distributed according to

$$\hat{m}_{ij} \sim N(m_{ij}, \sigma^2/K) \quad . \tag{10.125}$$

σ^2 is estimated in the model (10.122) by $\hat{\sigma}_\Omega^2$ (6.38). The estimates $\hat{m}_{ij}$ and $\hat{\sigma}^2$ change if restrictions are incorporated into the model. The restriction "no interaction" means $(\alpha\beta)_{ij} = 0$ (for all i, j) so that we get

$$m_{ij} = \mu + \alpha_i + \beta_j \tag{10.126}$$

and as ML estimates we have

$$\hat{m}_{ij} = y_{...} + (y_{i..} - y_{...}) + (y_{.j.} - y_{...}). \tag{10.127}$$

$\hat{\sigma}_\omega^2$ is obtained from (6.39). $H_0 : (\alpha\beta)_{ij} = 0$ is tested with the likelihood-ratio test.

10.4   Two-Way Contingency Tables

Suppose that we have a realization (sample) of two categorical variables with I and J categories with sample size n. This yields n_{ij} observations in $N = I \times J$ cells of the contingency table. The probabilities π_{ij} of the multinomial distribution form the joint distribution. Independence of the variables is equivalent to

$$\pi_{ij} = \pi_{i+} \pi_{+j} \quad \text{(for all } i, j). \tag{10.128}$$

If this is applied to the expected cell frequencies $m_{ij} = n\pi_{ij}$, the condition of independence is equivalent to

$$m_{ij} = n\pi_{i+}\pi_{+j} \quad . \tag{10.129}$$

The modelling of the $I \times J$-table is based on this relation as independence model on the logarithmic scale:

$$\ln(m_{ij}) = \ln n + \ln \pi_{i+} + \ln \pi_{+j} \quad . \tag{10.130}$$

Hence, the effects of the rows and columns on $\ln(m_{ij})$ are additive.

An alternative expression, following the models of analysis of variance of the form

$$y_{ij} = \mu + \alpha_i + \beta_j + \epsilon_{ij} , \quad \left(\sum \alpha_i = \sum \beta_j = 0 \right) \tag{10.131}$$

is given by

$$\ln(m_{ij}) = \mu + \lambda_i^X + \lambda_j^Y \tag{10.132}$$

with

$$\lambda_i^X = \ln(\pi_{i+}) - \frac{1}{I} \left(\sum_{k=1}^I \ln(\pi_{k+}) \right) , \tag{10.133}$$

$$\lambda_j^Y = \ln(\pi_{+j}) - \frac{1}{J} \left(\sum_{k=1}^J \ln(\pi_{+k}) \right) , \tag{10.134}$$

$$\mu = \ln n + \frac{1}{I} \left(\sum_{k=1}^I \ln(\pi_{k+}) \right) + \frac{1}{J} \left(\sum_{k=1}^J \ln(\pi_{+k}) \right) . \tag{10.135}$$

The parameters satisfy the constraints

$$\sum_{i=1}^I \lambda_i^X = \sum_{j=1}^J \lambda_j^Y = 0 \tag{10.136}$$

which make the parameters identifiable.

Remark: The λ_i^X are the deviations of $\ln(\pi_{i+})$ from their mean $\dfrac{1}{I} \displaystyle\sum_{i=1}^I \pi_{i+}$ so that

$$\sum_{i=0}^I \lambda_i^X = 0.$$

The model (10.132) is called *loglinear model of independence* in a two-way contingency table. The related *saturated model* contains the additional interaction parameters λ_{ij}^{XY}:

$$\ln(m_{ij}) = \mu + \lambda_i^X + \lambda_j^Y + \lambda_{ij}^{XY} . \tag{10.137}$$

This model describes the perfect fit. The interaction parameters satisfy

$$\sum_{i=1}^I \lambda_{ij}^{XY} = \sum_{j=1}^J \lambda_{ij}^{XY} = 0 . \tag{10.138}$$

Given λ_{ij}'s in the first $(I-1)(J-1)$ cells, these constraints determine the λ_{ij} in the last row or the last column. Thus, the saturated model contains

$$\underset{(\mu)}{1} \; + \; \underset{(\lambda_i^X)}{(I-1)} \; + \; \underset{(\lambda_j^Y)}{(J-1)} \; + \; \underset{(\lambda_{ij}^{XY})}{(I-1)(J-1)} \; = I \cdot J \qquad (10.139)$$

independent parameters (and hence 0 degrees of freedom). For the independence model, the number of independent parameters equals

$$1 + (I-1) + (J-1) = I + J - 1 \qquad (10.140)$$

(and hence $I \times J - I - J + 1 = (I-1)(J-1)$ degrees of freedom).

Interpretation of the Parameters

Loglinear models estimate the effects of rows and columns on $\ln(m_{ij})$. For this, no distinction is made between explanatory and response variables. The information of the rows and columns effect m_{ij} symmetrically.

Consider the simplest case – the $I \times 2$-table (independence model). According to (10.132), the logit of the binary variable Y equals

$$\begin{aligned}
\ln\left(\frac{\pi_{1/i}}{\pi_{2/i}}\right) &= \ln\left(\frac{m_{i1}}{m_{i2}}\right) \\
&= \ln(m_{i1}) - \ln(m_{i2}) \\
&= (\mu + \lambda_i^X + \lambda_1^Y) - (\mu + \lambda_i^X + \lambda_2^Y) \\
&= \lambda_1^Y - \lambda_2^Y. \qquad (10.141)
\end{aligned}$$

The logit is the same in every row, and hence independent of X or the categories $i = 1, \ldots, I$ respectively. For the constraints we have

$$\lambda_1^Y + \lambda_2^Y = 0 \qquad \Longrightarrow \qquad \lambda_1^Y = -\lambda_2^Y ,$$

$$\Longrightarrow \qquad \ln\left(\frac{\pi_{1/i}}{\pi_{2/i}}\right) = 2\lambda_1^Y \qquad (i = 1, \ldots, I).$$

Hence we obtain

$$\frac{\pi_{1/i}}{\pi_{2/i}} = \exp(2\lambda_1^Y) \qquad (i = 1, \ldots, I). \qquad (10.142)$$

In each category of X, the odds that Y is in category 1 rather than in category 2 is equal to $\exp(2\lambda_1^Y)$, if the independence model holds.

The following relationship exists between the odds ratio θ in a 2×2-table and the saturated loglinear model:

$$\begin{aligned}
\ln(\theta) &= \ln\left(\frac{m_{11}\, m_{22}}{m_{12}\, m_{21}}\right) \\
&= \ln(m_{11}) + \ln(m_{22}) - \ln(m_{12}) - \ln(m_{21}) \\
&= (\mu + \lambda_1^X + \lambda_1^Y + \lambda_{11}^{XY}) + (\mu + \lambda_2^X + \lambda_2^Y + \lambda_{22}^{XY}) \\
&\quad - (\mu + \lambda_1^X + \lambda_2^Y + \lambda_{12}^{XY}) - (\mu + \lambda_2^X + \lambda_1^Y + \lambda_{21}^{XY}) \\
&= \lambda_{11}^{XY} + \lambda_{22}^{XY} - \lambda_{12}^{XY} - \lambda_{21}^{XY}.
\end{aligned}$$

Because of $\sum\limits_{i=1}^{2} \lambda_{ij}^{XY} = \sum\limits_{j=1}^{2} \lambda_{ij}^{XY} = 0$, we have $\lambda_{11}^{XY} = \lambda_{22}^{XY} = -\lambda_{12}^{XY} = -\lambda_{21}^{XY}$ and thus $\ln\theta = 4\lambda_{11}^{XY}$. Hence the odds ratio in a 2×2-table equals

$$\theta = \exp(4\lambda_{11}^{XY}) , \tag{10.143}$$

and is dependent upon the association parameter in the saturated loglinear model. When there is no association (that is $\lambda_{ij}^{XY} = 0$) we have $\theta = 1$.

Example 10.1: In the following, we demonstrate the analysis of a two-way contingency table by loglinear models of the different types. The relationship of interest is that of the developement of tartar and tobacco consumption (cf. Tables 1.6 and 1.7).For a better overall view, the contingency table is shown again below as Table 10.1.

		no tartar	medium tartar	high-level tartar	
	j	1	2	3	$n_{i\cdot}$
i					
non-smoker	1	284	236	48	568
smoker, less than 6.5g per day	2	606	983	209	1798
smoker, more than 6.5g per day	3	1028	1871	425	3324
$n_{\cdot j}$		1918	3090	682	5690

Table 10.1: Contingency table tobacco consumption/tartar

For the analysis, the program LOGGY 1.0 (cf. Heumann and Jacobsen, 1993) is used. This program first assesses all models seperately by their G^2-value (Table 10.2). The models for one isolated main factor effect

$$\ln(m_{ij}) = \mu + \lambda_i^{\text{tartar}}$$

and

$$\ln(m_{ij}) = \mu + \lambda_j^{\text{smoking}}$$

are rejected. The model of independence (cf. (10.132)) with a G^2-value of 76.23 is rejected as well, although all parameter estimates are significant (Table 10.3). Hence, the only remaining model is the saturated model (10.137) with $G^2 = 0$ (perfect fit). The standardized parameter estimates of this model are shown in Table 1.7.

Even though all parameters are significant (except for smoking (low)/tartar (medium)) they cannot be interpreted seperately.

model	G^2	df	p-value
smoking	1740.29	6	0.00
tartar	2444.40	6	0.00
tartar+smoking	76.23	4	0.00
tartar*smoking	0.00	0	1.00

Table 10.2: Assessment of the four possible models tartar/tobacco consumption

parameter	estimate	estimate/ste	sterror
const	6.0420	313.30429	0.01928
tartar (non)	0.1857	8.94933	0.02075
tartar (medium)	0.6626	34.69686	0.01910
smoking (no)	−0.9730	−32.84409	0.02963
smoking (little)	0.1793	8.21489	0.02182

Table 10.3: Parameter estimation in the independence model

10.5 Three-Way Contingency Tables

Table 10.4 shows a three-way contingency table for the risk of an endodontic treatment having dependency on age of patients and type of prosthetic construction.

Age group	Type of construction	Endodontic treatment	
		yes	no
< 60	H	62	1041
	B	23	463
≥ 60	H	70	755
	B	30	215
Σ		185	2474

Table 10.4: $2 \times 2 \times 2$-table: Endodontic risk

In addition to the bivariate associations we want to model an overall association. The three variables are mutually independent if the following independence model for the cell frequencies m_{ijk} on a (logarithmic scale) holds:

$$\ln(m_{ijk}) = \mu + \lambda_i^X + \lambda_j^Y + \lambda_k^Z \quad . \tag{10.144}$$

(In the above example we have X: age group, Y: type of construction, Z: endodontic treatment).

The variable Z is independent of the joint distribution of X and Y (jointly independent) if

$$\ln(m_{ijk}) = \mu + \lambda_i^X + \lambda_j^Y + \lambda_k^Z + \lambda_{ij}^{XY} . \tag{10.145}$$

A third type of independence (conditional independence of two variables given a fixed category of the third variable) is expressed by the following model (j fixed !):

$$\ln(m_{ijk}) = \mu + \lambda_i^X + \lambda_j^Y + \lambda_k^Z + \lambda_{ij}^{XY} + \lambda_{jk}^{YZ} . \tag{10.146}$$

This is the approach for the conditional independence of X and Z at level j of Y. If this holds for all $j = 1, \ldots, J$, then X and Z are called conditionally independent given Y. Similarly, if X and Y are conditionally independent at level k of Z, the parameters λ_{ij}^{XY} and λ_{jk}^{YZ} in (10.146) are replaced by the parameters λ_{ik}^{XZ} and λ_{jk}^{YZ}. The parameters with two subscripts describe two-way interaction. The appropriate conditions for the cell probabilities are:

a) mutual independence of X, Y, Z:

$$\pi_{ijk} = \pi_{i++}\pi_{+j+}\pi_{++k} \quad \text{(for all } i, j, k) \tag{10.147}$$

b) joint independence:
   Y is jointly independent of X and Z if

$$\pi_{ijk} = \pi_{i+k}\pi_{+j+} \quad \text{(for all } i, j, k). \tag{10.148}$$

c) conditional independence:
   X and Y are conditionally independent of Z if

$$\pi_{ijk} = \frac{\pi_{i+k}\pi_{+jk}}{\pi_{++k}} \quad \text{(for all } i, j, k). \tag{10.149}$$

holds.

The most general loglinear model (saturated model) for three-way contingency tables is the following:

$$\ln(m_{ijk}) = \mu + \lambda_i^X + \lambda_j^Y + \lambda_k^Z + \lambda_{ij}^{XY} + \lambda_{ik}^{XZ} + \lambda_{jk}^{YZ} + \lambda_{ijk}^{XYZ} . \tag{10.150}$$

The last parameter describes the three-factor interaction. All association parameters that describe the deviation from the total mean μ, satisfy the constraints

$$\sum_{i=1}^{I} \lambda_{ij}^{XY} = \sum_{j=1}^{J} \lambda_{ij}^{XY} = \ldots = \sum_{k=1}^{K} \lambda_{ijk}^{XYZ} = 0 . \tag{10.151}$$

Similarly, we have for the main effects:

$$\sum_{i=1}^{I} \lambda_i^X = \sum_{j=1}^{J} \lambda_j^Y = \sum_{k=1}^{K} \lambda_k^Z = 0 .\tag{10.152}$$

From the general model (10.150) submodels can be constructed. For this, the hierarchical principle of construction is preferred. A model is called hierarchical if, in addition to significant higher-order effects, it contains all lower-order effects of the variables included in the higher-order effects, even if these parameter estimates are not statistically significant. For instance, if the model contains the association parameter λ_{ik}^{XZ}, it must also contain λ_i^X and λ_k^Z:

$$\ln(m_{ijk}) = \mu + \lambda_i^X + \lambda_k^Z + \lambda_{ik}^{XZ} .\tag{10.153}$$

A symbol is assigned to the various hierarchical models (Table 10.5). As an alternative to these symbols, the following notation is well-tried: a plus sign means independence, a '$*$ means association. Hence the following representations are equivalent:

(X,Y) $X + Y$
(XY,Z) $X * Y + Z$
(XY,XZ,YZ) $X * Y + X * Z + Y * Z$
(XYZ) $X * Y * Z$.

Similar to 2×2-tables, a close relationship exists between the parameters of the model and the odds ratio. Given a $2 \times 2 \times 2$-table under the constraints (10.151) and (10.152) for instance, we have

$$\frac{\theta_{11(1)}}{\theta_{11(2)}} = \frac{\frac{\pi_{111}\pi_{221}}{\pi_{211}\pi_{121}}}{\frac{\pi_{112}\pi_{222}}{\pi_{212}\pi_{122}}} = \exp(8\lambda_{111}^{XYZ}) .\tag{10.154}$$

This is the conditional odds ratio of X and Y given the levels $k = 1$ (numerator) and $k = 2$ (denominator) of Z. The same holds for X and Z under Y and for Y and Z under X. In the population, we thus have for the three-way interaction λ_{111}^{XYZ}

$$\frac{\theta_{11(1)}}{\theta_{11(2)}} = \frac{\theta_{1(1)1}}{\theta_{1(2)1}} = \frac{\theta_{(1)11}}{\theta_{(2)11}} = \exp(8\lambda_{111}^{XYZ}) .\tag{10.155}$$

In case of independence in the equivalent subtables, the odds ratios (of the populations) equal 1. The sample odds ratio gives an indication of a deviation from independence.

Consider the conditional odds ratio (10.154) for Table 10.4. We then have a value of 1.80. This indicates a positive tendency for an increased risk of endodontic treatment in comparing the following subtables

Loglinear model	Symbol
$\ln(m_{ij+}) = \mu + \lambda_i^X + \lambda_j^Y$	(X,Y)
$\ln(m_{i+k}) = \mu + \lambda_i^X + \lambda_k^Z$	(X,Z)
$\ln(m_{+jk}) = \mu + \lambda_j^Y + \lambda_k^Z$	(Y,Z)
$\ln(m_{ijk}) = \mu + \lambda_i^X + \lambda_j^Y + \lambda_k^Z$	(X,Y,Z)
$\ln(m_{ijk}) = \mu + \lambda_i^X + \lambda_j^Y + \lambda_k^Z + \lambda_{ij}^{XY}$	(XY,Z)
$\vdots$	$\vdots$
$\ln(m_{ijk}) = \mu + \lambda_i^X + \lambda_j^Y + \lambda_{ij}^{XY}$	(XY)
$\vdots$	$\vdots$
$\ln(m_{ijk}) = \mu + \lambda_i^X + \lambda_j^Y + \lambda_k^Z + \lambda_{ij}^{XY} + \lambda_{ik}^{XZ}$	(XY,XZ)
$\vdots$	$\vdots$
$\ln(m_{ijk}) = \mu + \lambda_i^X + \lambda_j^Y + \lambda_k^Z + \lambda_{ij}^{XY} + \lambda_{ik}^{XZ} + \lambda_{jk}^{YZ}$	(XY,XZ,YZ)
$\vdots$	$\vdots$
$\ln(m_{ijk}) = \mu + \lambda_i^X + \lambda_j^Y + \lambda_k^Z + \lambda_{ij}^{XY} + \lambda_{ik}^{XZ} + \lambda_{jk}^{YZ} + \lambda_{ijk}^{XYZ}$	(XYZ)

Table 10.5: Symbols of the hierarchical models for three-way contingency tables

	H	B
< 60	62	23
≥ 60	70	30

(endodontic treatment)

and

	H	B
< 60	1041	463
≥ 60	755	215

(no endodontic treatment)

The relationship (10.155) is also valid for the sample version. Thus a comparison of the following subtables

	Treatment yes	no
H	62	1041
B	23	463

(< 60)

and

	Treatment yes	no
H	70	755
B	30	215

(≥ 60)

or

	Treatment yes	no
< 60	62	1041
≥ 60	70	755

(H)

and

	Treatment yes	no
< 60	23	463
≥ 60	30	215

(B)

leads to the same sample value 1.80 and hence $\hat{\lambda}_{111}^{XYZ} = 0.073$. Calculations for Table 10.4:

$$\frac{\hat{\theta}_{11(1)}}{\hat{\theta}_{11(2)}} = \frac{\frac{n_{111}n_{221}}{n_{211}n_{121}}}{\frac{n_{112}n_{222}}{n_{212}n_{122}}} = \frac{\frac{62\cdot30}{70\cdot23}}{\frac{1041\cdot215}{755\cdot463}} = \frac{1.1553}{0.6403} = 1.80 \; ,$$

$$\frac{\hat{\theta}_{(1)11}}{\hat{\theta}_{(2)11}} = \frac{\frac{n_{111}n_{122}}{n_{121}n_{112}}}{\frac{n_{211}n_{222}}{n_{221}n_{212}}} = \frac{\frac{62\cdot463}{23\cdot1041}}{\frac{70\cdot215}{30\cdot755}} = \frac{1.1989}{0.6645} = 1.80 \; ,$$

$$\frac{\hat{\theta}_{1(1)1}}{\hat{\theta}_{1(2)1}} = \frac{\frac{n_{111}n_{212}}{n_{211}n_{112}}}{\frac{n_{121}n_{222}}{n_{221}n_{122}}} = \frac{\frac{62\cdot755}{70\cdot1041}}{\frac{23\cdot215}{30\cdot463}} = \frac{0.6424}{0.3560} = 1.80 \; .$$

10.6 Estimation of Parameters in Loglinear Models for Contingency Tables

For a chosen loglinear model and an accepted probability model of the population (Poisson or multinomial distribution), the expected cell frequencies m_{ijk} are to be estimated by $\hat{m}_{ijk}$, by means of the observed sample cell counts n_{ijk}. The ML estimates depend on the data n_{ijk} by the sufficient statistics only (marginal sums).

The most parsimonious model (X, Y, Z) without interaction requires for the estimation of λ_i^X, λ_j^Y, and λ_k^Z only the two-way marginal sums n_{i++}, n_{+j+}, and n_{++k}. Models with interaction require the appropriate one-way marginal sums (e.g., n_{i+k} for XZ-interaction). The ML estimates of the marginal expectations equal the marginal sums, as for example,

$$\hat{m}_{ij+} = n_{ij+}$$
$$\hat{m}_{i++} = n_{i++} \qquad \text{etc.}$$

The estimated $\hat{m}_{ijk}$'s have to satisfy these ML equations whereas the marginal conditions of the models are to be taken into account. In numerous submodels of the hierarchical model, the ML equations have an explicit solution, but in many others not. For this case iterative algorithms exist.

For the model of a three-way contingency table, exact solutions of the estimating equations for the $\hat{m}_{ijk}$ exist for all but one submodel (Table 10.6). Calculation of the $\hat{m}_{ijk}$ is done according to the usual rule

$$\text{ML estimate of } f(\alpha, \beta, \gamma) = f(\hat{\alpha}, \hat{\beta}, \hat{\gamma})$$

where $\hat{\alpha}, \hat{\beta}, \hat{\gamma}$ are the ML estimates of the parameters and with $\hat{m}_{ijk} = \hat{\pi}_{ijk} \cdot n$. For the model (XY, XZ, YZ) no explicit solution exists. The goodness-of-fit of the models (cf. (10.95)) is assessed by the statistic

$$G^2 = 2 \sum_{i,j,k} n_{ijk} \ln \left(\frac{n_{ijk}}{\hat{m}_{ijk}} \right) . \qquad (10.156)$$

Model	$\hat{m}_{ijk}$	Probability
(X, Y, Z)	$\dfrac{n_{i++}n_{+j+}n_{++k}}{n^2}$	$\pi_{ijk} = \pi_{i++}\pi_{+j+}\pi_{++k}$
		(independence model)
(XY, Z)	$\dfrac{n_{ij+}n_{++k}}{n}$	$\pi_{ijk} = \pi_{ij+}\pi_{++k}$
(XZ, Y)	$\dfrac{n_{i+k}n_{+j+}}{n}$	$\pi_{ijk} = \pi_{i+k}\pi_{+j+}$
(YZ, X)	$\dfrac{n_{+jk}n_{i++}}{n}$	$\pi_{ijk} = \pi_{+jk}\pi_{i++}$
(XY, XZ)	$\dfrac{n_{ij+}n_{i+k}}{n_{i++}}$	$\pi_{ijk} = \dfrac{\pi_{ij+}\pi_{i+k}}{\pi_{i++}}$
(XY, YZ)	$\dfrac{n_{ij+}n_{+jk}}{n_{+j+}}$	$\pi_{ijk} = \dfrac{\pi_{ij+}\pi_{+jk}}{\pi_{+j+}}$
(XZ, YZ)	$\dfrac{n_{i+k}n_{+jk}}{n_{++k}}$	$\pi_{ijk} = \dfrac{\pi_{i+k}\pi_{+jk}}{\pi_{++k}}$
(XYZ)	n_{ijk}	no approach (saturated model)

Table 10.6: ML estimates $\hat{m}_{ijk}$ (Agresti, 1990, p. 170)

This statistic G^2 is asymptotically χ^2-distributed with the following degrees of freedom:

df = total number of cells – number of independent parameters in the model.

Table 10.7 summarizes the degrees of freedom for a three-way contingency table and the submodels of the hierarchy.

Notice that the degrees of freedom in the saturated model (XYZ) is 0. In the independence model (X, Y, Z), we have

$$\underset{(\mu)}{1} \; + \; \underset{(\lambda_i^X)}{(I-1)} \; + \; \underset{(\lambda_j^Y)}{(J-1)} \; + \; \underset{(\lambda_k^Z)}{(K-1)}$$

independent parameters (with $\displaystyle\sum_{i=1}^{I}\lambda_i^X = \sum_{j=1}^{J}\lambda_j^Y = \sum_{k=1}^{K}\lambda_k^Z = 0$) and IJK cells, and hence

$$df = IJK - (1 + (I-1) + (J-1) + (K-1)) = IJK - I - J - K + 2 \,.$$

Model	df
(X, Y, Z)	$IJK - (I + J + K) + 2$
(XY, Z)	$(K - 1)(IJ - 1)$
(XZ, Y)	$(J - 1)(IK - 1)$
(YZ, X)	$(I - 1)(JK - 1)$
(XY, YZ)	$J(I - 1)(K - 1)$
(XZ, YZ)	$K(I - 1)(J - 1)$
(XY, XZ)	$I(J - 1)(K - 1)$
(XY, XZ, YZ)	$(I - 1)(J - 1)(K - 1)$
(XYZ)	0

Table 10.7: Degrees of freedom for the submodels of a three-way contingency table

This equals the number of parameters, which have to be equated to zero in the saturated model, in order to obtain the model (X, Y, Z):

$$\underset{(\lambda_{ij}^{XY})}{(I - 1)(J - 1)} + \underset{(\lambda_{ik}^{XZ})}{(I - 1)(K - 1)} + \underset{(\lambda_{jk}^{YZ})}{(J - 1)(K - 1)} +$$
$$\underset{(\lambda_{ijk}^{XYZ})}{(I - 1)(J - 1)(K - 1)} \quad ,$$

and hence

$$
\begin{aligned}
df &= IJ - I - J + 1 + IK - I - K + 1 + JK - J - K + 1 \\
&\quad + IJK - IK - JK - IJ + K + I + J - 1 \\
&= IJK - (I + J + K) + 2 \; .
\end{aligned}
$$

Model Choice

For testing H_0: *submodel is valid*, the test statistic

$$
\begin{aligned}
G^2(\text{submodel} \mid & \text{ large model}) \\
&= G^2(\text{submodel}) - G^2(\text{large model}) \quad (10.157)
\end{aligned}
$$

is used, which has the asymptotic distribution

$$\chi_{df}^2 \quad \text{with } df = df(\text{submodel}) - df(\text{large model}) \quad (10.158)$$

(cf. (10.60) and (10.98)).

10.6.1   The Special Case of Binary Response

If one of the variables is a binary response variable (in our example Z: endodontic treatment), these models lead to the already known logit model.

Given the independence model

$$\ln(m_{ijk}) = \mu + \lambda_i^X + \lambda_j^Y + \lambda_k^Z, \tag{10.159}$$

we then have for the logit of the response variable Z

$$\ln\left(\frac{m_{ij1}}{m_{ij2}}\right) = \lambda_1^Z - \lambda_2^Z. \tag{10.160}$$

With the constraint $\sum_{k=1}^{2} \lambda_k^Z = 0$, we have

$$\ln\left(\frac{m_{ij1}}{m_{ij2}}\right) = 2\lambda_1^Z \qquad (\text{for all } i, j) \,. \tag{10.161}$$

The higher the value of λ_1^Z, the higher the risk for category $Z = 1$ (endodontic treatment). If the other two variables are binary as well, implying a $2 \times 2 \times 2$-table, and considering the constraints

$$\lambda_2^X = -\lambda_1^X \quad , \quad \lambda_2^Y = -\lambda_1^Y \quad , \quad \lambda_2^Z = -\lambda_1^Z$$

the model (10.159) can be expressed as follows:

$$\begin{pmatrix} \ln(m_{111}) \\ \ln(m_{112}) \\ \ln(m_{121}) \\ \ln(m_{122}) \\ \ln(m_{211}) \\ \ln(m_{212}) \\ \ln(m_{221}) \\ \ln(m_{222}) \end{pmatrix} = \begin{pmatrix} 1 & 1 & 1 & 1 \\ 1 & 1 & 1 & -1 \\ 1 & 1 & -1 & 1 \\ 1 & 1 & -1 & -1 \\ 1 & -1 & 1 & 1 \\ 1 & -1 & 1 & -1 \\ 1 & -1 & -1 & 1 \\ 1 & -1 & -1 & -1 \end{pmatrix} \begin{pmatrix} \mu \\ \lambda_1^X \\ \lambda_1^Y \\ \lambda_1^Z \end{pmatrix}, \tag{10.162}$$

which is equivalent to

$$\ln(\boldsymbol{m}) = \boldsymbol{X}\boldsymbol{\beta} \,. \tag{10.163}$$

This corresponds to the effect coding of categorical variables. In the usual regression approach, the above representation is equivalent to the model

$$\boldsymbol{y} = \boldsymbol{X}\boldsymbol{\beta} + \boldsymbol{\epsilon} \quad | \quad \boldsymbol{r} = \boldsymbol{R}\boldsymbol{\beta}$$

where $\boldsymbol{r} = \boldsymbol{R}\boldsymbol{\beta}$ codes the exact restrictions for the parameters (cf. Chapter 3).

Assume $\boldsymbol{n}$ to be the vector of the observed cell counts n_{ijk}. The ML equation is then (cf. (10.92))

$$\boldsymbol{X}'\boldsymbol{n} = \boldsymbol{X}'\hat{\boldsymbol{m}} \,. \tag{10.164}$$

The estimated asymptotic covariance matrix for Poisson sampling reads as follows
(cf. (10.93))

$$\hat{V}(\hat{\beta}) = [\mathbf{X}'(\text{diag}(\hat{m}))\mathbf{X}]^{-1} \tag{10.165}$$

where $\text{diag}(\hat{m})$ has the elements of $\hat{m}$ on the main diagonal. The solution of
the normal equation (10.164) is obtained by Fisher-scoring or any other iterative
algorithm such as the IPF.

Example 10.2: In the following, we demonstrate model choice and parameter estimation for the three-way contingency table of endodontic treatment, age
group, and type of construction from Table 10.4. First, we examine the output
from LOGGY 1.0 for all possible models (Table 10.8). In case of three variables
X, Y and Z, the symbols denote:

$X + Y + Z$	:	independence model (mutual independence)
$X + Y * Z$	:	X jointly independent of Y and Z
$X * Y + X * Z$	:	for given X, Y and Z are independent
$X * Y * Z$	:	saturated model.

From Table 10.8, it is apparent that all models except the following are rejected
(H_0: *model valid* is tested with the appropriate G^2-value):

$$A * C + A * E \quad , \tag{10.166}$$
$$A * C + A * E + C * E \quad , \tag{10.167}$$
$$A * C * E \quad . \tag{10.168}$$

Here we use the abbreviations A: age, C: type of construction, and E: endodontic
treatment. We first examine the model (10.167) in comparison with the saturated
model.

The three-way interaction in the saturated model is estimated (standardized)
by $\hat{\lambda}_1^{ACE} = 1.73$. Hence, it is not significant (compared to the quantile 1.96). This
is an argument for proceeding to the next smaller model (10.167). An alternative
argument is possible with the model choice statistic $G^2(A * C + A * E + C * E|A *
C * E)$ (cf. (10.157)). We have (cf. Table 10.9)

$$G^2(A * C + A * E + C * E) - G^2(A * C * E) = 3.01 - 0 = \chi_1^2 \quad \text{(p-value 0.08)}.$$

Hence H_0: *submodel valid*, that is H_0: $\lambda_1^{A*C*E} = 0$, is not rejected.

In the model $A * C + A * E + C * E$, the parameters λ_1^{A*C} and λ_1^{A*E} are
significant (cf. Table 10.9, standardized parameter estimates), λ_1^{C*E} however is
not significant. The negative signs of $\hat{\lambda}_1^{A*C} = -0.0999$ and $\hat{\lambda}_1^{A*E} = -0.1527$
indicate that the construction type B and endodontic treatment occur less often
in the younger age group.

Model choice can now be continued by comparing (10.166) and (10.167):

$$G^2(A * C + A * E) - G^2(A * C + A * E + C * E) = 3.52 - 3.01 = 0.51 \quad ,$$
$$df = 2 - 1 = 1 \quad .$$

Because of $\chi_1^2 = 0.51$, H_0: $\hat{\lambda}_1^{C*E} = 0$ is not rejected. Hence, the submodel $A*C+A*E$ seems plausible. In this model, all parameter estimates are significant (cf. Table 10.10). For a given age, the type of construction and endodontic treatment are independent. Any further reduction (by so-called collapsing over the variable age) to the model $E + C$ is not allowed.

Model	G^2	df	p-value
Age	2940.050955	6	0.00000000+
Construct	2483.294331	6	0.00000000+
endo.treat.	698.847710	6	0.00000000+
Age + Construct	2381.339481	5	0.00000000+
Age + endo.treat.	596.892860	5	0.00000000+
Construct + endo.treat.	140.136236	5	0.00000000+
Age*Construct	2362.109194	4	0.00000000+
Age*endo.treat.	581.456952	4	0.00000000+
Construct*endo.treat.	140.003732	4	0.00000000+
Age + Construct + endo.treat.	38.181386	4	0.00000010+
endo.treat. + Age*Construct	18.951099	3	0.00027984+
Construct + Age*endo.treat.	22.745478	3	0.00004563+
Age + Construct*endo.treat.	38.048883	3	0.00000003+
Age*Construct + Age*endo.treat.	3.515190	2	0.17246353
Age*Construct + Construct*endo.treat.	18.818595	2	0.00008196+
Age*endo.treat. + Construct*endo.treat.	22.612974	2	0.00001229+
Age*Construct + Age*endo.treat. + Construct*endo.treat.	3.013646	1	0.08251487
Age*Construct*endo.treat.	0.000000	0	1.00000000

Table 10.8: Summary of models

10.6.2 Logistic Regression

In the following, we assume that the variable y is a binary response variable. We now model the probability for response 1, dependent on the explanatory variables (prognostic factors). Let $P(Y = 1) = \pi(\boldsymbol{x})$, where $\boldsymbol{x}' = (x_1, x_2, \ldots, x_p)$ denotes the vector of the prognostic factors. According to this approach, we have

$$\begin{aligned} \mathrm{E}(y) &= 1 \cdot \pi(\boldsymbol{x}) + 0 \cdot (1 - \pi(\boldsymbol{x})) = \pi(\boldsymbol{x}) \;, \\ \mathrm{E}(y^2) &= 1^2 \cdot \pi(\boldsymbol{x}) + 0^2 \cdot (1 - \pi(\boldsymbol{x})) = \pi(\boldsymbol{x}) \;, \\ \Longrightarrow \mathrm{Var}(y) &= \mathrm{E}(y^2) - (\mathrm{E}(y))^2 = \pi(\boldsymbol{x}) - \pi^2(\boldsymbol{x}) = \pi(\boldsymbol{x})(1 - \pi(\boldsymbol{x})) \;. \end{aligned}$$

For simplicity, we assume for the present that $p = 1$, that is we consider only one explanatory variable.

```
--------- New Model -----------------------------------------------
Age*Construct  +  Age*endo.treat. + Construct*endo.treat.
G^2  :  3.0136     Degrees of freedom:    1    p-Value: 0.08251487
Chi^2 :  3.0212     Degrees of freedom:    1    p-Value: 0.08213372
Algorithm: IPF
Max. Number of Iterations:     20
Criteria Eps: 0.0000010000
No. of IPF-Steps:      6
    i     estimate     est/ste      sterror      parameter
    1      5.0005     117.83801     0.04244      const
    2      0.1137       2.85358     0.03983      Age(<60)
    3      0.4827      11.35680     0.04250      Construct(H)
    4     -1.2729     -30.07448     0.04232      endo.treat.(yes)
    5     -0.0999      -4.38243     0.02279      Age(<60).Construct(H)
    6     -0.1527      -3.97057     0.03846      Age(<60).endo.treat.(yes)
    7     -0.0303      -0.71336     0.04250      Construct(H).endo.treat.(yes)

--- ML Estimates -----------------------
       Age    Construct  endo.treat. Orig.values    Expected  stand. Res.
       <60        H         yes          62          56.889     0.67761
       <60        H         no         1041        1046.111    -0.15802
       <60        B         yes          23          28.111    -0.96396
       <60        B         no          463         457.889     0.23884
      >=60        H         yes          70          75.111    -0.58972
      >=60        H         no          755         749.889     0.18664
      >=60        B         yes          30          24.889     1.02445
      >=60        B         no          215         220.111    -0.34449
```

Table 10.9: Model $A * C + A * E + C * E$

The model in this simplest case is given by

$$E\left(y_i\right) = \pi(x) = \alpha + \beta x. \tag{10.169}$$

If the y_i are independent, we obtain a GLM with identity link function. This model has a major structural defect: The probability $\pi(x)$ lies between 0 and 1, whereas $\alpha + \beta x$ can take values between $-\infty$ and ∞. This can lead to a possible contradiction. Additionally, we expect the values of x to influence $\pi(x)$ in a nonlinear rather than a monotone manner.

The model $\pi(x) = \alpha + \beta x$ is only valid for a specific range of x values. Another problem arises from the variance $V(y) = \pi(x)(1 - \pi(x))$, which is a function of x too, and hence not constant. This means, that the OLSE is not optimal. Since the y_i are not normally distributed, better estimators exist.

To avoid these difficulties, an approach is chosen that guarantees a monotonic course (S-curve) of the probability $\pi(x)$, under inclusion of the linear approach $\alpha + \beta x$ over the range of definition [0,1]:

```
-------- New Model ----------------------------------------------
Age*Construct + Age*endo.treat.
G^2  :  3.5152    Degrees of freedom:    2    p-Value: 0.17246353
Chi^2 : 3.6783    Degrees of freedom:    2    p-Value: 0.15895980
Algorithm: IPF
Max. Number of Iterations:    20
Criteria Eps: 0.0000010000
No. of IPF-Steps:    2
  i    estimate    est/ste    sterror     parameter
  1     4.9893    125.06720   0.03989     const
  2     0.1151      2.88606   0.03989     Age(<60)
  3     0.5084     22.37953   0.02272     Construct(H)
  4    -1.2863    -33.59284   0.03829     endo.treat.(yes)
  5    -0.0986     -4.34169   0.02272     Age(<60).Construct(H)
  6    -0.1503     -3.92448   0.03829     Age(<60).endo.treat.(yes)

--- ML Estimates -----------------------
     Age   Construct  endo.treat. Orig.values   Expected stand. Res.
     <60       H         yes          62         59.003      0.39023
     <60       H         no          1041      1043.997     -0.09277
     <60       B         yes          23         25.997     -0.58788
     <60       B         no           463       460.003      0.13976
    >=60       H         yes          70         77.103     -0.80890
    >=60       H         no           755       747.897      0.25972
    >=60       B         yes          30         22.897      1.48436
    >=60       B         no           215       222.103     -0.47660
```

Table 10.10: Model $A * C + A * E$

$$\pi(x) = \frac{\exp(\alpha + \beta x)}{1 + \exp(\alpha + \beta x)} \ . \tag{10.170}$$

At this point, the question arises, for which link function the logistic regression model is a GLM. In general, for binary variables the natural parameter is

$$Q(\pi) = \ln\left(\frac{\pi}{1 - \pi}\right) = \log \text{ odds} \ . \tag{10.171}$$

For this special situation, we now have

$$\frac{\pi(x)}{1 - \pi(x)} = \exp(\alpha + \beta x) = e^{\alpha}\left(e^{\beta}\right)^{x} \ , \tag{10.172}$$

that is, if x increases by one unit, the odds increases by e^{β}.

Thus, the use of the logit link: $\ln\left(\frac{\pi(x)}{1-\pi(x)}\right) = \alpha + \beta x$ provides a supplementary justification of the logistic regression model. An advantage of this link is that the effects of covariables can be estimated, no matter whether the study of interest is retrospective or prospective. The effects in the logistic model refer to the odds. For two different x values, $\frac{\alpha + \beta x_1}{\alpha + \beta x_2}$ is an odds ratio.

To find the appropriate form for the systematic component of the logistic regression, the sample logits are plotted against x (cf. Figure 10.1).

10.7 ML Estimates for the Logistic Regression

Let x be a vector of explanatory variables. The logistic regression model is then of the following form

$$\ln\left(\frac{\pi(x_i)}{1 - \pi(x_i)}\right) = \beta_0 + \beta_1 x_{i1} + \cdots + \beta_k x_{ik} \tag{10.173}$$

or equivalent with $\beta' = (\beta_0, \beta_1, \ldots, \beta_k)$, $x_i' = (x_{i0}, x_{i1}, \ldots, x_{ik})$, $x_{i0} \equiv 1$,

$$\pi(x_i) = \frac{\exp(x_i'\beta)}{1 + \exp(x_i'\beta)} \ . \tag{10.174}$$

For fixed x_i and n_i repetitions, response 1 is observed y_i-times. If the repetitions are independent, then y_i $(i = 1, \ldots, I)$ are independent binomial variables with $\mathrm{E}(y_i) = n_i \pi(x_i)$ and $n_1 + \cdots + n_I = N$.

Hence, the joint distribution of the $\{y_i\}$ is proportional to the product of the I binomial functions

$$\prod_{i=1}^{I} \pi(x_i)^{y_i}[1 - \pi(x_i)]^{n_i - y_i} = \prod_{i=1}^{I}[1 - \pi(x_i)]^{n_i} \prod_{i=1}^{I} \exp\left(\ln\left(\frac{\pi(x_i)}{1 - \pi(x_i)}\right)^{y_i}\right)$$

$$= \prod_{i=1}^{I}[1 - \pi(x_i)]^{n_i} \exp\left[\sum_{i=1}^{I} y_i \ln\left(\frac{\pi(x_i)}{1 - \pi(x_i)}\right)\right] . \tag{10.175}$$

Because of $\dfrac{\pi(x_i)}{1 - \pi(x_i)} = \exp(x_i'\beta)$, the ith logit is $\ln\left(\dfrac{\pi(x_i)}{1 - \pi(x_i)}\right) = x_i'\beta$, so that the last term in (10.175) is of the following form

$$\exp\left(\sum_{i=1}^{I} y_i x_i'\beta\right) = \exp\left(\sum_{j=0}^{k}\left(\sum_{i=0}^{I} y_i x_{ij}\right)\beta_j\right) . \tag{10.176}$$

Finally, with $1 - \pi(x_i) = \dfrac{1}{1 + \exp(x_i'\beta)}$ and using the identity $a = \exp(\ln a)$, after taking the logarithm, we get

$$l(\beta) = \sum_{j=0}^{k}\left(\sum_{i=1}^{I} y_i x_{ij}\right)\beta_j - \sum_{i=1}^{I} n_i \ln(1 + \exp(x_i'\beta)) . \tag{10.177}$$

Hence, the log-likelihood depends on the binomial variables y_i of the sample only by

$$\sum_{i=1}^{I} y_i x_{ij} \qquad (j = 0, \ldots, k). \tag{10.178}$$

We find the estimating equations as

$$\frac{\partial L}{\partial \beta_a} = \sum_{i=1}^{I} y_i x_{ia} - \sum_{i=1}^{I} n_i x_{ia} \left[\frac{\exp(\sum_{j=1}^{k} x_{ij} \beta_j)}{1 + \exp(\sum_{j=1}^{k} x_{ij} \beta_j)} \right] = 0 , \tag{10.179}$$

i.e.,

$$\sum_{i=1}^{I} y_i x_{ia} - \sum_{i=1}^{I} n_i x_{ia} \hat{\pi}_i = 0 \qquad a = 0, \ldots, k \tag{10.180}$$

with $\hat{\pi}_i = \dfrac{\exp(\sum_{j=1}^{k} x_{ij} \hat{\beta}_j)}{1 + \exp(\sum_{j=1}^{k} x_{ij} \hat{\beta}_j)}$ as the ML estimate of π_i, which is dependent on the ML estimate $\hat{\beta}$.

Let $\boldsymbol{X}$ be the $I \times (k+1)$-matrix of the x_{ij}. The ML equations can than be summarized in matrix notation as (cf. (10.92))

$$\boldsymbol{X}'(\boldsymbol{y} - \hat{\boldsymbol{m}}) = \boldsymbol{0} \tag{10.181}$$

with the components $\hat{m}_i = n_i \hat{\pi}_i$ of the vector $\hat{\boldsymbol{m}}$. The logistic model is based on the canonical link. The differentiated normal equation is typical for generalized linear models with the canonical link: it connects the sufficient statistic $\boldsymbol{X}'\boldsymbol{y}$ with the estimate of its expectation $\boldsymbol{X}'\hat{\boldsymbol{m}}$. The information matrix is the negative expectation of the matrix of the second derivatives of the log-likelihood. Under general conditions (cf. Section 10.1), the ML estimates are asymptotic normal with a covariance matrix equal to the inverse of the information matrix.

For the logistic model, we have:

$$\begin{aligned}
\frac{\partial^2 L}{\partial \beta_a \partial \beta_b} &= -\sum_{i=1}^{I} \frac{x_{ia} x_{ib} n_i \exp(\boldsymbol{x}_i' \boldsymbol{\beta})}{(1 + \exp(\boldsymbol{x}_i' \boldsymbol{\beta}))^2} \\
&= -\sum_{i=1}^{I} x_{ia} x_{ib} n_i \pi_i (1 - \pi_i) .
\end{aligned} \tag{10.182}$$

This is independent of y_i, so that the sample and population matrix of the second derivatives are identical. This holds for all GLM with a canonical link. Thus

$$V(\hat{\boldsymbol{\beta}}) = \left(-\frac{\partial^2 L}{\partial \beta_a \partial \beta_b} \right)^{-1} \tag{10.183}$$

and with (10.182) in matrix notation

$$V(\hat{\boldsymbol{\beta}}) = (\boldsymbol{X}'\mathrm{diag}(n_i\hat{\pi}_i(1-\hat{\pi}_i))\boldsymbol{X})^{-1} \ . \tag{10.184}$$

For fixed $\boldsymbol{x}_i$, the predicted logit is $\boldsymbol{x}_i'\hat{\boldsymbol{\beta}}$ with the estimated variance

$$\hat{\sigma}^2(\boldsymbol{x}_i'\hat{\boldsymbol{\beta}}) = \boldsymbol{x}_i'V(\hat{\boldsymbol{\beta}})\boldsymbol{x}_i \ . \tag{10.185}$$

Because of the asymptotic normal distribution, we have for large samples $\hat{\boldsymbol{\beta}} \overset{a.s.}{\sim} N(\boldsymbol{\beta}, V(\hat{\boldsymbol{\beta}}))$, so that

$$\boldsymbol{x}_i'\hat{\boldsymbol{\beta}} \pm u_{1-\alpha/2}\hat{\sigma}(\boldsymbol{x}_i'\hat{\boldsymbol{\beta}}) \tag{10.186}$$

leads to a $(1-\alpha)$-confidence interval. If the upper and lower limits L_u and L_l are retransformed according to

$$\hat{\pi} = \frac{\exp(\boldsymbol{x}_i'\hat{\boldsymbol{\beta}})}{1+\exp(\boldsymbol{x}_i'\hat{\boldsymbol{\beta}})} \tag{10.187}$$

we have a confidence interval

$$[\hat{\pi}_u, \hat{\pi}_o] \quad \text{for} \quad \pi(\boldsymbol{x}) \ .$$

The ML equation for $\hat{\boldsymbol{\beta}}$

$$\boldsymbol{X}'\boldsymbol{y} = \boldsymbol{X}'\hat{\boldsymbol{m}} \quad ,$$

with $\hat{m}_i = n_i\hat{\pi}_i$ and $\hat{\pi}_i = \dfrac{\exp(\boldsymbol{x}_i'\hat{\boldsymbol{\beta}})}{1+\exp(\boldsymbol{x}_i'\hat{\boldsymbol{\beta}})}$, is non-linear, hence iterative procedures have to be applied.

10.8   Logit Models for Categorical Data

10.8.1   Parameter Estimation

The explanatory variable X can be continuous or categorical. Assume X to be categorical and choose the logit link, then the logit models are equivalent to loglinear models (cf. Agresti, 1990, p. 152).

Logit Models for I×2-Tables

Let X be an explanatory variable with I categories. If response/nonresponse is the Y factor, then we have an $I \times 2$-table. In row i, the probability for response is $\pi_{1/i}$ and $\pi_{2/i}$ for nonresponse, with $\pi_{1/i} + \pi_{2/i} = 1$. This leads to the following logit model:

$$\ln\left(\frac{\pi_{1/i}}{\pi_{2/i}}\right) = \alpha + \beta_i \quad . \tag{10.188}$$

Here the x values are not included explicitly, but only through the category i. β_i describes the effect of category i on the response. If $\beta_i = 0$, there is no effect.

This model resembles the one-way analysis of variance and likewise, we have the constraints for identifiability $\sum \beta_i = 0$ or $\beta_I = 0$. Hence, $I - 1$ of the parameters $\{\beta_i\}$ suffice for characterization of the model. For the constraint $\sum \beta_i = 0$, α is the overall mean of the logits and β_i is the deviation from this mean for row i. The higher β_i, the higher the logit in row i, and the higher the value of $\pi_{1/i}$ ($=$ chance for response in category i).

If the factor X (in I categories) has no effect on the response variable, the model simplifies to the model of statistical independence of the factor and response:

$$\ln\left(\frac{\pi_{1/i}}{\pi_{2/i}}\right) = \alpha \quad \forall i \; .$$

We now have $\beta_1 = \beta_2 = \cdots = \beta_I$, and thus $\pi_{1/1} = \pi_{1/2} = \cdots = \pi_{1/I}$.

Logit Models for Higher Dimensions

As a generalization to two or more categorical factors that have an effect on the binary response, we now consider the two factors A and B with I and J levels. Let $\pi_{1/ij}$ and $\pi_{2/ij}$ denote the probabilities for response and nonresponse for the combination ij of factors, so $\pi_{1/ij} + \pi_{2/ij} = 1$. For the $I \times J \times 2$-table, the logit model

$$\ln\left(\frac{\pi_{1/ij}}{\pi_{2/ij}}\right) = \alpha + \beta_i^A + \beta_j^B \tag{10.189}$$

represents the effects of A and B without interaction. This model is equivalent to the two-way analysis of variance without interaction.

10.8.2 Goodness-of-Fit – Likelihood-Ratio Test

For a given model M, we can use the estimates of the parameters $(\widehat{\alpha + \beta_i})$ and $(\hat{\alpha}, \hat{\beta})$ respectively to predict the logits, to estimate the probabilities of response $\hat{\pi}_{1/i}$ and hence to calculate the expected cell frequencies $\hat{m}_{ij} = n_{i+}\hat{\pi}_{j/i}$. We can now test the goodness-of-fit of a model M with

$$G^2(M) = 2\sum_{i=1}^{I}\sum_{j=1}^{J} n_{ij} \ln\left(\frac{n_{ij}}{\hat{m}_{ij}}\right) . \tag{10.190}$$

The $\hat{m}_{ij}$ are calculated by using the estimated model parameters. The degrees of freedom equal the number of logits minus the number of independent parameters in the model M.

We now consider three models for binary response.

1. Independence model (I: independence):

$$\text{M=I:} \quad \ln\left(\frac{\pi_{1/i}}{\pi_{2/i}}\right) = \alpha \; . \tag{10.191}$$

Here we have I logits and one parameter, that is $I - 1$ degrees of freedom.

2. Logistic model:

$$M{=}L: \qquad \ln\left(\frac{\pi_{1/i}}{\pi_{2/i}}\right) = \alpha + \beta x_i \; . \qquad\qquad (10.192)$$

The number of degrees of freedom equals $I - 2$.

3. Logit model:

$$M{=}S: \qquad \ln\left(\frac{\pi_{1/i}}{\pi_{2/i}}\right) = \alpha + \beta_i \; . \qquad\qquad (10.193)$$

The model has I logits and I independent parameters. The number of degrees of freedom is 0, it has perfect fit. This model with equal number of parameters and observations is called saturated model.

The likelihood-ratio test compares a model M_2 with a simpler model M_1 (in which a few parameters equal zero). The test statistic then is (cf. (10.59) and (10.60))

$$\Lambda \;=\; \frac{L(M_1)}{L(M_2)} \qquad\qquad (10.194)$$

$$\text{or}\quad G^2\,(M_1|M_2) \;=\; 2\,(\ln L(M_2) - \ln L(M_1)) \; . \qquad\qquad (10.195)$$

The statistic $G^2(M)$ is a special case of this statistic, where $M_1 = M$ and M_2 is the saturated model. If we want to test the goodness of fit with $G^2(M)$, this is equivalent to testing whether all the parameters that are in the saturated model, but not in the model M, are equal to zero.

Let l_s denote the maximized log likelihood function for the saturated model. We then have

$$\begin{aligned}
G^2(M_1|M_2) \;&=\; 2\,(\ln L(M_2) - \ln L(M_1)) \\
&=\; 2\,(\ln L(M_2) - l_S) - [2(\ln L(M_1) - l_S)] \\
&=\; G^2(M_1) - G^2(M_2).
\end{aligned}$$

That is, the statistic $G^2(M_1|M_2)$ for comparing two models is identical to the difference of the goodness-of-fit statistics for the two models.

Example 10.3: In the example 10.4 "heart disease/blood pressure" we have for the logistic model:

	\multicolumn{4}{c}{Heart disease}
	\multicolumn{2}{c}{yes}	\multicolumn{2}{c}{no}
	observed	expected	observed	expected
1	3	5.2	153	150.8
2	17	10.6	235	241.4
3	12	15.0	272	269
4	16	18.0	255	253
5	12	11.5	127	127.5
6	8	8.8	77	76.2
7	16	14.1	83	84.9
8	8	8.3	35	34.7

$\implies G^2(\mathrm{L}) = 5.91$, $df = 8 - 2 = 6$.

For the independence model, we get $G^2(\mathrm{I}) = 30.02$ with $df = 7 = (I-1)(J-1) = (8-1)(2-1)$. The test statistic for testing H_0: $\beta = 0$ in the logistic model then is

$$G^2(\mathrm{I|L}) \;=\; G^2(\mathrm{I}) - G^2(\mathrm{L}) = 30.02 - 5.91 = 24.11 \quad , \quad df = 7 - 6 = 1 \; .$$

This value is significant, which means that the logistic model compared to the independence model holds.

10.8.3 Model Diagnosis

The statistics χ^2 and G^2 are measures for the overall fit of the chosen model. In case of a bad fit, additional means have to be applied in order to determine the cause, as for instance graphic analysis of residuals. For example, the estimated and observed proportions can be plotted against each other

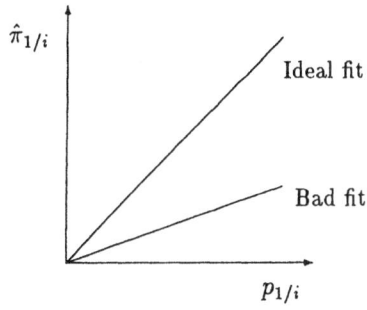

$\hat{\pi}_{1/i}$: from one model (not saturated, e.g., logistic regression)

or $\hat{\pi}(x)$

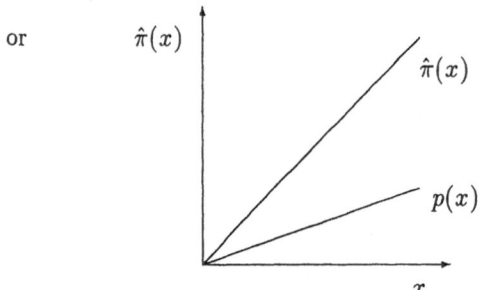

This helps to reveal whether the wrong link function was chosen or whether the regression model is non-linear.

Diagnosis Based on Residuals

Given n_i observations in category i, let y_i be the number of observations with response 1. Let $\hat{\pi}_{1/i}$ be the estimate according to a chosen binary response model. Then

$$e_i = \frac{y_i - n_i\hat{\pi}_{1/i}}{[n_i\hat{\pi}_{1/i}(1 - \hat{\pi}_{1/i})]^{1/2}} \qquad i = 1, \ldots, I \tag{10.196}$$

defines the ith residual [standardized with the variance of the binomial distribution $B(n_i; \hat{\pi}_{1/i})$].

If the estimate $\hat{\pi}_{1/i}$ in e_i is replaced by the true (unknown) parameter $\pi_{1/i}$, then e_i is a standardized binomial variable that converges to $N(0,1)$ for sufficiently large n_i. Hence, values of $| e_i | > 2$ will indicate a model error.

Since y_i is used in $\hat{\pi}_{1/i}$, the numerators in the e_i are smaller than the corresponding value $(y_i - n_i\pi_{1/i})$ of the population in general. This may lead to a smaller variation of the $\{e_i\}$ compared to standard normal values. Consequently, the fit may seem better than is actually the case. Hence, for single estimation of a residual, the two-sided 95%-quantile $u_{1-\alpha/2} = 1.96$ is not used; instead the value 2 is used.

Diagnosis Following the Determination Coefficient

In linear regression, R^2 and the adjusted $\overline{R}^2$ provide a measure for the goodness of fit of the model. We have a perfect fit if $R^2 = 1$. In analogy to R^2, a number of measures for contingency table models were developed.

Let l_M = $\max(\ln L)$ for the estimated (fitted) model M
       l_S = $\max(\ln L)$ for the saturated model S
       l_I = $\max(\ln L)$ for the independence model I.

In $\ln L$ is always non-positive, since the likelihood $\pi_i^{y_i}(1 - \pi)^{1-y_i}$ of a response probability lies between zero and one. The value of the likelihood cannot become smaller if the parameter space expands.

Because of the interlocking of the models, that is, the increase in complexity from

$$\ln\left(\frac{\pi_i}{1-\pi_i}\right) = \alpha \qquad \text{(independence model)}$$

$$\text{to} \quad \ln\left(\frac{\pi_i}{1-\pi_i}\right) = \alpha + \beta x_i \quad \text{(logistic regression model)}$$

$$\text{to} \quad \ln\left(\frac{\pi_i}{1-\pi_i}\right) = \alpha + \beta_i \quad \text{(saturated or logit model)},$$

we have

$$l_I \leq l_M \leq l_S. \tag{10.197}$$

Hence, the measure

$$L(\mathrm{I} \mid \mathrm{M}) = \frac{l_M - l_I}{l_S - l_I} \tag{10.198}$$

lies between 0 and 1. For $l_M = l_I$, we have $L(\mathrm{I} \mid \mathrm{M}) = 0$, i.e., the fitted model leads to no improvement compared to the independence model. For $l_M = l_S$ (perfect fit), we have $L(\mathrm{I} \mid \mathrm{M}) = 1$.

Let us consider this model more closely for the case of binary response for I categories ($I \times 2$-table). Let $\hat{\pi}_i$ be the estimated probability of response of the ith category and assume y_i is the binary response, both according to the chosen model. Suppose that we have a total of N observations of a binomial distribution. The maximized log-likelihood is then

$$\ln \prod_{i=1}^{N} \left[\hat{\pi}_i^{y_i}(1-\hat{\pi}_i)^{1-y_i}\right] = \sum_{i=1}^{N} [y_i \ln \hat{\pi}_i + (1-y_i)\ln(1-\hat{\pi}_i)]. \tag{10.199}$$

In the independence model, we have

$$\hat{\pi} = \bar{y} = \frac{1}{N}\sum y_i \tag{10.200}$$

and hence

$$l_I = N\left[\bar{y}\ln\bar{y} + (1-\bar{y})\ln(1-\bar{y})\right] . \tag{10.201}$$

In the saturated model, every observation leads to the corresponding ML estimate, that is, $\hat{\pi}_i = y_i$ ($i = 1,\ldots,N$) so that $l_S = 0$. This is evident because of

$$y_i \ln y_i + (1-y_i)\ln(1-y_i), \tag{10.202}$$

since y_i is either 0 or 1.

In the binomial model, the measure $L(\mathrm{I} \mid \mathrm{M})$ now simplifies to

$$D = \frac{l_I - l_M}{l_I} . \tag{10.203}$$

Considering the G^2-statistic once again, we have for the model of the $I \times 2$-table (that is in the model with I factor levels and binary response)

$$G^2(\text{M}) = -2(l_M - l_S) = -2l_M , \qquad (10.204)$$
$$G^2(\text{I}) = -2(L_I - l_S) = -2l_I , \qquad (10.205)$$

so that D may be written as

$$D^* = \frac{G^2(\text{I}) - G^2(\text{M})}{G^2(\text{I})} \qquad (10.206)$$

(Goodman, 1971, Theil, 1970). For values close to 1, this measure should indicate a good fit. However, D^* can become very large even if the fit is poor, since for instance $G^2(\text{I}) \to \infty$ for $N \to \infty$, while $G^2(\text{M})$ behaves like a χ^2-variable and stays limited. Hence, $D^* \to 1$ for $N \to \infty$, so that D^* is dependent on the sample size.

An alternative measure compares the prediction of y_i by $\hat{\pi}_i$ (model M) or by $\bar{y}$ (model I):

$$R^2 = 1 - \frac{\sum_{i=1}^{N}(y_i - \hat{\pi}_i)^2}{\sum_{i=1}^{N}(y_i - \bar{y})^2} \qquad (10.207)$$

This measure coincides with the usual R^2 of regression models if the linear probability model is estimated by OLSE.

10.8.4 Examples for Model Diagnosis

Example 10.4: We examine the risk (Y) for the loss of abutment teeth by extraction in dependence on the age (X) (Walther, 1992).

	Age	Loss		
i	group	yes	no	n_{i+}
1	< 40	4	70	74
2	40 − 50	28	147	175
3	50 − 60	38	207	245
4	60 − 70	51	202	253
5	> 70	32	92	124
	n_{+j}	153	718	871

Table 10.11: 5 × 2-table loss of abutment teeth/age groups

From Table 10.11, we calculate $\chi_4^2 = 15.56$ and $G^2 = 17.25$. Both values are significant ($\chi_{4;0.95}^2 = 9.49$). The decomposition of G^2 results in:

Loss	1	2	3	4	5
yes	4	28	38	51	32
no	70	147	207	202	92

1/2		
4	28	32
70	147	217
74	175	249

1+2/3		
32	38	70
217	207	424
249	245	494

1+2+3/4		
70	51	121
424	202	626
494	253	747

1+2+3+4/5		
121	32	153
626	92	718
747	124	871

This decomposition leads to $G^2 = \underline{6.00} + 0.72 + \underline{4.30} + \underline{6.22} = 17.25$, and thus three significant effects are found.

Modelling with the **logit model**

$$\ln\left(\frac{n_{1i}}{n_{2i}}\right) = \widehat{\alpha + \beta_i}$$

results in the following table:

i	Sample logits	$\hat{\pi}_{1/i} = \frac{n_{1i}}{n_{i+}}$
1	−2.86	0.054
2	−1.66	0.160
3	−1.70	0.155
4	−1.38	0.202
5	−1.06	0.258

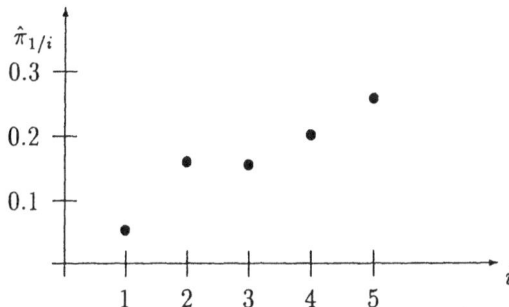

$\hat{\pi}_{1/i}$ is the estimated risk for loss of abutment teeth. It increases linearly with the age group. For instance, the age group 5 has 5 times the risk of age group 1.

Modelling with the **logistic regression**

$$\ln\left(\frac{\hat{\pi}_1(x_i)}{\hat{\pi}_2(x_i)}\right) = \alpha + \beta x_i$$

results in:

x_i	Sample logits	Fitted logits	$\hat{\pi}_1(x_i)$	Expected $n_{i+}\hat{\pi}_1(x_i)$	Observed n_{1i}
35	-2.86	-2.22	0.098	7.25	4
45	-1.66	-1.93	0.127	22.17	28
55	-1.70	-1.64	0.162	39.75	38
65	-1.38	-1.35	0.206	51.99	51
75	-1.06	-1.06	0.257	31.84	32

with the ML estimates

$$\hat{\alpha} = -3.233$$
$$\hat{\beta} = 0.029 \quad .$$

The hypothesis H_0: $\beta = 0$ is tested with the Wald statistic:

$$Z^2 = 13.06 > 3.84 = \chi^2_{1;0.95} \,,$$

which shows that the trend is significant.

The **LQ test** confirms this result:

n_{1i}	$\hat{m}_{1i}$	n_{2i}	$\hat{m}_{2i}$
4	7.25	70	66.75
28	22.17	147	152.83
38	39.75	207	205.25
51	51.99	202	201.01
32	31.84	92	92.16

Hence, we get

$$G^2(\text{I} \mid \text{L}) = G^2(\text{I}) - G^2(\text{L}) = 17.25 - 3.66 = 13.59 \,.$$

The number of degrees of freedom is calculated by

$$df: \quad 1 = 4 - 3 \,.$$

and $13.95 > \chi_{1;0.95} = 3.84$ leads to rejection of H_0 : $\beta = 0$. Thus, the logistic model is statistically significant compared to the independence model.

Summarizing the results, we have

i	logit $\hat{\pi}_{1/i}$	logistic $\hat{\pi}_1(x_i)$	n_{1i}	n_{i+}
1	0.054	0.098	4	74
2	0.160	0.127	28	175
3	0.155	0.162	38	245
4	0.202	0.206	51	253
5	0.258	0.257	32	124

From this we calculate the expected frequencies and residuals

$$e_i = \frac{n_{1i} - n_{i+}\hat{\pi}_1(x_i)}{(n_{i+}\hat{\pi}_1(x_i)(1 - \hat{\pi}_1(x_i)))^{1/2}}$$
(10.208)

logit	logistic		logit	logistic
$n_{i+}\hat{\pi}_{1/i}$	$n_{i+}\hat{\pi}_1(x_i)$		e_i	e_i
4	7.25		0	−1.271
28	22.17		0	1.325
38	39.75		0	−0.303
51	51.99		0	−0.154
32	31.84		0	0.033

We now calculate the analogue of the determination coefficient for this data set:

$$R^2 = 1 - \frac{\sum(n_{1i} - n_{i+}\hat{\pi}_{1/i})^2}{\sum(n_{1i} - n_{i+}\frac{n_{+1}}{n})^2}$$
(10.209)

n_{1i}	n_{i+}	$n_{i+}\dfrac{n_{+1}}{n}$	logistic $n_{i+}\hat{\pi}_1(x_i)$
4	74	12.99	7.25
28	175	30.73	22.17
38	245	43.02	39.75
51	253	44.43	51.99
32	124	21.77	31.84

This results in

$$\frac{n_{+1}}{n} = \frac{153}{871}$$

and hence

logit model : $R^2 = 1$,

logistic model: $R^2 = 1 - \dfrac{48.62}{261.29} = 0.814$.

The fit achieved by the logistic regression indicates the adequacy of the model. The value $R^2 = 1$ for the logit model is always to be expected because of the perfect fit.

10.9 Models for Correlated Categorical Response

10.9.1 Introduction

We now extend the problems of categorical response to the situations of correlation within the response values. These correlations are due to classification of the individuals into clusters of "related" elements.

Examples:

- Two or more implants or abutment teeth in dental reconstructions (Walther, 1992).

- Response of a patient in cross-over in case of significant carry-over effect (Chapter 8).

- Repeated categorical measurement of a response like function of the lungs, blood pressure, or performance in training (repeated measures design or panel data).

- Measurement of paired organs (eyes, kidneys, etc.)

- Response of members of a family.

Let y_{ij} be the categorical response of the jth individual in the ith cluster:

$$y_{ij}, \quad i = 1, \ldots, N, \quad j = 1, \ldots, n_i \quad . \tag{10.210}$$

We assume that the expectation of the response y_{ij} is dependent on prognostic variables (covariables) x_{ij} by a regression, i.e.,

$$E(y_{ij}) = \beta_0 + \beta_1 x_{ij} \quad . \tag{10.211}$$

Assume $\mathrm{Var}(y_{ij}) = \sigma^2$ and

$$\mathrm{Cov}(y_{ij}, y_{ij'}) = \sigma^2 \rho \quad (j \neq j'). \tag{10.212}$$

The response of individuals from different clusters is assumed to be uncorrelated. Let us assume that the covariance matrix for the response of every cluster equals

$$\mathrm{V} \begin{pmatrix} y_{i1} \\ \vdots \\ y_{in_i} \end{pmatrix} = \mathrm{V}(\boldsymbol{y}_i) = \sigma^2(1-\rho)\boldsymbol{I}_{n_i} + \sigma^2 \rho \boldsymbol{J}_{n_i} \tag{10.213}$$

and thus has a compound symmetric structure. Hence, the covariance matrix of the entire sample vector is block-diagonal

$$\boldsymbol{W} = \mathrm{V} \begin{pmatrix} \boldsymbol{y}_1 \\ \vdots \\ \boldsymbol{y}_N \end{pmatrix} = \mathrm{diag}(\mathrm{V}(\boldsymbol{y}_1), \ldots, \mathrm{V}(\boldsymbol{y}_N)) \quad . \tag{10.214}$$

Notice that the matrix $\boldsymbol{W}$ itself does not have a compound symmetric structure. Hence, we have a generalized regression model. The best linear unbiased estimate of $\boldsymbol{\beta} = (\beta_0, \beta_1)'$ is given by the Aitken estimate (3.251):

$$b = (\boldsymbol{X}'\boldsymbol{W}^{-1}\boldsymbol{X})^{-1}\boldsymbol{X}'\boldsymbol{W}^{-1}\boldsymbol{y} \quad , \tag{10.215}$$

and does not coincide with the OLS estimate, since the preconditions of Theorem 3.17 by McElroy are not fulfilled. The choice of an incorrect covariance structure leads, according to our remarks in Section 3.9.3, to a bias in the estimate of the variance. On the other hand, the unbiasedness or consistency of the estimate of β stays untouched even in case of incorrect choice of the covariance matrix. Liang and Zeger (1993) examined the bias of $\text{Var}(\hat{\beta}_1)$ for the wrong choice of $\rho = 0$. In case of positive correlation within the cluster, the variance is underestimated. This corresponds to the results of Goldberger (1964) for positive auto-correlation in econometric models.

The following problems arise in practice:

(i) identification of the covariance structure,

(ii) estimation of the correlation,

(iii) application of an Aitken-type estimate.

However, it is no more possible to assume the usual GLM approach, since this does not take the correlation structure into consideration. Various approaches were developed as extensions of the GLM approach, in order to be able to include the correlation structure in the response:

• marginal model,

• random effects model,

• observation driven models,

• conditional models.

For binary response, simplifications arise (Section 10.9.4). Liang and Zeger (1989) proved that the joint distribution of the y_{ij} can be descibed by n_i logistic models for y_{ij} given y_{ik} $(k \neq j)$. Rosner (1984) used this approach and developed beta-binomial models.

Modelling Approaches for Correlated Response

The modelling approaches can be ordered according to diverse criteria:

(A) Population-Averaged – Subject-Specific

The essential difference between population-averaged (PA) and subject-specific (SS) models lies in the answer of the question whether the regression coefficients vary for the individuals or not. In PA models, the β's are independent of the specific individual i. Examples are the marginal and conditional models. In SS models, the β's are dependent on the specific i and are therefore written as β_i.

An example for a SS model is the random-effects model.

(B) Marginal and Conditional Random-Effects Model
In the **marginal model**, the regression is modelled seperately from the dependence within the measurement in contrast to the two other approaches. The marginal expectation $E(y_{ij})$ is modelled as a function of the explanatory variables and is interpreted as the mean response over the population of individuals with the same x. Hence, marginal models are mainly suitable for the analysis of covariable effects in a population.

The **random-effects** model assumes that the covariable effects β are dependent on the indivual in contrast to the marginal model. Hence random-effects models are useful for estimates dependent on individuals and for population means (*mixed models*).

For the **conditional model** (observation-driven model), a time-dependent response y_{it} is modelled as a function of the covariables and of the past response values $y_{it-1}, \ldots, y_{i1}$. This is done by assuming a specific correlation structure between the response values. Conditional models are useful if the main point of interest is the conditional probability of a state or the transition of states.

10.9.2   Quasi-Likelihood Approach for Correlated Binary Response
The following sections are dedicated to binary response variables and especially the bivariate case (that is, cluster size $n_i = 2 \ \forall i$).

In case of a violation of independence or in case of a missing distribution assumption of the natural exponential family, the core of the ML method, namely the score function, may be used nevertheless for parameter estimation, as we have already mentioned in Section 10.1.7. We now want to specify the quasi-score function (10.79) for the binary response.

Let $y_i' = (y_{i1}, \ldots, y_{in_i})$ be the response vector of the ith cluster ($i = 1, \ldots, N$) with the true covariance matrix $\mathrm{Cov}(y_i)$ and let x_{ij} be the $p \times 1$-vector of the covariable corresponding to y_{ij}. Assume the variables y_{ij} are binary with the values 1 and 0 and assume $P(y_{ij} = 1) = \pi_{ij}$. We then have $\mu_{ij} = \pi_{ij}$. Let $\pi_i' = (\pi_{i1}, \ldots, \pi_{in_i})$. Suppose that the link function is $g(\cdot)$, that is,

$$g(\pi_{ij}) = \eta_{ij} = x_{ij}'\beta \quad .$$

Let $h(\cdot)$ be the inverse function, that is

$$\pi_{ij} = h(\eta_{ij}) = h(x_{ij}'\beta) \quad .$$

For the canonical link

$$\mathrm{Logit}(\pi_{ij}) = \ln\left(\frac{\pi_{ij}}{1 - \pi_{ij}}\right) = g(\pi_{ij})$$

we have

$$h(\eta_{ij}) = \frac{\exp(\eta_{ij})}{1 + \exp(\eta_{ij})} = \frac{\exp(\boldsymbol{x}'_{ij}\boldsymbol{\beta})}{1 + \exp(\boldsymbol{x}'_{ij}\boldsymbol{\beta})} \quad .$$

Hence

$$\boldsymbol{D} = \left(\frac{\partial \mu_{ij}}{\partial \boldsymbol{\beta}}\right) = \left(\frac{\partial \pi_{ij}}{\partial \boldsymbol{\beta}}\right) \quad .$$

We have

$$\frac{\partial \pi_{ij}}{\partial \boldsymbol{\beta}} = \frac{\partial \pi_{ij}}{\partial \eta_{ij}}\frac{\partial \eta_{ij}}{\partial \boldsymbol{\beta}} = \frac{\partial h(\eta_{ij})}{\partial \eta_{ij}}\boldsymbol{x}'_{ij} \quad ,$$

and hence, for $i = 1, \ldots, N$ and the $p \times n_i$-matrix $\boldsymbol{X}'_i = (\boldsymbol{x}_{i1}, \ldots, \boldsymbol{x}_{in_i})$

$$\boldsymbol{D}_i = \tilde{\boldsymbol{D}}_i \boldsymbol{X}_i \quad \text{with} \quad \tilde{\boldsymbol{D}}_i = \left(\frac{\partial h(\eta_{ij})}{\partial \eta_{ij}}\right) \quad .$$

For the quasi-score function for all N clusters, we now get

$$U(\boldsymbol{\beta}) = \sum_{i=1}^{N} \boldsymbol{X}'_i \tilde{\boldsymbol{D}}'_i \boldsymbol{V}_i^{-1}(\boldsymbol{y}_i - \boldsymbol{\pi}_i) \quad , \tag{10.216}$$

where $\boldsymbol{V}_i$ is the matrix of the working variances and covariances of the y_{ij} of the ith cluster. The solution of $U(\hat{\boldsymbol{\beta}}) = 0$ is found iteratively under further specifications which we will describe in the next section.

10.9.3   The GEE Method by Liang and Zeger

The variances are modelled as a function of the mean, that is

$$v_{ij} = \text{Var}(y_{ij}) = v(\pi_{ij})\phi \quad . \tag{10.217}$$

(In the binary case, the form of the variance of the binomial distribution is often chosen: $v(\pi_{ij}) = \pi_{ij}(1 - \pi_{ij})$.) With these, the following matrix is formed

$$\boldsymbol{A}_i = \text{diag}(v_{i1}, \ldots, v_{in_i}) \quad . \tag{10.218}$$

Since the structure of dependence is not known, a $n_i \times n_i$ *quasi-correlation matrix* $\boldsymbol{R}_i(\alpha)$ is chosen for the vector of the ith cluster $\boldsymbol{y}'_i = (y_{i1}, \ldots, y_{in_i})$ according to

$$\boldsymbol{R}_i(\alpha) = \begin{pmatrix} 1 & \rho_{i12}(\alpha) & \cdots & \rho_{i1n_i}(\alpha) \\ \rho_{i21}(\alpha) & 1 & \cdots & \rho_{i2n_i}(\alpha) \\ \vdots & & & \vdots \\ \rho_{in_i1}(\alpha) & \rho_{in_i2}(\alpha) & \cdots & 1 \end{pmatrix} \quad , \tag{10.219}$$

where the $\rho_{ikl}(\alpha)$ are the correlations as function of α, (α may be a scalar or a vector). $\boldsymbol{R}_i(\alpha)$ may vary for the clusters.

By multiplying the quasi-correlation matrix $\boldsymbol{R}_i(\alpha)$ with the diagonal matrix of the variances $\boldsymbol{A}_i$, we obtain a working covariance matrix

$$V_i(\boldsymbol{\beta}, \alpha, \phi) = \boldsymbol{A}_i^{1/2} \boldsymbol{R}_i(\alpha) \boldsymbol{A}_i^{1/2} \quad , \tag{10.220}$$

which is no longer completely specified by the expectations, as in the case of independent response. We have $V_i(\boldsymbol{\beta}, \alpha, \phi) = \mathrm{Cov}(\boldsymbol{y}_i)$ if and only if $\boldsymbol{R}_i(\alpha)$ is the true correlation matrix of $\boldsymbol{y}_i$.

If the matrices $\boldsymbol{V}_i$ in (10.216) are replaced by the matrices $V_i(\boldsymbol{\beta}, \alpha, \phi)$ from (10.220), we get the *generalized estimating equation (GEE)* by Liang and Zeger (1986), that is

$$U(\boldsymbol{\beta}, \alpha, \phi) = \sum_{i=1}^{N} \left(\frac{\partial \boldsymbol{\pi}_i}{\partial \boldsymbol{\beta}}\right)' V_i^{-1}(\boldsymbol{\beta}, \alpha, \phi)(\boldsymbol{y}_i - \boldsymbol{\pi}_i) = 0 \quad . \tag{10.221}$$

The solutions are denoted by $\hat{\boldsymbol{\beta}}_G$. For the quasi-Fisher matrix, we have

$$F_G(\boldsymbol{\beta}, \alpha) = \sum_{i=1}^{N} \left(\frac{\partial \boldsymbol{\pi}_i}{\partial \boldsymbol{\beta}}\right)' V_i^{-1}(\boldsymbol{\beta}, \alpha, \phi)\left(\frac{\partial \boldsymbol{\pi}_i}{\partial \boldsymbol{\beta}}\right) \quad . \tag{10.222}$$

To avoid the dependence of α in determining $\hat{\boldsymbol{\beta}}_G$, Liang and Zeger (1986) propose to replace α by a $N^{1/2}$-consistent estimate (cf. Lehmann, 1986, p. 422) $\hat{\alpha}(\boldsymbol{y}_1, \ldots, \boldsymbol{y}_N, \boldsymbol{\beta}, \phi)$ and ϕ by $\hat{\phi}$ (10.81) and to determine $\hat{\boldsymbol{\beta}}_G$ from $U(\boldsymbol{\beta}, \hat{\alpha}, \hat{\phi}) = 0$.

Remark:

- The iterative estimating procedure for GEE is described in detail in Liang and Zeger (1986). For the computational translation a SAS-macro by Karim and Zeger (1988, John Hopkins University, Baltimore) and a procedure of the Statistical Institute of the University of Munich (Heumann, 1993) exist.

- If $\boldsymbol{R}_i(\alpha) = \boldsymbol{I}_{n_i}$ for $i = 1, \ldots, N$ is chosen, then the GEE are reduced to the *independence estimating equations (IEE)*. The IEE are

$$U(\boldsymbol{\beta}, \phi) = \sum_{i=1}^{N} \left(\frac{\partial \boldsymbol{\pi}_i}{\partial \boldsymbol{\beta}}\right)' \boldsymbol{A}_i^{-1}(\boldsymbol{y}_i - \boldsymbol{\pi}_i) = 0 \tag{10.223}$$

  with $\boldsymbol{A}_i = \mathrm{diag}(v(\pi_{ij})\phi)$. The solution is denoted by $\hat{\boldsymbol{\beta}}_I$. Under some weak conditions, we have (Theorem 1 in Liang and Zeger, 1986)

  - $\hat{\boldsymbol{\beta}}_I$ is asymptotically consistent if the expectation $\pi_{ij} = h(\boldsymbol{x}'_{ij}\boldsymbol{\beta})$ is correctly specified and the dispersion parameter ϕ is consistantly estimated.

- The consistency is independent of the correct specification of the covariance.

- $\hat{\boldsymbol{\beta}}_I$ is asymptotically normal

$$\hat{\boldsymbol{\beta}}_I \stackrel{a.s.}{\sim} N(\boldsymbol{\beta}; F_Q^{-1}(\boldsymbol{\beta}, \phi) F_2(\boldsymbol{\beta}, \phi) F_Q^{-1}(\boldsymbol{\beta}, \phi)), \qquad (10.224)$$

where

$$F_Q^{-1}(\boldsymbol{\beta}, \phi) = \left[\sum_{i=1}^N \left(\frac{\partial \boldsymbol{\pi}_i}{\partial \boldsymbol{\beta}} \right)' \boldsymbol{A}_i^{-1} \left(\frac{\partial \boldsymbol{\pi}_i}{\partial \boldsymbol{\beta}} \right) \right]^{-1},$$

$$F_2(\boldsymbol{\beta}, \phi) = \sum_{i=1}^N \left(\frac{\partial \boldsymbol{\pi}_i}{\partial \boldsymbol{\beta}} \right)' \boldsymbol{A}_i^{-1} \mathrm{Cov}(\boldsymbol{y}_i) \boldsymbol{A}_i^{-1} \left(\frac{\partial \boldsymbol{\pi}_i}{\partial \boldsymbol{\beta}} \right)$$

and $\mathrm{Cov}(\boldsymbol{y}_i)$ is the true covariance matrix of $\boldsymbol{y}_i$.

- A consistent estimate for the variance of $\hat{\boldsymbol{\beta}}_I$ is found by replacing $\boldsymbol{\beta}_I$ by $\hat{\boldsymbol{\beta}}_I$, $\mathrm{Cov}(\boldsymbol{y}_i)$ by its estimate $(\boldsymbol{y}_i - \hat{\boldsymbol{\pi}}_i)(\boldsymbol{y}_i - \hat{\boldsymbol{\pi}}_i)'$, and ϕ by $\hat{\phi}$ from (10.81), if ϕ is an unknown nuisance parameter.

The advantages of $\hat{\boldsymbol{\beta}}_I$ are that $\hat{\boldsymbol{\beta}}_I$ is easy to calculate with the appropriate software, as for instance GLIM, and that in case of correct specification of the regression model, $\hat{\boldsymbol{\beta}}_I$ and $\mathrm{Cov}(\hat{\boldsymbol{\beta}}_I)$ are consistent estimates. However, $\hat{\boldsymbol{\beta}}_I$ looses in efficiency if the correlation between the clusters is large.

Properties of the GEE Estimate $\hat{\boldsymbol{\beta}}_G$

- Liang and Zeger (1986, p. 16) state in their Theorem 2, that under some weak assumptions and under the conditions

  (i) $\hat{\alpha}$ is $N^{1/2}$-consistent for α, given $\boldsymbol{\beta}$ and ϕ

  (ii) $\hat{\phi}$ is a $N^{1/2}$-consistent estimate for ϕ, given $\boldsymbol{\beta}$

  (iii) the derivation $\partial \hat{\alpha}(\boldsymbol{\beta}, \phi)/\partial \phi$ is independent of ϕ and α and is of the stochastic order $O_p(1)$

  the estimate $\hat{\boldsymbol{\beta}}_G$ is consistent and asymptotic normal:

$$\hat{\boldsymbol{\beta}}_G \stackrel{a.s.}{\sim} N(\boldsymbol{\beta}, \boldsymbol{V}_G) \qquad (10.225)$$

with the asymptotic covariance matrix

$$\boldsymbol{V}_G = F_Q^{-1}(\boldsymbol{\beta}, \alpha) F_2(\boldsymbol{\beta}, \alpha) F_Q^{-1}(\boldsymbol{\beta}, \alpha), \qquad (10.226)$$

where

$$F_Q^{-1}(\boldsymbol{\beta}, \alpha) = \left(\sum_{i=1}^N \left(\frac{\partial \boldsymbol{\pi}_i}{\partial \boldsymbol{\beta}} \right)' \boldsymbol{V}_i^{-1} \left(\frac{\partial \boldsymbol{\pi}_i}{\partial \boldsymbol{\beta}} \right) \right)^{-1},$$

$$F_2(\boldsymbol{\beta}, \alpha) = \sum_{i=1}^N \left(\frac{\partial \boldsymbol{\pi}_i}{\partial \boldsymbol{\beta}} \right)' \boldsymbol{V}_i^{-1} \mathrm{Cov}(\boldsymbol{y}_i) \boldsymbol{V}_i^{-1} \left(\frac{\partial \boldsymbol{\pi}_i}{\partial \boldsymbol{\beta}} \right)$$

and $\mathrm{Cov}(\boldsymbol{y}_i) = \mathrm{E}[(\boldsymbol{y}_i - \boldsymbol{\pi}_i)(\boldsymbol{y}_i - \boldsymbol{\pi}_i)']$ is the true covariance matrix of $\boldsymbol{y}_i$. A short outline of the proof may be found in the appendix of Liang and Zeger (1986).

The asymptotic properties only hold for $N \to \infty$. Hence, it should be remembered that the estimation procedure should only be used for a large number of clusters.

- An estimate $\hat{\mathbf{V}}_G$ for the covariance matrix $\mathbf{V}_G$ may be found by:

    - replacing $\boldsymbol{\beta}$, ϕ, α by their consistent estimates in (10.226),
    - replacing $\mathrm{Cov}(\boldsymbol{y}_i)$ by $(\boldsymbol{y}_i - \hat{\boldsymbol{\pi}}_i)(\boldsymbol{y}_i - \hat{\boldsymbol{\pi}}_i)'$.

- If the covariance structure is specified correctly so that $\mathbf{V}_i = \mathrm{Cov}(\boldsymbol{y}_i)$, then the covariance of $\hat{\boldsymbol{\beta}}_G$ is the inverse of the expected Fisher information matrix

$$\mathbf{V}_G = \left(\sum_{i=1}^{N} \left(\frac{\partial \boldsymbol{\pi}_i}{\partial \boldsymbol{\beta}} \right)' \mathbf{V}_i^{-1} \left(\frac{\partial \boldsymbol{\pi}_i}{\partial \boldsymbol{\beta}} \right) \right)^{-1} = F^{-1}(\boldsymbol{\beta}, \alpha).$$

The estimate of this matrix is more stable than that of (10.226), but it has a loss in efficiency if the correlation structure is specified incorrectly (cf. Prentice, 1988, p. 1040).

- The method of Liang and Zeger leads to an asymptotic variance of $\hat{\boldsymbol{\beta}}_G$ that is independent of the choice of the estimates $\hat{\alpha}$ and $\hat{\phi}$ within the class of the $N^{1/2}$-consistent estimates. This is true for the asymptotic distribution of $\hat{\boldsymbol{\beta}}_G$ as well.

- In case of correct specification of the regression model, the estimates $\hat{\boldsymbol{\beta}}_G$ and $\hat{\mathbf{V}}_G$ are consistent, independent of the choice of the quasi-correlation matrix $\boldsymbol{R}_i(\alpha)$. This means that even if $\boldsymbol{R}_i(\alpha)$ is specified incorrectly, $\hat{\boldsymbol{\beta}}_G$ and $\hat{\mathbf{V}}_G$ stay consistent as long as $\hat{\alpha}$ and $\hat{\phi}$ are consistent. This robustness of the estimates is important, since the admissibility of the working covariance matrix $\mathbf{V}_i$ is hard to check for small n_i.

- An incorrect specification of $\boldsymbol{R}_i(\alpha)$ can reduce the efficiency of $\hat{\boldsymbol{\beta}}_G$.

Remark:

1. If the identity matrix is assumed for $\boldsymbol{R}_i(\alpha)$, that is $\boldsymbol{R}_i(\alpha) = \boldsymbol{I}$, $i = 1, \cdots, N$, then the estimating equations for $\boldsymbol{\beta}$ are reduced to the IEE. If the variances of the binomial distribution are chosen, as is usually done in the binary case, then the IEE and the ML score function (with binomially distributed variables) lead to the same estimates for $\boldsymbol{\beta}$. However, the IEE method should be preferred in general, since the ML estimation procedure leads to incorrect variances for $\hat{\boldsymbol{\beta}}_G$ and hence incorrect *p-values*. This leads to incorrect conclusions, for instance, related to significance or non-significance of the covariables (cf. Liang and Zeger, 1993).

2. Diggle, Liang, and Zeger (1993, Chapter 7.5) propose to check the consistency of $\hat{\boldsymbol{\beta}}_G$ by fitting an appropriate model with various covariance structures. The estimates $\hat{\boldsymbol{\beta}}_G$ and their consistent variances are then compared. If these differ too much, the modelling of the covariance structure calls for more attention.

Efficiency of the GEE and IEE Methods

Liang and Zeger (1986) stated the following about the comparison of $\hat{\boldsymbol{\beta}}_I$ and $\hat{\boldsymbol{\beta}}_G$:

- $\hat{\boldsymbol{\beta}}_I$ is almost as efficient as $\hat{\boldsymbol{\beta}}_G$ if the true correlation α is small.

- $\hat{\boldsymbol{\beta}}_I$ is very efficient if α is small and the data are binary.

- If α is large, then $\hat{\boldsymbol{\beta}}_G$ is more efficient than $\hat{\boldsymbol{\beta}}_I$, and the efficiency of $\hat{\boldsymbol{\beta}}_G$ can be increased if the correlation matrix is specified correctly.

- In case of a high correlation within the blocks, the loss of efficiency of $\hat{\boldsymbol{\beta}}_I$ compared to $\hat{\boldsymbol{\beta}}_G$ is larger if the number of subunits n_i, $i = 1, \cdots, N$ varies between the clusters than if the clusters are all of the same size.

Choice of the Quasi-Correlation Matrix $\mathbf{R}_i(\alpha)$

The working correlation matrix $\boldsymbol{R}_i(\alpha)$ is chosen according to considerations like simplicity, efficiency, and amount of existing data. Furthermore, assumptions about the structure of the dependence between the data should be considered by the choice. As mentioned before, the importance of the correlation matrix is due to the fact that it influences the variance of the estimated parameters.

1. The simplest specification is the assumption, that the repeated observations of a cluster are uncorrelated, that is

$$\boldsymbol{R}_i(\alpha) = \boldsymbol{I}, \qquad i = 1, \cdots, N.$$

This assumption leads to the IEE equations for uncorrelated response variables.

2. Another special case, which is the most efficient according to Liang and Zeger (1986, §4) but may only be used if the number of observations per cluster is small and same for all clusters (e.g., equals n), is given by the choice

$$\boldsymbol{R}_i(\alpha) = \boldsymbol{R}(\alpha)$$

where $\boldsymbol{R}(\alpha)$ is left totally unspecified and may be estimated by the empirical correlation matrix. $n(n-1)/2$ parameters have to be estimated.

3. If it is assumed that the same pairwise dependencies exist between all response variables of one cluster, then the *exchangeable correlation structure*
   may be chosen:

$$\text{Corr}(y_{ik}, y_{il}) = \alpha, \qquad k \neq l, \quad i = 1, \dots, N \quad .$$

This corresponds to the correlation assumption in random effects models.

4. If

$$\text{Corr}(y_{ik}, y_{il}) = \alpha(|k - l|),$$

is chosen, then the correlations are stationary. The specific form $\alpha(|k-l|) = \alpha^{|l-k|}$ corresponds to the auto-correlation function of an AR(1)-process.

Further methods for parameter estimation in quasi-likelihood approaches are

- the GEE1 method by Prentice (1988) that estimates the α and β simultaneously from the GEE for α and β,

- the modified GEE1 method by Fitzmaurice and Laird (1993) based on
  conditional odds ratios and that by Lipsitz, Laird and Harrington (1991)
  and Liang, Zeger and Qaquish (1992) based on marginal odds ratios for
  modelling the cluster correlation,

- the GEE2 method by Liang, Zeger and Qaquish (1992) that estimates $\delta' = (\beta', \alpha)$ simultaneously as a joint parameter,

- the pseudo-ML method by Zhao and Prentice (1990, cf. Prentice and Zhao, 1991).

10.9.4   Bivariate Binary Correlated Response Variables

The previous sections introduced various methods developed for regression analysis of correlated binary data. They were described in a general form for N blocks
(clusters) of size n_i. These methods may of course be used for bivariate binary
data as well. This has the advantage that it simplifies the matter.

In this section, the methods GEE and IEE are developed for the bivariate
binary case. Afterwards an example demonstrates for the case of bivariate binary
data the difference between a naive ML estimate and the GEE method by Liang
and Zeger (1986).

We have: $\boldsymbol{y}_i = (y_{i1}, y_{i2})'$, $i = 1, \cdots, N$. Each response variable y_{ij}, $j = 1, 2$,
has its own vector of covariables $\boldsymbol{x}'_{ij} = (x_{ij1}, \cdots, x_{ijp})$. The chosen link function
for modelling the relationship between $\pi_{ij} = P(y_{ij} = 1)$ and $\boldsymbol{x}_{ij}$ is the logit link

$$\text{Logit}(\pi_{ij}) = \ln\left(\frac{\pi_{ij}}{1 - \pi_{ij}}\right) = \boldsymbol{x}'_{ij}\boldsymbol{\beta} \quad . \tag{10.227}$$

Let

$$\boldsymbol{\pi}'_i = (\pi_{i1}, \pi_{i2}) \ , \ \eta_{ij} = \boldsymbol{x}'_{ij}\boldsymbol{\beta} \ , \ \boldsymbol{\eta}' = (\eta_{i1}, \eta_{i2}) \ . \tag{10.228}$$

The logistic regression model has become the standard method for regression analysis of binary data.

The GEE Method

From Section 10.9.3 it can be seen that the form of the estimating equations for β is as follows:

$$U(\boldsymbol{\beta}, \alpha, \phi) = S(\boldsymbol{\beta}, \alpha) = \sum_{i=1}^{N} \left(\frac{\partial \boldsymbol{\pi}_i}{\partial \boldsymbol{\beta}} \right)' \boldsymbol{V}_i^{-1}(\boldsymbol{y}_i - \boldsymbol{\pi}_i) = 0 \ , \tag{10.229}$$

where $\boldsymbol{V}_i = \boldsymbol{A}_i^{1/2}\boldsymbol{R}_i(\alpha)\boldsymbol{A}_i^{1/2}$, $\boldsymbol{A}_i = \mathrm{diag}(v(\pi_{ij})\phi)$, $j = 1, 2$, and $\boldsymbol{R}_i(\alpha)$ is the working correlation matrix. Since only one correlation coefficient $\rho_i = \mathrm{Corr}(y_{i1}, y_{i2})$, $i = 1, \cdots, N$ has to be specified for bivariate binary data, and this is assumed to be constant, we have for the correlation matrix:

$$\boldsymbol{R}_i(\alpha) = \begin{pmatrix} 1 & \rho \\ \rho & 1 \end{pmatrix}, \qquad i = 1, \cdots, N \ . \tag{10.230}$$

For the matrix of derivatives we have:

$$\begin{aligned}
\left(\frac{\partial \boldsymbol{\pi}_i}{\partial \boldsymbol{\beta}} \right)' &= \left(\frac{\partial h(\boldsymbol{\eta}_i)}{\partial \boldsymbol{\beta}} \right)' = \left(\frac{\partial \boldsymbol{\eta}_i}{\partial \boldsymbol{\beta}} \right)' \left(\frac{\partial h(\boldsymbol{\eta}_i)}{\partial \boldsymbol{\eta}_i} \right)' \\
&= \begin{pmatrix} \boldsymbol{x}'_{i1} \\ \boldsymbol{x}'_{i2} \end{pmatrix}' \begin{pmatrix} \frac{\partial h(\eta_{i1})}{\partial \eta_{i1}} & 0 \\ 0 & \frac{\partial h(\eta_{i2})}{\partial \eta_{i2}} \end{pmatrix} .
\end{aligned}$$

Since $h(\eta_{i1}) = \pi_{i1} = \frac{\exp(\boldsymbol{x}'_{i1}\boldsymbol{\beta})}{1+\exp(\boldsymbol{x}'_{i1}\boldsymbol{\beta})}$ and $\exp(\boldsymbol{x}'_{i1}\boldsymbol{\beta}) = \frac{\pi_{i1}}{1-\pi_{i1}}$, we have $1 + \exp(\boldsymbol{x}'_{i1}\boldsymbol{\beta}) = 1 + \frac{\pi_{i1}}{1-\pi_{i1}} = \frac{1}{1-\pi_{i1}}$, and

$$\frac{\partial h(\eta_{i1})}{\partial \eta_{i1}} = \frac{\pi_{i1}}{1 + \exp(\boldsymbol{x}'_{i1}\boldsymbol{\beta})} = \pi_{i1}(1 - \pi_{i1}). \tag{10.231}$$

holds. Analogously we have:

$$\frac{\partial h(\eta_{i2})}{\partial \eta_{i2}} = \pi_{i2}(1 - \pi_{i2}). \tag{10.232}$$

If the variance is specified as $\mathrm{Var}(y_{ij}) = \pi_{ij}(1 - \pi_{ij})$, $\phi = 1$, then we get

$$\left(\frac{\partial \boldsymbol{\pi}_i}{\partial \boldsymbol{\beta}} \right)' = \boldsymbol{x}'_i \begin{pmatrix} \mathrm{Var}(y_{i1}) & 0 \\ 0 & \mathrm{Var}(y_{i2}) \end{pmatrix} = \boldsymbol{x}'_i \boldsymbol{\Delta}_i$$

with $\boldsymbol{x}'_i = (x_{i1}, x_{i2})$ and $\boldsymbol{\Delta}_i = \begin{pmatrix} \text{Var}(y_{i1}) & 0 \\ 0 & \text{Var}(y_{i2}) \end{pmatrix}$. For the covariance matrix $\boldsymbol{V}_i$ we have:

$$
\begin{aligned}
\boldsymbol{V}_i &= \begin{pmatrix} \text{Var}(y_{i1}) & 0 \\ 0 & \text{Var}(y_{i2}) \end{pmatrix}^{1/2} \begin{pmatrix} 1 & \rho \\ \rho & 1 \end{pmatrix} \begin{pmatrix} \text{Var}(y_{i1}) & 0 \\ 0 & \text{Var}(y_{i2}) \end{pmatrix}^{1/2} \\
&= \begin{pmatrix} \text{Var}(y_{i1}) & \rho(\text{Var}(y_{i1})\text{Var}(y_{i2}))^{1/2} \\ \rho(\text{Var}(y_{i1})\text{Var}(y_{i2}))^{1/2} & \text{Var}(y_{i2}) \end{pmatrix}
\end{aligned}
$$
(10.233)

and for the inverse of $\boldsymbol{V}_i$:

$$
\begin{aligned}
\boldsymbol{V}_i^{-1} &= \frac{1}{(1-\rho^2)\text{Var}(y_{i1})\text{Var}(y_{i2})} \\
&\quad \begin{pmatrix} \text{Var}(y_{i2}) & -\rho(\text{Var}(y_{i1})\text{Var}(y_{i2}))^{1/2} \\ -\rho(\text{Var}(y_{i1})\text{Var}(y_{i2}))^{1/2} & \text{Var}(y_{i1}) \end{pmatrix} \\
&= \frac{1}{1-\rho^2} \begin{pmatrix} [\text{Var}(y_{i1})]^{-1} & -\rho(\text{Var}(y_{i1})\text{Var}(y_{i2}))^{-1/2} \\ -\rho(\text{Var}(y_{i1})\text{Var}(y_{i2}))^{-1/2} & [\text{Var}(y_{i2})]^{-1} \end{pmatrix}.
\end{aligned}
$$
(10.234)

If $\boldsymbol{\Delta}_i$ is multiplied by $\boldsymbol{V}_i^{-1}$, we obtain

$$
\boldsymbol{W}_i = \boldsymbol{\Delta}_i \boldsymbol{V}_i^{-1} = \frac{1}{1-\rho^2} \begin{pmatrix} 1 & -\rho\left(\frac{\text{Var}(y_{i1})}{\text{Var}(y_{i2})}\right)^{1/2} \\ -\rho\left(\frac{\text{Var}(y_{i2})}{\text{Var}(y_{i1})}\right)^{1/2} & 1 \end{pmatrix}
$$
(10.235)

and for the GEE equations for $\boldsymbol{\beta}$ in the bivariate binary case:

$$
S(\boldsymbol{\beta}, \alpha) = \sum_{i=1}^{N} \boldsymbol{x}'_i \boldsymbol{W}_i (\boldsymbol{y}_i - \boldsymbol{\pi}_i) = 0.
$$
(10.236)

According to Theorem 2 by Liang and Zeger (1986), under some weak conditions and under the assumption that the correlation parameter was consistently estimated, the solution $\hat{\boldsymbol{\beta}}_G$ is consistent and asymptotic normal with the expectation $\boldsymbol{\beta}$ and the covariance matrix (10.226).

The IEE Method

If it is assumed that the response variables of each of the blocks are independent, that is, $\boldsymbol{R}_i(\alpha) = \boldsymbol{I}$ and $\boldsymbol{V}_i = \boldsymbol{A}_i$, then the GEE equations are reduced to the IEE equations (10.224)

$$
U(\boldsymbol{\beta}, \phi) = S(\boldsymbol{\beta}) = \sum_{i=1}^{N} \left(\frac{\partial \boldsymbol{\pi}_i}{\partial \boldsymbol{\beta}}\right)' \boldsymbol{A}_i^{-1}(\boldsymbol{y}_i - \boldsymbol{\pi}_i) = 0.
$$
(10.237)

As we just showed, we have for the bivariate binary case:

$$\left(\frac{\partial \pi_i}{\partial \beta}\right)' = x_i' \Delta_i = x_i' \left(\begin{array}{cc} \mathrm{Var}(y_{i1}) & 0 \\ 0 & \mathrm{Var}(y_{i2}) \end{array} \right) \tag{10.238}$$

with $\mathrm{Var}(y_{ij}) = \pi_{ij}(1 - \pi_{ij})$, $\phi = 1$, and $A_i^{-1} = \left(\begin{array}{cc} [\mathrm{Var}(y_{i1})]^{-1} & 0 \\ 0 & [\mathrm{Var}(y_{i2})]^{-1} \end{array} \right)$.

The IEE equations then simplify to

$$S(\beta) = \sum_{i=1}^{N} x_i'(y_i - \pi_i) = 0. \tag{10.239}$$

The solution $\hat{\beta}_I$ is consistent and asymptotic normal, according to Theorem 1 by Liang and Zeger (1986).

10.9.5 An Example from the Field of Dentistry

In this section, we want to demonstrate the procedure of the GEE method by means of a 'twin' data set, that was documented by Dr. W. Walther of the Dental Clinic in Karlsruhe (Walther, 1992). The focal point is to show the difference between a robust estimate (GEE method) that takes the correlation of the response variables into account and the naive ML estimate. For the parameter estimation with the GEE method, a SAS macro is available (Karim and Zeger, 1988), as well as a procedure by Heumann (1993).

Description of the 'Twin' Data Set

During the examined interval, 331 patients were provided with two conical crowns each in the Dental Clinic Karlsruhe, Germany. Since 50 conical crowns showed missing values and since the SAS macro for the GEE method needs complete data sets, these patients were excluded. Hence, for estimation of the regression parameters, the remaining 612 completely observed twin data were used. In this example, the twin pairs make up the clusters and the twins themselves (1.twin, 2.twin) are the subunits of the clusters.

The Response Variable

For all twin pairs in this study, the life-time of the conical crowns was recorded in days. This life-time is chosen as response and is transformed into a binary response variable y_{ij} of the jth twin ($j = 1, 2$) in the ith cluster with

$$y_{ij} = \left\{ \begin{array}{ll} 1 & , \quad \text{if the conical crown is in function longer than } x \text{ days} \\ 0 & , \quad \text{if the conical crown is not in function longer than } x \text{ days.} \end{array} \right.$$

Different values may be defined for x. In the example, the values 360 (c. 1 year), 1100 (c. 3 years) and 2000 (c. 5 years) were chosen. Since the response variable is binary, the response probability of y_{ij} is modelled by the logit link (logistic regression). The model for the log-odds (i.e., the logarithm of the odds $\pi_{ij}/(1 - \pi_{ij})$ of the response $y_{ij} = 1$) is linear in the covariables and in the model for the odds itself, the covariables have a multiplicative effect on the odds. Aim of the analysis is to find out whether the prognostic factors have a significant influence on the response probability.

Prognostic Factors

The covariables that were included in the analysis with the SAS macro, are

- age (in years)

- sex (1: male, 2: female)

- jaw (1: upper jaw, 2: lower jaw)

- type (1: dentoalveolar design, 2: transversal design)

All covariables, except for the covariable age, are dichotomous. The two types of conical crowns construction, dentoalveolar and transversal design, are distinguished as follows (cf. Walther, 1992):

- the dentoalveolar design connects all abutments exclusively by a rigid connection that runs on the alveolar ridge;

- the transversal design is used if the parts of reconstruction have to be connected by a transversal bar. This is the case if teeth in the front area are not included in the construction.

292 conical crowns were included in a dentoalveolar and 320 in a transversal design. 258 conical crowns were placed in the upper jaw, 354 in the lower jaw.

The GEE Method

A problem that arises for the twin data, is that the twins of a block are correlated. If this correlation is not taken into account, then the estimates $\hat{\beta}$ stay unchanged but the variance of the $\hat{\beta}$ is underestimated. In case of positive correlation in a cluster, we have:

$$\text{Var}(\hat{\beta})_{\text{naive}} < \text{Var}(\hat{\beta})_{\text{robust}}.$$

Because of

$$\frac{\hat{\beta}}{\sqrt{\text{Var}(\hat{\beta})_{\text{naive}}}} > \frac{\hat{\beta}}{\sqrt{\text{Var}(\hat{\beta})_{\text{robust}}}}$$

this leads to incorrect tests and possibly to significant effects that might not be significant in a correct analysis (e.g., GEE). For this reason, appropriate methods that estimate the variance correctly should be chosen if the response variables are correlated.

The following regression model without interaction is assumed:

$$\ln \frac{P(\text{Life-time} \geq x)}{P(\text{Life-time} < x)} = \beta_0 + \beta_1 \cdot \text{Age} + \beta_2 \cdot \text{Sex} + \beta_3 \cdot \text{Jaw} + \beta_4 \cdot \text{Type}. \quad (10.240)$$

Additionally, we assume that the dependencies between the twins are identical and hence the exchangeable correlation structure is suitable for describing the dependencies.

In order to demonstrate the effect of various correlation assumptions on the estimation of the parameters, the following logistic regression models, that only differ in the assumed association parameter, are compared:

Model 1: naive (incorrect) ML estimation

Model 2: robust (correct) estimation, where independence is assumed
$(R_i(\alpha) = I)$

Model 3: robust estimation with exchangeable correlation structure
$(\rho_{ikl} = \text{Corr}(y_{ik}, y_{il}) = \alpha, \ k \neq l)$

Model 4: robust estimation with unspecified correlation structure
$(R_i(\alpha) = R(\alpha))$.

As a test statistic (z-naive and z-robust) the ratio of estimate and standard error is calculated.

Results

Table 10.13 summarizes the estimated regression parameters, the standard errors, the z-statistics, and the p-values of the models 2, 3, and 4 of the response variables

$$y_{ij} = \begin{cases} 1 & , \quad \text{if the conical crown is longer in function than 360 days} \\ 0 & , \quad \text{if the conical crown is not longer in function than 360 days.} \end{cases}$$

It turns out that the $\hat{\beta}$-values and the z-statistics are identical, independent of the choice of R_i, eventhough a high correlation between the twins exists. The exchangeable correlation model yields the value 0.9498 for the estimated correlation parameter $\hat{\alpha}$. In the model with the unspecified correlation structure, ρ_{i12} and ρ_{i21} were estimated as 0.9498 as well. The fact that the estimates of the models 2, 3, and 4 coincide was observed in the analyses of the response variables with $x=1100$ and $x=2000$ as well. This means that the choice of R_i has no influence on the estimation procedure in case of bivariate binary response. The GEE method is robust with respect to various correlation assumptions.

Table 10.14 compares the results of model 1 and 2. A striking difference between the two methods is that the covariable age in case of a naive ML estimation (model 1) is significant on the 10%-level, eventhough this significance does not turn up if the robust method with the assumption of independence (model 2) is used. In case of coinciding estimated regression parameters, the robust variances of $\hat{\beta}$ are larger and, accordingly, the robust z-statistics are smaller than the naive z-statistics. This result shows clearly that the ML method, which is incorrect in this case, underestimates the variances of $\hat{\beta}$ and hence leads to an incorrect age effect.

≥ 360	Model 2 (independence assump.)	Model 3 (exchangeable)	Model 4 (unspecified)
Age	$0.017^{1)}$ $(0.012)^{2)}$ $1.33^{3)}$ $0.185^{4)}$	0.017 (0.012) 1.33 0.185	0.017 (0.012) 1.33 0.185
Sex	-0.117 (0.265) -0.44 0.659	-0.117 (0.265) -0.44 0.659	-0.117 (0.265) -0.44 0.659
Jaw	0.029 (0.269) 0.11 0.916	0.029 (0.269) 0.11 0.916	0.029 (0.269) 0.11 0.916
Type	-0.027 (0.272) -0.10 0.920	-0.027 (0.272) -0.10 0.920	-0.027 (0.272) -0.10 0.920

Table 10.12: Results of the robust estimates for the models 2, 3, and 4 ($x=360$).
[1]: estimated regression values $\hat{\beta}$; [2]: standard errors of $\hat{\beta}$; [3]: z-statistic; [4]: p-value

≥ 360	Model 1 (naive)			Model 2 (robust)		
	σ	z	p-value	σ	z	p-value
Age	0.008	1.95	0.051*	0.012	1.33	0.185
Sex	0.190	-0.62	0.538	0.265	-0.44	0.659
Jaw	0.192	0.15	0.882	0.269	0.11	0.916
Type	0.193	-0.14	0.887	0.272	-0.10	0.920

Table 10.13: Comparison of the standard errors, the z-statistics and the p-values of models 1 and 2 ($x=360$). *: significant on the 10%-level

Tables 10.14 and 10.15 summarize the results with the x-values 1100 and 2000. Table 10.14 shows that if the response variable is modelled with $x=1100$, then none of the observed covariables is significant. As before, the estimated correlation parameter $\hat{\alpha} = 0.9578$ indicates a strong dependency between the twins.

≥1100	$\hat{\beta}$	Model 1 (naive)				Model 2 (robust)		
		σ	z	p-value		σ	z	p-value
Age	0.0006	0.008	0.08	0.939		0.010	0.06	0.955
Sex	-0.0004	0.170	-0.00	0.998		0.240	-0.00	0.999
Jaw	0.1591	0.171	0.93	0.352		0.240	0.66	0.507
Type	0.0369	0.172	0.21	0.830		0.242	0.15	0.878

Table 10.14: Comparison of the standard errors, the z-statistics and the p-values of models 1 and 2 (x=1100).

≥2000	$\hat{\beta}$	Model 1 (naive)				Model 2 (robust)		
		σ	z	p-value		σ	z	p-value
Age	-0.0051	0.013	-0.40	0.691		0.015	-0.34	0.735
Sex	-0.2177	0.289	-0.75	0.452		0.399	-0.55	0.586
Jaw	0.0709	0.287	0.25	0.805		0.412	0.17	0.863
Type	0.6531	0.298	2.19	0.028*		0.402	1.62	0.104

Table 10.15: Comparison of the standard errors, the z-statistics and the p-values of models 1 and 2 (x=2000). *: significant on the 10%-level

In Table 10.15, the covariable 'type' has a significant influence in case of the naive estimation. In case of the GEE method ($\boldsymbol{R} = \boldsymbol{I}$) it might be significant with a p-value=0.104 (10%-level). The result $\hat{\beta}_{\text{type}} = 0.6531$ indicates that a dentoalveolar design significantly increases the log-odds of the response variable

$$y_{ij} = \begin{cases} 1 & , \quad \text{if the conical crown is longer in function than 2000 days} \\ 0 & , \quad \text{if the conical crown is not longer in function than 2000 days.} \end{cases}$$

Assuming the model

$$\frac{P(\text{Life-time} \geq 2000)}{P(\text{Life-time} < 2000)} = \exp(\beta_0 + \beta_1 \cdot \text{Age} + \beta_2 \cdot \text{Sex} + \beta_3 \cdot \text{Jaw} + \beta_4 \cdot \text{Type})$$

the odds $\frac{P(\text{Life-time} \geq 2000)}{P(\text{Life-time} < 2000)}$ for a dentoalveolar design is higher than the odds for a transversal design by the factor $\exp(\beta_4) = \exp(0.6531) = 1.92$, or alternatively: the odds ratio equals 1.92. The correlation parameter yields the value 0.9035.

Summarizing, it can be said that 'age' and 'type' are significant but not time-dependent covariables. The robust estimation yields no significant interaction and a high correlation α exists between the twins of a pair.

Problems

The GEE estimations, which were carried out stepwise, have to be compared with caution, since they are not independent due to the time effect in the response variables. In this context, time-adjusted GEE methods that could be applied in this example are still missing. Therefore, further efforts are necessary in the field of survivorship analysis, in order to be able to complement the standard procedures, as for instance Kaplan-Meier estimate and log-rank test, which are based on the independence of the response variables.

10.10   Exercises and Questions

10.10.1   Let two models be defined by their design matrices X_1 and $X_2 = (X_1, X_3)$. Name the test statistic for testing H_0 : 'Model X_1 holds' and its distribution.

10.10.2   What is meant by overdispersion?
How is it parametrized in case of a binomial distribution?

10.10.3   Why would a quasi-log-likelihood approach be chosen?
How is the correlation in cluster data parametrized?

10.10.4   Compare the models of two-way classification for continuous, normal data (ANOVA) and for categorical data.
What are the reparametrization conditions in each case?

10.10.5   Given the following G^2-analysis of a two-way model with all submodels:

Model	G^2	p-value
A	200	0.00
B	100	0.00
A+B	20	0.10
A*B	0	1.00

Which model is valid?

10.10.6   Given the following $I \times 2$-table for X : age group and Y : binary response.

	1	0
< 40	10	8
40–50	15	12
50–60	20	12
60–70	30	20
> 70	30	25

Analyse the trend of the sample logits.

A Matrix Algebra

There are numerous books on matrix albegra that contain results useful for the discussion of linear models. See for instance books by Graybill (1961), Mardia et al. (1979), Searle (1982), Rao (1973), Rao and Mitra (1971) to mention a few. In this Appendix, we collected some of the important results for ready reference. Proofs are generally omitted. References to original sources are given wherever necessary.

A.1 Introduction

Definition A.1 *An $m \times n$-matrix A is a rectangular array of elements in m rows and n columns.*

In the context of the material treated in the book and in this Appendix, the elements of a matrix are taken as real numbers.

We refer to an $m \times n$-matrix of type (or order) $m \times n$ and indicate this by writing $A : m \times n$ or $\underset{m,n}{A}$.

Let a_{ij} be the element in the ith row and the jth column of A. Then A may be represented as

$$A = \begin{pmatrix} a_{11} & a_{12} & \cdots & a_{1n} \\ a_{21} & a_{22} & \cdots & a_{2n} \\ \vdots & \vdots & & \cdots \\ a_{m1} & a_{m2} & \cdots & a_{mn} \end{pmatrix} = (a_{ij}).$$

A matrix with $n = m$ rows and columns is called square matrix. A square matrix having zeros as elements below (above) the diagonal is called an upper (lower) triangular matrix.

Definition A.2 *The transpose $A' : n \times m$ of a matrix $A : m \times n$ is given by interchanging the rows and columns of A. Thus*

$$A' = (a_{ji}).$$

We then have the following rules

$$(A')' = A, \quad (A + B)' = A' + B', \quad (AB)' = B'A'.$$

Definition A.3 *A square matrix is called symmetric, if $A' = A$.*

Definition A.4 *An $m \times 1$–matrix a is said to be an m-vector and is written as a column*

$$a = \begin{pmatrix} a_1 \\ \vdots \\ a_m \end{pmatrix}.$$

Definition A.5 *A $1 \times n$–matrix a' is said to be a row vector*

$$a' = (a_1, \cdots, a_n).$$

Hence, a matrix $A : m \times n$ may be written alternatively as

$$A = (a_{(1)}, \ldots, a_{(n)}) = \begin{pmatrix} a'_1 \\ \vdots \\ a'_m \end{pmatrix}$$

with

$$a_{(j)} = \begin{pmatrix} a_{1j} \\ \vdots \\ a_{mj} \end{pmatrix}, \quad a_i = \begin{pmatrix} a_{i1} \\ \vdots \\ a_{in} \end{pmatrix}.$$

Definition A.6 *The $n \times 1$ row vector $(1, \cdots, 1)'$ is denoted by $\mathbf{1}'_n$ or $\mathbf{1}'$.*

Definition A.7 *The matrix $A : m \times m$ with $a_{ij} = 1$ (for all i,j) is denoted by the symbol J_m, i.e.,*

$$J_m = \begin{pmatrix} 1 & \cdots & 1 \\ \vdots & & \vdots \\ 1 & \vdots & 1 \end{pmatrix} = \mathbf{1}_m \mathbf{1}'_m.$$

Definition A.8 *The n-vector*

$$e_i = (0, \cdots, 0, 1, 0, \cdots, 0)'$$

whose ith component is one and whose remaining components are zero, is called the ith unit vector.

Definition A.9 *An $n \times n$ (square) matrix with elements 1 on the main diagonal and zeros off the diagonal is called the identity matrix I_n.*

Definition A.10 *A square matrix $A : n \times n$ with zeros off the diagonal is called a diagonal matrix. We write*

$$A = \mathrm{diag}\,(a_{11}, \cdots, a_{nn}) = \mathrm{diag}\,(a_{ii}) = \begin{pmatrix} a_{11} & & 0 \\ & \ddots & \\ 0 & & a_{nn} \end{pmatrix}.$$

Definition A.11 *A matrix A is said to be partitioned if its elements are arranged in submatrices.*

Examples are

$$\underset{m,n}{A} = (\underset{m,r}{A_1}, \underset{m,s}{A_2}) \quad \text{with} \quad r+s=n$$

or

$$\underset{m,n}{A} = \left(\begin{array}{cc} \underset{r,n-s}{A_{11}} & \underset{r,s}{A_{12}} \\ \underset{m-r,n-s}{A_{21}} & \underset{m-r,s}{A_{22}} \end{array} \right).$$

For partitioned matrices we get the transpose as

$$A' = \left(\begin{array}{c} A_1' \\ A_2' \end{array} \right), \quad A' = \left(\begin{array}{cc} A_{11}' & A_{21}' \\ A_{12}' & A_{22}' \end{array} \right).$$

A.2   Trace of a Matrix

Definition A.12 *Let $a_{11}, \ldots, a_{nn}$ be the elements on the main diagonal of a square matrix $A : n \times n$. The trace of A is then defined as the sum*

$$\text{tr}(A) = \sum_{i=1}^{n} a_{ii}.$$

Theorem A.13 *Let A and B be square $n \times n$ matrices and let c be a scalar factor. We then have the following rules*

(i) $\text{tr}(A \pm B) = \text{tr}(A) \pm \text{tr}(B)$,

(ii) $\text{tr}(A') = \text{tr}(A)$,

(iii) $\text{tr}(cA) = c\,\text{tr}(A)$,

(iv) $\text{tr}(AB) = \text{tr}(BA)$,

(v) $\text{tr}(AA') = \text{tr}(A'A) = \sum_{i,j} a_{ij}^2$.

(vi) *If $a = (a_1, \cdots, a_n)'$ is an n-vector, then its squared norm may be written as*

$$\| a \|^2 = a'a = \sum_{i=1}^{n} a_i^2 = \text{tr}(aa').$$

Note, that the rules *(iv)* and *(v)* also hold for the case $A : n \times m$ and $B : m \times n$.

A.3   Determinant of a Matrix

Definition A.14 *Let $n > 1$ be a positive integer. The determinant of a square matrix $\boldsymbol{A} : n \times n$ is defined by*

$$|\boldsymbol{A}| = \sum_{i=1}^{n} (-1)^{i+j} a_{ij} |\boldsymbol{M}_{ij}| \quad \text{(for any } j, \; j \text{ fixed)},$$

with $|\boldsymbol{M}_{ij}|$ being the minor of the element a_{ij}. $|\boldsymbol{M}_{ij}|$ is the determinant of the remaining $(n-1) \times (n-1)$ matrix when the ith row and the jth column of $\boldsymbol{A}$ are deleted. $\boldsymbol{A}_{ij} = (-1)^{i+j} |\boldsymbol{M}_{ij}|$ is called the cofactor of a_{ij}.

Examples: $n = 2$:

$$|\boldsymbol{A}| = a_{11} a_{22} - a_{12} a_{21}$$

$n = 3$: First column ($j = 1$) fixed:

$$\boldsymbol{A}_{11} = (-1)^2 \begin{vmatrix} a_{22} & a_{23} \\ a_{32} & a_{33} \end{vmatrix}$$

$$\boldsymbol{A}_{21} = (-1)^3 \begin{vmatrix} a_{12} & a_{13} \\ a_{32} & a_{33} \end{vmatrix}$$

$$\boldsymbol{A}_{31} = (-1)^4 \begin{vmatrix} a_{12} & a_{13} \\ a_{22} & a_{23} \end{vmatrix}$$

$$\Rightarrow |\boldsymbol{A}| = a_{11} \boldsymbol{A}_{11} + a_{21} \boldsymbol{A}_{21} + a_{31} \boldsymbol{A}_{31}.$$

Remark: As an alternative, one may fix a row and develop the determinant of $\boldsymbol{A}$ according to

$$|\boldsymbol{A}| = \sum_{j=1}^{n} (-1)^{i+j} a_{ij} |\boldsymbol{M}_{ij}| \quad \text{(for any } i, \; i \text{ fixed)}.$$

Definition A.15 *A square matrix $\boldsymbol{A}$ is said to be regular or non-singular if $|\boldsymbol{A}| \neq 0$. Otherwise, $\boldsymbol{A}$ is said to be singular.*

Theorem A.16 *Let $\boldsymbol{A}$ and $\boldsymbol{B}$ be $n \times n$ square matrices and let c be a scalar. Then we have*

(i) $|\boldsymbol{A}'| = |\boldsymbol{A}|$

(ii) $|c\boldsymbol{A}| = c^n |\boldsymbol{A}|$

(iii) $|\boldsymbol{A}\boldsymbol{B}| = |\boldsymbol{A}||\boldsymbol{B}|$

(iv) $|\boldsymbol{A}^2| = |\boldsymbol{A}|^2$

(v) If A is diagonal or triangular then

$$|A| = \prod_{i=1}^{n} a_{ii}.$$

(vi) For $D = \begin{pmatrix} A_{n,n} & C_{n,m} \\ 0_{m,n} & B_{m,n} \end{pmatrix}$ we have

$$\begin{vmatrix} A & C \\ 0 & B \end{vmatrix} = |A||B|,$$

and analogously

$$\begin{vmatrix} A' & 0' \\ C' & B' \end{vmatrix} = |A||B|.$$

(vii) If A is partitioned with $A_{11} : p \times p$ and $A_{22} : q \times q$ square and non-singular, then

$$\begin{vmatrix} A_{11} & A_{12} \\ A_{21} & A_{22} \end{vmatrix} = |A_{11}||A_{22} - A_{21}A_{11}^{-1}A_{12}|$$
$$= |A_{22}||A_{11} - A_{12}A_{22}^{-1}A_{21}|.$$

Proof: Define the following matrices

$$Z_1 = \begin{pmatrix} I & -A_{12}A_{22}^{-1} \\ 0 & I \end{pmatrix} \quad and \quad Z_2 = \begin{pmatrix} I & 0 \\ -A_{22}^{-1}A_{21} & I \end{pmatrix}$$

where $|Z_1| = |Z_2| = 1$ by (vi). Then we have

$$Z_1 A Z_2 = \begin{pmatrix} A_{11} - A_{12}A_{22}^{-1}A_{21} & 0 \\ 0 & A_{22} \end{pmatrix}$$

and [using (iii) and (iv)]

$$|Z_1 A Z_2| = |A| = |A_{22}||A_{11} - A_{12}A_{22}^{-1}A_{21}|.$$

(viii) $\begin{vmatrix} A & x \\ x' & c \end{vmatrix} = |A|(c - x'A^{-1}x)$ where x is an $(n, 1)$–vector.

Proof: Use (vii) with A instead of A_{11} and c instead of A_{22}.

(ix) Let $B : p \times n$ and $C : n \times p$ be any matrices and $A : p \times p$ a non–singular matrix. Then

$$|A + BC| = |A||I_p + A^{-1}BC|$$
$$= |A||I_n + CA^{-1}B|.$$

Proof: The first relationship follows from (iii) and

$$(A + BC) = A(I_p + A^{-1}BC),$$

immediately.

The second relationship is a consequence of (vii) applied to the matrix

$$\begin{vmatrix} I_p & -A^{-1}B \\ C & I_n \end{vmatrix} = |I_p||I_n + CA^{-1}B|$$

$$= |I_n||I_p + A^{-1}BC|.$$

(x) $|A + aa'| = |A|(1 + a'A^{-1}a)$, if A is non–singular.

(xi) $|I_p + BC| = |I_n + CB|$, if $B : (p, n)$ and $C : (n, p)$.

A.4 Inverse of a Matrix

Definition A.17 *The inverse of a square matrix $A : n \times n$ is written as A^{-1}. The inverse exists, if and only if, A is non–singular. The inverse A^{-1} is unique and characterized by*

$$AA^{-1} = A^{-1}A = I.$$

Theorem A.18 *If all the inverses exist, we have*

(i) $(cA)^{-1} = c^{-1}A^{-1}$

(ii) $(AB)^{-1} = B^{-1}A^{-1}$

(iii) *If $A : p \times p$, $B : p \times n$, $C : n \times n$, and $D : n \times p$ then*

$$(A + BCD)^{-1} = A^{-1} - A^{-1}B(C^{-1} + DA^{-1}B)^{-1}DA^{-1}.$$

(iv) *If $1 + b'A^{-1}a \neq 0$, we get from (iii)*

$$(A + ab')^{-1} = A^{-1} - \frac{A^{-1}ab'A^{-1}}{1 + b'A^{-1}a}.$$

(v) $|A^{-1}| = |A|^{-1}$.

Theorem A.19 (Inverse of a partitioned matrix)
For a partitioned regular A

$$A = \begin{pmatrix} E & F \\ G & H \end{pmatrix},$$

where $\underset{n_1,n_1}{E}$, $\underset{n_1,n_2}{F}$, $\underset{n_2,n_1}{G}$ *and* $\underset{n_2,n_2}{H}$ *with $n_1 + n_2 = n$ are such that E and $D = H - GE^{-1}F$ are regular, the partitioned inverse is given by*

$$A^{-1} = \begin{pmatrix} E^{-1}(I + FD^{-1}GE^{-1}) & -E^{-1}FD^{-1} \\ -D^{-1}GE^{-1} & D^{-1} \end{pmatrix} = \begin{pmatrix} A^{11} & A^{12} \\ A^{21} & A^{22} \end{pmatrix}.$$

Proof: Check that the product of A and A^{-1} reduces to the identity matrix, i.e.,

$$AA^{-1} = A^{-1}A = I.$$

A.5 Orthogonal Matrices

Definition A.20 *A square matrix $A : n \times n$ is said to be orthogonal if $AA' = I = A'A$. For orthogonal matrices we have*

(i) $A' = A^{-1}$

(ii) $|A| = \pm 1$

(iii) Let δ_{ij} denote the Kronecker symbol, with $\delta_{ij} = 1$ for $i = j$ and $\delta_{ij} = 0$ for $i \neq j$. The row vectors a_i and the column vectors $a_{(i)}$ of A then satisfy the conditions

$$a_i a_j' = \delta_{ij}, \quad a_{(i)}' a_{(j)} = \delta_{ij}.$$

(iv) AB is orthogonal, if A and B are orthogonal.

Theorem A.21 *For $\underset{n,n}{A}$ and $\underset{n,n}{B}$ symmetric, an orthogonal matrix H exists such that $H'AH$ and $H'BH$ become diagonal if and only if A and B commute, i.e.,*

$$AB = BA.$$

A.6 Rank of a Matrix

Definition A.22 *The rank of $A : m \times n$ is the maximum number of linearly independent rows (or columns) of A. We write $\mathrm{rank}\left(\underset{m,n}{A}\right) = p$.*

Theorem A.23 (Rules for ranks)

(i) $0 \leq \mathrm{rank}\,(A) \leq \min(m, n)$

(ii) $\mathrm{rank}\,(A) = \mathrm{rank}\,(A')$

(iii) $\mathrm{rank}\,(A + B) \leq \mathrm{rank}\,(A) + \mathrm{rank}\,(B)$

(iv) $\mathrm{rank}\,(AB) \leq \min\{\mathrm{rank}\,(A), \mathrm{rank}\,(B)\}$

(v) $\mathrm{rank}\,(AA') = \mathrm{rank}\,(A'A) = \mathrm{rank}\,(A) = \mathrm{rank}\,(A')$

(vi) For $\underset{m,m}{B}$ and $\underset{n,n}{C}$ regular, we have $\mathrm{rank}\,(BAC) = \mathrm{rank}\,(A)$.

(vii) For $\underset{n,n}{A}$, $\mathrm{rank}\,(A) = n$ if and only if A is regular.

(viii) If $A = \mathrm{diag}\,(a_i)$, then $\mathrm{rank}\,(A)$ equals the number of the $a_i \neq 0$.

A.7   Range and Null Space
Definition A.24

(i) *The range $\mathcal{R}(A)$ of a matrix $\underset{m,n}{A}$ is the vector space spanned by the column vectors of A, i.e.,*

$$\mathcal{R}(A) = \{z : z = Ax = \sum_{i=1}^{n} a_{(i)}x_i, \quad x \in \Re^n\} \subset \Re^m$$

where $a_{(1)}, \ldots, a_{(n)}$ are the column vectors of A.

(ii) *The null space $\mathcal{N}(A)$ is the vector space defined by*

$$\mathcal{N}(A) = \{x \in \Re^n \text{ and } Ax = 0\} \subset \Re^n.$$

Theorem A.25

(i) $\text{rank}(A) = \dim \mathcal{R}(A)$, *where $\dim V$ denotes the number of basis vectors of a vector space V.*

(ii) $\dim \mathcal{R}(A) + \dim \mathcal{N}(A) = n$

(iii) $\mathcal{N}(A) = \{\mathcal{R}(A')\}^{\perp}$.
     ($V^{\perp}$ the orthogonal complement of a vector space V is defined by $V^{\perp} = \{x : x'y = 0 \quad \text{for all} \quad y \in V\}$).

(iv) $\mathcal{R}(AA') = \mathcal{R}(A)$.

(v) $\mathcal{R}(AB) \subseteq \mathcal{R}(A)$ *for any A and B*

(vi) *For $A \geq 0$ and any B, $\mathcal{R}(BAB') = \mathcal{R}(BA)$.*

A.8   Eigenvalues and Eigenvectors
Definition A.26 *If $A : p \times p$ is a square matrix then*

$$q(\lambda) = |A - \lambda I|$$

is a pth order polynomial in λ. The p roots $\lambda_1, \ldots, \lambda_p$ of the characteristic equation $q(\lambda) = |A - \lambda I| = 0$ are called eigenvalues or characteristic roots of A.

The eigenvalues may be complex numbers. Since $|A - \lambda_i I| = 0$, $A - \lambda_i I$ is a singular matrix. Hence, a non–zero vector $\gamma_i \neq 0$ exists, satisfying $(A - \lambda_i I)\gamma_i = 0$, i.e.,

$$A\gamma_i = \lambda_i \gamma_i.$$

γ_i is called (right) eigenvector of A for the eigenvalue λ_i. If λ_i is complex, then γ_i may have complex components. An eigenvector γ with real components is called standardized if $\gamma'\gamma = 1$.

Theorem A.27

   (i) If x and y are non–zero eigenvectors of A for λ_i and if α and β are any real numbers, then $\alpha x + \beta y$ also is an eigenvector for λ_i, i. e.

$$A(\alpha x + \beta y) = \lambda_i(\alpha x + \beta y).$$

   Thus, the eigenvectors for any λ_i span a vector space that is called eigen-space of A for λ_i.

   (ii) The polynomial $q(\lambda) = |A - \lambda I|$ has the normal form in terms of the roots

$$q(\lambda) = \prod_{i=1}^{p}(\lambda_i - \lambda).$$

   Hence, $q(0) = \prod_{i=1}^{p} \lambda_i$ and

$$|A| = \prod_{i=1}^{p} \lambda_i.$$

   (iii) Matching the coefficients of λ^{n-1} in $q(\lambda) = \prod_{i=1}^{p}(\lambda_i - \lambda)$ and $|A - \lambda I|$ gives

$$\operatorname{tr}(A) = \sum_{i=1}^{p} \lambda_i.$$

   (iv) Let $C : p \times p$ be a regular matrix. Then A and CAC^{-1} have the same eigenvalues λ_i. If γ_i is an eigenvector for λ_i, then $C\gamma_i$ is an eigenvector of CAC^{-1} for λ_i.

Proof: As C is non–singular, it has an inverse C^{-1} with $CC^{-1} = I$. We have $|C^{-1}| = |C|^{-1}$ and

$$
\begin{aligned}
|A - \lambda I| &= |C||A - \lambda C^{-1}C||C^{-1}| \\
&= |CAC^{-1} - \lambda I|.
\end{aligned}
$$

Thus, A und CAC^{-1} have the same eigenvalues. Let $A\gamma_i = \lambda_i\gamma_i$ and multiply from the left by C

$$CAC^{-1}C\gamma_i = (CAC^{-1})(C\gamma_i) = \lambda_i(C\gamma_i).$$

   (v) The matrix $A + \alpha I$ with α a real number has the eigenvalues $\tilde{\lambda}_i = \lambda_i + \alpha$, and the eigenvectors of A and $A + \alpha I$ coincide.

   (vi) Let λ_1 denote any eigenvalue of $\underset{p,p}{A}$ with eigenspace H of dimension r. If k denotes the multiplicity of λ_1 in $q(\lambda)$, then

$$1 \le r \le k.$$

Remark:

(a) For symmetric matrices A we have $r = k$.

(b) If A is not symmetric, then it is possible that $r < k$.

Example: $A = \begin{pmatrix} 0 & 1 \\ 0 & 0 \end{pmatrix}$, $A \neq A'$

$$|A - \lambda I| = \begin{vmatrix} -\lambda & 1 \\ 0 & -\lambda \end{vmatrix} = \lambda^2 = 0.$$

The multiplicity of the eigenvalue $\lambda_{1,2} = 0$ is $k = 2$.

The eigenvectors for $\lambda = 0$ are $\gamma = \alpha \begin{pmatrix} 1 \\ 0 \end{pmatrix}$ and generate an eigenspace of dimension 1.

(c) If for any particular eigenvalue λ, $\dim(H) = r = 1$, then the standardized eigenvector for λ is unique (except for the sign).

Theorem A.28 Let $\underset{n,p}{A}$ and $\underset{p,n}{B}$, with $n \geq p$, be any two matrices. Then from A.16 (vii)

$$\begin{vmatrix} -\lambda I_n & -A \\ B & I_p \end{vmatrix} = (-\lambda)^{n-p}|BA - \lambda I_p| = |AB - \lambda I_n|.$$

Hence the n eigenvalues of AB are equal to the p eigenvalues of BA plus the eigenvalue 0 with multiplicity $n - p$. Suppose that $x \neq 0$ is an eigenvector of AB for any particular $\lambda \neq 0$. Then $y = Bx$ is an eigenvector of BA for this λ, and we have $y \neq 0$, too.

Corollary: A matrix $A = aa'$ with $a \neq 0$ has the eigenvalues 0 and $\lambda = a'a$ and the eigenvector a.

Corollary: The non–zero eigenvalues of AA' are equal to the non-zero eigenvalues of $A'A$.

Theorem A.29 If A is symmetric, then all the eigenvalues are real.

A.9   Decomposition of Matrices

Theorem A.30 (Spectral decomposition theorem)

Any symmetric matrix $\underset{p,p}{A}$ can be written as

$$A = \Gamma \Lambda \Gamma' = \sum \lambda_i \gamma_{(i)} \gamma'_{(i)}$$

where $\Lambda = \operatorname{diag}(\lambda_1, \cdots, \lambda_p)$ is the diagonal matrix of the eigenvalues of A and $\Gamma = (\gamma_{(1)}, \cdots, \gamma_{(p)})$ is the matrix of the standardized eigenvectors $\gamma_{(i)}$. Γ is orthogonal:

$$\Gamma\Gamma' = \Gamma'\Gamma = I.$$

Theorem A.31 *Suppose A is symmetric and $A = \Gamma\Lambda\Gamma'$. Then*

(i) *A and Λ have the same eigenvalues (with the same multiplicity).*

(ii) *From $A = \Gamma\Lambda\Gamma'$ we get $\Lambda = \Gamma'A\Gamma$.*

(iii) *If $A : p \times p$ is a symmetric matrix, then for any integer n we have $A^n = \Gamma\Lambda^n\Gamma'$ and $\Lambda^n = \operatorname{diag}(\lambda_i^n)$. If the eigenvalues of A are positive, then we can define the rational powers*

$$A^{r/s} = \Gamma\Lambda^{r/s}\Gamma' \quad \text{with } \Lambda^{r/s} = \operatorname{diag}(\lambda_i^{r/s})$$

for integers $s > 0$ and r. Important special cases are (for $\lambda_i > 0$)

$$A^{-1} = \Gamma\Lambda^{-1}\Gamma' \quad \text{with } \Lambda^{-1} = \operatorname{diag}(\lambda_i^{-1}),$$

the symmetric square root decomposition of A (for $\lambda_i \geq 0$)

$$A^{1/2} = \Gamma\Lambda^{1/2}\Gamma' \quad \text{with } \Lambda^{1/2} = \operatorname{diag}(\lambda_i^{1/2})$$

and for $\lambda_i > 0$

$$A^{-1/2} = \Gamma\Lambda^{-1/2}\Gamma' \quad \text{with } \Lambda^{-1/2} = \operatorname{diag}(\lambda_i^{-1/2}).$$

(iv) *For any square matrix A the rank of A equals the number of non–zero eigenvalues.*

Proof: According to Theorem A.23 (vi) we have $\operatorname{rank}(A) = \operatorname{rank}(\Gamma\Lambda\Gamma') = \operatorname{rank}(\Lambda)$. But $\operatorname{rank}(\Lambda)$ equals the number of non–zero λ_i's.

(v) *A symmetric matrix A is uniquely determined by its distinct eigenvalues and the corresponding eigenspaces. If the distinct eigenvalues λ_i are ordered as $\lambda_1 \geq \cdots \geq \lambda_p$, then the matrix Γ is unique (except for the sign).*

(vi) *$A^{1/2}$ and A have the same eigenvectors. Hence, $A^{1/2}$ is unique.*

(vii) Let $\lambda_1 \geq \lambda_2 \geq \cdots \geq \lambda_k > 0$ be the non–zero eigenvalues and $\lambda_{k+1} = \cdots = \lambda_p = 0$. Then we have

$$A = (\Gamma_1 \Gamma_2) \begin{pmatrix} \Lambda_1 & 0 \\ 0 & 0 \end{pmatrix} \begin{pmatrix} \Gamma_1' \\ \Gamma_2' \end{pmatrix} = \Gamma_1 \Lambda_1 \Gamma_1'$$

with $\Lambda_1 = \mathrm{diag}\,(\lambda_1, \cdots, \lambda_k)$ and $\Gamma_1 = (\gamma_{(1)}, \cdots, \gamma_{(k)})$, whereas $\Gamma_1'\Gamma_1 = I_k$ holds so that Γ_1 is column–orthogonal.

(viii) A symmetric matrix A is of rank 1 if and only if $A = aa'$, where $a \neq 0$.

Proof: If $\mathrm{rank}\,(A) = \mathrm{rank}\,(\Lambda) = 1$, then $\Lambda = \begin{pmatrix} \lambda & 0' \\ 0 & 0 \end{pmatrix}$, $A = \lambda\gamma\gamma' = aa'$ with $a = \sqrt{\lambda}\gamma$. If $A = aa'$, then by A.23 (iv) we have $\mathrm{rank}\,(A) = \mathrm{rank}\,(a) = 1$.

Theorem A.32 (Singular value decomposition of a rectangular matrix)
Let $\underset{n,p}{A}$ be a rectangular matrix of rank r. We then have

$$\underset{n,p}{A} = \underset{n,r}{U}\,\underset{r,r}{L}\,\underset{r,p}{V'}$$

with $U'U = I_r$, $V'V = I_r$ and $L = \mathrm{diag}\,(l_1, \cdots, l_r)$, $l_i > 0$.
For a proof see Rao (1973, p.42).

Theorem A.33 If $A : p \times q$ has $\mathrm{rank}\,(A) = r$, then A contains at least one non-singular (r,r)-submatrix X, such that A has the so-called normal presentation

$$\underset{p,q}{A} = \begin{pmatrix} \underset{r,r}{X} & \underset{r,q-r}{Y} \\ \underset{p-r,r}{Z} & \underset{p-r,q-r}{W} \end{pmatrix}.$$

All square submatrices of type $(r+s, r+s)$ with $(s \geq 1)$ are singular.
Proof: As $\mathrm{rank}\,(A) = \mathrm{rank}\,(X)$ holds, the first r rows of (X, Y) are linearly independent. Then, the $p - r$ rows (Z, W) are linear combinations of (X, Y), i.e., a matrix F exists, such that

$$(Z, W) = F(X, Y).$$

Analogously, a matrix H exists that satisfies

$$\begin{pmatrix} Y \\ W \end{pmatrix} = \begin{pmatrix} X \\ Z \end{pmatrix} H.$$

Hence, we get $W = FY = FXH$ and

$$A = \begin{pmatrix} X & Y \\ Z & W \end{pmatrix} = \begin{pmatrix} X & XH \\ FX & FXH \end{pmatrix}$$

$$= \begin{pmatrix} I \\ F \end{pmatrix} X(I, H)$$

$$= \begin{pmatrix} X \\ FX \end{pmatrix} (I, H) = \begin{pmatrix} I \\ F \end{pmatrix} (X, XH).$$

As X is non–singular, the inverse X^{-1} exists. We obtain $F = ZX^{-1}$, $H = X^{-1}Y$ and $W = ZX^{-1}Y$ and

$$A = \begin{pmatrix} X & Y \\ Z & W \end{pmatrix} = \begin{pmatrix} I \\ ZX^{-1} \end{pmatrix} X(I, X^{-1}Y)$$

$$= \begin{pmatrix} X \\ Z \end{pmatrix}(I, X^{-1}Y)$$

$$= \begin{pmatrix} I \\ ZX^{-1} \end{pmatrix}(X\ Y).$$

Theorem A.34 (Full rank factorization)

(i) If $\underset{p,q}{A}$ has $\operatorname{rank}(A) = r$, then A may be written as

$$\underset{p,q}{A} = \underset{p,r}{K}\ \underset{r,q}{L}$$

with K of full column rank r and L of full row rank r.
Proof: Theorem A.33.

(ii) If $\underset{p,q}{A}$ has $\operatorname{rank}(A) = p$ then A may be written as

$$A = M(I, H) \quad \text{where } \underset{p,p}{M} \text{ is regular.}$$

Proof: Theorem A.34 (i).

A.10 Definite Matrices and Quadratic Forms

Definition A.35 *Suppose $\underset{n,n}{A}$ is symmetric and $\underset{n,1}{x}$ is any vector. The quadratic form in x is then defined as the function*

$$Q(x) = x'Ax = \sum_{i,j} a_{ij}x_i x_j.$$

Clearly $Q(0) = 0$.

Definition A.36 *The matrix A is called positive definite (p.d.) if $Q(x) > 0$ for all $x \neq 0$. We write $A > 0$.*

Note: If $A > 0$, then $(-A)$ is called negative definite.

Definition A.37 *The quadratic form $x'Ax$ (and the matrix A, also) is called positive semi–definite (p.s.d.), if $Q(x) \geq 0$ for all x and $Q(x) = 0$ for at least one $x \neq 0$.*

Definition A.38 *The quadratic form $x'Ax$ (and A) is called nonnegative definite (n.n.d.), if it is either p.d. or p.s.d., i.e., if $x'Ax \geq 0$ for all x. If A is n.n.d., we write $A \geq 0$.*

Theorem A.39 *Let the* $n \times n$ *matrix* $A > 0$. *Then*

(i) A *has all eigenvalues* $\lambda_i > 0$

(ii) $x'Ax > 0$ *for any* $x \neq 0$

(iii) A *is non-singular and* $|A| > 0$

(iv) $A^{-1} > 0$

(v) $\operatorname{tr}(A) > 0$

(vi) *Let* $\underset{n,m}{P}$ *be of rank* $(P) = m \leq n$. *Then* $P'AP > 0$ *and in particular* $P'P > 0$, *choosing* $A = I$.

(vii) *Let* $\underset{n,m}{P}$ *be of rank* $(P) < m \leq n$. *Then* $P'AP \geq 0$ *and* $P'P \geq 0$.

Theorem A.40 *Given* $\underset{n,n}{A}$ *and* $\underset{n,n}{B}$ *with* $A > 0$ *and* $B \geq 0$. *Then*

(i) $C = A + B > 0$

(ii) $A^{-1} - (A + B)^{-1} \geq 0$

(iii) $|A| \leq |A + B|$

Theorem A.41 *Let* $A \geq 0$. *Then*

(i) $\lambda_i \geq 0$

(ii) $\operatorname{tr}(A) \geq 0$

(iii) $A = A^{1/2}A^{1/2}$ *with* $A^{1/2} = \Gamma\Lambda^{1/2}\Gamma'$

(iv) *For any matrix* $\underset{n,m}{C}$ *we have* $C'AC \geq 0$.

(v) *For any matrix* C *we have* $C'C \geq 0$ *and* $CC' \geq 0$.

Theorem A.42 *For any matrix* $A \geq 0$ *we have* $0 \leq \lambda_i \leq 1$ *if and only if* $(I - A) \geq 0$.
Proof: Write the symmetric matrix A in its spectral form as $A = \Gamma\Lambda\Gamma'$. We then have

$$(I - A) = \Gamma(I - \Lambda)\Gamma' \geq 0$$

if and only if

$$\Gamma'\Gamma(I - \Lambda)\Gamma'\Gamma = I - \Lambda \geq 0.$$

(a) If $I - \Lambda \geq 0$, then for the eigenvalues of $I - A$ we have $1 - \lambda_i \geq 0$, i.e., $0 \leq \lambda_i \leq 1$.

(b) If $0 \leq \lambda_i \leq 1$, then for any $x \neq 0$

$$x'(I - \Lambda)x = \sum x_i^2(1 - \lambda_i) \geq 0,$$

i.e., $I - \Lambda \geq 0$.

Theorem A.43 (Theobald, 1974)

Let $\underset{n,n}{D}$ be symmetric. Then $D \geq 0$ if and only if $\operatorname{tr}\{CD\} \geq 0$ for all $C \geq 0$.

Proof: D is symmetric, so that

$$D = \Gamma\Lambda\Gamma' = \sum \lambda_i \gamma_i \gamma_i'$$

and hence,

$$\begin{aligned} \operatorname{tr}\{CD\} &= \operatorname{tr}\{\sum \lambda_i C \gamma_i \gamma_i'\} \\ &= \sum \lambda_i \gamma_i' C \gamma_i. \end{aligned}$$

(a) Let $D \geq 0$, and hence, $\lambda_i \geq 0$ for all i. Then $\operatorname{tr}(CD) \geq 0$ if $C \geq 0$.

(b) Let $\operatorname{tr}\{CD\} \geq 0$ for all $C \geq 0$. Choose $C = \gamma_i \gamma_i'$ $(i = 1, \ldots, n,\ i$ fixed) so that

$$\begin{aligned} 0 \leq \operatorname{tr}\{CD\} &= \operatorname{tr}\{\gamma_i \gamma_i'(\sum_j \lambda_j \gamma_j \gamma_j')\} \\ &= \lambda_i \qquad (i = 1, \cdots, n) \end{aligned}$$

and $D = \Gamma\Lambda\Gamma' \geq 0$.

Theorem A.44 *Let $\underset{n,n}{A}$ be symmetric with eigenvalues $\lambda_1 \geq \cdots \geq \lambda_n$. Then*

$$\sup_x \frac{x'Ax}{x'x} = \lambda_1, \qquad \inf_x \frac{x'Ax}{x'x} = \lambda_n.$$

Proof: See Rao (1973, p. 62).

Theorem A.45 *Let $\underset{n,r}{A} = (\underset{n,r_1}{A_1}, \underset{n,r_2}{A_2})$ and $\operatorname{rank}(A) = r = r_1 + r_2$.*

Define the orthogonal projectors $M_1 = A_1(A_1'A_1)^{-1}A_1'$ and $M = A(A'A)^{-1}A'$. Then

$$M = M_1 + (I - M_1)A_2(A_2'(I - M_1)A_2)^{-1}A_2'(I - M_1).$$

Proof: M_1 and M are symmetric idempotent matrices fulfilling $M_1 A_1 = 0$ and $MA = 0$. Using A.19 for partial inversion of $A'A$, i.e.,

$$(A'A)^{-1} = \begin{pmatrix} A_1'A_1 & A_1'A_2 \\ A_2'A_1 & A_2'A_2 \end{pmatrix}^{-1}$$

and using the special form of the matrix D defined in A.19, i.e.,

$$D = A_2'(I - M_1)A_2,$$

straightforward calculation concludes the proof.

Theorem A.46 *Let $A : n \times m$ with* $\mathrm{rank}\,(A) = m \leq n$ *and $B : m \times m$ be any symmetric matrix. Then*

$$ABA' \geq 0 \quad \text{if and only if } B \geq 0.$$

Proof:
(i) $B \geq 0 \Rightarrow ABA' \geq 0$ for all A.
(ii) Let $\mathrm{rank}\,(A) = m \leq n$ and assume $ABA' \geq 0$, so that $x'ABA'x \geq 0$ for all $x \in \Re^n$.
We have to prove that $y'By \geq 0$ for all $y \in \Re^m$. As $\mathrm{rank}\,(A) = m$, the inverse $(A'A)^{-1}$ exists. Setting $z = A(A'A)^{-1}y$, we have $A'z = y$ and $y'By = z'ABA'z \geq 0$ so that $B \geq 0$.

Definition A.47 *Let $\underset{n,n}{A}$ and $\underset{n,n}{B}$ be any matrices. The roots $\lambda_i = \lambda_i^B(A)$ of the equation*

$$|A - \lambda B| = 0$$

are called the eigenvalues of A in the metric of B. For $B = I$, we obtain the usual eigenvalues defined in A.26. (cf. Dhrymes, 1974, p. 581).

Theorem A.48 *Let $B > 0$ and $A \geq 0$. Then $\lambda_i^B(A) \geq 0$.*
Proof: $B > 0$ is equivalent to $B = B^{1/2}B^{1/2}$ with $B^{1/2}$ non–singular and unique (A.31 (iii)). We may write

$$0 = |A - \lambda B| = |B^{1/2}|^2 |B^{-1/2}AB^{-1/2} - \lambda I|$$

and $\lambda_i^B(A) = \lambda_i^I(B^{-1/2}AB^{-1/2}) \geq 0$, as $B^{-1/2}AB^{-1/2} \geq 0$.

Theorem A.49 (Simultaneous diagonalization)
Let $B > 0$ and $A \geq 0$ and denote by $\Lambda = \mathrm{diag}\,(\lambda_i^B(A))$ the diagonal matrix of the eigenvalues of A in the metric of B. Then a non–singular matrix W exists, such that
$$B = W'W \quad \text{and} \quad A = W'\Lambda W.$$

Proof: From the proof of A.48 we know that the roots $\lambda_i^B(A)$ are the usual eigenvalues of the matrix $B^{-1/2}AB^{-1/2}$. Let X be the matrix of the corresponding eigenvectors:
$$B^{-1/2}AB^{-1/2}X = X\Lambda,$$

i.e.,

$$A = B^{1/2}X\Lambda X'B^{1/2} = W'\Lambda W$$

with $W' = B^{1/2}X$ regular and

$$B = W'W = B^{1/2}XX'B^{1/2} = B^{1/2}B^{1/2}.$$

Theorem A.50 *Let $A > 0$ (or $A \geq 0$) and $B > 0$. Then*

$$B - A > 0 \quad \text{if and only if} \quad \lambda_i^B(A) < 1.$$

Proof: Using A.49 we may write

$$B - A = W'(I - \Lambda)W,$$

i.e.,

$$
\begin{aligned}
x'(B - A)x &= x'W'(I - \Lambda)Wx \\
&= y'(I - \Lambda)y \\
&= \sum(1 - \lambda_i^B(A))y_i^2
\end{aligned}
$$

with $y = Wx$, W regular, and hence $y \neq 0$ for $x \neq 0$. Then $x'(B - A)x > 0$ holds if and only if

$$\lambda_i^B(A) < 1.$$

Theorem A.51 *Let $A > 0$ (or $A \geq 0$) and $B > 0$. Then*

$$A - B \geq 0$$

if and only if

$$\lambda_i^B(A) \leq 1.$$

Proof: Similar to A.50.

Theorem A.52 *Let $A > 0$ and $B > 0$. Then*

$$B - A > 0 \quad \text{if and only if} \quad A^{-1} - B^{-1} > 0.$$

Proof: From A.49 we have

$$B = W'W, \quad A = W'\Lambda W.$$

Since W is regular we have

$$B^{-1} = W^{-1}W'^{-1}, \quad A^{-1} = W^{-1}\Lambda^{-1}W'^{-1},$$

i.e.,

$$A^{-1} - B^{-1} = W^{-1}(\Lambda^{-1} - I)W'^{-1} > 0,$$

as $\lambda_i^B(A) < 1$ and, hence, $\Lambda^{-1} - I > 0$.

Theorem A.53 *Let $B - A > 0$. Then $|B| > |A|$ and $\operatorname{tr}(B) > \operatorname{tr}(A)$.*
If $B - A \geq 0$, then $|B| \geq |A|$ and $\operatorname{tr}(B) \geq \operatorname{tr}(A)$.
Proof: From A.49 and A.16 (iii),(v) we get

$$
\begin{aligned}
|B| &= |W'W| = |W|^2, \\
|A| &= |W'\Lambda W| = |W|^2|\Lambda| = |W|^2 \prod \lambda_i^B(A),
\end{aligned}
$$

i.e.,

$$|A| = |B| \prod \lambda_i^B(A).$$

For $B - A > 0$ we have $\lambda_i^B(A) < 1$, i.e., $|A| < |B|$.
For $B - A \geq 0$ we have $\lambda_i^B(A) \leq 1$, i.e., $|A| \leq |B|$.
$B - A > 0$ implies $\operatorname{tr}(B - A) > 0$, and $\operatorname{tr}(B) > \operatorname{tr}(A)$. Analogously, $B - A \geq 0$
implies $\operatorname{tr}(B) \geq \operatorname{tr}(A)$.

Theorem A.54 (Cauchy–Schwarz Inequality)
Let x, y be real vectors of same dimension. Then

$$(x'y)^2 \leq (x'x)(y'y),$$

with equality if and only if x and y are linearly dependent.

Theorem A.55 *Let x, y be n-vectors and $A > 0$. We then have the following results:*

(i) $(x'Ay)^2 \leq (x'Ax)(y'Ay)$.

(ii) $(x'y)^2 \leq (x'Ax)(y'A^{-1}y)$.

Proof: (i) $A \geq 0$ is equivalent to $A - BB$ with $B - A^{1/2}$ (A.41 (iii)). Let
$Bx = \tilde{x}$ and $By = \tilde{y}$. Then (i) is a consequence of A.54.
(ii) $A > 0$ is equivalent to $A = A^{1/2}A^{1/2}$ and $A^{-1} = A^{-1/2}A^{-1/2}$. Let $A^{1/2}x = \tilde{x}$
and $A^{-1/2}y = \tilde{y}$, then (ii) is a consequence of A.54.

Theorem A.56 *Let $A > 0$ and T any square matrix. Then*

(i) $\sup_{x \neq 0} \dfrac{(x'y)^2}{x'Ax} = y'A^{-1}y$

(ii) $\sup_{x \neq 0} \dfrac{(y'Tx)^2}{x'Ax} = y'TA^{-1}T'y.$

Proof: Use A.55 (ii).

Theorem A.57 *Let* $\underset{n,n}{\mathbf{I}}$ *be the identity matrix and* $\mathbf{a}$ *an n–vector. Then*

$$\mathbf{I} - \mathbf{a}\mathbf{a}' \geq 0 \quad \text{if and only if} \quad \mathbf{a}'\mathbf{a} \leq 1.$$

Proof: The matrix $\mathbf{a}\mathbf{a}'$ is of rank 1 and $\mathbf{a}\mathbf{a}' \geq 0$. The spectral decomposition is $\mathbf{a}\mathbf{a}' = \mathbf{C}\mathbf{\Lambda}\mathbf{C}'$ with $\mathbf{\Lambda} = \operatorname{diag}(\lambda, 0, \cdots, 0)$ and $\lambda = \mathbf{a}'\mathbf{a}$. Hence, $\mathbf{I} - \mathbf{a}\mathbf{a}' = \mathbf{C}(\mathbf{I} - \mathbf{\Lambda})\mathbf{C}' \geq 0$ if and only if $\lambda = \mathbf{a}'\mathbf{a} \leq 1$ (see A 42).

Theorem A.58 *Assume* $\mathbf{M}\mathbf{M}' - \mathbf{N}\mathbf{N}' \geq 0$. *Then a matrix* $\mathbf{H}$ *exists, such that* $\mathbf{N} = \mathbf{M}\mathbf{H}$.

Proof: (Milliken and Akdeniz, 1977) Let $\underset{n,r}{\mathbf{M}}$ of rank $(\mathbf{M}) = s$ and let $\mathbf{x}$ be any vector $\in \mathcal{R}(\mathbf{I} - \mathbf{M}\mathbf{M}^-)$, implying $\mathbf{x}'\mathbf{M} = \mathbf{0}$ and $\mathbf{x}'\mathbf{M}\mathbf{M}'\mathbf{x} = 0$. As $\mathbf{N}\mathbf{N}'$ and $\mathbf{M}\mathbf{M}' - \mathbf{N}\mathbf{N}'$ (by assumption) are n.n.d., we may conclude that $\mathbf{x}'\mathbf{N}\mathbf{N}'\mathbf{x} \geq 0$ and

$$\mathbf{x}'(\mathbf{M}\mathbf{M}' - \mathbf{N}\mathbf{N}')\mathbf{x} = -\mathbf{x}'\mathbf{N}\mathbf{N}'\mathbf{x} \geq 0,$$

so that $\mathbf{x}'\mathbf{N}\mathbf{N}'\mathbf{x} = 0$ and $\mathbf{x}'\mathbf{N} = 0$. Hence, $\mathbf{N} \subset \mathcal{R}(\mathbf{M})$ or, equivalently, $\mathbf{N} = \mathbf{M}\underset{r,k}{\mathbf{H}}$ for some matrix $\mathbf{H}$.

Theorem A.59 *Let* $\mathbf{A}$ *be an* $n \times n$–*matrix and assume* $(-\mathbf{A}) > 0$. *Let* $\mathbf{a}$ *be an* n–*vector. In case of* $n \geq 2$, *the matrix* $\mathbf{A} + \mathbf{a}\mathbf{a}'$ *is never n.n.d..*

Proof: (Guilkey and Price, 1981) The matrix $\mathbf{a}\mathbf{a}'$ is of rank ≤ 1. In case of $n \geq 2$, a non–zero vector $\mathbf{w}$ exists, such that $\mathbf{w}'\mathbf{a}\mathbf{a}'\mathbf{w} = 0$, implying $\mathbf{w}'(\mathbf{A} + \mathbf{a}\mathbf{a}')\mathbf{w} = \mathbf{w}'\mathbf{A}\mathbf{w} < 0$.

A.11 Idempotent Matrices

Definition A.60 *A square matrix* $\mathbf{A}$ *is called idempotent, if it satisfies*

$$\mathbf{A}^2 = \mathbf{A}\mathbf{A} = \mathbf{A}.$$

An idempotent matrix $\mathbf{A}$ *is called an orthogonal projector if* $\mathbf{A} = \mathbf{A}'$. *Otherwise* $\mathbf{A}$ *is called an oblique projector.*

Theorem A.61 *Let* $\underset{n,n}{\mathbf{A}}$ *be idempotent with* rank $(\mathbf{A}) = r \leq n$. *Then we have:*

(i) *The eigenvalues of* $\mathbf{A}$ *are 1 oder 0.*

(ii) tr $(\mathbf{A}) = $ rank $(\mathbf{A}) = r$.

(iii) *If* $\mathbf{A}$ *is of full rank* n, *then* $\mathbf{A} = \mathbf{I}_n$.

(iv) *If* $\mathbf{A}$ *and* $\mathbf{B}$ *are idempotent and if* $\mathbf{A}\mathbf{B} = \mathbf{B}\mathbf{A}$, *then* $\mathbf{A}\mathbf{B}$ *is also idempotent.*

(v) *If* $\mathbf{A}$ *is idempotent and* $\mathbf{P}$ *is orthogonal, then* $\mathbf{P}\mathbf{A}\mathbf{P}'$ *is also idempotent.*

(vi) If A is idempotent, then $I - A$ is idempotent and

$$A(I - A) = (I - A)A = 0.$$

Proof:
(i) The characteristic equation

$$Ax = \lambda x$$

multiplied by A gives

$$AAx = Ax = \lambda Ax = \lambda^2 x.$$

Multiplication of both the equations by x' yields

$$x'Ax = \lambda x'x = \lambda^2 x'x,$$

i.e.,

$$\lambda(\lambda - 1) = 0.$$

(ii) From the spectral decomposition

$$A = \Gamma \Lambda \Gamma'$$

we obtain

$$\text{rank}(A) = \text{rank}(\Lambda) = \text{tr}(\Lambda) = r,$$

where r is the number of characteristic roots with value 1.
(iii) Let $\text{rank}(A) = \text{rank}(\Lambda) = n$, then $\Lambda = I_n$ and

$$A = \Gamma \Lambda \Gamma' = I_n.$$

(iv) – (vi) follow from the definition of an idempotent matrix.

A.12   Generalized Inverse

Definition A.62 *Let A be an $m \times n$ matrix. Then a matrix $A^- : n \times m$ is said to be a generalized inverse of A if*

$$AA^- A = A$$

holds.

Theorem A.63 *A generalized inverse always exists although it is not unique in general.*
Proof: Assume $\text{rank}(A) = r$. According to A.32, we may write

$$\underset{m,n}{A} = \underset{m,r}{U}\ \underset{r,r}{L}\ \underset{r,n}{V'}$$

with $U'U = I_r$ and $V'V = I_r$ and

$$L = \operatorname{diag}(l_1, \cdots, l_r), \quad l_i > 0.$$

Then

$$A^- = V \begin{pmatrix} L^{-1} & X \\ Y & Z \end{pmatrix} U',$$

where X, Y, Z are arbitrary matrices (of suitable dimensions), is a g-inverse. Using Theorem A.33, i.e.,

$$A = \begin{pmatrix} X & Y \\ Z & W \end{pmatrix}$$

with X non-singular, we have

$$A^- = \begin{pmatrix} X^{-1} & 0 \\ 0 & 0 \end{pmatrix}$$

as a special g-inverse.

For details on g-inverses, the reader is referred to Rao and Mitra (1991).

Definition A.64 (Moore-Penrose Inverse)
A matrix A^+ satisfying the following conditions is called Moore–Penrose inverse of A:

$$\begin{array}{ll} (i) \quad AA^+A = A & (ii) \quad A^+AA^+ = A^+ \\ (iii) \quad (A^+A)' = A^+A & (iv) \quad (AA^+)' = AA^+. \end{array}$$

A^+ *is unique.*

Theorem A.65 *For any matrix $\underset{m,n}{A}$ and any g-inverse $\underset{m,n}{A^-}$ we have*

(i) A^-A and AA^- are idempotent

(ii) $\operatorname{rank}(A) = \operatorname{rank}(AA^-) = \operatorname{rank}(A^-A)$

(iii) $\operatorname{rank}(A) \leq \operatorname{rank}(A^-)$.

Proof:
(i) Using the definition of a g-inverse

$$(A^-A)(A^-A) = A^-(AA^-A) = A^-A.$$

(ii) According to A.23 (iv), we get

$$\operatorname{rank}(A) = \operatorname{rank}(AA^-A) \leq \operatorname{rank}(A^-A) \leq \operatorname{rank}(A),$$

i.e., $\operatorname{rank}(A^-A) = \operatorname{rank}(A)$. Analogously, we see that $\operatorname{rank}(A) = \operatorname{rank}(AA^-)$.
(iii) $\operatorname{rank}(A) = \operatorname{rank}(AA^-A) \leq \operatorname{rank}(AA^-) \leq \operatorname{rank}(A^-)$.

Theorem A.66 *Let A be an $m \times n$ matrix. Then*

(i) A *regular* $\Rightarrow A^+ = A^{-1}$

(ii) $(A^+)^+ = A$

(iii) $(A^+)' = (A')^+$

(iv) $\operatorname{rank}(A) = \operatorname{rank}(A^+) = \operatorname{rank}(A^+A) = \operatorname{rank}(AA^+)$

(v) A *an orthogonal projector* $\Rightarrow A^+ = A$

(vi) $\operatorname{rank}\left(\underset{m,n}{A}\right) = m \Rightarrow A^+ = A'(AA')^{-1}$ *and* $AA^+ = I_m$

(vii) $\operatorname{rank}\left(\underset{m,n}{A}\right) = n \Rightarrow A^+ = (A'A)^{-1}A'$ *and* $A^+A = I_n$

(viii) *If* $\underset{m,m}{P}$ *and* $\underset{n,n}{Q}$ *are orthogonal* $\Rightarrow \quad (PAQ)^+ = Q^{-1}A^+P^{-1}$

(ix) $(A'A)^+ = A^+(A')^+ \quad$ *and* $\quad (AA')^+ = (A')^+A^+$

(x) $A^+ = (A'A)^+A' = A'(AA')^+$

Theorem A.67 (Baksalary et al., 1983)
Let $\underset{n,n}{M} \geq 0$ *and* $\underset{m,n}{N}$ *be any matrices. Then*

$$M - N'(NM^+N')^+N \geq 0$$

if and only if

$$\mathcal{R}(N'NM) \subset \mathcal{R}(M).$$

Theorem A.68 *Let A be any square $n \times n$ matrix and a be an n-vector with $a \notin \mathcal{R}(A)$. Then a g-inverse of $A + aa'$ is given by*

$$(A + aa')^- = A^- - \frac{A^-aa'U'U}{a'U'Ua} - \frac{VV'aa'A^-}{a'VV'a} + \phi\frac{VV'aa'U'U}{(a'U'Ua)(a'VV'a)},$$

with A^- any g-inverse of A and

$$\phi = 1 + a'A^-a, \quad U = I - AA^-, \quad V = I - A^-A.$$

Proof: straightforward, by checking $AA^-A = A$. $(A + aa')(A + aa')^-(A + aa') = (A + aa')$.

Theorem A.69 *Let A be a square $n \times n$ matrix. We then have the following results*

(i) Assume a, b to be vectors with $a, b \in \mathcal{R}(A)$ and let A be symmetric. Then the bilinear form $a' A^- b$ is invariant to the choice of A^-.

(ii) $A(A'A)^- A'$ is invariant to the choice of $(A'A)^-$.

Proof:
(i) $a, b \in \mathcal{R}(A) \Rightarrow a = Ac$ and $b = Ad$.
Using the symmetry of A gives

$$
\begin{aligned}
a' A^- b &= c' A' A^- A d \\
&= c' A d.
\end{aligned}
$$

(ii) Using the row–wise representation of A as $A = \begin{pmatrix} a'_1 \\ \vdots \\ a'_n \end{pmatrix}$ gives

$$
A(A'A)^- A' = (a'_i (A'A)^- a_j).
$$

As $A'A$ is symmetric, we may conclude from (i) that all bilinear forms $a'_i(A'A)a_j$ are invariant to the choice of $(A'A)^-$ and hence, (ii) is proved.

Theorem A.70 Let $\underset{n,n}{A}$ be symmetric, $a \in \mathcal{R}(A)$, $b \in \mathcal{R}(A)$ and assume $1 + b' A^+ a \neq 0$. Then

$$
(A + ab')^+ = A^+ - \frac{A^+ ab' A^+}{1 + b' A^+ a}.
$$

Proof: straightforward, using A.68 and A.69.

Theorem A.71 Let $\underset{n,n}{A}$ symmetric, a an n-vector and $\alpha > 0$ any scalar. Then the following statements are equivalent

(i) $\alpha A - aa' \geq 0$

(ii) $A \geq 0$, $a \in \mathcal{R}(A)$ and $a' A^- a \leq \alpha$, with A^- being any g-inverse of A.

Proof:
(i) $\Rightarrow$ (ii): $\alpha A - aa' \geq 0 \Rightarrow \alpha A = (\alpha A - aa') + aa' \geq 0 \Rightarrow A \geq 0$. Using
A.31 for $\alpha A - aa' \geq 0$ we have $\alpha A - aa' = BB$ and, hence,

$$
\begin{aligned}
\alpha A = BB + aa' &= (B, a)(B, a)'. \\
\Rightarrow \quad \mathcal{R}(\alpha A) &= \mathcal{R}(A) = \mathcal{R}(B, a) \\
\Rightarrow \quad a &\in \mathcal{R}(A) \\
\Rightarrow \quad a &= Ac \quad \text{with} \quad c \in \mathfrak{R}^n. \\
\Rightarrow \quad a' A^- a &= c' A c.
\end{aligned}
$$

As $\alpha A - aa' \geq 0 \quad \Rightarrow$

$$x'(\alpha A - aa')x \geq 0$$

for any vector x. Choosing $x = c$ we have

$$\alpha c'Ac - c'aa'c = \alpha c'Ac - (c'Ac)^2 \geq 0,$$

$$\Rightarrow \quad c'Ac \leq \alpha.$$

(ii) $\Rightarrow$ (i): Let $x \in \Re^n$ be any vector. Then, using A.54,

$$\begin{aligned} x'(\alpha A - aa')x &= \alpha x'Ax - (x'a)^2 \\ &= \alpha x'Ax - (x'Ac)^2 \\ &\geq \alpha x'Ax - (x'Ax)(c'Ac) \end{aligned}$$

$$\Rightarrow \quad x'(\alpha A - aa')x \geq (x'Ax)(\alpha - c'Ac).$$

In (ii) we have assumed $A \geq 0$ and $c'Ac = a'A^- a \leq \alpha$. Hence, $\alpha A - aa' \geq 0$.

(Note: This Theorem is due to Baksalary and Kala (1983). The version given here and the proof are formulated by G. Trenkler, cf. Büning et al., 1995.)

Theorem A.72 *For any matrix A we have*

$$A'A = 0 \quad \text{if and only if } A = 0.$$

Proof:
(i) $A = 0 \quad \Rightarrow \quad A'A = 0$.
(ii) Let $A'A = 0$ and let $A = (a_{(1)}, \cdots, a_{(n)})$ be the column–wise presentation. Then

$$A'A = (a'_{(i)}a_{(j)}) = 0,$$

so that all the elements on the diagonal are zero: $a'_{(i)}a_{(i)} = 0 \quad \Rightarrow \quad a_{(i)} = 0$ and $A = 0$.

Theorem A.73 *Let $X \neq 0$ be an $m \times n$ matrix and A an $n \times n$ matrix. Then*

$$X'XAX'X = X'X \quad \Rightarrow \quad XAX'X = X \quad \text{and} \quad X'XAX' = X'$$

Proof: As $X \neq 0$ and $X'X \neq 0$, we have

$$X'XAX'X - X'X = (X'XA - I)X'X = 0 \quad \Rightarrow$$

$$\begin{aligned} (X'XA - I) &= 0 \quad \Rightarrow \\ 0 &= (X'XA - I)(X'XAX'X - X'X) \\ &= (X'XAX' - X')(XAX'X - X) = Y'Y, \end{aligned}$$

so that (by A.72) $Y = 0$ and, hence, $XAX'X = X$.

Corollary: *Let $X \neq 0$ be an (m,n)-matrix and A and B (n,n)-matrices. Then*

$$AX'X = BX'X \quad \Longleftrightarrow \quad AX' = BX'$$

Theorem A.74 (Albert's Theorem)

Let $A = \begin{pmatrix} A_{11} & A_{12} \\ A_{21} & A_{22} \end{pmatrix}$ symmetric. Then

(a) $A \geq 0$ *if and only if*

    (i) $A_{22} \geq 0$

    (ii) $A_{21} = A_{22}A_{22}^{-}A_{21}$

    (iii) $A_{11} \geq A_{12}A_{22}^{-}A_{21}$

    ((ii) and (iii) are invariant of the choice of A_{22}^{-}).

(b) $A > 0$ *if and only if*

    (i) $A_{22} > 0$

    (ii) $A_{11} > A_{12}A_{22}^{-1}A_{21}$.

Proof: (Bekker and Neudecker, 1989)

(a) Assume $A \geq 0$.

    (i) $A \geq 0 \quad \Rightarrow \quad x'Ax \geq 0$ for any x. Choosing $x' = (0', x_2')$,
    $\Rightarrow x'Ax = x_2'A_{22}x_2 \geq 0$ for any $x_2 \Rightarrow A_{22} \geq 0$

    (ii) Let $B' = (0, I - A_{22}A_{22}^{-}) \quad \Rightarrow,$

$$\begin{aligned} B'A &= \left((I - A_{22}A_{22}^{-})A_{21}, A_{22} - A_{22}A_{22}^{-}A_{22} \right) \\ &= \left((I - A_{22}A_{22}^{-})A_{21}, 0 \right) \end{aligned}$$

and $B'AB = B'A^{1/2}A^{1/2}B = 0, \quad \Rightarrow \quad B'A^{1/2} = 0$
$$(A.72)$$

$$\Rightarrow \quad B'A^{1/2}A^{1/2} = B'A = 0$$
$$\Rightarrow \quad (I - A_{22}A_{22}^{-})A_{21} = 0$$

This proves (ii).

    (iii) Let $C' = (I, -(A_{22}^{-}A_{21})')$. As $A \geq 0 \quad \Rightarrow,$

$$\begin{aligned} 0 \leq C'AC &= A_{11} - A_{12}(A_{22}^{-})'A_{21} - A_{12}A_{22}^{-}A_{21} \\ &\quad + A_{12}(A_{22}^{-})'A_{22}A_{22}^{-}A_{21} \\ &= A_{11} - A_{12}A_{22}^{-}A_{21} \end{aligned}$$

(as A_{22} is symmetric, we have $(A_{22}^{-})' = A_{22}^{-}$).

Assume now (i), (ii), and (iii). Then

$$D = \begin{pmatrix} A_{11} - A_{12}A_{22}^{-}A_{21} & 0 \\ 0 & A_{22} \end{pmatrix} \geq 0,$$

as the submatrices are n.n.d. by (i) and (ii). Hence,

$$A = \begin{pmatrix} I & A_{12}(A_{22}^{-}) \\ 0 & I \end{pmatrix} D \begin{pmatrix} I & 0 \\ A_{22}^{-}A_{21} & I \end{pmatrix} \geq 0.$$

(b) Proof as in (a) if A_{22}^{-} is replaced by A_{22}^{-1}.

Theorem A.75 *If $\underset{n,n}{A}$ and $\underset{n,n}{B}$ are symmetric, then*

(a) $0 \leq B \leq A$ *if and only if*

   (i) $A \geq 0$

   (ii) $B = AA^{-}B$

   (iii) $B \geq BA^{-}B$

(b) $0 < B < A$ *if and only if* $0 < A^{-1} < B^{-1}$.

Proof: Apply Theorem A.74 to the matrix $\begin{pmatrix} B & B \\ B & A \end{pmatrix}$.

Theorem A.76 *Let A be symmetric and $c \in \mathcal{R}(A)$. Then the following statements are equivalent*

(i) $\operatorname{rank}(A + cc') = \operatorname{rank}(A)$

(ii) $\mathcal{R}(A + cc') = \mathcal{R}(A)$

(iii) $1 + c'A^{-}c \neq 0$.

Corollary: *Assume (i) or (ii) or (iii) to hold, then*

$$(A + cc')^{-} = A^{-} - \frac{A^{-}cc'A^{-}}{1 + c'A^{-}c}$$

for any choice of A^{-}.

Corollary: *Assume (i) or (ii) or (iii) to hold, then*

$$c'(A + cc')^{-}c = c'A^{-}c - \frac{(c'A^{-}c)^2}{1 + c'A^{-}c}$$

$$= 1 - \frac{1}{1 + c'A^{-}c} \quad .$$

Moreover, as $c \in \mathcal{R}(A + cc')$, this is seen to be invariant of the special choice of the g-inverse.

Proof: $c \in \mathcal{R}(A) \quad \Leftrightarrow \quad AA^-c = c \Rightarrow$

$$\mathcal{R}(A + cc') = \mathcal{R}(AA^-(A + cc')) \subset \mathcal{R}(A).$$

Hence, (i) and (ii) become equivalent.

Proof of (iii): Consider the following product of matrices

$$\begin{pmatrix} 1 & 0 \\ c & A + cc' \end{pmatrix} \begin{pmatrix} 1 & -c \\ 0 & I \end{pmatrix} \begin{pmatrix} 1 & 0 \\ -A^-c & I \end{pmatrix} = \begin{pmatrix} 1 + c'A^-c & -c \\ 0 & A \end{pmatrix}.$$

The left–hand side has the rank

$$1 + \operatorname{rank}(A + cc') = 1 + \operatorname{rank}(A)$$

(see (i) or (ii)). The right–hand side has the rank $1 + \operatorname{rank}(A)$ if and only if $1 + c'A^-c \neq 0$.

Theorem A.77 *Assume $\underset{n,n}{A}$ to be a symmetric and non–singular matrix and $c \notin \mathcal{R}(A)$. Then we have*

(i) $c \in \mathcal{R}(A + cc')$

(ii) $\mathcal{R}(A) \subset \mathcal{R}(A + cc')$

(iii) $c'(A + cc')^-c = 1$

(iv) $A(A + cc')^- A = A$

(v) $A(A + cc')^-c = 0$.

Proof: As A is assumed to be non–singular, the equation $Al = 0$ has a non–trivial solution $l \neq 0$ which may be standardized as $\left(\frac{l}{c'l}\right)$, such that $c'l = 1$. Then we have $c = (A + cc')l \in \mathcal{R}(A + cc')$ and, hence, (i) is proven. Relation (ii) holds as $c \notin \mathcal{R}(A)$. Relation (i) is seen to be equivalent to

$$(A + cc')(A + cc')^-c = c.$$

Therefore, (iii) follows:

$$\begin{aligned} c'(A + cc')^-c &= l'(A + cc')(A + cc')^-c \\ &= l'c = 1. \end{aligned}$$

From

$$c = (A + cc')(A + cc')^- c$$
$$= A(A + cc')^- c + cc'(A + cc')^- c$$
$$= A(A + cc')^- c + c$$

we have (v).

(iv) is a consequence of the general definition of a g-inverse and of (iii) and (iv):

$$A + cc' = (A + cc')(A + cc')^-(A + cc')$$
$$= A(A + cc')^- A$$
$$+ cc'(A + cc')^- cc' \qquad [= cc' \text{ using (iii)}]$$
$$+ A(A + cc')^- cc' \qquad [= 0 \text{ using (v)}]$$
$$+ cc'(A + cc')^- A \qquad [= 0 \text{ using (v)}].$$

Theorem A.78 *We have* $A \geq 0$ *if and only if*

(i) $A + cc' \geq 0$

(ii) $(A + cc')(A + cc')^- c = c$

(iii) $c'(A + cc')^- c \leq 1$.

Assume $A \geq 0$, *then*

(a) $c = 0 \Longleftrightarrow c'(A + cc')^- c = 0$

(b) $c \in \mathcal{R}(A) \Longleftrightarrow c'(A + cc')^- c < 1$

(c) $c \notin \mathcal{R}(A) \Longleftrightarrow c'(A + cc')^- c = 1.$

Proof: $A \geq 0$ is equivalent to

$$0 \leq cc' \leq A + cc'.$$

Straightforward application of Theorem A.75 gives (i)-(iii).
Proof of (a): $A \geq 0 \quad \Rightarrow \quad A + cc' \geq 0$. Assume

$$c'(A + cc')^- c = 0$$

and replace c by (ii) $\Rightarrow$

$$c'(A + cc')^-(A + cc')(A + cc')^- c = 0 \Rightarrow$$
$$(A + cc')(A + cc')^- c = 0$$

as $(A + cc') \geq 0$. Assuming $c = 0 \quad \Rightarrow \quad c'(A + cc')c = 0.$

Proof of (b): Assume $A \geq 0$ and $c \in \mathcal{R}(A)$, and use Theorem A.76 (Corollary 2) $\Rightarrow$

$$c'(A + cc')^- c = 1 - \frac{1}{1 + c'A^- c} < 1.$$

The opposite direction of (b) is a consequence of (c).
Proof of (c): Assume $A \geq 0$ and $c \notin \mathcal{R}(A)$, and use Theorem A.77 (iii) $\Rightarrow$

$$c'(A + cc')^- c = 1.$$

The opposite direction of (c) is a consequence of (b).

Note: The proofs of Theorems A74 – A78 are given in Bekker and Neudecker (1989).

Theorem A.79 *The linear equation $Ax = a$ has a solution if and only if*

$$a \in \mathcal{R}(A) \quad or \quad AA^- a = a$$

for any g-inverse A.
If this condition holds, then all solutions are given by

$$x = A^- a + (I - A^- A)w$$

where w is an arbitrary m-vector. Furthermore, $q'x$ has a unique value for all solutions of $Ax = a$ if and only if $q'A^- A = q'$, or $q \in \mathcal{R}(A')$.
For a proof see Rao (1973, p. 25).

A.13 Projectors

Consider the range $\mathcal{R}(A)$ of the matrix $A : m \times n$ with rank r. Then $\mathcal{R}(A)^\perp$ exists, which is the orthogonal complement of $\mathcal{R}(A)$ with dimension $m - r$. Any vector $x \in \Re^m$ has the unique decomposition

$$x = x_1 + x_2, \quad x_1 \in \mathcal{R}(A) \quad and \quad x_2 \in \mathcal{R}(A)^\perp,$$

of which the component x_1 is called the orthogonal projection of x on $\mathcal{R}(A)$. The component x_1 can be computed as Px, where

$$P = A(A'A)^- A',$$

which is called the projection operator on $\mathcal{R}(A)$. Note that P is unique for any choice of the g-inverse $(A'A)^-$.

Theorem A.80 *For any $P : n \times n$, the following statements are equivalent:*

(i) P is an orthogonal projection operator.

(ii) **P** *is symmetric and idempotent.*

For proofs and other details the reader is referred to Rao (1973) and Rao and Mitra (1991).

Theorem A.81 *Let* X *be a matrix of order* $T \times K$ *with rank* $r < K$ *and* $U :$ $(K - r) \times K$ *be such that* $\mathcal{R}(X') \cap \mathcal{R}(U') = \{0\}$. *Then*

(i) $X(X'X + U'U)^{-1}U' = 0$

(ii) $X'X(X'X + U'U)^{-1}X'X = X'X$, *i.e.,* $(X'X + U'U)^{-1}$ *is a g-inverse of* $X'X$

(iii) $U'U(X'X + U'U)^{-1}U'U = U'U$, *i.e.,* $(X'X + U'U)^{-1}$ *is also a g-inverse of* $U'U$

(iv) $U(X'X + U'U)^{-1}U'u = u$ *if* $u \in \mathcal{R}(U)$.

Proof: Since $X'X + U'U$ is of full rank, a matrix A exists, such that

$$(X'X + U'U)A = U'$$
$$\Rightarrow \quad X'XA = U' - U'UA \quad \Rightarrow \quad XA = 0 \text{ and } U' = U'UA$$

since $\mathcal{R}(X')$ and $\mathcal{R}(U')$ are disjoint.
Proof of (i):

$$X(X'X + U'U)^{-1}U' = X(X'X + U'U)^{-1}(X'X + U'U)A = XA = 0$$

Proof of (ii):

$$X'X(X'X + U'U)^{-1}(X'X + U'U - U'U) = X'X - X'X(X'X + U'U)^{-1}U'U$$
$$= X'X.$$

The result (iii) follows on the same lines as result (ii).
Proof of (iv):

$$U(X'X + U'U)^{-1}U'u = U(X'X + U'U)^{-1}U'Ua = Ua = u$$

since $u \in \mathcal{R}(U)$.

A.14 Functions of Normally Distributed Variables

Let $\boldsymbol{x}' = (x_1, \cdots, x_p)$ be a p–dimensional random vector. Then $\boldsymbol{x}$ is p–dimensional normally distributed with expectation vector $\boldsymbol{\mu}$ and covariance matrix $\boldsymbol{\Sigma} > 0$, i.e., $\boldsymbol{x} \sim N_p(\boldsymbol{\mu}, \boldsymbol{\Sigma})$ if the joint density is

$$f(\boldsymbol{x}; \boldsymbol{\mu}, \boldsymbol{\Sigma}) = \{(2\pi)^p |\boldsymbol{\Sigma}|\}^{-1/2} \exp\{-\frac{1}{2}(\boldsymbol{x} - \boldsymbol{\mu})' \boldsymbol{\Sigma}^{-1}(\boldsymbol{x} - \boldsymbol{\mu})\}.$$

Theorem A.82 *Assume* $\boldsymbol{x} \sim N_p(\boldsymbol{\mu}, \boldsymbol{\Sigma})$, *and* $\underset{p,p}{\boldsymbol{A}}$ *and* $\underset{p,1}{\boldsymbol{b}}$ *nonstochastic. Then*

$$\boldsymbol{y} = \boldsymbol{A}\boldsymbol{x} + \boldsymbol{b} \sim N_q(\boldsymbol{A}\boldsymbol{\mu} + \boldsymbol{b}, \boldsymbol{A}\boldsymbol{\Sigma}\boldsymbol{A}') \quad with \ q = \mathrm{rank}\,(\boldsymbol{A}).$$

Theorem A.83 *If* $\boldsymbol{x} \sim N_p(\boldsymbol{0}, \boldsymbol{I})$, *then*

$$\boldsymbol{x}'\boldsymbol{x} \sim \chi_p^2$$

(central χ^2-distribution with p degrees of freedom).

Theorem A.84 *If* $\boldsymbol{x} \sim N_p(\boldsymbol{\mu}, \boldsymbol{I})$, *then*

$$\boldsymbol{x}'\boldsymbol{x} \sim \chi_p^2(\lambda)$$

has a non central χ^2-distribution with noncentrality parameter

$$\lambda = \boldsymbol{\mu}'\boldsymbol{\mu} = \sum_{i=1}^p \mu_i^2.$$

Theorem A.85 *If* $\boldsymbol{x} \sim N_p(\boldsymbol{\mu}, \boldsymbol{\Sigma})$, *then*

(i) $\boldsymbol{x}'\boldsymbol{\Sigma}^{-1}\boldsymbol{x} \sim \chi_p^2(\boldsymbol{\mu}'\boldsymbol{\Sigma}^{-1}\boldsymbol{\mu})$

(ii) $(\boldsymbol{x} - \boldsymbol{\mu})'\boldsymbol{\Sigma}^{-1}(\boldsymbol{x} - \boldsymbol{\mu}) \sim \chi_p^2.$

Proof: $\boldsymbol{\Sigma} > 0 \ \Rightarrow \ \boldsymbol{\Sigma} = \boldsymbol{\Sigma}^{1/2}\boldsymbol{\Sigma}^{1/2}$ with $\boldsymbol{\Sigma}^{1/2}$ regular and symmetric. Hence, $\boldsymbol{\Sigma}^{-1/2}\boldsymbol{x} = \boldsymbol{y} \sim N_p(\boldsymbol{\Sigma}^{-1/2}\boldsymbol{\mu}, \boldsymbol{I}) \ \Rightarrow$

$$\boldsymbol{x}'\boldsymbol{\Sigma}^{-1}\boldsymbol{x} = \boldsymbol{y}'\boldsymbol{y} \sim \chi_p^2(\boldsymbol{\mu}'\boldsymbol{\Sigma}^{-1}\boldsymbol{\mu})$$

and

$$(\boldsymbol{x} - \boldsymbol{\mu})'\boldsymbol{\Sigma}^{-1}(\boldsymbol{x} - \boldsymbol{\mu}) = (\boldsymbol{y} - \boldsymbol{\Sigma}^{-1/2}\boldsymbol{\mu})'(\boldsymbol{y} - \boldsymbol{\Sigma}^{-1/2}\boldsymbol{\mu}) \sim \chi_p^2.$$

Theorem A.86 *If $Q_1 \sim \chi_m^2(\lambda)$ and $Q_2 \sim \chi_n^2$ and Q_1 and Q_2 are independent, then*

(i) The ratio

$$F = \frac{Q_1/m}{Q_2/n}$$

has a noncentral $F_{m,n}(\lambda)$-distribution.

(ii) If $\lambda = 0$, then $F \sim F_{m,n}$ (the central F–distribution).

(iii) If $m = 1$, then $\sqrt{F}$ has a noncentral $t_n(\sqrt{\lambda})$ distribution or a central t_n distribution if $\lambda = 0$.

Theorem A.87 *If $x \sim N_p(\mu, \mathbf{I})$ and $\underset{p,p}{A}$ is a symmetric idempotent matrix with rank $(A) = r$, then*

$$x'Ax \sim \chi_r^2(\mu'A\mu).$$

Proof: We have $A = P\Lambda P'$ (Theorem A.30) and without loss of generality (Theorem A.61 (i)) we may write $\Lambda = \begin{pmatrix} \mathbf{I}_r & 0 \\ 0 & 0 \end{pmatrix}$, i.e., $P'AP = \Lambda$ with P orthogonal. Let $P = (\underset{p,r}{P_1}, \underset{p,(p-r)}{P_2})$ and

$$P'x = y = \begin{pmatrix} y_1 \\ y_2 \end{pmatrix} = \begin{pmatrix} P_1'x \\ P_2'x \end{pmatrix}.$$

Therefore

$$\begin{aligned} y &\sim N_p(P'\mu, \mathbf{I}_p) \qquad \text{(Theorem A.82)} \\ y_1 &\sim N_r(P_1'\mu, \mathbf{I}_r) \\ \text{and} \quad y_1'y_1 &\sim \chi_r^2(\mu'P_1P_1'\mu) \quad \text{(Theorem A.84)}. \end{aligned}$$

As P is orthogonal, we have

$$\begin{aligned} A &= (PP')A(PP') = P(P'AP)P \\ &= (P_1\,P_2)\begin{pmatrix} \mathbf{I}_r & 0 \\ 0 & 0 \end{pmatrix}\begin{pmatrix} P_1' \\ P_2' \end{pmatrix} = P_1P_1' \end{aligned}$$

and therefore

$$x'Ax = x'P_1P_1'x = y_1'y_1 \sim \chi_r^2(\mu'A\mu).$$

Theorem A.88 *Assume $x \sim N_p(\mu, \mathbf{I})$, $\underset{p,p}{A}$ idempotent of rank r, $\underset{p,n}{B}$ any matrix. Then the linear form Bx is independent of the quadratic form $x'Ax$ if and only if $AB = 0$.*

Proof: Let P be the matrix as in Theorem A.87. Then $BPP'AP = BAP = 0$, as $BA = 0$ was assumed. Let $BP = D = (D_1, D_2) = (BP_1, BP_2)$, then

$$BPP'AP = (D_1, D_2)\begin{pmatrix} \mathbf{I}_r & 0 \\ 0 & 0 \end{pmatrix} = (D_1, 0) = (0, 0),$$

so that $D_1 = 0$. This gives

$$Bx = BPP'x = Dy = (0, D_2)\begin{pmatrix} y_1 \\ y_2 \end{pmatrix} = D_2y_2$$

where $y_2 = P_2'x$. Since P is orthogonal and, hence, regular we may conclude that all the components of $y = P'x$ are independent $\Rightarrow$ $Bx = D_2y_2$ and $x'Ax = y_1'y_1$ are independent.

Theorem A.89 *Let $x \sim N_p(0, I)$ and assume A and B to be idempotent $p \times p$ matrices with rank $(A) = r$ and rank $(B) = s$. Then the quadratic forms $x'Ax$ and $x'Bx$ are independent if and only if $BA = 0$.*

Proof: If we use P from Theorem A.87 and set $C = P'BP$ (C symmetric), we get with the assumption $BA = 0$

$$CP'AP = P'BPP'AP$$
$$= P'BAP = 0.$$

Using

$$C = \begin{pmatrix} P_1 \\ P_2 \end{pmatrix} B(P_1' \ P_2')$$

$$= \begin{pmatrix} C_1 & C_2 \\ C_2' & C_3 \end{pmatrix} = \begin{pmatrix} P_1BP_1' & P_1BP_2' \\ P_2BP_1' & P_2BP_2' \end{pmatrix}$$

this relation may be written as

$$CP'AP = \begin{pmatrix} C_1 & C_2 \\ C_2' & C_3 \end{pmatrix} \begin{pmatrix} I_r & 0 \\ 0 & 0 \end{pmatrix} = \begin{pmatrix} C_1 & 0 \\ C_2' & 0 \end{pmatrix} = 0 \quad .$$

Hence, $C_1 = 0$ and $C_2 = 0$ and

$$x'Bx = x'(PP')B(PP')x$$
$$= x'P(P'BP)P'x$$
$$= x'PCP'x$$
$$= (y_1', y_2') \begin{pmatrix} 0 & 0 \\ 0 & C_3 \end{pmatrix} \begin{pmatrix} y_1 \\ y_2 \end{pmatrix} = y_2'C_3y_2 \quad .$$

As shown in Theorem A.87, we have $x'Ax = y_1'y_1$ and therefore the quadratic forms $x'Ax$ and $x'Bx$ are independent.

A.15 Differentiation of Scalar Functions of Matrices

Definition A.90 *If $f(X)$ is a real function of an $m \times n$ matrix $X = (x_{ij})$, then the partial differential of f with respect to X is defined as the $m \times n$ matrix of partial differentials $\frac{\partial f}{\partial x_{ij}}$:*

$$\frac{\partial f(X)}{\partial X} = \begin{pmatrix} \frac{\partial f}{\partial x_{11}} & \cdots & \frac{\partial f}{\partial x_{1n}} \\ \vdots & & \vdots \\ \frac{\partial f}{\partial x_{m1}} & \cdots & \frac{\partial f}{\partial x_{mn}} \end{pmatrix} .$$

Theorem A.91 *Let x be an n-vector and A a symmetric $n \times n$ matrix. Then*

$$\frac{\partial}{\partial x} x'Ax = 2Ax.$$

Proof:

$$x'Ax = \sum_{r,s=1}^{n} a_{rs}x_r x_s$$

$$\frac{\partial f}{\partial x_i}x'Ax = \sum_{\substack{s=1 \\ (s\neq i)}}^{n} a_{is}x_s + \sum_{\substack{r=1 \\ (r\neq i)}}^{n} a_{ri}x_r + 2a_{ii}x_i$$

$$= 2\sum_{s=1}^{n} a_{is}x_s \qquad (\text{as } a_{ij}=a_{ji})$$

$$= 2a_i'x \qquad (a_i': i\text{th row vector of } A).$$

According to definition A.90 we get

$$\frac{\partial x'Ax}{\partial x} = \begin{pmatrix} \frac{\partial}{\partial x_1} \\ \vdots \\ \frac{\partial}{\partial x_n} \end{pmatrix}(x'Ax) = 2\begin{pmatrix} a_1' \\ \vdots \\ a_n' \end{pmatrix}x = 2Ax.$$

Theorem A.92 *If x is an n-vector, y is an m-vector and C an $n \times m$ matrix, then*

$$\frac{\partial}{\partial C}x'Cy = xy'.$$

Proof:

$$x'Cy = \sum_{r=1}^{m}\sum_{s=1}^{n} x_s c_{sr} y_r,$$

$$\frac{\partial}{\partial c_{k\lambda}}x'Cy = x_k y_\lambda \qquad (\text{the } (k,\lambda)\text{th element of } xy'),$$

$$\frac{\partial}{\partial C}x'Cy = (x_k y_\lambda) = xy'.$$

Theorem A.93 *Let x be a K-vector, A a symmetric $T \times T$-matrix, C a $T \times K$-matrix. Then*

$$\frac{\partial}{\partial C}x'C'ACx = 2ACxx'.$$

Proof: We have

$$x'C' = \left(\sum_{i=1}^{K} x_i c_{1i}, \cdots, \sum_{i=1}^{K} x_i c_{Ti}\right),$$

$$\frac{\partial}{\partial c_{k\lambda}}x'C' = (0,\cdots,0,x_\lambda,0,\cdots,0) \qquad (x_\lambda \text{ is an element of the } k\text{th column}).$$

Using the product rule yields

$$\frac{\partial}{\partial c_{k\lambda}}x'C'ACx = \left(\frac{\partial}{\partial c_{k\lambda}}x'C'\right)ACx + x'C'A\left(\frac{\partial}{\partial c_{k\lambda}}Cx\right).$$

Since

$$\boldsymbol{x}'\boldsymbol{C}'\boldsymbol{A} = \left(\sum_{t=1}^{T}\sum_{i=1}^{K} x_i c_{ti} a_{t1}, \cdots, \sum_{t=1}^{T}\sum_{i=1}^{K} x_i c_{ti} a_{Tt}\right)$$

we get

$$
\begin{aligned}
\boldsymbol{x}'\boldsymbol{C}'\boldsymbol{A}\left(\frac{\partial}{\partial c_{k\lambda}}\boldsymbol{C}\boldsymbol{x}\right) &= \sum_{t,i} x_i x_\lambda c_{ti} a_{kt} \\
&= \sum_{t,i} x_i x_\lambda c_{ti} a_{tk} \qquad \text{(as } \boldsymbol{A} \text{ is symmetric)} \\
&= \left(\frac{\partial}{\partial c_{k\lambda}}\boldsymbol{x}'\boldsymbol{C}'\right)\boldsymbol{A}\boldsymbol{C}\boldsymbol{x}.
\end{aligned}
$$

But $\sum_{t,i} x_i x_\lambda c_{ti} a_{tk}$ is just the (k, λ)th element of the matrix $\boldsymbol{A}\boldsymbol{C}\boldsymbol{x}\boldsymbol{x}'$.

Theorem A.94 *Assume $\boldsymbol{A} = \boldsymbol{A}(x)$ to be an $n \times n$-matrix, with elements $a_{ij}(x)$ that are real functions of a scalar x. Let $\boldsymbol{B}$ be an $n \times n$-matrix, such that its elements are independent of x. Then*

$$\frac{\partial}{\partial x}\text{tr}\,(\boldsymbol{A}\boldsymbol{B}) = \text{tr}\,\left(\frac{\partial \boldsymbol{A}}{\partial x}\boldsymbol{B}\right).$$

Proof:

$$
\begin{aligned}
\text{tr}\,(\boldsymbol{A}\boldsymbol{B}) &= \sum_{i=1}^{n}\sum_{j=1}^{n} a_{ij} b_{ji}, \\
\frac{\partial}{\partial x}\text{tr}\,(\boldsymbol{A}\boldsymbol{B}) &= \sum_{i}\sum_{j}\frac{\partial a_{ij}}{\partial x} b_{ji} \\
&= \text{tr}\,\left(\frac{\partial \boldsymbol{A}}{\partial x}\boldsymbol{B}\right),
\end{aligned}
$$

where $\frac{\partial \boldsymbol{A}}{\partial x} = \left(\frac{\partial a_{ij}}{\partial x}\right)$.

Theorem A.95 *For the differential of the trace, we have the following rules*

	y	$\partial y/\partial \boldsymbol{X}$
(i)	$\text{tr}\,(\boldsymbol{A}\boldsymbol{X})$	$\boldsymbol{A}'$
(ii)	$\text{tr}\,(\boldsymbol{X}'\boldsymbol{A}\boldsymbol{X})$	$(\boldsymbol{A} + \boldsymbol{A}')\boldsymbol{X}$
(iii)	$\text{tr}\,(\boldsymbol{X}\boldsymbol{A}\boldsymbol{X})$	$\boldsymbol{X}'\boldsymbol{A} + \boldsymbol{A}'\boldsymbol{X}'$
(iv)	$\text{tr}\,(\boldsymbol{X}\boldsymbol{A}\boldsymbol{X}')$	$\boldsymbol{X}(\boldsymbol{A} + \boldsymbol{A}')$
(v)	$\text{tr}\,(\boldsymbol{X}'\boldsymbol{A}\boldsymbol{X}')$	$\boldsymbol{A}\boldsymbol{X}' + \boldsymbol{X}'\boldsymbol{A}$
(vi)	$\text{tr}\,(\boldsymbol{X}'\boldsymbol{A}\boldsymbol{X}\boldsymbol{B})$	$\boldsymbol{A}\boldsymbol{X}\boldsymbol{B} + \boldsymbol{A}'\boldsymbol{X}\boldsymbol{B}'$

Differentiation of Inverse Matrices

Theorem A.96 *Let $T = T(x)$ be a regular matrix, such that its elements depend on a scalar x. Then*

$$\frac{\partial T^{-1}}{\partial x} = -T^{-1}\frac{\partial T}{\partial x}T^{-1}.$$

Proof: We have $T^{-1}T = I$, $\frac{\partial I}{\partial x} = 0$,

$$\frac{\partial(T^{-1}T)}{\partial x} = \frac{\partial T^{-1}}{\partial x}T + T^{-1}\frac{\partial T}{\partial x} = 0.$$

Theorem A.97 *For non–singular X, we have*

$$\frac{\partial \mathrm{tr}\,(AX^{-1})}{\partial X} = -(X^{-1}AX^{-1})'$$

$$\frac{\partial \mathrm{tr}\,(X^{-1}AX^{-1}B)}{\partial X} = -(X^{-1}AX^{-1}BX^{-1} + X^{-1}BX^{-1}AX^{-1})'$$

Proof: Use Theorems A95, A96 and the product rule.

Differentiation of a Determinant

Theorem A.98 *For a non–singular matrix Z, we have*

(i) $\frac{\partial}{\partial Z}|Z| = |Z|(Z')^{-1}$

(ii) $\frac{\partial}{\partial Z}\log|Z| = (Z')^{-1}$.

A.16   Miscellaneous Results, Stochastic Convergence

Theorem A.99 (Kronecker Product)

Let $\underset{m,n}{A} = (a_{ij})$ and $\underset{p,q}{B} = (b_{rs})$ be any matrices. Then the Kronecker product of A and B is defined as

$$\underset{mp,nq}{C} = \underset{m,n}{A} \otimes \underset{p,q}{B} = \begin{pmatrix} a_{11}B & a_{12}B & \cdots & a_{1n}B \\ \vdots & \vdots & & \vdots \\ a_{m1}B & a_{m2}B & \cdots & a_{mn}B \end{pmatrix}$$

and the following rules hold:

(i) $c(A \otimes B) = (cA) \otimes B = A \otimes (cB)$ (*c being a scalar*),

(ii) $A \otimes (B \otimes C) = (A \otimes B) \otimes C$,

(iii) $A \otimes (B + C) = (A \otimes B) + (A \otimes C)$,

(iv) $(A \otimes B)' = A' \otimes B'$.

Theorem A.100 (Tschebyschev's Inequality)

For any n-dimensional random vector $\boldsymbol{x}$ and a given scalar $\epsilon > 0$ we have

$$\mathbb{P}\left\{|\boldsymbol{x}| \geq \epsilon\right\} \leq \frac{\mathrm{E}\,|\boldsymbol{x}|^2}{\epsilon^2}\ .$$

Proof: Let $F(\boldsymbol{x})$ be the joint distribution function of $\boldsymbol{x} = (x_1, \ldots, x_n)$. Then

$$
\begin{aligned}
\mathrm{E}\,|\boldsymbol{x}|^2 &= \int |\boldsymbol{x}|^2 dF(\boldsymbol{x}) \\
&= \int_{\{\boldsymbol{x}:|\boldsymbol{x}|\geq\epsilon\}} |\boldsymbol{x}|^2 dF(\boldsymbol{x}) + \int_{\{\boldsymbol{x}:|\boldsymbol{x}|<\epsilon\}} |\boldsymbol{x}|^2 dF(\boldsymbol{x}) \\
&\geq \epsilon^2 \int_{\{\boldsymbol{x}:|\boldsymbol{x}|\geq\epsilon\}} dF(\boldsymbol{x}) = \epsilon^2 \mathbb{P}\left\{|\boldsymbol{x}| \geq \epsilon\right\}\ .
\end{aligned}
$$

Definition A.101 *Let $\{\boldsymbol{x}(t)\}$, $t = 1, 2, \ldots$ be a multivariate stochastic process.*

(i) Weak convergency:
   If

$$\lim_{t\to\infty} \mathbb{P}\left\{|\boldsymbol{x}(t) - \tilde{\boldsymbol{x}}| \geq \delta\right\} = 0$$

   where $\delta > 0$ is any given scalar and $\tilde{\boldsymbol{x}}$ is a finite vector, then $\tilde{\boldsymbol{x}}$ is called the probability limit of $\{\boldsymbol{x}(t)\}$ and we write

$$p \lim \boldsymbol{x} = \tilde{\boldsymbol{x}}.$$

(ii) Strong convergency:
   Assume that $\{\boldsymbol{x}(t)\}$ is defined on a probability space $(\Omega, \Sigma, \mathbb{P})$. Then $\{\boldsymbol{x}(t)\}$ is said to be strongly convergent to $\tilde{\boldsymbol{x}}$, i.e.,

$$\{\boldsymbol{x}(t)\} \ \to \ \tilde{\boldsymbol{x}} \quad almost\ sure\ (a.s.)$$

   if there exists a set $T \in \Sigma$, $\mathbb{P}(T) = 0$ and $\boldsymbol{x}_\omega(t) \to \tilde{\boldsymbol{x}}_\omega$, as $T \to \infty$, for each $\omega \in \Omega - T$ (M.M. Rao, 1984, p. 45).

Theorem A.102 (Slutsky's Theorem)

Using Definition A.101, we have

(i) if $p \lim \boldsymbol{x} = \tilde{\boldsymbol{x}}$ then $\lim_{t\to\infty} \mathrm{E}\{\boldsymbol{x}(t)\} = \bar{\mathrm{E}}(\boldsymbol{x}) = \tilde{\boldsymbol{x}}$,

(ii) if $\boldsymbol{c}$ is a vector of constants then $p \lim \boldsymbol{c} = \boldsymbol{c}$,

(iii) (Slutsky's Theorem) if $p \lim \boldsymbol{x} = \tilde{\boldsymbol{x}}$ and $\boldsymbol{y} = f(\boldsymbol{x})$ is any continuous vector function of $\boldsymbol{x}$, then $p \lim \boldsymbol{y} = f(\tilde{\boldsymbol{x}})$,

(iv) *if **A** and **B** are random matrices, and assuming that the following limits exist, then:*

$$p \lim(\boldsymbol{AB}) = p \lim \boldsymbol{A} \; p \lim \boldsymbol{B}$$

and

$$p \lim(\boldsymbol{A}^{-1}) = (p \lim \boldsymbol{A})^{-1} \quad ,$$

(v) *if* $p \lim \left[\sqrt{T}(\boldsymbol{x}(t) - \mathrm{E}\,\boldsymbol{x}(t)) \right]' \left[\sqrt{T}(\boldsymbol{x}(t) - \mathrm{E}\,\boldsymbol{x}(t)) \right] = \boldsymbol{V}$ *then the asymptotic covariance matrix is*

$$\bar{V}(\boldsymbol{x}, \boldsymbol{x}) = \bar{\mathrm{E}} \left[\boldsymbol{x} - \bar{\mathrm{E}}\,(\boldsymbol{x}) \right]' \left[\boldsymbol{x} - \bar{\mathrm{E}}(\boldsymbol{x}) \right] = T^{-1}\boldsymbol{V} \quad .$$

Definition A.103 *If* $\{\boldsymbol{x}(t)\}, t = 1, 2, \dots$ *is a multivariate stochastic process satisfying*

$$\lim_{t \to \infty} \mathrm{E}\,|\boldsymbol{x}(t) - \tilde{\boldsymbol{x}}|^2 = 0$$

then $\{\boldsymbol{x}(t)\}$ *is called convergent in the quadratic mean, and we write*

$$l.i.m. \; \boldsymbol{x} = \tilde{\boldsymbol{x}} \quad .$$

Theorem A.104 *If* $l.i.m. \; \boldsymbol{x} = \tilde{\boldsymbol{x}}$, *then* $p \lim \boldsymbol{x} = \tilde{\boldsymbol{x}}$.
Proof: Using Theorem A.100 we get

$$0 \leq \lim_{t \to \infty} \mathbb{P}\,(|\boldsymbol{x}(t) - \tilde{\boldsymbol{x}}| \geq \epsilon) \leq \lim_{t \to \infty} \frac{\mathrm{E}\,|\boldsymbol{x}(t) - \tilde{\boldsymbol{x}}|^2}{\epsilon^2} = 0 \quad .$$

Theorem A.105 *If* $l.i.m. \; (\boldsymbol{x}(t) - \mathrm{E}\,\boldsymbol{x}(t)) = \boldsymbol{0}$ *and* $\lim_{t \to \infty} \mathrm{E}\,\boldsymbol{x}(t) = \boldsymbol{c}$, *then* $p \lim \boldsymbol{x}(t) = \boldsymbol{c}$.
Proof:

$$\begin{aligned}
\lim_{t \to \infty} \mathbb{P}\,(|\boldsymbol{x}(t) - \boldsymbol{c}| \geq \epsilon) \; &\leq \; \epsilon^{-2} \lim_{t \to \infty} \mathrm{E}\,|\boldsymbol{x}(t) - \boldsymbol{c}|^2 \\
&= \; \epsilon^{-2} \lim_{t \to \infty} \mathrm{E}\,|\boldsymbol{x}(t) - \mathrm{E}\,\boldsymbol{x}(t) + \mathrm{E}\,\boldsymbol{x}(t) - \boldsymbol{c}|^2 \\
&= \; \epsilon^{-2} \lim_{t \to \infty} \mathrm{E}\,|\boldsymbol{x}(t) - \mathrm{E}\,\boldsymbol{x}(t)|^2 + \epsilon^{-2} \lim_{t \to \infty} |\mathrm{E}\,\boldsymbol{x}(t) - \boldsymbol{c}|^2 \\
&\quad + 2\epsilon^{-2} \lim_{t \to \infty} \{(\mathrm{E}\,\boldsymbol{x}(t) - \boldsymbol{c})'(\boldsymbol{x}(t) - \mathrm{E}\,\boldsymbol{x}(t))\} \\
&= \; 0 \quad .
\end{aligned}$$

Theorem A.106 $l.i.m. \; \boldsymbol{x} = \boldsymbol{c}$ *if and only if*

$$l.i.m.(\boldsymbol{x}(t) - \mathrm{E}\,\boldsymbol{x}(t)) = \boldsymbol{0} \text{ and } \lim_{t \to \infty} \mathrm{E}\,\boldsymbol{x}(t) = \boldsymbol{c} \quad .$$

Proof: As in Theorem A.105, we may write

$$\begin{aligned}
\lim_{t \to \infty} \mathrm{E}\,|\boldsymbol{x}(t) - \boldsymbol{c}|^2 \; &= \; \lim_{t \to \infty} \mathrm{E}\,|\boldsymbol{x}(t) - \mathrm{E}\,\boldsymbol{x}(t)|^2 \\
&\quad + \lim_{t \to \infty} |\mathrm{E}\,\boldsymbol{x}(t) - \boldsymbol{c}|^2 \\
&\quad + 2 \lim_{t \to \infty} \mathrm{E}\,(\mathrm{E}\,\boldsymbol{x}(t) - \boldsymbol{c})'(\boldsymbol{x}(t) - \mathrm{E}\,\boldsymbol{x}(t)) \\
&= \; 0
\end{aligned}$$

Theorem A.107 *Let $\boldsymbol{x}(t)$ be an estimator of a parameter vector $\boldsymbol{\theta}$. Then we have the result*

$$\lim_{t \to \infty} \mathrm{E}\,\boldsymbol{x}(t) = \boldsymbol{\theta} \quad \text{if} \quad l.i.m.(\boldsymbol{x}(t) - \boldsymbol{\theta}) = \boldsymbol{0} \quad .$$

That is, $\boldsymbol{x}(t)$ is an asymptotically unbiased estimator for $\boldsymbol{\theta}$ if $\boldsymbol{x}(t)$ converges to $\boldsymbol{\theta}$ in the quadratic mean.

Proof: Use A.106.

B Distributions and Tables

x	0.0	0.02	0.04	0.06	0.08
0.0	0.3989	0.3989	0.3986	0.3982	0.3977
0.2	0.3910	0.3894	0.3876	0.3857	0.3836
0.4	0.3814	0.3653	0.3621	0.3589	0.3555
0.6	0.3332	0.3292	0.3251	0.3209	0.3166
0.8	0.2897	0.2850	0.2803	0.2756	0.2709
1.0	0.2419	0.2371	0.2323	0.2275	0.2226
1.2	0.1942	0.1895	0.1849	0.1804	0.1758
1.4	0.1497	0.1456	0.1415	0.1374	0.1334
1.6	0.1109	0.1074	0.1039	0.1006	0.0973
1.8	0.0789	0.0761	0.0734	0.0707	0.0681
2.0	0.0539	0.0519	0.0498	0.0478	0.0459
2.2	0.0355	0.0339	0.0325	0.0310	0.0296
2.4	0.0224	0.0213	0.0203	0.0194	0.0184
2.6	0.0136	0.0167	0.0122	0.0116	0.0110
2.8	0.0059	0.0075	0.0071	0.0067	0.0063
3.0	0.0044	0.0024	0.0012	0.0006	0.0003

Table B.1: Probability density function $\phi(x)$ of the $N(0,1)$–distribution

u	0.00	0.01	0.02	0.03	0.04
0.0	0.500000	0.503989	0.507978	0.511966	0.515953
0.1	0.539828	0.543795	0.547758	0.551717	0.555670
0.2	0.579260	0.583166	0.587064	0.590954	0.594835
0.3	0.617911	0.621720	0.625516	0.629300	0.633072
0.4	0.655422	0.659097	0.662757	0.666402	0.670031
0.5	0.691462	0.694974	0.698468	0.701944	0.705401
0.6	0.725747	0.729069	0.732371	0.735653	0.738914
0.7	0.758036	0.761148	0.764238	0.767305	0.770350
0.8	0.788145	0.791030	0.793892	0.796731	0.799546
0.9	0.815940	0.818589	0.821214	0.823814	0.826391
1.0	0.841345	0.843752	0.846136	0.848495	0.850830
1.1	0.864334	0.866500	0.868643	0.870762	0.872857
1.2	0.884930	0.886861	0.888768	0.890651	0.892512
1.3	0.903200	0.904902	0.906582	0.908241	0.909877
1.4	0.919243	0.920730	0.922196	0.923641	0.925066
1.5	0.933193	0.934478	0.935745	0.936992	0.938220
1.6	0.945201	0.946301	0.947384	0.948449	0.949497
1.7	0.955435	0.956367	0.957284	0.958185	0.959070
1.8	0.964070	0.964852	0.965620	0.966375	0.967116
1.9	0.971283	0.971933	0.972571	0.973197	0.973810
2.0	0.977250	0.977784	0.978308	0.978822	0.979325
2.1	0.982136	0.982571	0.982997	0.983414	0.983823
2.2	0.986097	0.986447	0.986791	0.987126	0.987455
2.3	0.989276	0.989556	0.989830	0.990097	0.990358
2.4	0.991802	0.992024	0.992240	0.992451	0.992656
2.5	0.993790	0.993963	0.994132	0.994297	0.994457
2.6	0.995339	0.995473	0.995604	0.995731	0.995855
2.7	0.996533	0.996636	0.996736	0.996833	0.996928
2.8	0.997445	0.997523	0.997599	0.997673	0.997744
2.9	0.998134	0.998193	0.998250	0.998305	0.998359
3.0	0.998650	0.998694	0.998736	0.998777	0.998817

Table B.2: Probability function $\Phi(u)$ of the $N(0,1)$-distribution

u	0.05	0.06	0.07	0.08	0.09
0.0	0.519939	0.523922	0.527903	0.531881	0.535856
0.1	0.559618	0.563559	0.567495	0.571424	0.575345
0.2	0.598706	0.602568	0.606420	0.610261	0.614092
0.3	0.636831	0.640576	0.644309	0.648027	0.651732
0.4	0.673645	0.677242	0.680822	0.684386	0.687933
0.5	0.708840	0.712260	0.715661	0.719043	0.722405
0.6	0.742154	0.745373	0.748571	0.751748	0.754903
0.7	0.773373	0.776373	0.779350	0.782305	0.785236
0.8	0.802337	0.805105	0.807850	0.810570	0.813267
0.9	0.828944	0.831472	0.833977	0.836457	0.838913
1.0	0.853141	0.855428	0.857690	0.859929	0.862143
1.1	0.874928	0.876976	0.879000	0.881000	0.882977
1.2	0.894350	0.896165	0.897958	0.899727	0.901475
1.3	0.911492	0.913085	0.914657	0.916207	0.917736
1.4	0.926471	0.927855	0.929219	0.930563	0.931888
1.5	0.939429	0.940620	0.941792	0.942947	0.944083
1.6	0.950529	0.951543	0.952540	0.953521	0.954486
1.7	0.959941	0.960796	0.961636	0.962462	0.963273
1.8	0.967843	0.968557	0.969258	0.969946	0.970621
1.9	0.974412	0.975002	0.975581	0.976148	0.976705
2.0	0.979818	0.980301	0.980774	0.981237	0.981691
2.1	0.984222	0.984614	0.984997	0.985371	0.985738
2.2	0.987776	0.988089	0.988396	0.988696	0.988989
2.3	0.000613	0.000863	0.991106	0.991344	0.991576
2.4	0.992857	0.993053	0.993244	0.993431	0.993613
2.5	0.994614	0.994766	0.994915	0.995060	0.995201
2.6	0.995975	0.996093	0.996207	0.996319	0.996427
2.7	0.997020	0.997110	0.997197	0.997282	0.997365
2.8	0.997814	0.997882	0.997948	0.998012	0.998074
2.9	0.998411	0.998462	0.998511	0.998559	0.998605
3.0	0.998856	0.998893	0.998930	0.998965	0.998999

Table B.3: Probability function $\Phi(u)$ of the $N(0,1)$-distribution

	Type I error probabilities α					
df	0.99	0.975	0.95	0.05	0.025	0.01
1	0.0001	0.001	0.004	3.84	5.02	6.62
2	0.020	0.051	0.103	5.99	7.38	9.21
3	0.115	0.216	0.352	7.81	9.35	11.3
4	0.297	0.484	0.711	9.49	11.1	13.3
5	0.554	0.831	1.15	11.1	12.8	15.1
6	0.872	1.24	1.64	12.6	14.4	16.8
7	1.24	1.69	2.17	14.1	16.0	18.5
8	1.65	2.18	2.73	15.5	17.5	20.1
9	2.09	2.70	3.33	16.9	19.0	21.7
10	2.56	3.25	3.94	18.3	20.5	23.2
11	3.05	3.82	4.57	19.7	21.9	24.7
12	3.57	4.40	5.23	21.0	23.3	26.2
13	4.11	5.01	5.89	22.4	24.7	27.7
14	4.66	5.63	6.57	23.7	26.1	29.1
15	5.23	6.26	7.26	25.0	27.5	30.6
16	5.81	6.91	7.96	26.3	28.8	32.0
17	6.41	7.56	8.67	27.6	30.2	33.4
18	7.01	8.23	9.39	28.9	31.5	34.8
19	7.63	8.91	10.1	30.1	32.9	36.2
20	8.26	9.59	10.9	31.4	34.2	37.6
25	11.5	13.1	14.6	37.7	40.6	44.3
30	15.0	16.8	18.5	43.8	47.0	50.9
40	22.2	24.4	26.5	55.8	59.3	63.7
50	29.7	32.4	34.8	67.5	71.4	76.2
60	37.5	40.5	43.2	79.1	83.3	88.4
70	45.4	48.8	51.7	90.5	95.0	100.4
80	53.5	57.2	60.4	101.9	106.6	112.3
90	61.8	65.6	69.1	113.1	118.1	124.1
100	70.1	74.2	77.9	124.3	129.6	135.8

Table B.4: Quantiles of the χ^2-distribution

One-sided type I error probabilities α			
0.05	0.025	0.01	0.005
Two-sided type I error probabilities α			
df 0.10	0.05	0.02	0.01
1 6.31	12.71	31.82	63.66
2 2.92	4.30	6.97	9.92
3 2.35	3.18	4.54	5.84
4 2.13	2.78	3.75	4.60
5 2.01	2.57	3.37	4.03
6 1.94	2.45	3.14	3.71
7 1.89	2.36	3.00	3.50
8 1.86	2.31	2.90	3.36
9 1.83	2.26	2.82	3.25
10 1.81	2.23	2.76	3.17
11 1.80	2.20	2.72	3.11
12 1.78	2.18	2.68	3.05
13 1.77	2.18	2.65	3.01
14 1.76	2.14	2.62	2.98
15 1.75	2.13	2.60	2.95
16 1.75	2.12	2.58	2.92
17 1.74	2.11	2.57	2.90
18 1.73	2.10	2.55	2.88
19 1.73	2.09	2.54	2.86
20 1.73	2.09	2.53	2.85
30 1.70	2.04	2.46	2.75
40 1.68	2.02	2.42	2.70
60 1.67	2.00	2.39	2.66
∞ 1.64	1.96	2.33	2.58

Table B.5: Quantiles of the t-distribution

$$df_1$$

df_2	1	2	3	4	5	6	7	8	9
1	161	200	216	225	230	234	237	239	241
2	18.51	19.00	19.16	19.25	19.30	19.33	19.36	19.37	19.38
3	10.13	9.55	9.28	9.12	9.01	8.94	8.88	8.84	8.81
4	7.71	6.94	6.59	6.39	6.26	6.16	6.09	6.04	6.00
5	6.61	5.79	5.41	5.19	5.05	4.95	4.88	4.82	4.78
6	5.99	5.14	4.76	4.53	4.39	4.28	4.21	4.15	4.10
7	5.59	4.74	4.35	4.12	3.97	3.87	3.79	3.73	3.68
8	5.32	4.46	4.07	3.84	3.69	3.58	3.50	3.44	3.39
9	5.12	4.26	3.86	3.63	3.48	3.37	3.29	3.23	3.18
10	4.96	4.10	3.71	3.48	3.33	3.22	3.14	3.07	3.02
11	4.84	3.98	3.59	3.36	3.20	3.09	3.01	2.95	2.90
12	4.75	3.88	3.49	3.26	3.11	3.00	2.92	2.85	2.80
13	4.67	3.80	3.41	3.18	3.02	2.92	2.84	2.77	2.72
14	4.60	3.74	3.34	3.11	2.96	2.85	2.77	2.70	2.65
15	4.54	3.68	3.29	3.06	2.90	2.79	2.70	2.64	2.59
20	4.35	3.49	3.10	2.87	2.71	2.60	2.52	2.45	2.40
30	4.17	3.32	2.92	2.69	2.53	2.42	2.34	2.27	2.21

Table B.6: Quantiles of F_{df_1, df_2}-distribution for $\alpha = 0.05$

				df_1				
df_2	10	11	12	14	16	20	24	30
1	242	243	244	245	246	248	249	250
2	19.39	19.40	19.41	19.42	19.43	19.44	19.45	19.46
3	8.78	8.76	8.74	8.71	8.69	8.66	8.64	8.62
4	5.96	5.93	5.91	5.87	5.84	5.80	5.77	5.74
5	4.74	4.70	4.68	4.64	4.60	4.56	4.53	4.50
6	4.06	4.03	4.00	3.96	3.92	3.87	3.84	3.81
7	3.63	3.60	3.57	3.52	3.49	3.44	3.41	3.38
8	3.34	3.31	3.28	3.23	3.20	3.15	3.12	3.08
9	3.13	3.10	3.07	3.02	2.98	2.93	2.90	2.86
10	2.97	2.94	2.91	2.86	2.82	2.77	2.74	2.70
11	2.86	2.82	2.79	2.74	2.70	2.65	2.61	2.57
12	2.76	2.72	2.69	2.64	2.60	2.54	2.50	2.46
13	2.67	2.63	2.60	2.55	2.51	2.46	2.42	2.38
14	2.60	2.56	2.53	2.48	2.44	2.39	2.35	2.31
15	2.55	2.51	2.48	2.43	2.39	2.33	2.29	2.25
20	2.35	2.31	2.28	2.23	2.18	2.12	2.08	2.04
30	2.16	2.12	2.00	2.04	1.99	1.93	1.89	1.84

Table B.7: Quantiles of the F_{df_1,df_2}-distribution for $\alpha = 0.05$

Bibliography

[1] A. Agresti. *Categorical data analysis.* Wiley, New York, 1990.

[2] J. Aitchison and S. D. Silvey. Maximum likelihood estimation of parameters subject to restraints. *Ann. Math. Statist.*, 29:813–828, 1958.

[3] A. Albert. *Regression and the Moore-Penrose pseudoinverse.* Academic Press, New York, 1972.

[4] T. Amemiya. *Advanced econometrics.* Blackwell, Oxford, 1985.

[5] J. K. Baksalary, R. Kala, and K. Klaczynski. The matrix inequality $M \geq B^*MB$. *Linear Algebra and its Applications*, 54:77–86, 1983.

[6] M. S. Bartlett. Some examples of statistical methods of research in agriculture and applied botany. *J. Royal Statist. Soc. B*, 4:137–170, 1937.

[7] R. J. Beckmann and H. J. Trussel. The distribution of an arbitrary studentized residual and the effects of updating in multiple regression. *J. Amer. Statist. Assoc.*, 69:199–201, 1974.

[8] P. A. Bekker and H. Neudecker. Albert's theorem applied to problems of efficiency and MSE superiority. *Statistica Neerlandica*, 43:157–167, 1989.

[9] M. W. Birch. Maximum likelihood in three-way contingency tables. *J. Royal. Statist. Soc. B*, 25:220–233, 1963.

[10] Y. M. M. Bishop, S. E. Fienberg, and P. W. Holland. *Discrete multivariate analysis: theory and practice.* MIT Press, Cambridge, 1975.

[11] R. J. Boik. A priori tests in repeated measures designs: effects of nonsphericity. *Psychometrika*, 46(3):241–255, 1981.

[12] K. Bosch. *Statistik-Taschenbuch.* Oldenbourg, München, 1992.

[13] G. E. P. Box. A general distribution theory for a class of likelihood criteria. *Biometrics*, 36:317–346, 1949.

[14] R. J. Brook and G. C. Arnold. *Applied regression analysis and experimental design.* Dekker, New York, 1985.

[15] B. W. Brown. The crossover experiment. *Biometrics*, 36:69–79, 1980.

[16] H. Büning and G. Trenkler. *Nichtparametrische Statistische Methoden.* de Gruyter, Berlin, 1978.

[17] S. L. Campbell and C. D. Meyer. *Generalized inverses of linear transformations.* Pitman, London, 1979.

[18] Christensen. *Loglinear models.* Springer, New York, 1990.

[19] W. Cochran and D. R. Cox. *Experimental design.* Wiley, New York, 1957.

[20] D. R. Cox. *The analysis of binary data.* Chapman and Hall, London, 1970.

[21] D. R. Cox. The analysis of multivariate binary data. *Appl. Statist.*, 21:113–120, 1972.

[22] D.R. Cox and D.V. Hinkley. *Theoretical Statistics.* Chapman and Hall, London, 1974.

[23] M. J. Crowder and D. J. Hand. *Analysis of repeated measures.* Chapman and Hall, London, 1990.

[24] A. P. Dempster, N. M. Laird, and D. B. Rubin. Maximum likelihood from incomplete data via the EM algorithm. *J. Royal Statist. Soc. B*, 43:1–22, 1977.

[25] P. J. Dhrymes. *Indroductory econometrics.* Springer, New York, 1978.

[26] P. J. Diggle, K.-Y. Liang, and S. L. Zeger. *Analysis of longitudinal data.* London (to appear), 1993.

[27] N. Draper and H. Smith. *Applied regression analysis.* Wiley, New York, 1966.

[28] D. B. Duncan. t-tests and intervals for comparisons suggested by the data. *Biometrics*, 31:339–359, 1975.

[29] O. J. Dunn. Multiple comparisons using rank sums. *Technometrics*, 6:241–252, 1964.

[30] O. J. Dunn and V. A. Clark. *Applied statistics. Analysis of variance and regression.* Wiley, New York, 1987.

[31] C. W. Dunnett. A multiple comparison procedure for comparing treatments with a control. *J. Amer. Statist. Assoc.*, 50:1096–1121, 1955.

[32] C. W. Dunnett. New tables for multiple comparisons with a control. *Biometrics*, 20:482–491, 1964.

[33] L. Fahrmeir and A. Hamerle. *Multivariate statistische Verfahren*. de Gruyter, Berlin, 1984.

[34] L. Fahrmeir and H. Kaufmann. Consistency and asymptotic normality of the maximum likelihood estimator in generalized linear models. *Annals of Statistics*, 13:342–368, 1985.

[35] L. Fahrmeir and G. Tutz. *Multivariate statistical modelling based on generalized linear models*. Springer, Berlin, 1994.

[36] G. M. Fitzmaurice and N. M. Laird. A likelihood-based method for analysing longitudinal binary responses. *Biometrika*, 80:141–151, 1993.

[37] J. L. Fleiss. A critique of recent research in the two-treatment crossover design. *Controlled Clinical Trials*, 10:237–243, 1989.

[38] M. Friedman. The use of ranks to avoid the assumption of normality implicit in the analysis of variance. *J. Amer. Statist. Assoc.*, 32:675–701, 1937.

[39] J. J. Gart. An exact test for comparing matched proportions in crossover designs. *Biometrika*, 56(1):75–80, 1969.

[40] U. Gather and I. Pigeot. Multiples Testen. *Skript, Universität Dortmund, Fachbereich Statistik*, 1990.

[41] E. R. Girden. *ANOVA – repeated measures*. Sage Publications, New York, 1992.

[42] A. S. Goldberger. *Econometric theory*. Wiley, New York, 1964.

[43] L. A. Goodman. The analysis of multidimensional contingency tables: stepwise procedures and direct estimation methods for building models for multiple classifications. *Technometrics*, 13:33–61, 1971.

[44] F. A. Graybill. *An introduction to linear statistical models*. McGraw-Hill, New York, 1961.

[45] S. W. Greenhouse and S. Geisser. On methods in the analysis of profile data. *Psychometrika*, 24(2):95–112, 1959.

[46] A. P. Grieve. The two-period changeover design in clinical trials. letter to the editor. *Biometrics*, 38:517, 1982.

[47] A. P. Grieve. *Crossover versus parallel designs. In D. A. Berry (ed.): Statistical methodology in the pharmaceutical sciences.* Dekker, New York, 1990.

[48] J. E. Grizzle. The two-period change-over design and its use in clinical trials. *Biometrics*, 21, 1965.

[49] J. E. Grizzle, C. F. Starmer, and G. G. Koch. Analysis of categorical data by linear models. *Biometrics*, 25:489–504, 1969.

[50] D. K. Guilkey and J. M. Price. On comparing restricted least squares estimators. *Journal of Econometrics*, 15:397–404, 1981.

[51] P. D. Haaland. *Experimental design in biotechnology.* Dekker, New York, 1989.

[52] Y. Haitovsky. Missing data in regression analysis. *J. Royal Statist. Soc. B*, 34:67–82, 1968.

[53] W. L. Hays. *Statistics.* Holt, Rinehart and Winston, New York, 4th ed. edition, 1988.

[54] Ch. Heumann. GEE1-procedure for categorical correlated response. *University of Munich, Institute of Statistics, Research Report*, 1993.

[55] Ch. Heumann and M. Jacobsen. *LOGGY 1.0 – Ein Programm zur Analyse von loglinearen Modellen und Kontingenztafeln.* Ch. Heumann, Ludwig-Dill-Str. 17, 85221 Dachau, 1993.

[56] Ch. Heumann, M. Jacobsen, and H. Toutenburg. Rechnergestützte grafische Analyse von ordinalen Kontingenztafeln – eine Alternative zum Pareto-Prinzip. *Universität München, Institut für Statistik*, 1993.

[57] M. Hills and P. Armitage. The two-period cross-over clinical trial. *British Journal of clinical Pharmacology*, 8:7–20, 1979.

[58] R. R. Hocking. A discussion of the two-way mixed models. *The American Statistician*, 27(4):148–152, 1973.

[59] M. Hollander and D. A. Wolfe. *Nonparametric statistical methods.* Wiley, New York, 1973.

[60] H. Huynh and L. S. Feldt. Conditions under which mean square ratios in repeated measurements designs have exact *F*-distribution. *J. Amer. Statist. Assoc.*, 65:1582–1589, 1970.

[61] H. Huynh and G. K. Mandeville. Validity conditions in repeated measures designs. *Psychological Bulletin*, 86(5):964–973, 1979.

[62] K. Ishihawa. *Guide to quality control.* Unipub, New York, 1976.

[63] J. Johnston. *Econometric methods.* McGraw-Hill, New York, 1972.

[64] B. Jones and M. G. Kenward. *Design and Analysis of Crossover Trials.* Chapman and Hall, London, 1989.

[65] G. G. Judge, W. E. Griffiths, R. C. Hill, and T. C. Lee. *The theory and practice of econometrics.* Wiley, New York, 1980.

[66] G. G. Judge, W. E. Griffiths, R. C. Hill, H. Lütkepohl, and T. C. Lee. *The theory and practice of econometrics.* Wiley, New York, 2nd edition, 1985.

[67] M. R. Karim and S. L. Zeger. SAS-macro for GEE estimation. *John Hopkins University*, 1988.

[68] J. Kmenta. *Elements of econometrics.* Macmillan, New York, 1971.

[69] G. G. Koch. Some aspects of the statistical analysis of split-plot experiments in completely randomized layouts. *J. Amer. Statist. Assoc.*, 64:485–505, 1969.

[70] G. G. Koch. The use of nonparametric methods in the analysis of the two period change-over design. *Biometrics*, 28:577–584, 1972.

[71] G. G. Koch, J. R. Landis, D. H. Freemanand, J. L. Freeman, and R. G. Lehnen. A general methodology for the analysis of experiments with repeated measurements of categorical data. *Biometrics*, 33:133–158, 1977.

[72] H. V. Kres. *Statistical tables for multivariate analysis.* Springer, New York, 1983.

[73] W. H. Kruskal and W. A. Wallis. Use of ranks in one-criterion variance analysis. *J. Amer. Statist. Assoc.*, 47:583–621, 1952.

[74] W. Lehmacher. *Verlaufskurven und Crossover (Med. Informatik und Statistik 67).* Springer, Berlin Heidelberg New York, 1987.

[75] W. Lehmacher. Analyse von K Stichproben von Verlaufskurven; in: Bauer, Hommel und Sonnemann (Hrsg.): Multiple Hypothesenprüfung. *Med. Informatik und Statistik*, 70:33–47, 1988.

[76] W. Lehmacher. Verlaufskurvenanalyse und Crossover-Pläne in der Therapieforschung; in: Giani und Repges (Hrsg.): Biometrie und Informatik – neue Erkenntnisgewinnung in der Medizin. *Med. Informatik und Statistik*, 72:72–77, 1990.

[77] W. Lehmacher. Analysis of the crossover design in the presence of residual effects. *Statistics in Medicine*, 10:891–899, 1991.

[78] W. Lehmacher and K. D. Wall. A new nonparametric approach to the comparison of k independent samples of response curves. *Biometrical Journal*, 20(3):261–273, 1978.

[79] E. C. Lehmann. *Testing statistical hypotheses*. Wiley, New York, 2nd ed. edition, 1986.

[80] K.-Y. Liang and S. L. Zeger. Longitudinal data analysis using generalized linear models. *Biometrika*, 73:13–22, 1986.

[81] K.-Y. Liang and S. L. Zeger. Regression analysis for correlated data. *Annu. Rev. Pib. Health*, 14:43–68, 1993.

[82] K.-Y. Liang, S. L. Zeger, and B. Quaqish. Multivariate regression analysis for categorical data. *J. Royal. Statist. Soc. B*, 54:3–40, 1992.

[83] G. A. Lienert. *Verteilungsfreie Methoden in der Biostatistik.* Hain, Königstein/Ts., 1986.

[84] S. R. Lipsitz, N. M. Laird, and D. P. Harrington. Maximum likelihood regression methods for paired binary data. *Statistics in Medicine*, 9:1517–1525, 1990.

[85] R. J. A. Little and D. B. Rubin. *Statistical analysis with missing data.* Wiley, New York, 1987.

[86] K. V. Mardia, J. T. Kent, and J. M. Bibby. *Multivariate analysis.* Academic Press, London, 1979.

[87] J. W. Mauchly. Significance test for sphericity of a normal n-variate distribution. *Annals Math. Statistics*, 11:204–209, 1940.

[88] P. McCullagh and J. A. Nelder. *Generalized Linear Models.* Chapmann and Hall, London, 1989.

[89] F. W. McElroy. A necessary and sufficient condition that ordinary least-squares estimators be best linear unbiased. *J. Amer. Statist. Assoc.*, 62:1302–1304, 1967.

[90] D. McFadden. Conditional logit analysis of qualitative choice; in: P. Zarembka (Hrsg.): Frontiers in econometrics. *Academic Press, New York*, pages 105–142, 1974.

[91] J. Michaelis. Schwellenwerte des Friedman-Tests. *Biometr. Zeitschr.*, 13:118–122, 1971.

[92] R. G. Miller, Jr. *Simultaneous statistical inference.* Springer, New York, 1981.

[93] G. A. Milliken and F. Akdediz. A theorem on the difference of the generalized inverse of two nonnegative matrices. *Comm. Statist. A*, 6:73–79, 1977.

[94] G. A. Milliken and D. E. Johnson. *Designed experiments.* Van Nostrand Reinhold, New York, 1984.

[95] H. C. Mitzel and P. A. Games. Circularity and multiple comparisons in repeated measure designs. *British Journal of Math. and Stat. Psychology*, 34:253–259, 1981.

[96] D. C. Montgomery. *Design and analysis of experiments.* Wiley, New York, 1976.

[97] D. F. Morrison. A test for equality of means of correlated variates with missing data on one response. *Biometrika*, 60:101–105, 1973.

[98] D. F. Morrison. *Multivariate statistical methods.* Mc Graw Hill, New York, 1984.

[99] J. Nelder and R. W. M. Wedderburn. Generalized linear models. *J Royal Statist. Soc. A*, 135:370–384, 1972.

[100] W. Oberhofer and J. Kmenta. A general procedure for obtaining maximum likelihood estimates in generalized regression models. *Econometrica*, 42:579–590, 1974.

[101] H. I. Patel. Analysis of incomplete data in a two-period crossover design. *Biometrika*, 72:411–18, 1985.

[102] M. S. Pepe and T. R. Fleming. A nonparametric method for dealing with mismeasured covariate data. *J. Amer. Statist. Assoc.*, 86:108–113, 1991.

[103] R. G. Petersen. *Design and analysis of experiments.* Dekker, New York, 1985.

[104] D. S. G. Pollock. *The algebra of econometrics.* Wiley, New York, 1979.

[105] R. L. Prentice. Correlated binary regression with covariates specific to each binary observation. *Biomerics*, 44:1033–1048, 1988.

[106] R. L. Prentice and L. P. Zhao. Estimation equations for parameters in means and covariance of multivariate discrete and continous responses. *Biomerics*, 47:825–839, 1991.

[107] R. J. Prescott. The comparison of success rates in cross-over trials in the presence of an order effect. *Applied Statistics*, 30(1):9–15, 1981.

[108] M. L. Puri and P. K. Sen. *Nonparametric methods in multivariate analysis*. Wiley, New York, 1971.

[109] C. R. Rao. *Linear statistical inference and its applications*. Wiley, New York, 2ne ed. edition, 1973.

[110] C. R. Rao. Methodology based on the l_1-norm in statistical inference. *Sankhya*, 50:289–313, 1988.

[111] C. R. Ráo and P. K. Mitra. *Generalized inverse of matrices and its applications*. Wiley, New York, 1971.

[112] D. A. Ratkowsky, M. A. Evans, and J. R. Alldredge. *Cross-over experiments*. Dekker, New York, 1993.

[113] B. Rosner. Multivariate metjods in ophtalmology with application to paired-data situations. *Biometrics*, 40:961–971, 1984.

[114] H. Rouanet and D. Lepine. Comparison between treatments in a repeated-measurement design: ANOVA and multivariate methods. *The British Journal of Math. and Stat. Psychology*, 23(2):147–163, 1970.

[115] S. N. Roy. On a heuristic method of test construction and its use in multivariate analysis. *Ann. Math. Statistics*, 24:220–238, 1953.

[116] S. N. Roy. *Some aspects of multivariate analysis*. Wiley, New York, 1957.

[117] D. B. Rubin. Inference and missing data. *Biometrika*, 63:581–592, 1976.

[118] D. B. Rubin. *Multiple imputation for nonresponse in surveys*. Wiley, New York, 1987.

[119] B. Rüger. *Induktive Statistik: Einführung für Wirtschafts- und Sozialwissenschaftler, 2. Auflage*. Oldenbourg, München, 1988.

[120] L. Sachs. *Angewandte Statistik*. Springer, Berlin, 1974.

[121] H. Scheffé. A method for judging all contrasts in the analysis of variance. *Biometrika*, 40:87–104, 1953.

[122] H. Scheffé. A 'mixed model' for the analysis of variance. *Ann. Math. Statistics*, 27:23–26, 1956.

[123] H. Scheffé. *The analysis of variance*. Wiley, New York, 1959.

[124] S. R. Searle. *Matrix algebra useful for statistics.* Wiley, New York, 1982.

[125] S. R. Searle, G. Casella, and C. E. McCullock. *Variance components.* Wiley, New York, 1992.

[126] S. D. Silvey. Multicollinearity and imprecise estimation. *J. Royal Statist. Soc. B*, 35:67–75, 1969.

[127] W. Y. Tan. Note on an extension of the GM-theorem to multivariate linear regression models. *SIAM J. Appl. Math.*, 1:24–28, 1971.

[128] H. Theil. On the estimation of relationships involving qualitative variables. *Amer. J. Sociol.*, 76:103–154, 1970.

[129] C. M. Theobald. Generalizations of mean square error applied to ridge regression. *J. Royal Statist. Soc. B*, 36:103–106, 1974.

[130] N. H. Timm. *Multivariate analysis.* Brooks/Coole Publishing Company, Monterey (California), 1975.

[131] H. Toutenburg. *Lineare Modelle.* Physica, Heidelberg, 1992.

[132] H. Toutenburg, Ch. Heumann, A. Fieger, and S. H. Park. Missing values in regression: mixed and weighted mixed estimation. *(to appear)*, 1995.

[133] H. Toutenburg, S. Toutenburg, and W. Walther. *Datenanalyse und Statistik für Zahnmediziner.* Hanser, München, Wien, 1991.

[134] H. Toutenburg and W. Walther. Statistische Behandlung unvollständiger Datensätze. *Dtsch. Zahnärztl. Z.*, 47:104–106, 1992.

[135] S. Toutenburg. Eine Methode zur Berechnung des Betreungsgrades in der prothetischen und konservierenden Zahnmedizin auf der Basis von Arbeitsablaufstudien, Arbeitszeitmessungen und einer Morbiditätsstudie. *Med. Diss. Berlin*, 1977.

[136] J. W. Tukey. The problem of multiple comparisons. Technical report, Department of Mathematics, Princeton University, 1953.

[137] W. Vach and M. Blettner. Biased estimation of the odds ratio in case-control studies due to the use of ad-hoc methods of correcting for missing values in confounding variables. *American Journal of Epidemiology*, 134:895–907, 1991.

[138] W. Vach and M. Schumacher. Logistic regression with incompletely observed categorial covariates – a comparison of three approaches. *Biometrika*, 80:353–62, 1992.

[139] R. A. Waller and D. B. Duncan. A bayes rule for the symmetric multiple comparison problem. *J. Amer. Statist. Assoc.*, 67:253–255, 1972.

[140] W. Walther. *Ein Modell zur Erfassung und statistischen Bewertung klinischer Therapieverfahren – entwickelt durch Evaluation des Pfeilerverlustes bei Konuskronenersatz.* PhD thesis, Med. Habil., Universität Homburg, 1992.

[141] W. Walther and H. Toutenburg. Datenverlust bei klinischen Studien. *Dtsch. Zahnärztl. Z.*, 46:219–222, 1991.

[142] R. W. M. Wedderburn. Quasi-likelihood functions, generalized linear models, and the gauss-newton method. *Biometrika*, 61:439–447, 1974.

[143] R. W. M. Wedderburn. On the existence and uniqueness of the maximum likelihood estimates for certain generalized linear models. *Biometrika*, 63:27–32, 1976.

[144] S. Weisberg. *Applied linear regression.* Wiley, New York, 1980.

[145] S. S. Wilks. Moments and distributions of estimates of population parameters from fragmentary samples. *Ann. Math. Statist.*, 3:163–195, 1932.

[146] R. F. Woolson. *Statistical methods for the analysis of biomedical data.* Wiley, New York, 1987.

[147] F. Yates. The analysis of replicated experiments when the field results are incomplete. *Emp. J. Exp. Agric.*, 1:129–142, 1933.

[148] L. P. Zhao and R. L. Prentice. Correlated binary regression using a generalized quadratic model. *Biometrika*, 77:642–648, 1990.

[149] H. Zimmermann and W. Rahlfs. Testing hypotheses in the two period change-over with binary data. *Biometrical Journal*, 20(2):133–141, 1978.

Index

Springer-Verlag
and the Environment

We at Springer-Verlag firmly believe that an international science publisher has a special obligation to the environment, and our corporate policies consistently reflect this conviction.

We also expect our business partners – paper mills, printers, packaging manufacturers, etc. – to commit themselves to using environmentally friendly materials and production processes.

The paper in this book is made from low- or no-chlorine pulp and is acid free, in conformance with international standards for paper permanency.